D1669520

Industrial Metrology

Springer
London
Berlin
Heidelberg
New York
Barcelona
Hong Kong
Milan
Paris
Singapore
Tokyo

http://www.springer.de/phys/

Previous books for Springer-Verlag by the author

Advanced Machining: The Handbook of Cutting Technology (1989)

CNC Machining Technology series:

Book 1: Design, Development and CIM Strategies
Book 2: Cutting, Fluids and Workholding Technologies
Book 3: Part Programming Techniques (1993)

CNC Machining Technology: Library Edition (1993)

Graham T. Smith

Industrial Metrology

Surfaces and Roundness

Graham T. Smith, MPhil (Brunel), PhD (Birmingham), CEng, FIMechE, FIEE
Faculty of Technology
Southampton Institute
East Park Terrace
Southampton SO14 0RD
U.K.

British Library Cataloguing in Publication Data
Smith, Graham T., 1947–
Industrial metrology: surfaces and roundness
1. Mensuration 2. Surfaces (Technology) 3. Roundness measurement
I. Title
620'0044
ISBN 1852335076

Library of Congress Cataloging-in-Publication Data
Smith, Graham T., 1947–
Industrial metrology: surfaces and roundness/Graham T. Smith.
p. cm
Includes bibliographical references and index.
ISBN 1-85233-507-6 (alk. paper)
1. Surfaces (Technology) – Measurement. 2. Roundness measurement. I. Title

TA418.7.S55 2001
620'.44–dc 21001032028

ISBN 1-85233-507-6 Springer-Verlag London Berlin Heidelberg
a member of BertelsmannSpringerScience + Business Media GmbH
http://www.springer.co.uk

Printed in Great Britain

Typeset by Florence Production Ltd, Stoodleigh, Devon
Printed and bound by Kyodo Printing Co (S'Pore) Pte Ltd, Singapore
69/3830–543210 Printed on acid-free paper SPIN 10835643

To my wife Brenda, who has supported me throughout our married and my academic life, with much love and affection.

Foreword

The subject of this book is surface metrology, in particular two major aspects: surface texture and roundness. It has taken a long time for manufacturing engineers and designers to realise the usefulness of these features in quality of conformance and quality of design. Unfortunately this awareness has come at a time when engineers versed in the use and specification of surfaces are at a premium.

Traditionally surface metrology usage has been dictated by engineers who have served long and demanding apprenticeships, usually in parallel with studies leading to technician-level qualifications. Such people understood the processes and the achievable accuracies of machine tools, thereby enabling them to match production capability with design requirements. This synergy, has been made possible by the understanding of adherence to careful metrological procedures and a detailed knowledge of surface measuring instruments and their operation, in addition to wider inspection room techniques.

With the demise in the UK of polytechnics and technical colleges, this source of skilled technicians has all but dried up. The shortfall has been made up of semi-skilled craftsmen, or inexperienced graduates who cannot be expected to satisfy traditional or new technology needs. Miniaturisation, for example, has had a profound effect. Engineering parts are now routinely being made with nanometre surface texture and flatness. At these molecular and atomic scales, the engineer has to be a physicist.

This book is intended to bridge the gap between technology and training. Dr Smith has approached the subject from the point of view of a precision engineer, but in addition has made the reader aware, whenever appropriate, of new techniques and instruments. He is admirably suited to the task, having served an apprenticeship in industry which has led to him becoming a master toolmaker. He has balanced these practical skills by in-depth studies leading to the degree of Ph.D in manufacturing engineering. Taking these complementary paths has given him a unique pedigree that has enabled him to write with authority.

Dr Smith is recognised as an expert in his subject of manufacturing and has lectured extensively throughout Great Britain and North America. This book is a must for those involved in inspection and quality control and could well be used as a core for training courses in metrology.

Professor David J. Whitehouse
Formerly Chief Scientist,
School of Engineering,
University of Warwick

Preface

As the UK's national measurement institute the National Physical Laboratory (NPL) has been at the forefront of research in the field of metrology – measurement science – since its inception in 1900. NPL has always been involved in measurements at the highest levels of accuracy in order to suppport its role as custodian of the national measurement standards and to address the UK industry's most demanding measurement problems. NPL also plays a key role in the dissemination of measurement good practice to industry through its wide range of knowledge transfer activities. It was therefore a pleasure for myself and members of the Dimensional and Optical Metrology Team at NPL to contribute to this important book, which I believe provides a rich source of information for anyone working in the field of surface texture and roundness – areas which have been and will continue to be among the most important but sometimes least appreciated areas of dimensional metrology in industry.

A measure of the importance of surface texture and roundness can be gauged by the work carried out at NPL in this area over the past 70 years or so. As early as the 1930s NPL was increasingly involved in developing instrumentation for surface texture and roundness metrology. In 1940, NPL was working on the quest for a single number or parameter that would define the texture of a surface and hence enable comparisons between different surfaces to be made. Around this time NPL was also producing standard artefacts for calibrating surface and roundness instrumentation.

In the 1970s much of the surface texture work carried out at NPL related to the manufacture of X-ray optical components that required control of roundness to within a few tens of nanometers, profile to a few nanometres and surface texture often to sub-nanometres tolerances. In order to control the manufacture of these optical components, measurement techniques with sufficient accuracy had to be developed. This led to an NPL-designed instument, subsequently marketed by Rank Taylor Hobson, called "Nanostep", which was the first instrument of its time to have sub-nanometre vertical accuracy with a large horizontal movement range.

In recent years one of NPL's key objectives has been to undertake research work to support UK industry by providing calibrated references artefacts for surface texture measurements. This work culminated in 2000 in a new NPL-designed instrument, "NanoSurf IV", capable of calibrating surface texture reference artefacts with fully traceable measurements in all the axes, leading to a combined expanded uncertainty of ± 1.3 nm.

Roundness measurement was revolutionised by the introduction of the "Talyrond" in 1949 and has progressed since then. More recently Taylor Hobson and NPL have also developed a roundness-measuring facility that has a measurement capability of 5 nm, making it undoubtedly among the most accurate instruments of its type.

A view of NPL's new laboratory complex

With the ever-increasing importance of surface measurements to industries operating in diverse areas ranging from aerospace to automotive and from medical to environmental, it should be no surprise to learn that NPL plans to continue to strive for further improvements in surface measurements in its future work programmes. It is expected that this work will be helped by the imminent move of the whole of the National Physical Laboratory into a completely new purpose-built laboratory complex. The move is expected to be completed during 2002 and will provide NPL scientists with new and improved facilities to enable them to provide the underpinning research and traceable calibration services required by industry in this important field well into the new century.

Professor Graham N. Peggs
Science Leader, Dimensional and Optical Metrology Team
Centre for Basic, Thermal and Length Metrology

Contents

Acknowledgements

A book that leans heavily on current industrial practices cannot be successful without the proactive support of both relevant industrial companies and research-based organisations, to whom I am particularly indebted. It would be impossible to name all of the people who have contributed to this book, and if I do not thank you in name then please accept my apologies. However, certain companies have provided unstinting support to me in this work and I would like first to thank Taylor Hobson Limited for their major support of information on surface texture and roundness instrumentation, Jeol (UK) Limited for information on surface microscopy equipment and, likewise, Nikon (UK) Limited, VG Scientific Limited, Zygo Photometrics Limited and Rubert & Co. Limited, for additional information on their products. The National Physical Laboratory must be singled out for their significant contribution to the book; I am particularly indebted to them for information on the current standards.

First and foremost, I would like to offer my sincere gratitude to Dr Richard K. Leach at the National Physical Laboratory, who read the appropriate sections on surfaces and standards and suggested significant ways to improve the text. Also from the National Physical Laboratory, I would like to show my appreciation to Professor Graham N. Peggs for agreeing to write the Preface, to Mr David Flack, who read the roundness sections and supplied considerable information on roundness standards, and to Dr Bob Angus, who along with Dr Andrew Lewis has shown significant enthusiasm for this project.

At Taylor Hobson Limited, I am truly indebted to Dr Paul Scott, who supplied me with information on current working standards and for numerous discussions on their current instrumentation. Similarly, I must thank Mr Pat Kilbane (Manager) and his calibration staff in the UKAS Calibration Laboratory for information on certain calibration procedures. Other notable Taylor Hobson personnel that I would like to express my thanks to include Mr Darian P.L. Mauger (Manager) in their Centre of Excellence, Mr Mick Garton, Mr Jon Gardiner and Mr Steve Foster (Technical Liaison Engineers); to Dr Mike Hills (Chief Engineer) and from the Sales and Marketing Department Mr Julian Shaw (Director of Sales and Marketing), Ms Louise O'Reilly (formerly Marketing Manager), Mr Peter Atkinson (Sales Executive Technical); and to Mr Bruce P. Wilson (Managing Director) and to all the Taylor Hobson staff I have neglected to mention.

Jeol (UK) Limited were of considerable help and support in supplying information and producing photomicrographs for the chapter on surface microscopy. Most notably, I would like to express my thanks to Dr Larry Stoter (Sales Executive) for supplying instrumentation photographs and relevant information, Mr John Critchall (Applications Manager) and Mr Andy Yarwood (Application Specialist) for undertaking SEM/AFM photomicrographs and to Mr Chris Walker (General Manager) for further technical assistance.

Other notable support came from companies that have contributed information and photographic support to the book, including Zygo/Lamda Photometrics Limited – Mr Joe Armstrong (Technical Sales Engineer); VG scientific – Dr John Wolstenholme (Marketing Manager): Nikon UK Limited – Mr Robert Forster (General Manager) and Keith Poulton (Product Manager); also Rubert & Co. Limited – Mr Paul Rubert (Managing Director).

This book would not have been possible without the help and guidance of some of Great Britain's leading academics and I would like to single out several people for special mention. In the first instance, an old friend Professor David J. Whitehouse (formerly Chief Scientist at the University of Warwick). For conversations on David's immense knowledge on the topics discussed here and for previous instruction on all manner of aspects concerning both surface texture and roundness, furthermore, I would like to express my sincere thanks for him kindly agreeing to write the Foreword for this book. Likewise to Dr Brian J. Griffiths (Reader at Brunel University), my previous mentor and dear friend, for his instruction and support on all topics appertaining to surface integrity; and to Professor Tom R. Thomas (Chalmers University, Sweden) for conversations concerning surface texture and amplitude-wavelength analysis.

Numerous people at my publishers Springer-Verlag have been of great support in this current work, most notably Ms Francesca Warren (Editorial Assistant), Mrs Beverly Ford (Editorial Director), Mr Roger Dobbing (Production Manager) and Mr David S. Anderson (Sales and Business Development Director). Finally, I would like to acknowledge a long and interesting acquaintanceship with Mr Nick Pinfield (fomerly of Springer-Verlag), who gave me considerable encouragement during the period of the book's development and for some interesting conversations ranging from engineering to historical topics.

Lastly, a considerable number of graphs, line and assembly diagrams were drawn by me, and if there are any misinterpretations in their execution then they are a reflection on the author. Further, in the body of the text some manuscript errors may have inadvertently crept in, and if this is the case they are not to be associated with any of the companies or individuals that have supported the book.

Dr Graham T. Smith
West End,
Southampton

1 Surface texture: two-dimensional

"Errors, like straws, upon the surface flow;
He who would search for pearls must dive below."

(*All for Love*, Prologue, 25; John Dryden 1631–1700)

Surface texture: an overview

The concept of *surface technology* has previously varied according to whether a scientist, technologist, or engineer has inspected a surface. An individual's viewpoint differs depending upon their specific interest and technical emphasis. Therefore, a surface cannot be thought of in isolation from other features, because each is valid and necessary and needs to be considered to gain an overall impression of its industrial, or scientific usage.

Surface technology encompasses a wide range of disciplines that includes: metrology; metallurgy; materials science; physics; chemistry; tribology and mechanical design. Moreover, it has become apparent that this topic, and associated characteristics that relate to surface technology needs to be recognised as a subject in its own right.

Surface technology provides an important and valuable insight into the practical and theoretical applications of a manufactured surface, most notably for the following reasons:

- the surface is the final link from the original design concept through to its manufacture, with the industrial engineer being the last person to add value to the product prior to shipment;
- industry is continually attempting to improve component power-to-weight ratios, drive down manufacturing costs and find efficient ways of either producing or improving surfaces – this has become an established goal of world-class competing companies today;
- internationally, there is growing demand to recognise the legal implications of a product's performance. Product liability is directly related to the production process technique and, hence, to a surface's quality;
- technical data that in the past was considered valid, has of late been found to be either unreliable or may give insufficient detailed information on surface phenomena. By way of an example, originally it was thought that the surface texture was directly related to a component's fatigue life; however, it has been shown that surface topography must be extended to include subsurface metallurgical transformations, namely surface integrity;
- the production process selected for the manufacture of a part has been shown to have some influence on the likely in-service reliability of components.

1.1 Introduction

On the macro scale the natural world mimics surfaces found in engineering. Typical of these is that shown in Figure 1, where the desert sand is comprised of sand grains – *roughness*; ripples in the surface – *waviness*; together with the undulating nature of the land – *profile*. From an engineering point of view, surfaces are boundaries between two distinct media, namely, the component and its working environment. When a designer develops a feature for a part – whether on a computer-aided design (CAD) system, or drawing board – neat lines delineate the desired surface condition, which is further specified by its specific geometric tolerance. In reality, this theoretical surface condition cannot exist, as it results from process-induced surface texture modifications. Regardless of the method of manufacture, an engineering surface must have

Figure 1. The natural world – the desert – can exhibit a large-scale combination of surface characteristics, for instance:
- agglomeration (i.e. clusters) of sand – *roughness;*
- ripples – *waviness;* and
- the undulating nature of the land – *profile.*

(Source: James Smith, *Prayer in the desert*, circa 1920s.)

some form of texture associated with it, this being a combination of several interrelated factors, such as:

- the influence of the material's microstructure;
- the surface generation method, the tool's cutting action, tool geometry, cutting speed, feedrate and the effect of cutting fluid;
- instability during the manufacturing process, promoted by induced chatter – poor loop-stiffness between the machine-tool–workpiece system, or imbalance in a grinding wheel;
- inherent residual stresses within the part, promoted by stress patterns causing deformations in the component.

Within the limitations resulting from the component's manufacture, a designer must select a functional surface condition that will satisfy the operational constraints, which might be a requirement for either a "smooth" or "rough" surface. This then begs the question posed many years ago: "How smooth is smooth?" This is not as superficial a statement as it might initially seem, because unless we can quantify a surface accurately, we can only hope that it will function correctly in service. In fact, a surface's texture, as illustrated in Figure 2, is a complex condition resulting from a combination of:

- *roughness* – comprising of irregularities that occur due to the mechanism of the material removal production process; tool geometry, wheel grit, or the EDM spark;
- *waviness* – that component of the surface texture upon which roughness is superimposed, resulting from factors such as machine or part deflections, vibrations and chatter, material strain and extraneous effects;

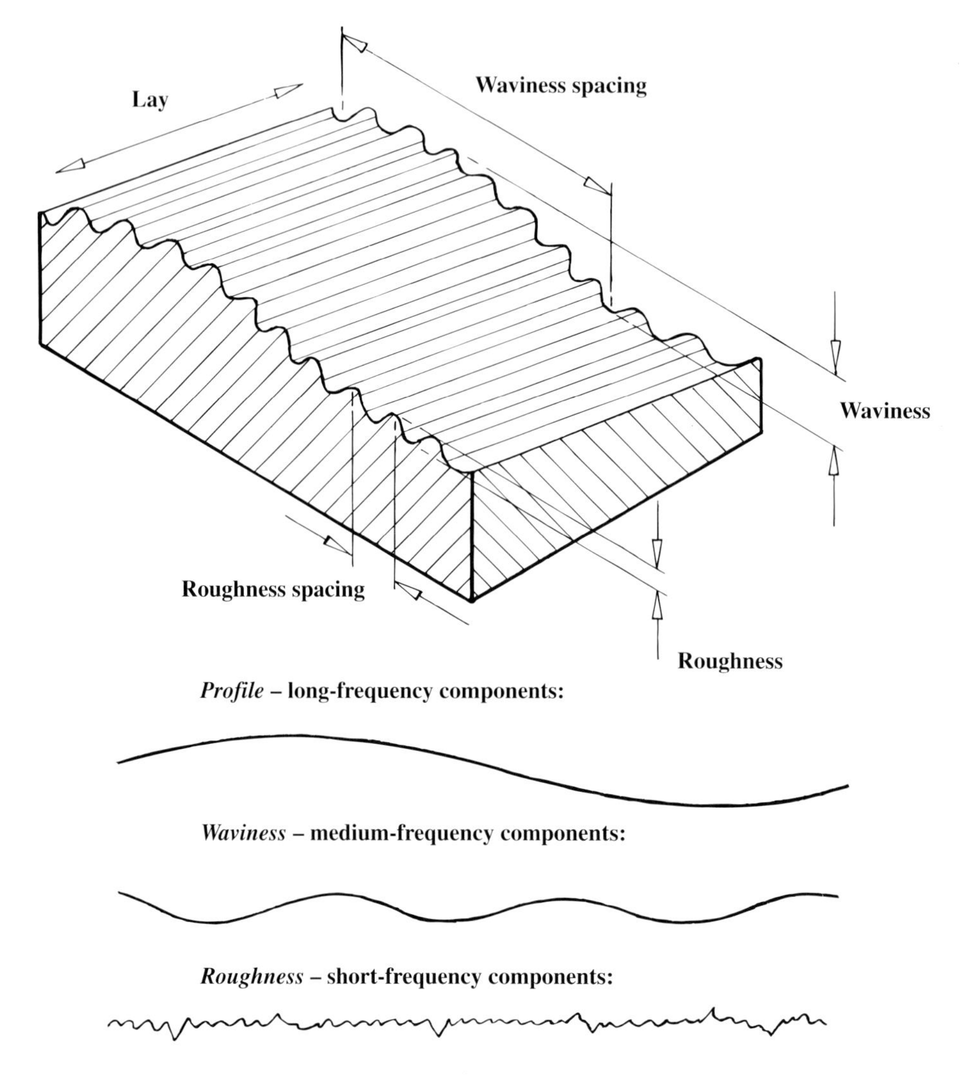

Figure 2. The major components that constitute a typical *surface texture*, also exhibiting some degree of directionality (lay).

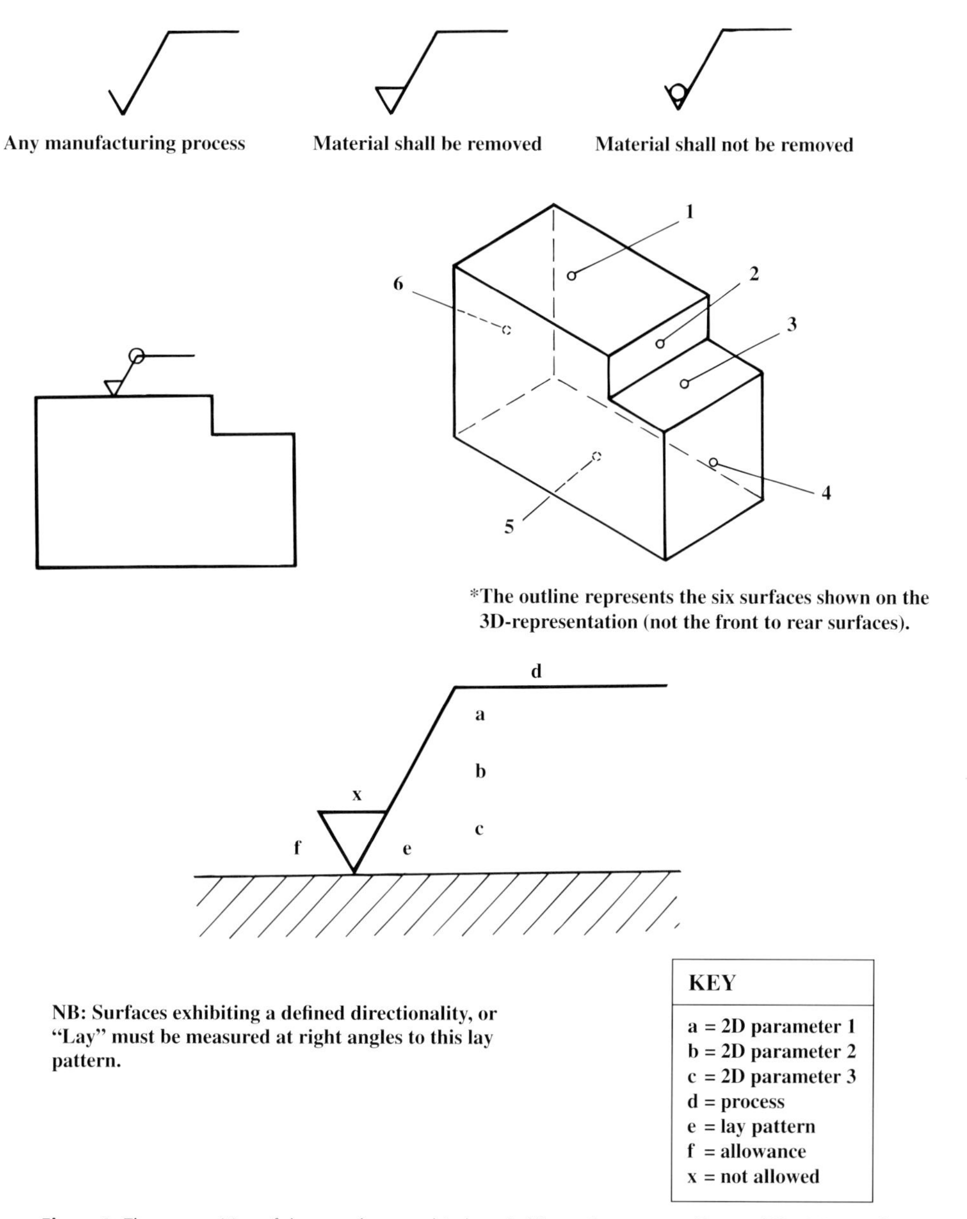

Figure 3. The composition of the complete graphical symbol for surface texture. [Source: ISO 1302: 2001]

- *profile* – the overall shape of the surface – ignoring roughness and waviness variations – is caused by errors in machine tool slideways.

Such surface distinctions tend to be qualitative – not expressible as a number – yet have considerable practical importance, being an established procedure that is functionally sound (Figure 3). The combination of these surface texture conditions, together with the surface's associated "lay", are idealistically shown in Figure 4, where the lay of the surface can be defined as the *direction of the dominant pattern*. Figure 4 also depicts the classification of lay, which tends to be either *anisotropic* – having directional properties such as feed marks, or *isotropic* – devoid of a predominant lay direction. The lay of any surface is important when attempting to characterise its potential functional performance. If the direction of the trace being produced by a stylus instrument – record of the stylus motion over the assessed surface topography – is not taken into account, then a totally misrepresentative reading will result from an anisotropic surface. This is not the case when measuring an isotropic surface, for example a "multi-directional lay" – as indicated in Figure 4. Under this isotropic condition, perhaps three different measurement paths could be scanned and the worst roughness trace would be utilised to

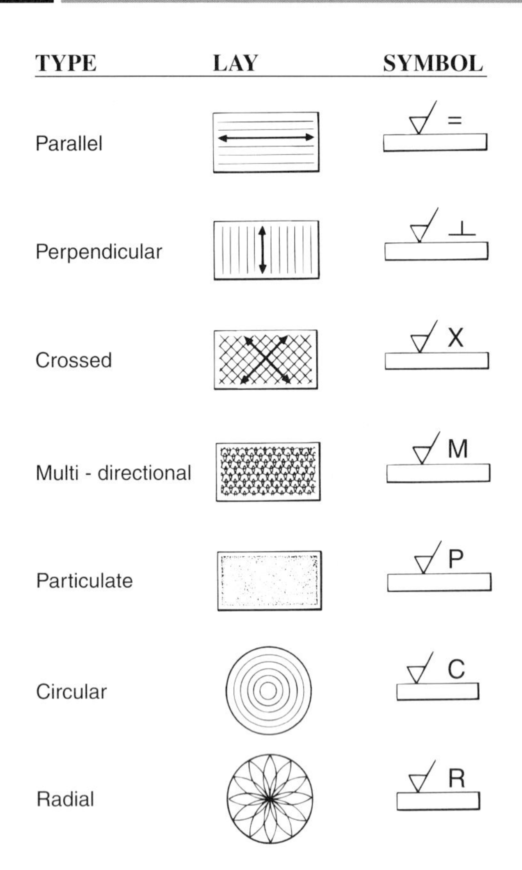

Figure 4. The indication of surface *'lay'* as denoted on engineering drawings. [Source: ISO 1302: 2001]

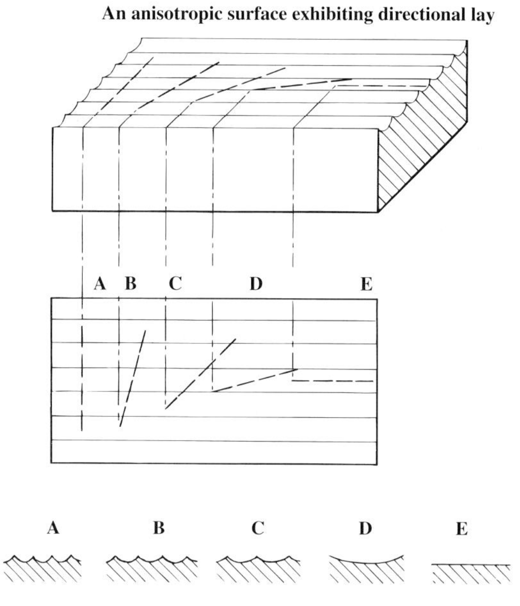

Figure 5. Effect of the relative direction of lay and its associated influence on the measurement of profile shape. (Courtesy of Taylor Hobson.)

specify the surface. The designer should be aware that the lay condition and other information can be incorporated into the surface texture symbol (ISO 1302: 2001) – which is often misused – see Figure 3 for the positioning of this data and other surface texture descriptors. The surface texture symbol shown in Figure 4, indicated by "surface texture ticks", highlights the potential process-related lay type and associated directionality (ISO 1302: 2001).

Returning to the theme of an isotropic lay condition, if this surface is assessed at a direction not at right angles to the lay, then a totally unrepresentative surface profile will result, which is graphically illustrated in Figure 5. As indicated in this figure, as the trace angle departs to greater obliquity from condition A then the surface profile becomes flat once the direction has reached condition E. This would give a totally false impression of the true surface topography. If employed in a critical and perhaps highly stressed in-service state, with the user thinking that this incorrectly assessed surface was flat, then, potentially, a premature failure state could arise. Normally, if the production process necessitates, for example, a cross-honing operation, the lay condition would be "crossed" – as shown in Figure 4. Under this roughness state, it is usual to measure the surface at 45°, which has the effect of averaging out the influence of the two directions imparted by the cross-honing operation. Where surfaces are devoid of a lay direction – as in the case of "particulate lay" (see Figure 4) from either shot-blasting or the sintering process – then under such circumstances a trace will produce the same surface texture reading, irrespective of the direction of measurement.

The production of certain milled surfaces can sometimes exhibit roughness and waviness conditions at 90° to each other – this is not the same condition as "crossed lay" – because the relative heights and spacings of the two lays differ markedly. Hence, the cusp and chatter marks of the milled topography should be measured in opposing directions. When a turned component has had a "facing-off" operation undertaken, and a "circular lay" condition occurs (see Figure 4), then under these circumstances it is normal to measure the surface texture in a radial direction, otherwise an inappropriate reading will result. Conversely, if a "radial lay" occurs – resulting from "cylindrically grinding" the end face – it is more usual to measure surface roughness at a series of tangent positions with respect to the circumferential direction.When measuring surface texture in the majority of practical situations it can be achieved by direct measurement of the profile, positioning the workpiece in the correct manner to that of the stylus of the surface texture instrument. When the part is

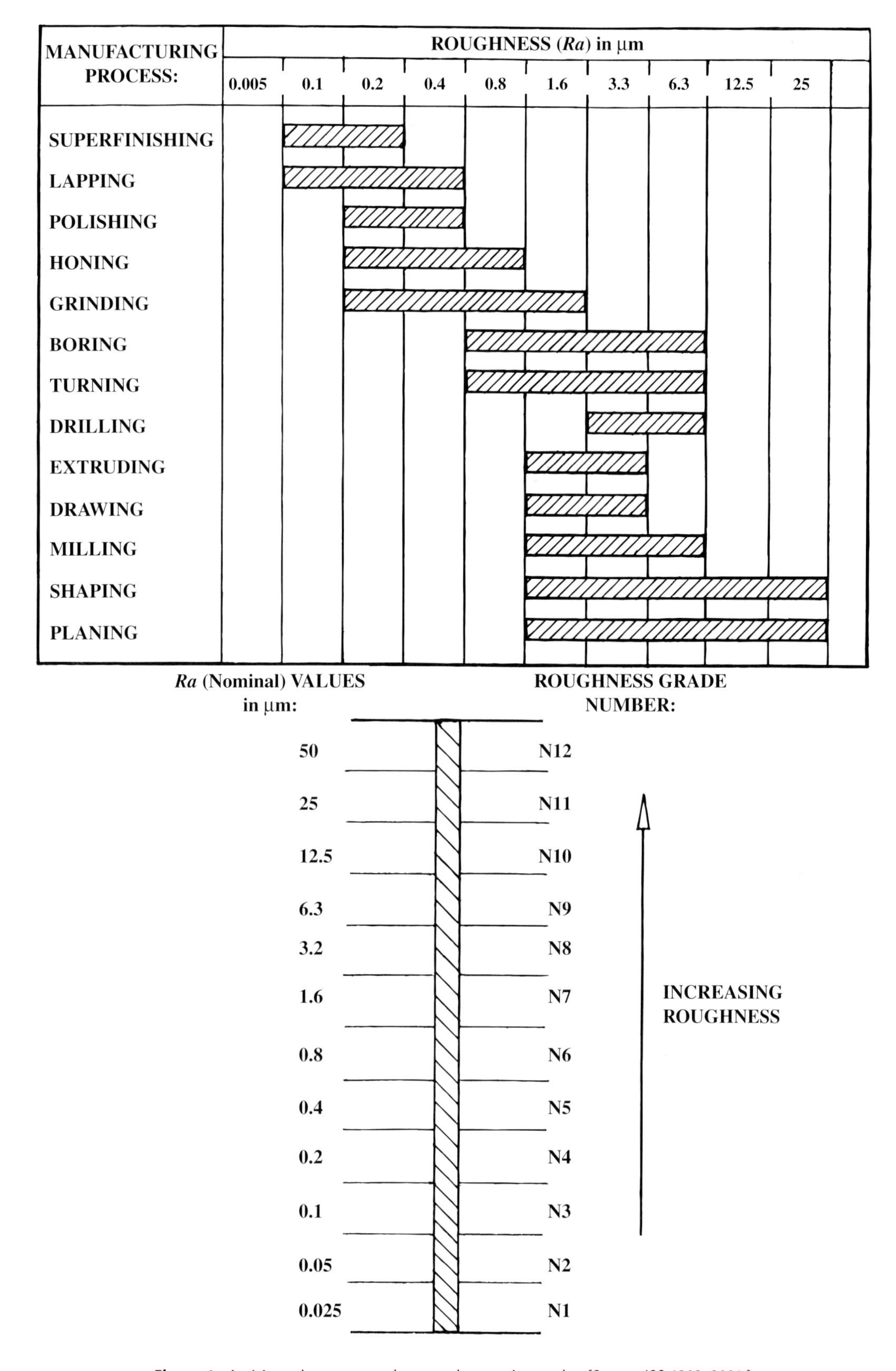

Figure 6. Anticipated process roughness and respective grades. [Source: ISO 1302: 2001.]

either too large or this is impracticable – due to either stylus access to the surface or surface inclination – then placing a small portable instrument on the surface can allow a measurement. Occasions arise when even this is not an option and under these circumstances it may be possible to take a "cast replica" of the surface, then present this to the instrument. Some surface texture measuring instrument manufacturers offer "replica kits" for just this purpose. It should be remembered that this replica of the surface is an inverted image.

1.2 Establishing the *Ra* numerical value of surface texture from the production process

Numerical data (ISO 1302: 2001) to define the roughness grade for surface texture has been established (see Figure 6) which can be related to the method of production. Caution should be made when using these values for control of the surface condition, because they can misrepresent the actual state of the surface topography, being based solely on a derived numerical value for height. More will be said on this later. Moreover, the "N-number" has been used to establish the arithmetic roughness *Ra* value, this being just one number to cover a spread of potential *Ra* values for that production process. Nevertheless, this single numerical value has merit, in that it defines a "global" roughness (*Ra*) value and accompanying "N-roughness grade", that can be used by a designer to specify a particularly desired surface condition being related to the manufacturing process. The spread of the roughness for a specific production process has been established from experimental data over the years – covering the maximum expected variance – which can be modified depending upon whether a fine, medium or coarse surface texture is required. Due to the variability in the production process and its stochastic output, such surface texture values do not reflect the likely in-service performance of the part. Neither the surface topography nor its associated integrity has been quantified by assigning to a surface representative numerical parameters. In many instances, "surface engineering" is utilised to enhance specific component in-service conditions, but more will be said on this subject toward the end of Chapter 5.

Previously, it was mentioned that with many in-service applications that the accompanying surface texture is closely allied to functional performance, particularly when one or more surfaces are in motion with respect to an adjacent surface. This suggests that the smoother the surface the better, but this is not necessarily true if the surfaces in question are required to maintain an efficient lubrication film. The apparent roughness of one of these surfaces with respect to the other enables it to retain a "holding film" in its associated roughness valleys. Yet another factor which may limit the designer from selecting a smooth surface is related to its production cost (see Figure 7). If a smooth surface is the requirement then this takes a significantly longer time to produce than an apparently rougher surface, which is exacerbated if this is allied to very close dimensional tolerancing.

1.3 Surface texture roughness comparison blocks and precision reference specimens

In order to gain a basic "feel" for actual replica surfaces in conjunction with their specific numerical value for surface roughness. Then the use of comparison blocks allows an efficient and useful means of interpreting the production process with their expected surface condition. A designer (Figure 8a) can select a surface using a visual and tactile method to reflect the required workpiece condition. Correspondingly, the machine tool operator can utilise similar comparison blocks (see Figure 8b) to establish the same surface condition, without the need to break down the set-up and inspect the part with a suitable surface texture measurement instrument. This is a low-cost solution to on-machine inspection and additionally enables each operator to simply act as an initial source for a decision on the part's surface texture condition. Such basic visual and tactile inspection might at first glance seem inappropriate for any form of inspection today, but within its limitations it offers a quick solution to surface roughness assessment. Tactile and visual sensing is very sensitive to minute changes in the surface roughness, but is purely subjective in nature and is open to some degree of variation from one person to another. Comparison block surface assessment is purely a subjective form of attribute sampling. Therefore, not only will there be some divergence of opinion between two operators on the actual surface's numerical *Ra* value due to their differing sensual assessments, but also this is only a very rudimentary form of inspection.

Precision reference specimens are manufactured to exacting and consistent tolerances, particularly

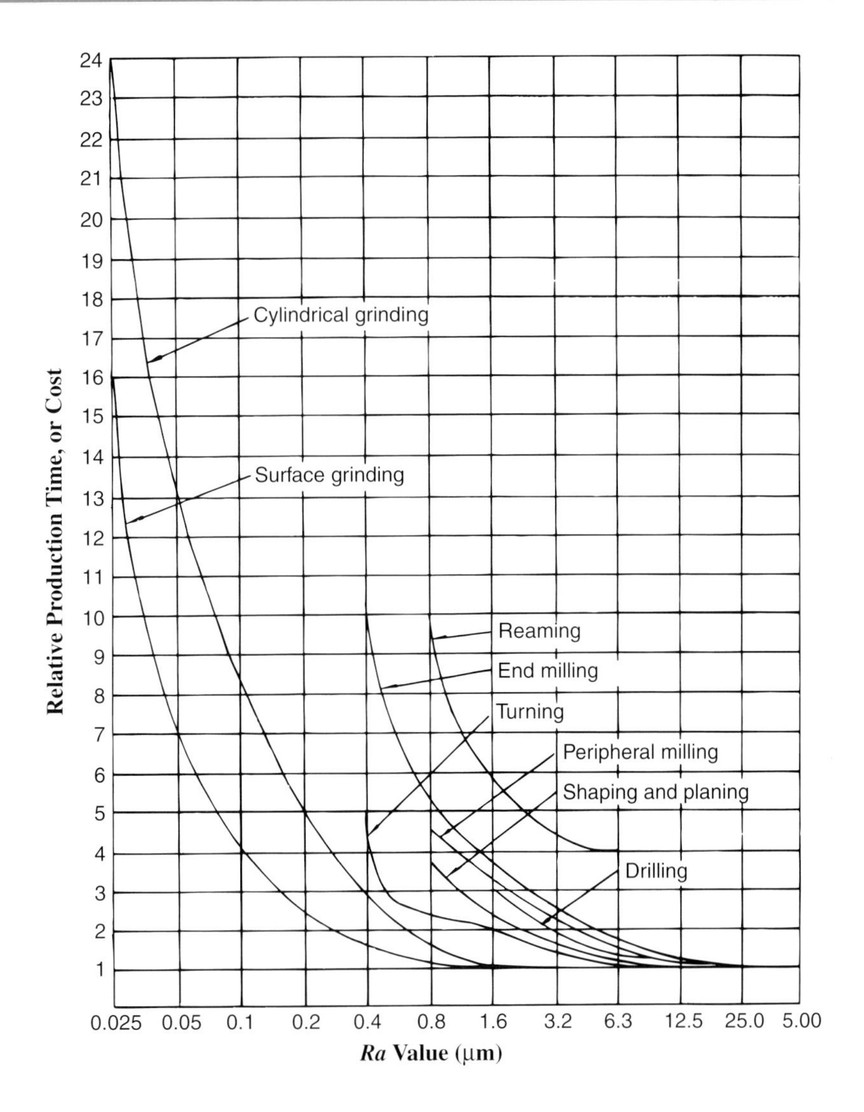

Figure 7. Relative cost, in production time, necessary to produce surface finishes by different production processes. (Courtesy of Taylor Hobson.)

when used to calibrate stylus-based surface texture measuring instruments. A range of standard reference specimens can be purchased for this purpose and will act as a "health check" on both the current stylus condition – whether its point is worn or partially broken – and indicate if the allied instrumentation/electronics has "drifted" since the last calibration check. This enables the inspector to speedily and efficiently remedy the situation and bring the instrument back into calibration. Most of these calibration reference specimens are manufactured from replicas of an original master production surface by nickel electroforming; this enables the block to be a reproduction of the whole surface, with the finest details being approximately 1 μm in size. Each reference specimen offers exacting uniformity of profile shape. In order to minimise wear on these reference specimens from frequent use a hard boron nitride layer can be specified.

1.4 The basic operating principle of the pick-up, its stylus and skid

Prior to describing some of the more sophisticated instruments by which the surface is measured, it is worth describing the basic function of a typical stylus instrument. In Figure 9, a schematic illustration is depicted of the basic components for a typical surface texture measuring instrument. The

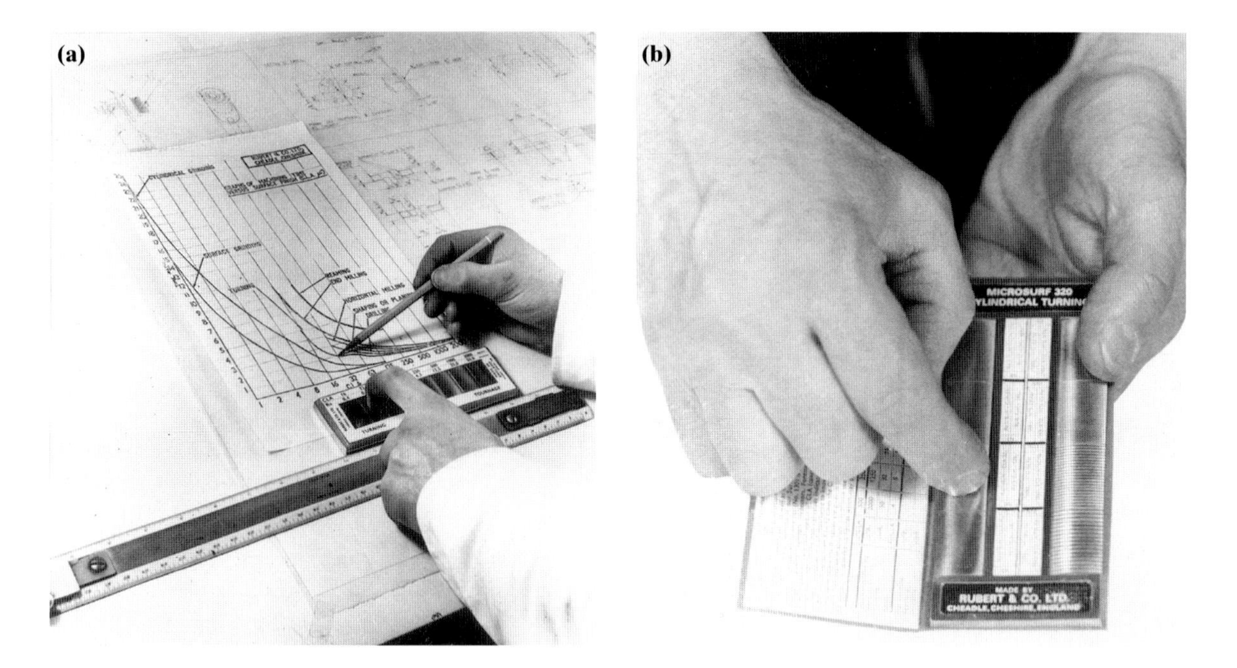

Figure 8. (a) Utilising a comparator gauge to determine the surface finish for a specific manufacturing process. (b) A typical comparator gauge, for both visual and tactile assessment. (Courtesy of Rubert & Co Ltd.)

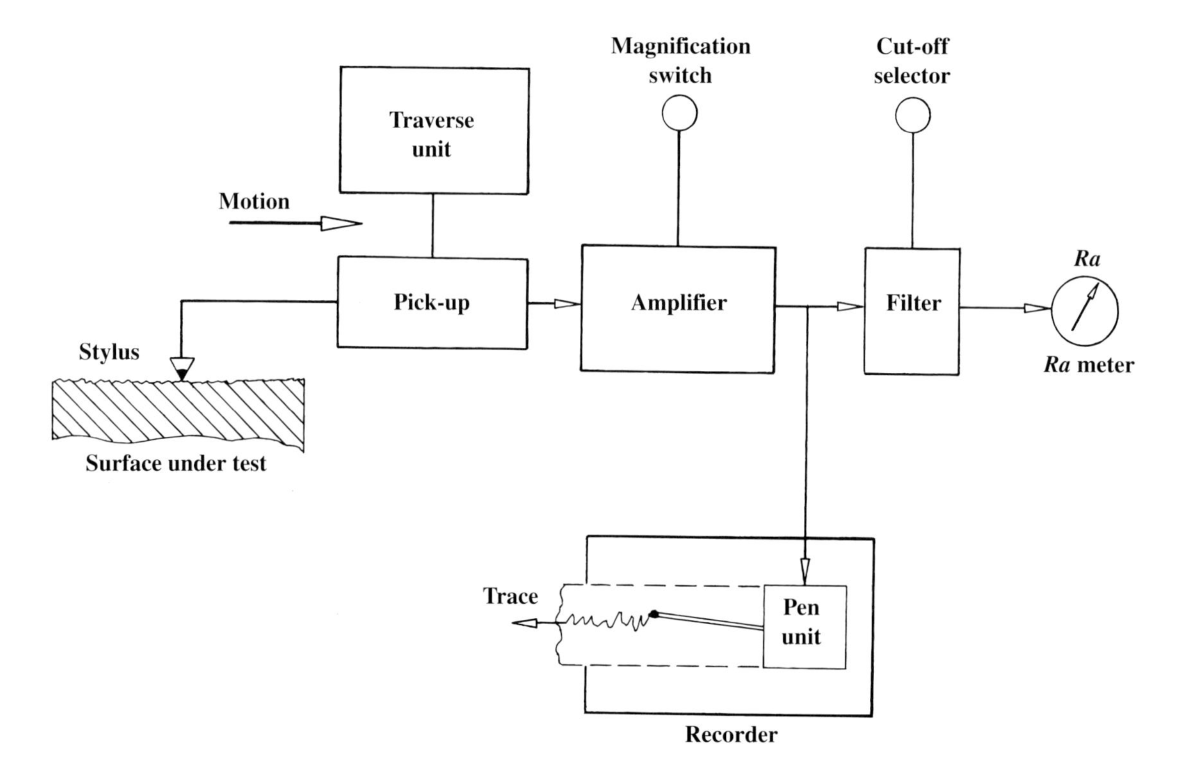

Figure 9. Schematic diagram illustrating the major constituents of a stylus-type of surface texture measuring instrument. (Courtesy of Taylor Hobson.)

Figure 10. Scanning electron photomicrograph of a stylus (5 μm point radius) on a surface. [Courtesy of Hommelwerke GmbH.]

stylus traverses across the surface and the transducer converts its vertical movements into an electrical signal. This signal is then amplified for subsequent processing, or output to operate a pen recorder. The required parameter value is subsequently derived from the filtered signal, having previously been displayed on screen. In general, the profile is the result of the stylus tracing the *movement* across the surface under test, the contacting of consecutive points of the profile being spaced in *time*. Therefore this relationship between movement and time is closely associated with its "cut-off". "Cut-off" refers to the limiting wavelength at which components of the profile are passed nominally unchanged by a filter.

The stylus is normally either conical with a spherical tip (see Figure 10), or a four-sided pyramid with a truncated flat tip (Figure 11a). The conical type of styli (Figure 10) have a cone angle of either 60° or 90°, with a tip radius ranging from less than 0.1 μm to 12 μm in size. A pyramidal stylus tends to be approximately 2 μm wide in the traverse direction, but is normally wider transversely to the direction of travel, giving the tip greater strength (see Figure 11a left). The sharp-pointed 0.1 μm wide pyramidal stylus (Figure 11a right), is utilised on the Taylor Hobson "Talystep" and "Nanostep" instruments. However, in use the edges of a pyramidal tip tend to become rounded with wear; conversely, the spherical tip version develops a flat, so the distinction between the two profile geometries becomes less marked with time. If a stylus tip becomes damaged during use, this results in a considerable increase in its width, leading to potentially very serious measuring errors.

When a comparison is made of the scale of an actual surface under test to that of the stylus tip (Figure 10) against a vertical magnification (V_v) of ×5000 and horizontal magnification (V_h) of ×100, then in this situation the respective size of the stylus is almost indistinguishable from that of a (vertical) straight line or point contact (see Figure 11b). This significant size differential enables the stylus to penetrate into quite narrow valleys, although its finite size affects the accuracy with which the surface profile can be traced. These stylus profile and size limitations influence the following:

- *distortion of peak shape* – as the spherical stylus traverses over a sharp peak, the point of contact shifts across the stylus – from one side to the other (Figure 11c). This results in the stylus having to follow a path which is more rounded than the peak, but due to the stylus being raised to its full height when at the crest the true peak height is measured;
- *penetration into valleys* – due to the tip's size, it may not be able to completely penetrate to deep and narrow valley bottoms (see Figure 12);
- *re-entrant features cannot be traced* – whenever a stylus encounters a re-entrant feature (Figure 12 detail) the stylus tip loses partial contact with the profile and as a result will remove this feature from the trace. Typical surfaces that exhibit such re-entrant characteristics are powder metallurgy (PM) compacts and other porous parts, together with certain cast irons. This means that misleading surface texture parameter measurements may occur – although recently DIN 4776: 1988 has described a method of obtaining "distortion-free" assessment for PM parts. If transverse sections are taken and positioned within a scanning electron microscope re-entrant features can be visually revealed.

If the slope of a surface texture feature has a steeper angle than that of the half angle of the side of a conical stylus tip, then it will lose contact with the profile, having the same effect as a re-entrant angle on the resulting measurement. The stylus is the only active contact between the surface and the instrument and it has been shown above that its shape and size affect profile trace accuracy. If the force exerted by the stylus is too great, it may scratch or deform the surface, leading to erroneous results. Conversely, the stylus force must be sufficient to ensure that the stylus maintains continuous contact with the surface at the adopted traverse speed. Metrological instrument manufacturers incorporate this into the transducer design, with most surface texture equipment having a tip radius of between 0.1 and 12 μm, with a stylus force of ≈0.75 mN. This stylus force changes as the tip becomes progressively more blunt as it bears on the surface. An instrument with very fine tip radius (Figure 10a

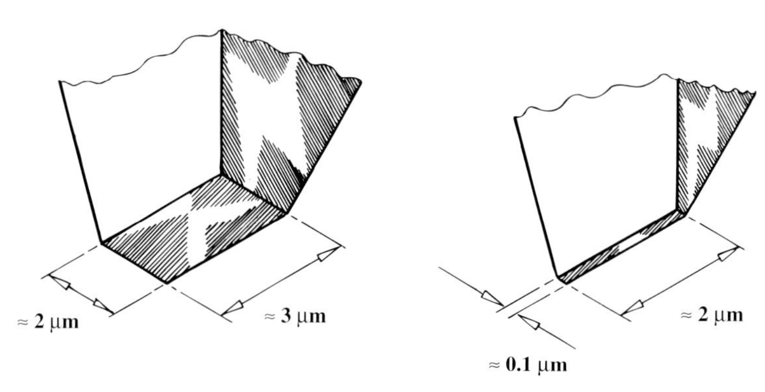
(a) Typical stylus geometries
≈ 2 μm
≈ 3 μm
≈ 2 μm
≈ 0.1 μm

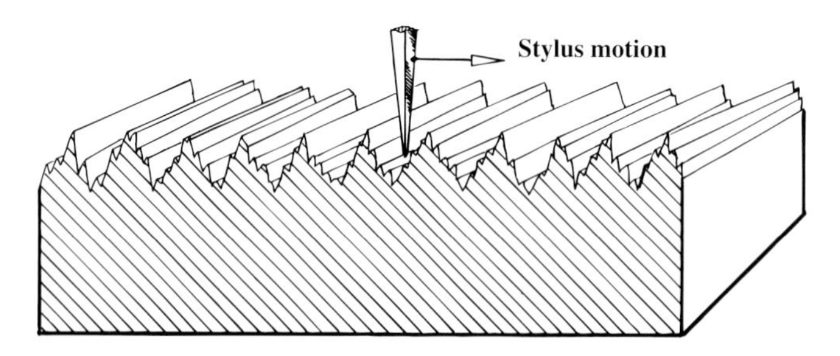
(b) Relative size of the stylus against the surface topography
Stylus motion

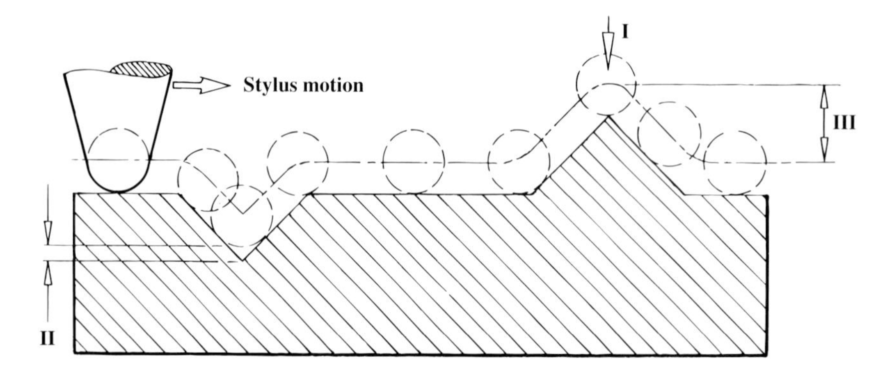
(c) The curve tends to round the peaks (I) and reduce the depth of the valleys (II), although the peak height is not affected (III)
Stylus motion
I
II
III

Figure 11. Stylus geometry and its mechanical filtering effect on surface texture. (Courtesy of Taylor Hobson.)

right) must utilise an exceedingly low force, typically $\approx 3 \times 10^{-5}$ N or less. The transducer assembly can incorporate both a stylus and skid, with the skid offering a local datum for this transducer and with that of the surface. Moreover, the skid provides a local datum for the stylus with respect to its vertical and horizontal motion. Therefore, with the skid's large curved radius to that of the relatively small surface under test, it rides along the surface being measured (Figure 13), providing a "local datum". As long as the skid's radius is greater than the peak spacing, the apparent line of movement of this skid will be virtually a straight line. Under such surface transducer conditions as the vertical skid moves from crest to crest, its relative horizontal skid motion with respect to the test surface can be ignored, as the skid's vertical motion is virtually insignificant. However, once the crests in the surface under test become more widely spaced, this horizontal crest spacing tends to introduce a significant

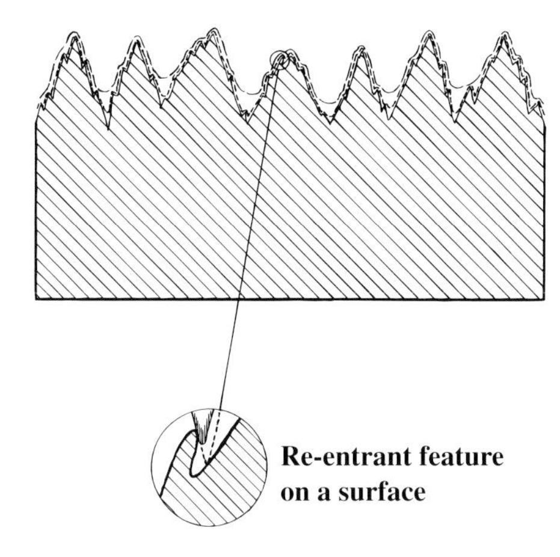

Figure 12. Illustrating how the use of a larger-radius stylus reduces the apparent amplitude of closely spaced irregularities. (Courtesy of Taylor Hobson.)

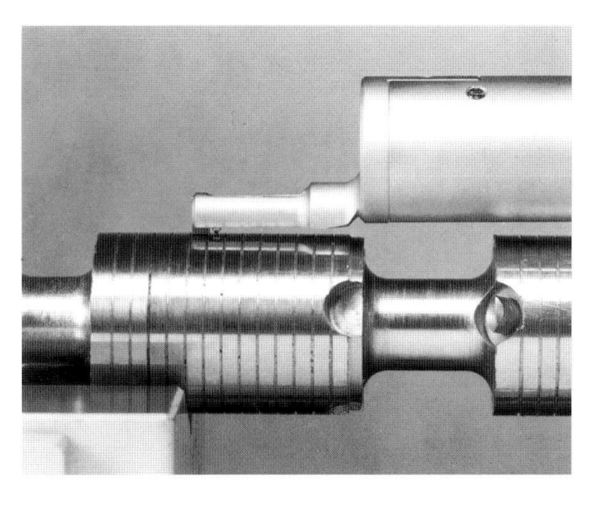

Figure 13. Surface texture of component being inspected using a skid. (Courtesy of Taylor Hobson.)

amount of vertical skid motion. If the skid's vertical motion affects the subsequent surface texture results, it is then preferable to use the instrument in the "skidless mode", as instruments employing a skid should be used for the measurement of roughness parameters (ISO 3274: 1996). To achieve skidless operation and minimise the unwanted affects of vertical skid motion which might otherwise interfere with the results, the skid must be removed, then by utilising the instrument's in-built straight edge as a "sliding datum" the surface assessment can be undertaken, as illustrated in Figure 14. More will be mentioned on the influence of the skid and its associated "phase effect" later.

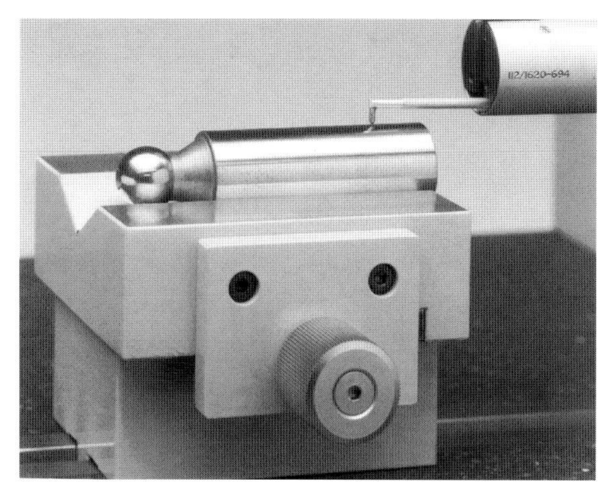

Figure 14. A high-quality component's surface texture being assessed without a skid which tends to be the "approved method", of late. (Courtesy of Taylor Hobson.)

1.5 Filters and cut-off

It was previously suggested that surface geometry is comprised of roughness, waviness and profile (see Figure 2). These interrelated factors also tend to have different relationships to the performance of the component. It is normally the case that they are separated out during analysis. To obtain this separation on surface texture measuring instruments, selectable filters for roughness and waviness are applied. In this section, only a superficial treatment of filtering and its effects will be mentioned, as this will be dealt with in greater depth later in the chapter.

In general, selection of the desired filtering can be established by the following "rules of thumb":

- *roughness filtering* would be applied if control is required of workpiece performance (for example: resistance to stress failure; wear resistance; friction; reflectivity and lubrication properties);
- *waviness filtering* might be selected if control is necessary for workpiece or machine tool performance (for example, vibration or noise generation).

Once the type of filter has been chosen then the filter cut-offs should be selected (see ISO 11562: 1996 for more detailed information) with a range, notably: "... mm; 0.08 mm; 0.25 mm; 0.8 mm; 2.5 mm; 8.0 mm; ..." (ISO 3274: 1996). However, this can only be undertaken when the specific features needing to be measured are known. Generally, the following criteria can be applied:

- a 0.08 mm cut-off length should be selected for high-quality optical components;
- a 0.8 mm cut-off length should be chosen for general engineering components;
- longer cut-offs are necessary when very rough components are to be measured, or if surfaces for cosmetic appearance are important.

NB A practical guideline is that the cut-off chosen should be of the order of 10 times the length of the feature spacings under test. If there is any doubt about the selection of cut-off, initially measure the surface in an unfiltered manner and then make a qualified decision from this test.

So far, the term cut-off has been liberally used, but what is it? As an example of cut-off and assuming a perfect filter characteristic, if a *high-pass filter* has a cut-off of 0.8 mm, this will only allow wavelengths below 0.8 mm to be assessed, with wavelengths above this value being removed. Conversely, when a *low-pass filter* of 0.8 mm cut-off has been selected, this will only allow wavelengths above 0.8 mm to be assessed. As previously mentioned (ISO 3274: 1996), the internationally recognised cut-off lengths for surface roughness measurement are given in Table 1. One basic feature of stylus-based instruments is that the profile is traced by *movement* across the test surface, with the consecutive points of the profile in contact being spaced in *time*. Hence, this relationship between movement and time is closely associated with cut-off.

Table 1. Typical cut-offs for variously manufactured surfaces

Typical finishing process	Meter cut-off (mm)				
	0.25	0.8	2.5	8.0	25.0
Milling		X	X	X	
Boring		X	X	X	
Turning		X	X		
Grinding	X	X	X		
Planing			X	X	X
Reaming		X	X		
Broaching		X	X		
Diamond boring	X	X			
Diamond turning	X	X			
Honing	X	X			
Lapping	X	X			
Superfinishing	X	X			
Buffing	X	X			
Polishing	X	X			
Shaping		X	X	X	
Electro-discharge machining	X	X			
Burnishing		X	X		
Drawing		X	X		
Extruding		X	X		
Moulding		X	X		
Electro-polishing		X	X		

NB The cut-off selected must be one that will give a valid assessment of the surface characteristic of interest. For example, although a cut-off of 0.8 mm can validly be used for almost all of these surfaces, it may not necessarily be suitable for assessing a particular feature of the texture. This entails examining the surface and considering the purpose of the measurement before selecting the cut-off. (Courtesy of Taylor Hobson.)

The frequencies present are dependent upon two factors: traverse speed and spacing irregularities. In Figure 15 (A and B), the frequencies of 100 Hz and 50 Hz represent irregularity spacings of 0.01 mm and 0.02 mm respectively. At the same traverse speed, frequencies of 4.0 Hz and 1.25 Hz would be generated by spacings of 0.25 mm and 0.8 mm. If a high-pass electrical filter is used, this will suppress any frequency below 4.0 Hz, enabling only those irregularities having spacings of 0.25 mm or less to be represented in the filtered profile. Therefore, this relationship is ideal to obtain a sampling length of 0.25 mm, hence the term "cut-off" length – since the response now cuts off at irregularity spacings of 0.25 mm, denoted by the international symbol, (λc) for cut-off length. By introducing different filters, the most suitable cut-off for the surface can be selected, where Table 1 acts as a guide to the selection of suitable cut-offs for various machining processes.

Until 1990, standard roughness and waviness filters were analogue or 2RC (two resistors and two capacitors) types. Since 1996, the standard (ISO 11562: 1996) has specified the digital filter for surface texture measurements. For example, if a 0.8 mm cut-off value is utilised, then there is a 50% transmission of the profile equal to this 0.8 mm. This type of filtering is analogous to a sieve, only allowing values below the cut-off to be passed through. The main problem with utilising an analogue filter is that it assumes that the surface is basically sinusoidal in nature, but this rarely is the case. This mismatch with the "real" surface under test causes the so-called "Gibb's phenomenon". The Gibb's phenomenon results in an overshooting of the filtered profile when analysing certain extreme forms of profile. The "Gibb's phenomenon" can be explained by the following example.

Filtering is used to determine the mean line of the roughness profile – being identical to the waviness profile. An analogue filter determines a profile's mean line by a "moving average" technique. Hence, at any point in time the filter averages the profile previously traversed by the stylus. This has been equated to a person walking backwards! Such a filter cannot anticipate what lies ahead of it in the remainder of the profile, such as abrupt changes in the topography. Under these conditions, it reacts after a significant change has occurred; this means that the true profile is no longer represented by the mean line, and as a result the filtered profile will be distorted vertically

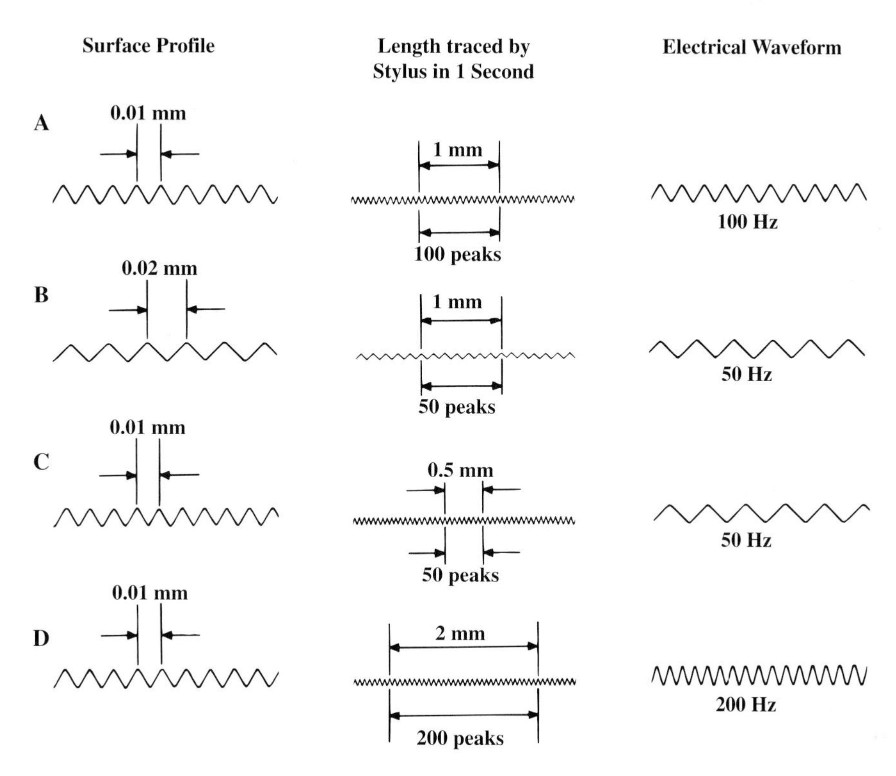

Figure 15. Signal frequency depends on irregularity spacing (compare A and B) and traverse speed (compare A, C and D). (Courtesy of Taylor Hobson.)

and horizontally shifted. Such an action by the analogue filter may lead to serious errors for either plateau honed types of surface or PM (porous) types of topographies.

If a digital filter is employed, such as that shown on the graph in Figure 16(a), then the vertical distortions of the profile caused by abrupt changes in the height of the topography are minimised. The principle operation of the digital filter can be explained as below.

Digital profiles are no longer represented by a smooth curve, but instead, are described by a series of ordinates (numbers) relating to profile heights at regular intervals. In order to estimate the profile's mean line the filter's action is to calculate the average height at a given point, as the arithmetic average of the ordinate points in its vicinity. Hence, if the profile is divided up into "windows" the mean profile height for each window can be estimated. A line joins these window averages together, which represents the profile mean line.

Such a digital filter allows up to 100% transmission of the profile, representing wavelengths shorter than the actual cut-off length, together with zero transmission of wavelengths greater than this cut-off. Figure 16(bi) shows how the unfiltered profile equates to an electrically filtered one – Figure 16(bii) – for the same value of cut-off, but related to 50% of its depth. The main difference between the (PC) digital filter and the analogue filter is that the former type evaluates the profile mean around a point – enabling it to anticipate changes in the profile height – whereas the latter (analogue) filter cannot. It is possible to emulate the analogue filter's characteristics by employing a "weighting function" for each ordinate height when a determination of the arithmetic average or mean point is required. However, unless there is a large variation in the surface topography, or special processing conditions apply – such as the "Gibb's phenomenon" – then the analogue-to-digital differences remain slight.

1.6 Measuring lengths

In order to determine a workpiece's surface texture, three characteristic lengths are associated with the profile (ISO 4287: 1997). These are:

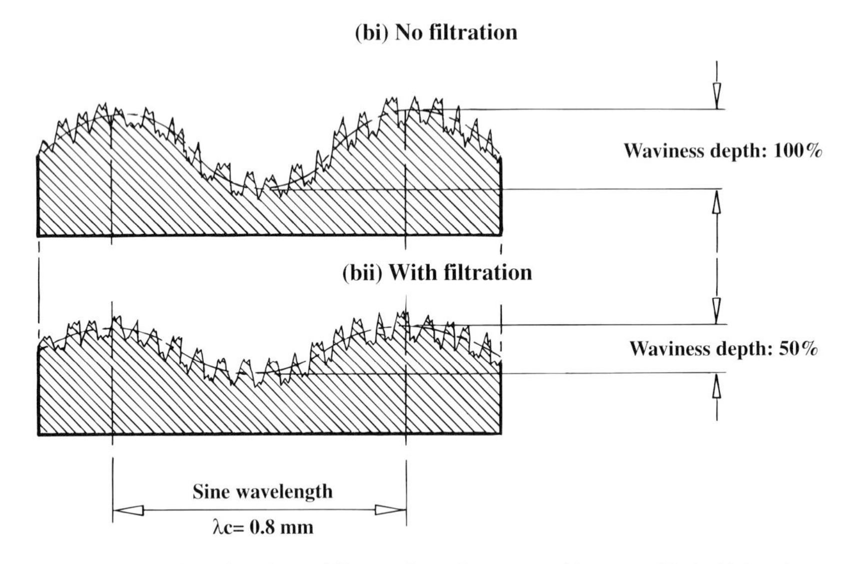

Figure 16. Effect of an electrical filter on the surface texture. (Courtesy of Taylor Hobson.)

- *sampling length*;
- *evaluation (or assessment) length*;
- *traverse length.*

Sampling length (*lp*, *lr* and *lw*)

The surface texture profile illustrated in Figure 17(a) has both roughness and waviness present, with both patterns being generally repetitive in nature. Therefore the roughness pattern A over the distance L might be typical for the whole length of the surface, only differing in minor detail from that indicated at B over a similar distance.

NB If a shorter distance T was selected, the roughness patterns at C and D would not be identical. This would give a misleading picture of their respective roughness heights.

What was true for the reduced distances also applies to waviness, neglecting the roughness (Figure 17b). This is repetitive over distance U, although if the examination is confined to a shorter distance V then the waviness at E and F appear to be markedly different in character. Such a variation demonstrates why it is necessary to select the length of the profile over which a parameter is to be determined, termed its "sampling length", to suit the surface under test. Yet another feature concerning

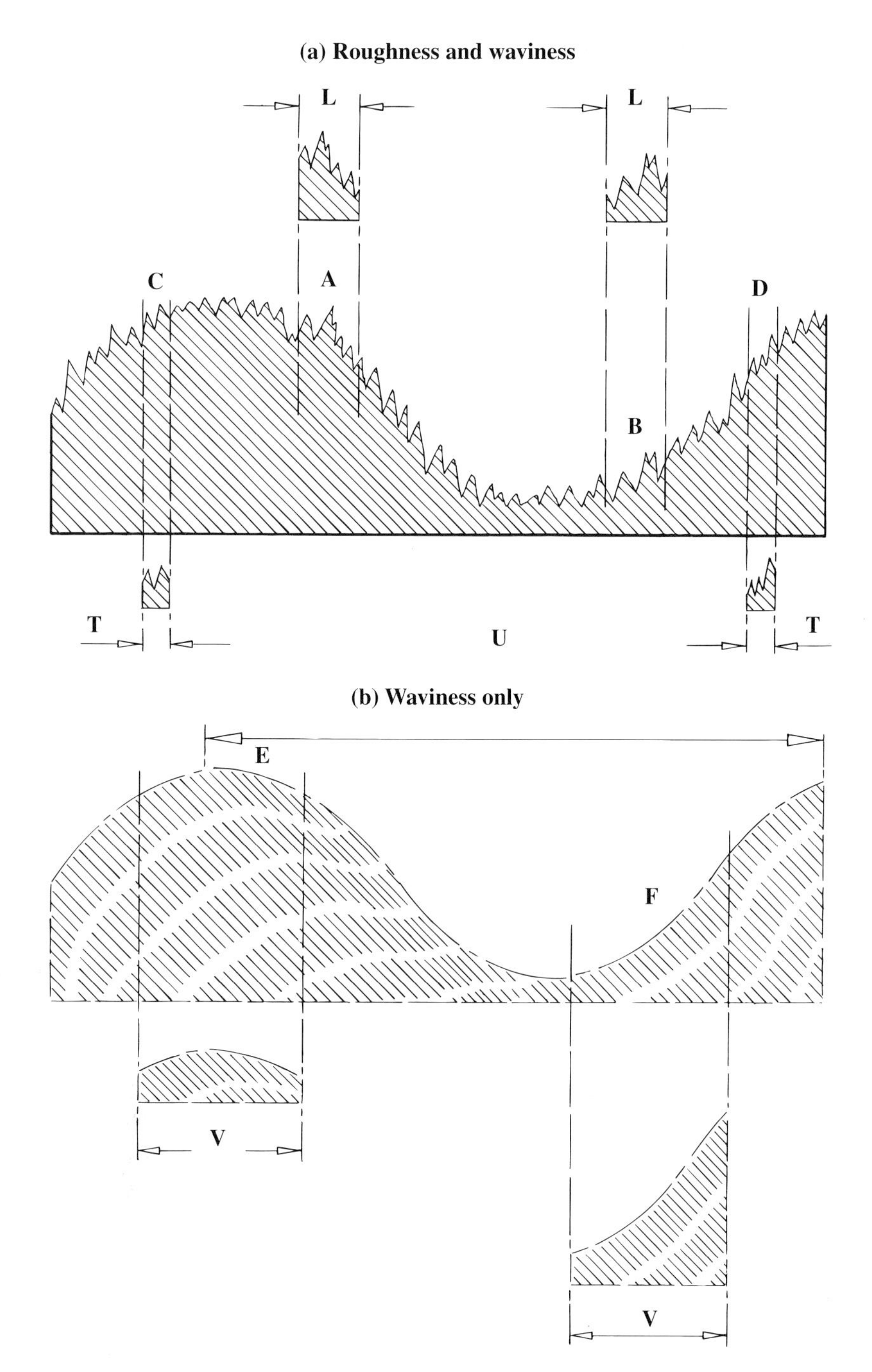

Figure 17. Effect of different sampling lengths. (Courtesy of Taylor Hobson.)

selection of appropriate lengths can be demonstrated in Figure 17(a). Figure17(a) indicates that while the sampling length L is sufficient to reveal the whole of the roughness pattern, the waviness has very little influence over this length. Even if the sampling length was increased, it would simply include more roughness detail but would have little, if any, effect on a roughness parameter value. In this case, the waviness would play a greater role and for this reason it is undesirable to increase the sampling length much beyond that required to obtain a representative assessment of roughness.

The sampling length can be defined as *the length in the direction of the* ***x-axis*** *used for identifying the irregularities that characterise the profile under evaluation*. By specifying a sampling length, this implies

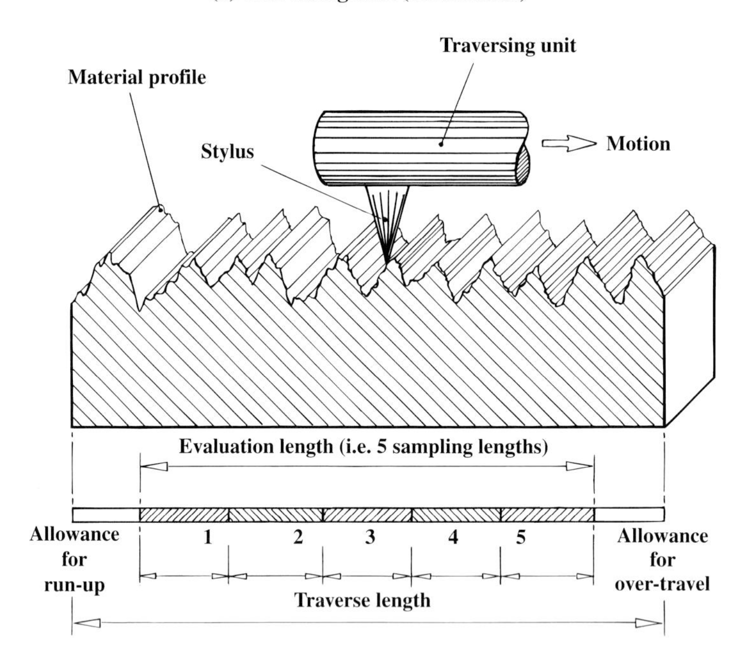

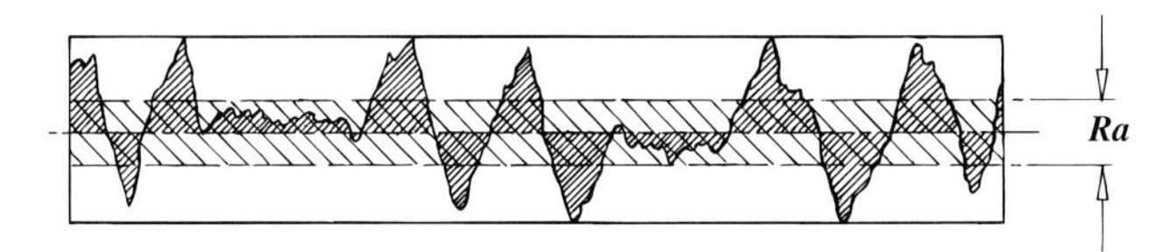

Figure 18. Determining surface texture analysis roughness parameter Ra, together with its arithmetic derivation.

that structure in the profile occurring over longer lengths will not be applicable in this particular evaluation. The sampling length for *lr* – roughness – is equal (numerically) to the wavelength of the profile λc, whereas the sampling length for *lw* – waviness – equates to that of the profile filter λf. Virtually all the parameters necessitate evaluation over the sampling length; however, reliability is improved by taking an average of them from several sampling lengths as depicted by the evaluation length in Figure 18(a).

Evaluation length (*ln*)

A simplistic diagram of the relationship of the stylus to the material profile/topography for a pre-selected traverse length is indicated in Figure 18(a). The evaluation length can be defined as *the total length in the* ***x-axis*** *used for the assessment of the profile under evaluation* (see Figure 18a). As shown, this length may include several sampling lengths – typically five – being the normal practice in evaluating roughness and waviness profiles. The evaluation length measurement is the sum of the individual sampling lengths. If a different number is used for the roughness parameter assessment, ISO 4287: 1997 advocates that this number be included in the symbol, for example *Ra6*. In the case of waviness, no default value is recommended. From a practical viewpoint, the selection for the correct filter can normally be ensured when at least 2.5 times the peak spacing occurs, with two peaks and valleys within each one. Typically, this would mean that an evaluation length of 0.8 mm would be selected, but there are occasions when either a larger or smaller evaluation length might be preferable for the surface under test. The metrologist's experience and judgement will come into play here.

Traverse length

The traverse length can be defined as *the total length of the surface traversed by the stylus in making a measurement*. It will normally be greater than the evaluation length, due to the necessity of allowing run-up and over-travel at each end of the evaluation length to ensure that any mechanical and electrical transients, together with filter edge effects, are excluded from the measurement (see Figure 18a).On shorter surfaces, it may be necessary to confine the measurements to one sampling length, under these circumstances the sampling and evaluation lengths are identical, but the over-travel necessary to contribute to the traverse length must be retained.

1.7 Filtering effects (λs, λc and λf)

In Figure 19 a flow chart is illustrated showing the manner in which surface assessment occurs, with the primary profile broken down into filtered elements to obtain the waviness and roughness profiles of the workpiece surface. Algorithms enable suitable characteristic functions and parameters to be obtained, scaled to an appropriate size. As has been mentioned previously in Section 1.5, filters are vital to any form of surface texture analysis. Today, most forms of filtering are normally electrical or computational, although they can occasionally be mechanical for analysis of the range of structure in the total profile, which can be adjudged to be of most significance in a particular situation. Conversely, the effects of filtering can be considered as a means of removing irrelevant information, such as instrument noise and imperfections. Therefore, filters can select or reject surface topographical structure depending upon its scale in the x-axis, in terms of spatial frequencies, or wavelengths. The terms *low-pass* and *high-pass filters* refer to whether they either reject or preserve surface data. For example, a *low-pass filter* will reject short wavelengths while retaining longer ones, whereas a *high-pass filter* preserves shorter-wavelength features while rejecting longer ones. The term *bandpass filter* refers to the combination of a low-/high-pass filter selected to restrict the range of wavelengths, with both high- and low-pass regions being rejected. Filter attenuation (rejection) should be somewhat gradual, otherwise significantly differing results occur from almost identical part surfaces, the exception being when a significant surface feature causes a slight wavelength shift.

The term *cut-off* refers to the 50% transmission (and rejection) wavelength filter, this being specific to the topic of surface texture. In the case of the large majority of surface texture assessments, it is suggested that when the width of a specific surface feature is significant, but its size may only be 1% of the overall width, this will make it less important. Under these circumstances that affect feature transmission/rejection, it is suggested that *bandpass*

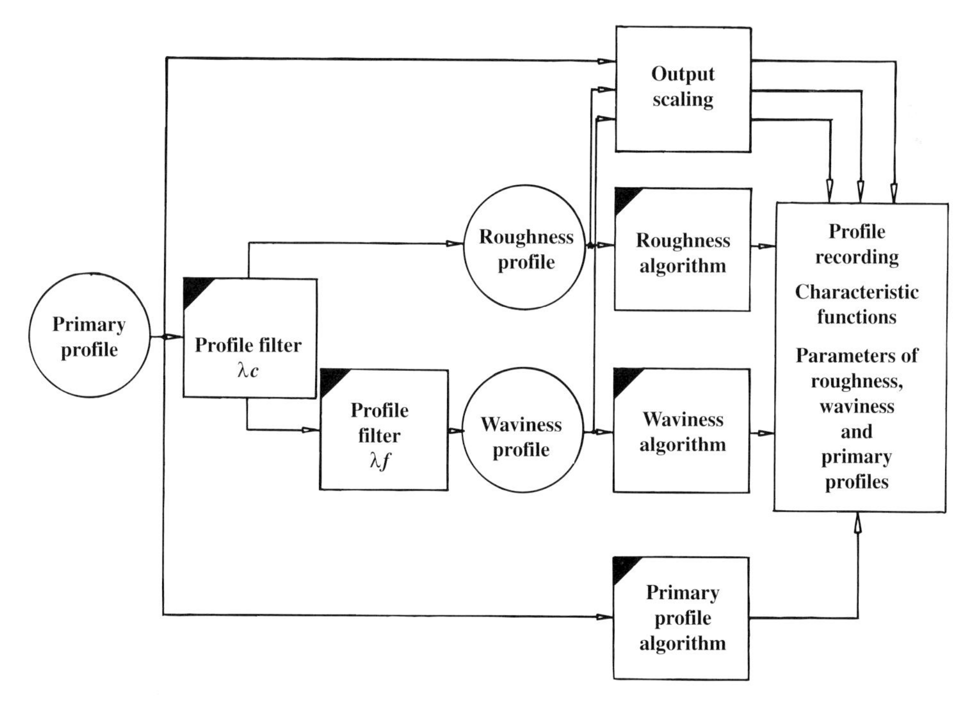

Figure 19. A flowchart for surface assessment. [Source: ISO 4287: 1997.]

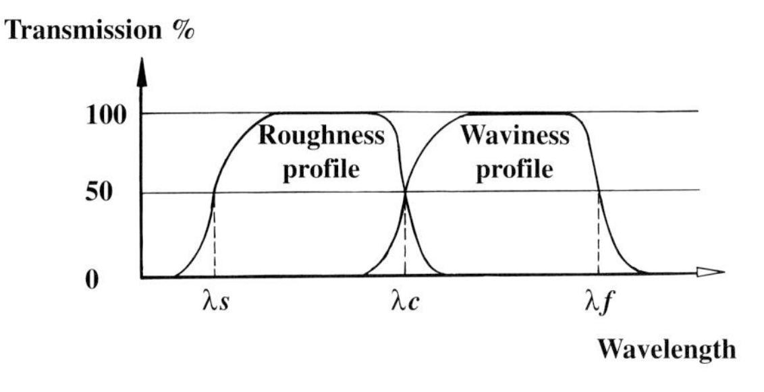

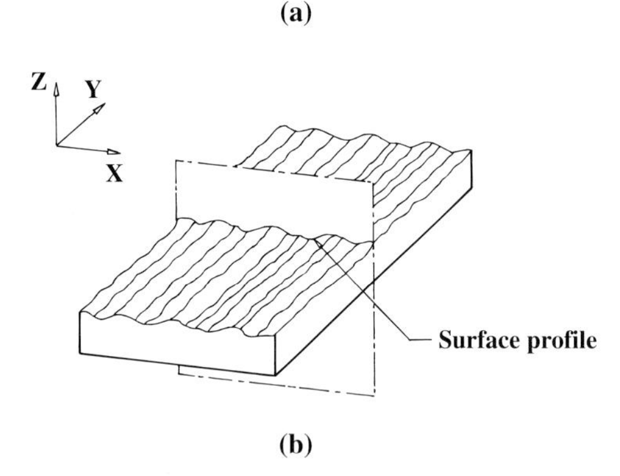

Figure 20. (a) Transmission characteristic of roughness and waviness profiles. (b) Surface profile. [Source: ISO 4287: 1997]

filtering is vital to any potential analysis of surface texture.

Profile filter

The "profile filter" can be defined (ISO 4287: 1997) as a *filter that separates profiles into long- and short-wave components* (see Figure 20a). In particular, three filters can be employed for surface measurement (Figure 20a):

- *profile filter* (λs) – this filter defines where the intersection occurs between the roughness and the presence of shorter-wavelength components in a surface;
- *profile filter* (λc) – this filter defines the intersection position between the components of roughness and waviness;
- *profile filter* (λf) – filters of this type define the intersection point between waviness and the presence of any longer wavelengths in a surface.

Primary profile

The *primary profile* forms the basis for primary profile parameter evaluations, being defined as *the total profile after application of a short-wavelength* (low-pass) *filter, with a cut-off* λs. The finite size of a stylus will eliminate any very short wavelengths, which practically is a form of mechanical filtering, being in the main utilised as a default for the λs filter. As the size of styli will vary and the instrument can introduce vibration and/or supplementary noise into the profile signal (having equivalent wavelengths to that of the stylus dimensions) under these circumstances, it is recommended to ignore any λs filtration along the total profile.

Roughness profile

The *roughness profile* can be defined as *the profile derived from the primary profile by suppressing the long-wave component using a long-wavelength* (high-pass) *filter, with a cut-off* λc. This roughness profile provides the foundation for the evaluation of the roughness profile parameters and will automatically include λf profile filter information, because it is derived from the primary profile.

Waviness profile

The derivation of this waviness profile is the result of the application of the bandpass filter to select the surface structure at somewhat longer wavelengths to that of roughness. The *filter* λf will suppress any longer-wave components, namely the *profile* component, while the *filter* λc suppresses the shorter-wave components, such as any of the *roughness* components. Hence, the waviness profile forms the basis for the evaluation of the waviness profile parameters.

Roughness profile mean line

The roughness profile mean line is a reference line for parameter calculation, which corresponds to the suppression of the long-wave profile component by the profile filter λc.

Waviness profile mean line

The waviness profile mean line is a reference line for parameter calculation, which corresponds to the suppression of the long-wave profile component by the profile filter λf.

Primary profile mean line

The primary profile mean line is a reference line for parameter calculation, being a line determined by fitting a least-squares line of nominal form through the primary profile.

Surface profile

In ISO 4287: 1997, the coordinate system is defined by making use of the rectangular coordinate of a right-handed Cartesian set for the surface profile (Figure 20b). Here, the *x-axis* provides the *direction* of the trace; the *y-axis* nominally lies on the *real surface*, with the *z-axis* facing in an *outward direction* from the material to the surrounding medium. Thus, the *real surface* can be defined as *the surface limiting the body and separating it from the surrounding medium.* The *lay* previously alluded to in Section 1.1 is normally functionally significant for the workpiece, so it is important to specify it on engineering drawings in terms of the type of *lay* and its direction (ISO 1302: 2001 - Appendix B).

1.8 Geometrical parameters

The following discussion is concerned with the calculation of parameters from the measurement data. These parameters are all derived *after* the *form* has been removed. It should be stated that any parameters selected should be appropriate for a given application and that not all of them would be necessary in all circumstances. On an engineering drawing, a designer may have specified a parameter to define the part's functional behaviour, which means that this parameter is outside the control of the user of the instrument. When surface texture parameters have been specified, the user needs to have a clear and unambiguous understanding of the manner in which they are calculated.

In understanding and evaluating surfaces, the philosophy behind the reasons why *peaks* and *valleys* are significant is important, although just how and what represents a peak/valley is not always clear (see further discussion on this topic in the Appendix). When a surface topography shows the presence of a peak or valley that is not significant, the decision concerning its importance for in-service functional characteristics is not always a simple matter to establish. In the surface profile shown in Figure 21(a), for example, are both the peaks here of importance, or even significant? In order to minimise any confusion for the user, the latest standards have introduced the important concept that the *profile element* consists of a *peak and valley event.* Moreover, associated with this *profile element* is a discrimination that prevents any minute, unreliable measurement features from affecting the detection of these elements. Several of these surface texture-related elements and associated features are described in ISO 4287: 1997, consisting of:

- *profile element* - the section of a profile that crosses the mean line to the point at which it next crosses this mean line in the same direction - for example, from below to above the mean line;
- *profile peak* - the section of the profile element that is above the mean line, namely, the profile by which it crosses the mean line in the positive direction until it once again crosses this mean line, but now in the negative direction;
- *profile valley* - as described for the "peak" above, but with the direction reversed;
- *discrimination level* - within a profile it may be likely that a minute fluctuation occurs which takes the profile trace across the mean line, then back again almost immediately. In such a condition the departure of the trace from the mean line is not considered to be a real profile peak or valley. In order to prevent automatic systems from counting small trace departures, allowing them to only consider those larger surface features, they specify particular heights and widths to be counted. Hence, default levels are established in the absence of other specifications. These default levels are set such that their profile peak heights or valley depths must exceed 10% of the Rz, Wz or Pz parameter value. Furthermore, the width of either a profile peak or valley must exceed 1% of the sampling length, with both criteria being simultaneously met;
- *ordinate value* $Z(x)$ - this represents the height of the assessed profile at any position x, its height being regarded as negative when the ordinate lies below the x-axis, and otherwise positive;
- *profile element width* Xs - this is concerned with the x-axis segment intersecting with the profile element (see Figure 21a);
- *profile peak height* Zp - this represents the distance between the mean line on the x-axis and the highest point of the largest profile peak (see Figure 21a);
- *profile valley depth* Zv - this represents is the distance between the mean line on the x-axis and the lowest point of the deepest valley (see Figure 21a);
- *profile element height* Zt - this equates to the sum of the peak height and valley depth of the profile element, namely, the sum of Zp and Zv (see Figure 21a);
- *local slope* dZ/dX - this represents the slope of the assessed profile at position x_i (see Figure 21b).

The local slope's numerical value critically depends upon the ordinate spacing and thus influences $P\Delta q$, $R\Delta q$ and $W\Delta q$;

- *material length of profile at level "c" Ml(c)* – this represents the section lengths obtained when intersecting with the profile element by a parallel line to the x-axis at a given level "c" (see Figure 21c).

1.9 Surface profile parameters

The following examples of defined parameters can be calculated from any profile. The designation of letters follows the logic that the parameter symbol's first capital letter denotes the type of profile under evaluation, for example the *Ra* is calculated from the roughness profile, while *Wa* has it origin from the waviness profile, with the *Pa* occurring from the primary profile. In the subsequent sections, the definitions and schematic representations of many of these surface profile parameters are described along with examples of their usage.

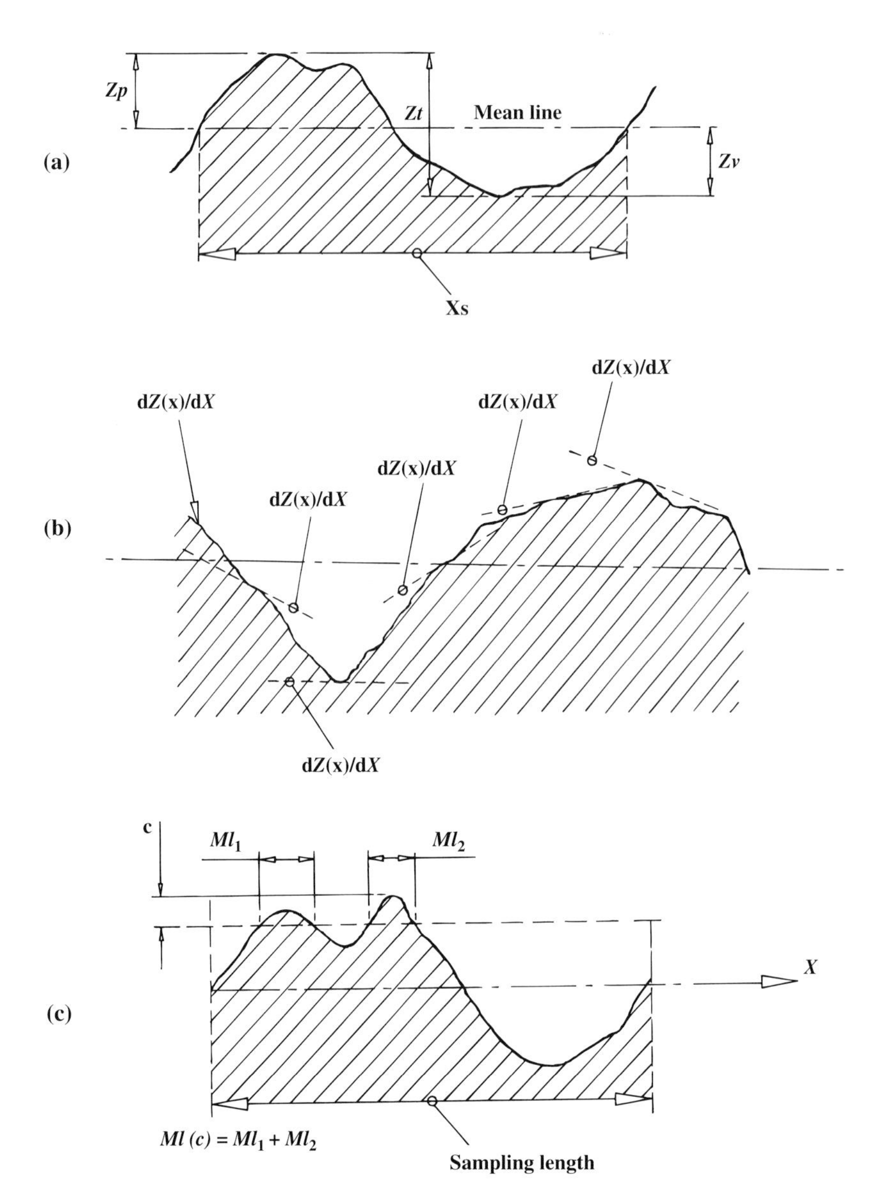

Figure 21. (a) Profile length, (b) local slope and (c) material length of a surface trace. [Source: ISO 4287: 1997]

1.9.1 Amplitude parameters (peak-to-valley)

Maximum profile peak height Rp, Wp and Pp

This parameter is represented by the largest peak height (*Zp*) within the sampling length, with its height being measured from the mean line to the highest point (see Figure 22a). The *Rp* parameter (equating to roughness) is generally less favoured, with preference given to parameters based on the total peak-to-valley height. Often *Rp* and its associated parameters *Wp and Pp* are referred to as extreme-value parameters, being somewhat unrepresentative of the overall surface, because their numerical value can vary between respective samples. In order to minimise variation in *Rp*, it is feasible to average readings over consecutive sampling lengths, although in most cases the value obtained is numerically too large to offer practical assistance. The *Rp* and its *Wp* and *Pp* derivatives are not without some merit, as they can establish unusual surface features such as either a burr on the surface or sharp spike, and will indicate the presence of scratches or cracks – possibly indicative of low-grade material or poor production processing.

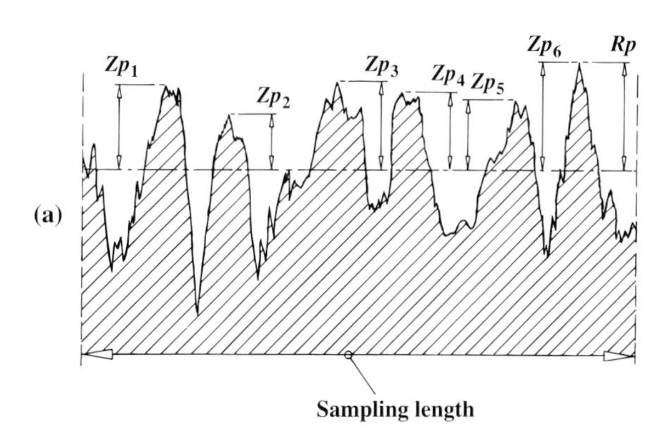

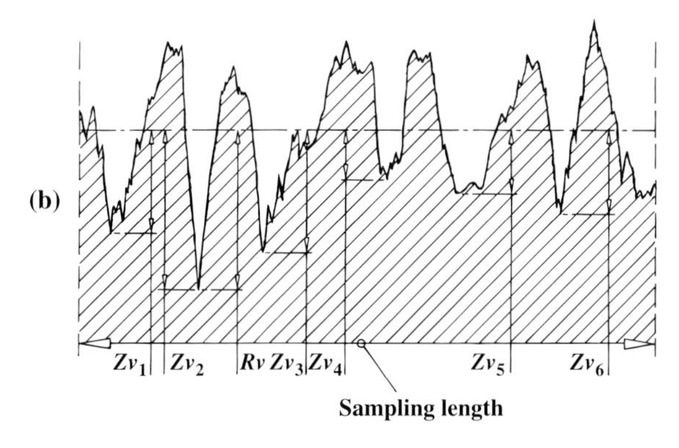

Figure 22. (a) Maximum profile peak height and (b) maximum profile valley depth (examples of roughness profiles). [Source: ISO 4287: 1997]

Maximum profile valley depth Rv, Wv and Pv

This parameter is represented by the largest valley depth (*Zv*) within the sampling length, with its absolute value *Rp* (equating to roughness) being obtained from the lowest point on the profile from the mean line (see Figure 22b). As in the case above for peaks, the maximum profile valley depths are extreme-value parameters. In a similar manner to *Rp*, *Rv* can establish whether there is a tendency to workpiece cracking, spikes and burrs, detecting such features on the surface.

Maximum height of profile Rz, Wz and Pz

The maximum height parameter (for example, *Rz* – for roughness) is the sum of the height of the largest profile peak height (*Zp*), together with the largest profile valley (*Zv*) within the sampling length (see Figure 23a). In isolation, the *Rz* does not provide too much useful surface information, and therefore it is often divided into *Rp* and *Rv*, previously described. In fact, in ISO 4287: 1984, the *Rz* symbol indicated the *ten point height irregularities*, with many of the previous surface texture measuring instruments measuring this *Rz* parameter – see the Appendix for more information regarding this previously utilised parameter.

Mean height of profile elements Rc, Wc and Pc

Parameters of this type evaluate surface profile element heights (*Zt*) within the sampling length (see Figure 23b) and to obtain this parameter it requires both height and spacing discrimination, previously mentioned. When these values are not specified then the default height discrimination used shall be 10% of *Rz*, *Wz* and *Pz*, respectively, with the default spacing discrimination being stated as 1% of the sampling length. Both of these height and spacing requirements must be met. Normally in practice it is extremely rare to utilise this parameter because of its difficulty in interpretation and, as a result, it would rarely be used on an engineering drawing.

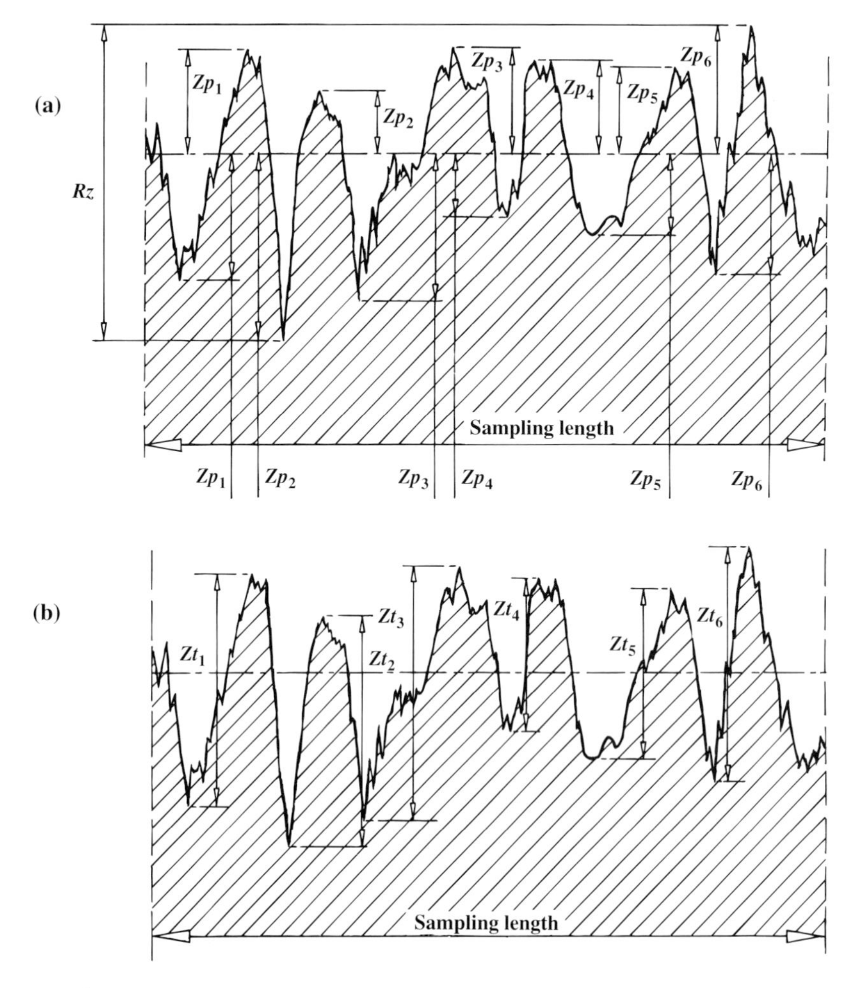

Figure 23. (a) Maximum height of profile and (b) height of profile elements (examples of roughness profiles). [Source: ISO 4287: 1997]

Total height of profile Rt, Wt and Pt

This parameter relates to the sum of the height of the largest peak height (Zp) and the largest profile valley depth (Zv) within the evaluation length. It is defined over the evaluation length rather than the sampling length and in this manner has no averaging effect. This lack of averaging means that any surface contamination (dirt) or scratches present will have a direct effect on the numerical value of, say, Rt.

1.9.2 Amplitude parameters (average of ordinates)

Arithmetical mean deviation of the assessed profile Ra, Wa and Pa

The arithmetic mean parameter is the absolute ordinate value $Z(x)$ within the sampling length The numerical value of Ra is able to vary somewhat without unduly influencing the performance of a surface. Often on engineering drawings a designer will specify a tolerance band, or alternatively a maximum value for Ra that is acceptable. The mathematical expression of this parameter is given below for these absolute ordinate values as:

$$Za = \frac{1}{N}\sum_{i=1}^{N} |Z_i|$$

where N = number of measured points in a sampling length (see Figure 24b).

In Figure 24(a) is illustrated the graphical derivation of this parameter within the sampling length, with the shaded areas of the graph below the centre line in A being repositioned above this centre line in B. Here, the *Ra* value is the mean line of the resulting profile indicated in C.

The absolute average roughness over one sampling length is given by the *Ra* value. This means that the influence of a discrete but non-typical peak or valley on the numerical value of *Ra* is not really significant. Good metrological practice is to ensure that a number of assessments of *Ra* occur over consecutive sampling lengths, enabling the acceptance of an average of these values. This sampling strategy ensures that the numerical value of *Ra* is typical for the inspected surface. As has been previously mentioned on several occasions, if a defined "unidirectional lay" is present on the surface (Figure 4), then any measurements should be undertaken perpendicularly to this "lay" if misinterpretation of the surface is to be avoided (Figure 5).

The numerical value of *Ra* does not impart any information regarding the surface shape, nor of any irregularities present. In fact, it is quite possible to have identical *Ra* values for surfaces that are markedly different in their profiles (Figure A in the Appendix) and this will affect their potential in-service performance, with engineering drawings often quoting the type of production process (Figure 3c). Historically, in Great Britain the most widely used of all the surface texture parameters has been *Ra*, but this should not discourage industrial and research organisations from using others, as they may offer more in terms of information regarding a surface's functional performance.

The mathematical derivation of both the commonly utilised parameters *Ra* and *Rq* are given below (see Figure 24b):

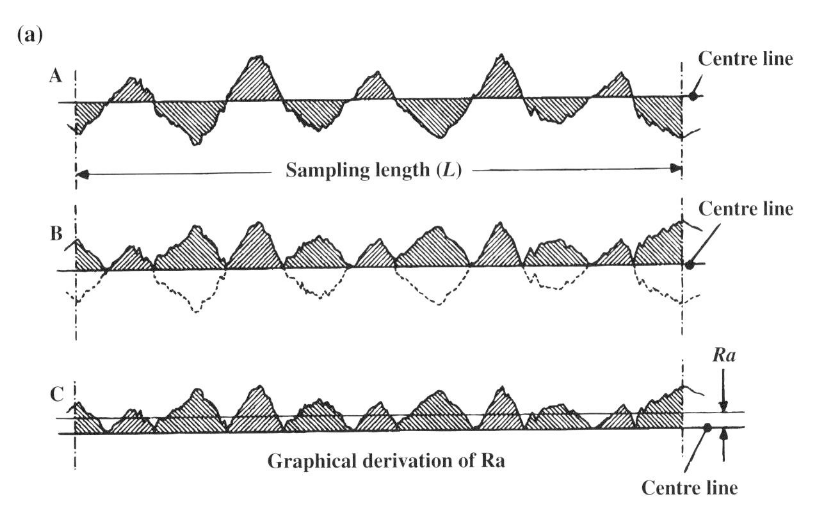

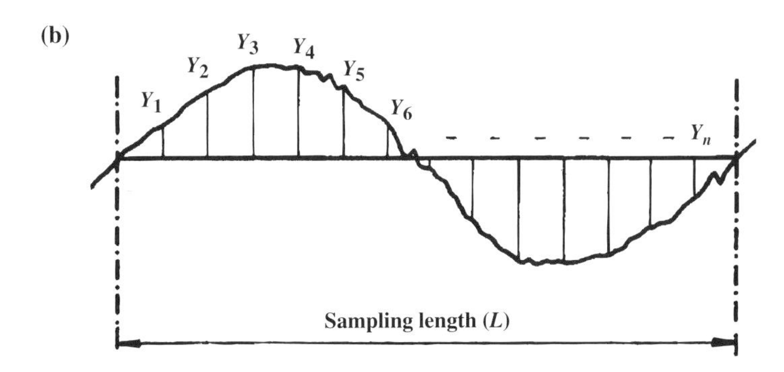

Figure 24. Determination of *Ra* and *Rq*. (Courtesy of Taylor Hobson.)

$$Ra = |Y_1| + |Y_2| + |Y_3| \ldots + |Y_n| / n$$

$$Rq = \sqrt{(Y_1^2 + Y_2^2 + Y_3^2 \ldots + Y_n^2)} / n$$

Root mean square deviation from the assessed profile *Rq*, *Wq* and *Pq*

The root mean square of the ordinate values *Z(x)* within the sampling length is established by the departures from mean line of the profile and can be mathematically expressed in the following way:

$$Zq = \sqrt{\frac{1}{N} \sum_{i=1}^{N} Z_i^2}$$

If a comparison is made between the arithmetic average and the root mean square parameters, the latter has the effect of providing additional weighting to the numerically greater values of surface height. One of the reasons for the harmony between both the *Ra* and *Rq* parameters is mainly from a historical perspective. The *Ra* parameter is much easier to determine graphically from a profile recording and for this reason, it was primarily adopted prior to automatic surface measurement instruments being developed. Once the roughness parameters utilising automatic surface texture instrumentation were made available, the parameter *Rq* had the advantage of being able to neglect the *phase effect* from the electrical filters, while the *Ra* parameter using arithmetic average will be affected by these phase effects and cannot be disregarded. To compound the problem, *Ra* has been virtually universally adopted for machining specifications to the detriment of *Rq*. The *Rq* parameter is still employed in many optical applications for the assessment of lenses, mirrors and the like for surface optical quality.

Skewness of the assessed profile *Rsk*, *Wsk* and *Psk*

This parameter represents the quotient of the mean cube value of the ordinate values *Z*(*x*) and the cube of *Pq*, *Rq* or *Wq*, respectively, within the sampling length (Figure 25). The skewness is derived from the amplitude distribution curve, representing the symmetry about the mean line. This parameter cannot distinguish if profile spikes are evenly distributed above or below the mean line, being considerably influenced by any isolated peaks or valleys present within the sampling length. Skewness can be expressed in mathematical form with *Rsk*, equating to:

$$Rsk = \frac{1}{Rq^3} \frac{1}{N} \sum_{i=1}^{N} Z_i^3$$

The skewness parameter of an amplitude distribution curve as depicted in Figure 25(a) indicates a certain amount of bias that might be either in an upward or downward direction. The amplitude distribution curve shape is very informative as to the overall construction of the surface topography. If this curve is symmetrical in nature, then it indicates symmetry of the surface profile; conversely, an unsymmetrical surface profile will be indicative of a skewed amplitude distribution curve (Figure 25b and c). As can be seen in Figure 25(b) and (c), the direction of skew is dependent upon whether the bulk material appears above the mean line – termed *negative skewness* (Figure 25c) – or below the mean line – *positive skewness* (Figure 25b). Utilising this skewness parameter distinguishes between profiles having similar if not identical *Ra* values.

Further, one can deal with this amplitude or height data statistically, in the same manner as one might physically measure anthropometric data such as a person's average stature, within a specific population range. As with the statistical dispersion of population stature, engineering surfaces can exhibit a broad range of profile heights.

For example, a boring operation with a relatively long length-to-diameter ratio may cause deflection (elastic deformation of the boring bar) and occasion the cutting insert to deflect, producing large peak-to-valley undulations along the bore (waviness). Superimposed onto these longer wavelengths are small-amplitude cyclical peaks (periodic oscillations) indicating vibrations resulting from the cutting process. The resultant surface profile for the bored hole would depict the interactions from the boring bar deformations and any harmonic oscillations. This boring bar motion reflected in the profile trace would exhibit low average profile height, but with a large range of height values.

Surface texture data can be statistically manipulated, beginning with the profile trace's height, or amplitude distribution curve – this being a graphical representation of the distribution of height ordinates over the total depth of the profile. The characteristics of amplitude distribution curves can be defined mathematically by several terms called "moments", but more will mentioned concerning this aspect in the Appendix.

A numerical value can be given to *Rsk* and in the case of Figure 25(b) the "positive skewness" may eventually obtain an adequate bearing surface, although it is unlikely to have oil-retaining abilities. This type of surface can typically be exploited for adhesive-bonding applications. The surface charac-

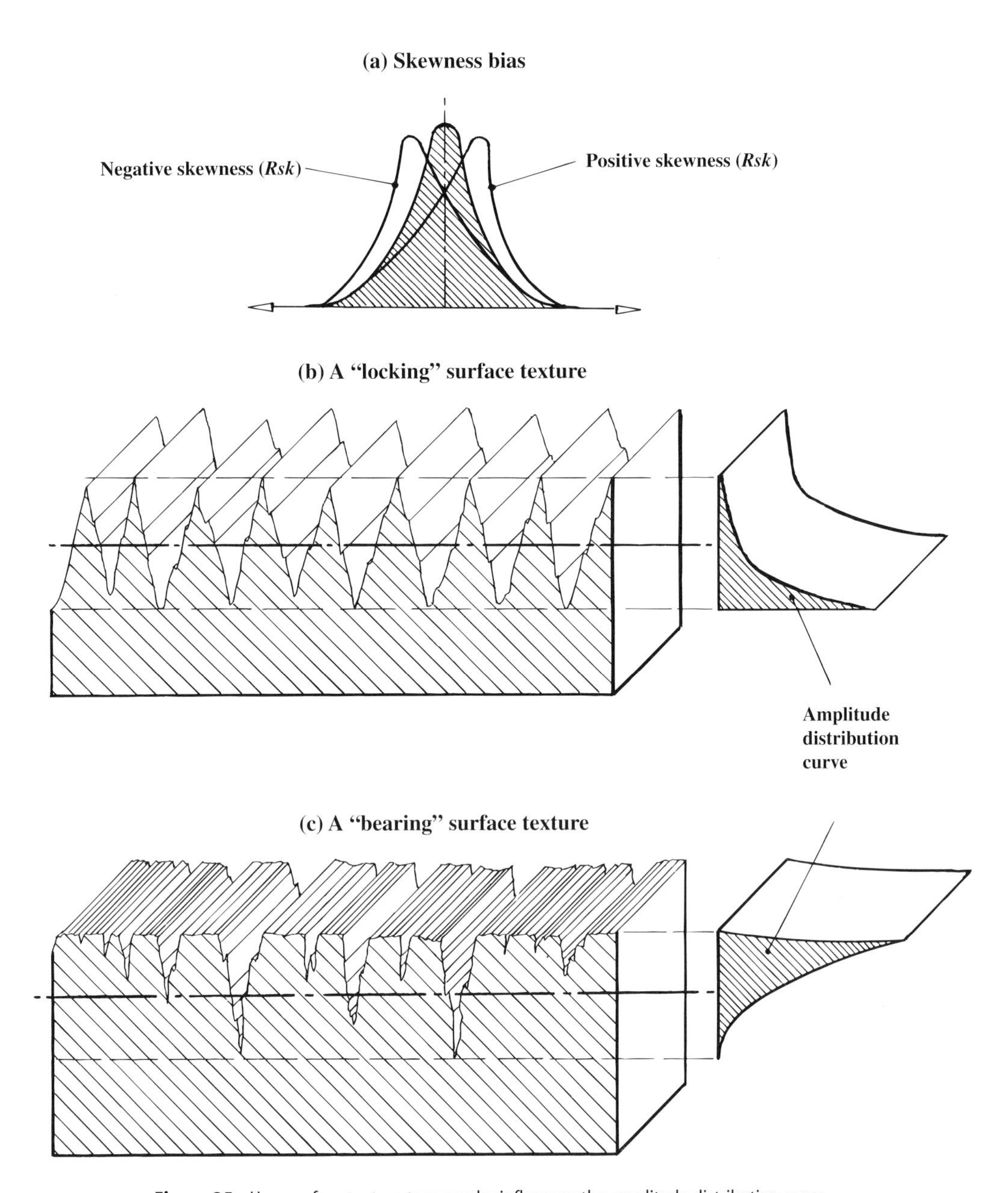

Figure 25. How surface texture topography influences the amplitude distribution curve.

terised by Figure 25(c) might occur in the cases of porous, sintered or cast-iron surface topography – having comparatively large numerical values of negative skewness. The surface is sensitive to extreme ordinate values within the profile under test; this is due to *Rsk* being a function of the cube of the ordinate height. As a result of this peak sensitivity, it is a hindrance when attempting to inspect plateau-type surfaces, although the *Rsk* parameter indicates a reasonable correlation with a component's potential load carrying ability, or its porosity.

The shape, or "spikeness" of the amplitude distribution curve can also relay useful information about the dispersion, or "randomness" of the surface profile, which can be quantified by means of a parameter known as kurtosis (*Rku*).

Kurtosis of the assessed profile Rku, Wku and Pku

The kurtosis parameter, typified by *Rku*, is the mean quadratic value of the ordinate values *Z(x)* and the

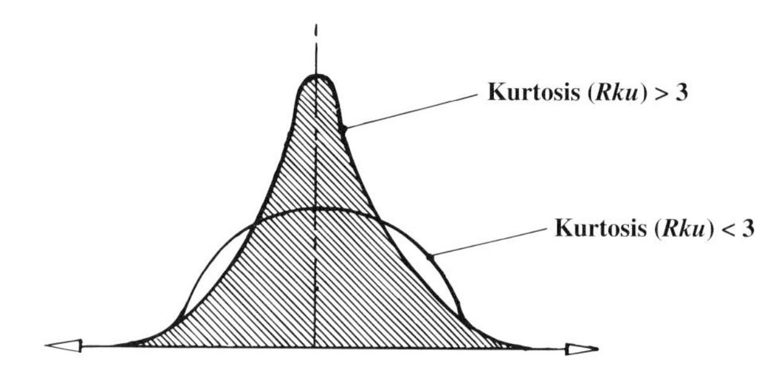
(a) Kurtosis is influenced by the distribution shape
Kurtosis (Rku) > 3
Kurtosis (Rku) < 3

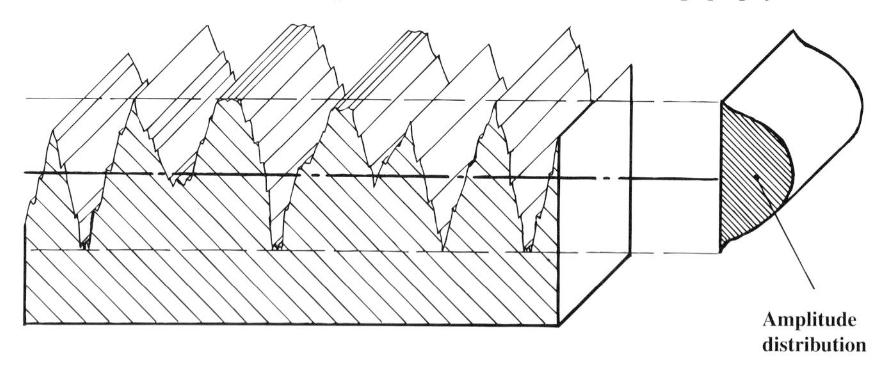
(b) Material distributed evenly across the whole of the surface topography
Amplitude
distribution

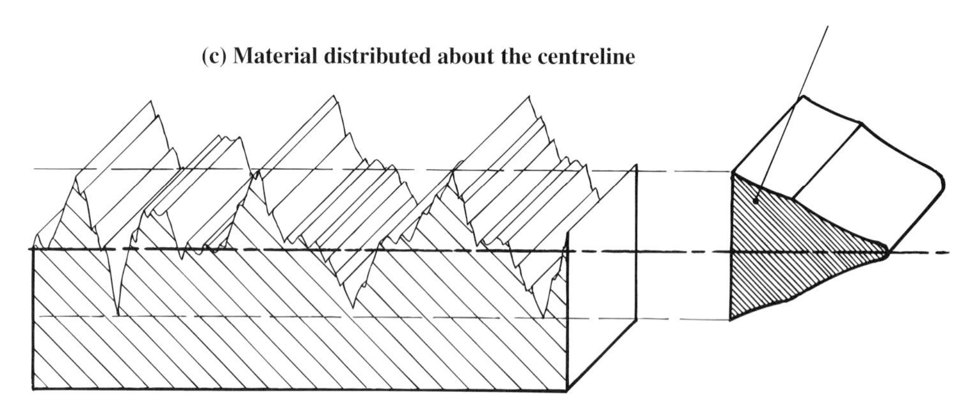
(c) Material distributed about the centreline

Figure 26. Variation in surface topography influences the shape and height of the amplitude distribution curve.

fourth power of *Pq*, *Rq* or *Wq*, respectively, within the sampling length. However, unlike the skewness parameters previously described, namely *Rsk* and its derivatives, *Rku* can detect if the profile peaks are distributed in an even manner across the sampling length trace, as well as providing information on the profile's sharpness (see Figure 26).

Kurtosis provides a means of measuring the *sharpness* of the profile, with a "spiky" surface exhibiting a high numerical value of *Rku* (Figure 26c); alternatively, a "bumpy" surface topography will have a low *Rku* value (Figure 26b). As a consequence of this ability to distinguish variations in the actual surface topography, *Rku* is a useful parameter in the prediction of in-service component performance with respect to lubricant retention and subsequent wear behaviour.

It should be mentioned that kurtosis cannot convey differences between either peaks or valleys in the assessed profile.

The kurtosis parameter indicative of *Rku*, can be mathematically expressed in the following manner:

$$Rku = \frac{1}{Rq^4}\frac{1}{N}\sum_{i=1}^{N} Z_i^4$$

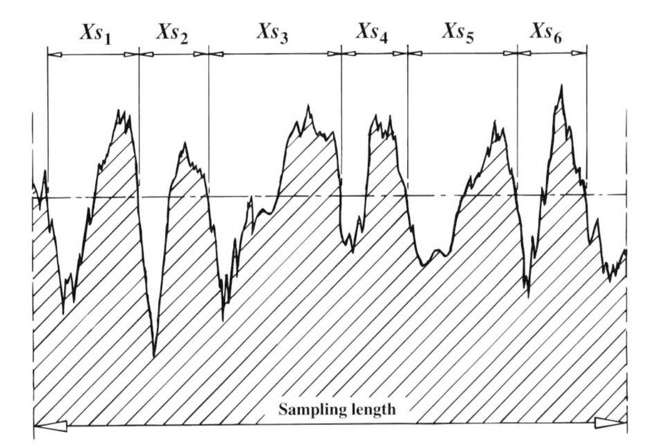

Figure 27. Width of profile elements. [Source: ISO 4287: 1997]

Later, the parameters *Rsk* and *Rku* will be applied to specific manufacturing processes to produce *manufacturing process envelopes* describing the functional aspects of a particular generated surface condition.

1.9.3 Spacing parameters

Mean width of profile elements *Rsm*, *Wsm* and *Psm*

The mean width parameter of the profile element widths *Xs* within the sampling length relates to the average value of the length of the mean line section that contains adjacent profile peaks and valleys (Figure 27). The parameter needs both height and spacing discrimination, and if not specified the default height bias utilised shall be 10% of *Pz*, *Rz* or *Wz*, respectively. Moreover, the default spacing discrimination shall be 1% of the sampling length, with both of these conditions requiring to be met.

The spacing parameters are particularly useful in determining the feedmarks from a specific machining operation, as they relate very closely to that of the actual feed per revolution of either the cutter or workpiece, depending on which production process was selected.

NB More is mentioned on spacing parameters and their affects on the surface topography in the Appendix.

1.9.4 Hybrid parameters

Root mean square slope of the assessed profile $R\Delta q$, $W\Delta q$ and $P\Delta q$

The root mean square slope parameter refers to the value of the ordinate slope dZ/dX within the sampling length, depending upon both amplitude and spacing information, hence the term *hybrid parameter*. The slope of the profile is the angle it makes with a line that is parallel to the mean line, with the mean of the slopes at all points in the profile being termed the average slope within the sampling length. By way of illustration of its use, it might be necessary to determine the developed or actual profile length, namely the length occupied if all the peaks and valleys were laid out along a single straight line. Hence, the steeper the slope the longer will be the actual surface length. In practical industrial situations, the parameter might be utilised in either a plating or painting operation, where the surface length for *keying* of the coating is a critical feature. Moreover, the average slope can be related to certain mechanical properties such as hardness, elasticity or more generally to the *crushability* of a surface. Further, if the root mean square value is small, this is an indication that the surface would have a good optical reflection property.

NB In the Appendix are listed some of the previous and current techniques for the assessment for these *hybrid parameters*.

1.9.5 Curves and related parameters

Curves and their related parameters are defined over the evaluation length rather than the sampling length.

Material ratio of the profile *Rmr(c)*, *Wmr(c)* and *Pmr(c)*

The material ratio of a profile refers to the ratio of the bearing length to the evaluation length and is represented as a percentage. This bearing length can be found by the sum of the section lengths obtained by cutting the profile line – termed *slice level* – that is drawn parallel to the mean line at a specified level. The ratio is assumed to be 0% if the slice level is at the highest peak; conversely, at the deepest valley this would represent 100%. Parameters *Rmr(c)*, *Wmr(c)* and *Pmr(c)* will determine the percentage of each bearing length ratio of a single slice level, or say, 19 slice levels that are drawn at equal intervals within the *Rt*, *Wt* or *Pt*, respectively.

NB More is mentioned on this topic in the Appendix.

Material ratio of the curve of the profile (Abbot–Firestone or bearing ratio curve)

The material ratio curve represents the profile as a function of level. Specifically, by plotting the bearing ratio at a range of depths in the profile, the manner by which the bearing ratio changes with depth provides a method of characterising differing shapes present on the profile (Figure 28a). The bearing area fraction can be defined as *the sum of the lengths of individual plateaux at a particular height, normalised by the total assessment length* – this parameter is designated by *Rmr* (Figure 28a). The values of *Rmr* are occasionally specified on engineering drawings, although this may lead to large uncertainties (see Chapter 6, Section 6.4) if the bearing area curve is referred to by the highest and lowest profile points.

In the majority of circumstances mating surfaces require specific tribological functions; these are the direct result of particular machining operational sequences. Normally, the initial production operation will establish the general shape of the surface – by roughing-out – providing a somewhat coarse finish, with subsequent operations necessary to improve the finish, resulting in the desired design properties. This machining strategy provides the operational sequence that will invariably remove surface peaks from the original process, but often leaves any deep valleys intact. This roughing and finishing machining technique leads to a surface texture type that is often termed a *stratified surface*. In such cases the height distributions are negatively skewed; therefore this would make it otherwise difficult for an average parameter such as *Ra* to represent the surface effectively for either its specification or in the case of quality control matters.

NB This topic is discussed in further detail in the Appendix.

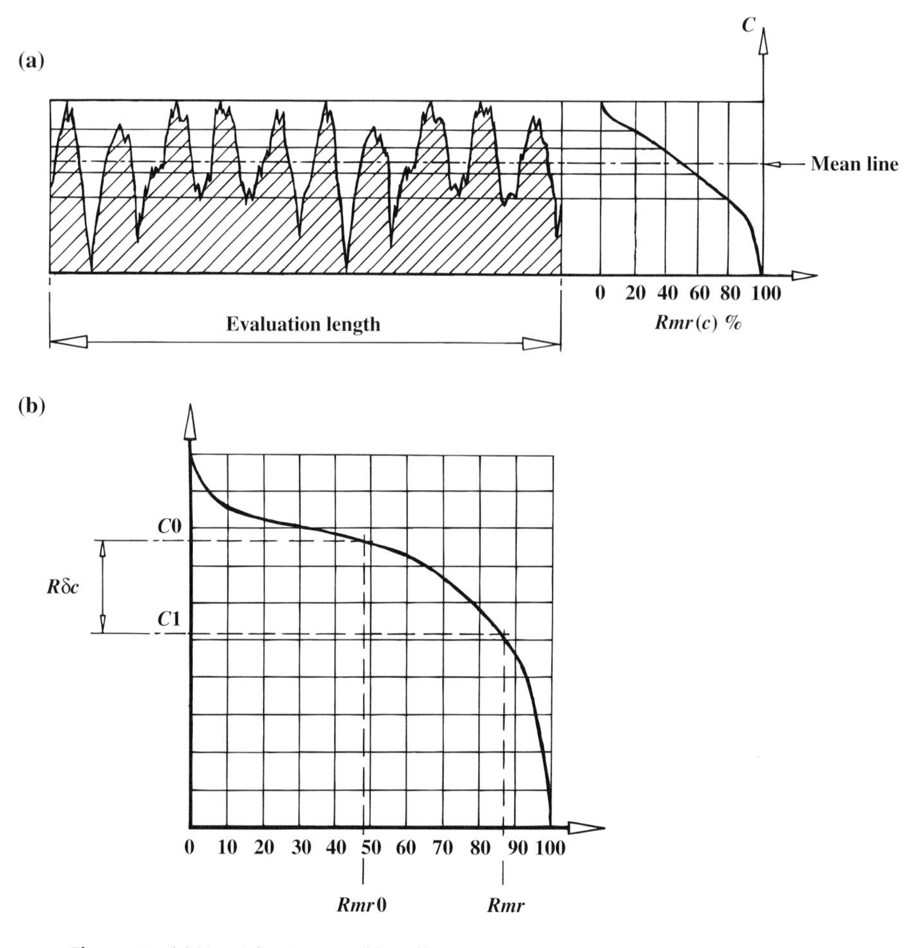

Figure 28. (a) Material ratio curve. (b) Profile section level separation. [Source: ISO 4287: 1997]

Profile section height difference $R\delta c$, $W\delta c$ and $P\delta c$

This parameter can be represented as the vertical distance between two section levels of a given material ratio curve (see Figure 28b).

Relative material ratio Rmr, Wmr and Pmr

The relative material ratio can be established at a profile section level $R\delta c$, being related to a reference $C0$ (see Figure 28b), where:

$C1 = C0 - R\delta c$ or ($W\delta c$ or $P\delta c$)

$C0 = C(Rmr0, Wmr0, Pmr0)$

Rmr refers to the bearing ratio at a predetermined height (see Figure 28b). One method of specifying the height is to move over to a certain percentage – reference percentage – on the bearing ratio curve, then move down to a certain depth (slice depth), with the bearing ratio at the corresponding position here being the *Rmr* (Figure 28b). The reason for the reference percentage is to eliminate any spurious peaks from the assessment, as they would tend to be worn away during the initial *burn-in/running-in* period. The slice depth will then correspond to a satisfactory surface roughness, or to an acceptable level of surface wear.

Profile height amplitude curve

This represents the sample probability density function of the ordinate $Z(x)$ within the evaluation length. The amplitude distribution curve, as it is usually known, is a probability function giving the probability that a profile of the surface has a particular height at a certain position. Like many probability distributions, the curve normally follows the contours of a Gaussian – *bell-shaped* – distribution (Figure 29a). The amplitude distribution curve informs the user of how much of the profile is situated at a particular height, from the aspect of a histogram-like sense.

This curve illustrates the relative total lengths

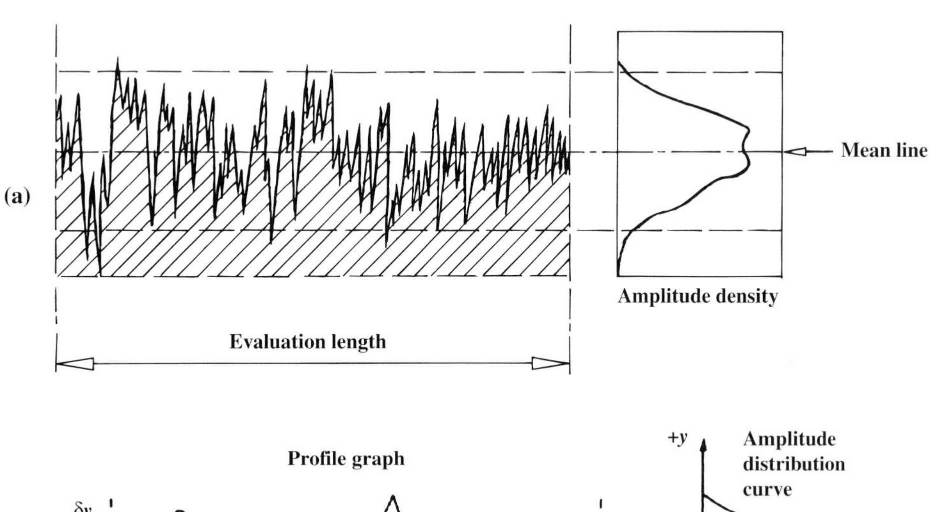

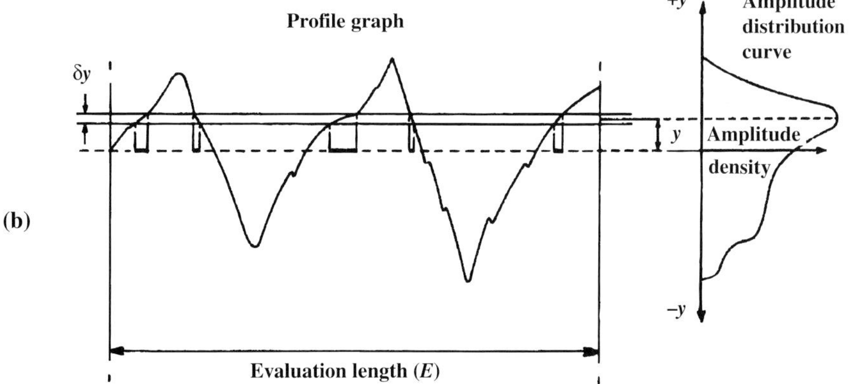

Figure 29. Profile height and amplitude distribution curve. [Source: ISO 4287: 1997]

over which the profile graph will attain any selected range of heights that are above or below the mean line (see Figure 29b). The profile's horizontal lengths are included within a narrow zone width δy at a height z and are represented in Figure 29(b) by "a, b, c, d and e". Therefore, by expressing the sum of these lengths as a percentage of the evaluation length, this will give a measure of the relative amount of the profile at height z.

This graph is known as the amplitude distribution at height z, so by plotting density against height the amplitude over the whole profile can be determined – producing the amplitude density distribution curve.

1.9.6 Overview of parameters

In Table 2 are shown the more recent and previously utilised parameters, together with information concerning whether the parameter has to be calculated over the sampling or evaluation length.

1.10 Auto-correlation function

Every day of one's life, the technique of correlation is applied to compare sights, sounds, tastes, objects – the list is almost limitless – this being a process of comparison. The auto-correlation function (**ACF**) is essentially a process of determining the relationship of any point on the profile to all other points. Some manufactured surfaces clearly indicate visible repetitive marks on either the material itself or indirectly via the profile graph, while on other surfaces it becomes difficult to distinguish any repetitive irregularities from random occurrences. This visual difficulty is particularly true when the amplitudes of the repetitive irregularities are less than the random occurrences. Moreover, the presence of repetitive surface features – however small – can indicate factors such as tool wear, machine vibration or deficiencies in the machine, with identification of them being an important factor. Any random pattern occurring on the surface can reveal, for example, whether a built-up edge condition has transpired on the cutting tool resulting in a degree of tearing of the machined surface – this tends to be of a random nature and is not predictable.

The extent of this randomness of the surface can be monitored and assessed by isolating random from repetitive textural patterns, this being achieved by auto-correlation. When a profile is perfectly periodic in nature – typified by a sine wave – the relationship of a particular group of points repeats itself at a distance equal to the wavelength. Conversely, if the profile under inspection is comprised entirely from random irregularities, the precise relationship between any specific points will not occur at any position along the trace length, hence any repetitive feature or group of features can be identified. Computers equipped with fast digital processors have significantly reduced the tedious task of determining a surface profile's auto-correlation.

The technique exploited by auto-correlation is to compare different parts of the surface profile; in this

Table 2. Past and present parameters for surface texture

Parameters 1997 edition	1984 edition	1997 edition	Determined within:	
			Evaluation length	Sampling length
Maximum profile height	*Rp*	*Rp*		X
Maximum profile valley depth	*Rm*	*Rv*		X
Maximum height of profile	*Ry*	*Rz*		X
Mean height of profile	*Rc*	*Rc*		X
Total height of profile	–	*Rt*	X	
Arithmetical mean deviation of the assessed profile	*Ra*	*Ra*		X
Root mean square deviation of the assessed profile	*Rq*	*Rq*		X
Skewness of the assessed profile	*Sk*	*Rsk*		X
Kurtosis of the assessed profile	–	*Rku*		X
Mean width of profile elements	*Sm*	*RSm*		X
Root mean square slope of the assessed profile	Δq	$R\Delta q$		X
Material ratio of the profile		*RMr(c)*	X	
Profile section height difference	–	$R\delta c$	X	
Relative material ratio	*tp*	*Rmr*	X	
Ten-point height (deleted as an ISO parameter)	*Rz*	–		

Source: ISO 4287: 1997 (E/F).

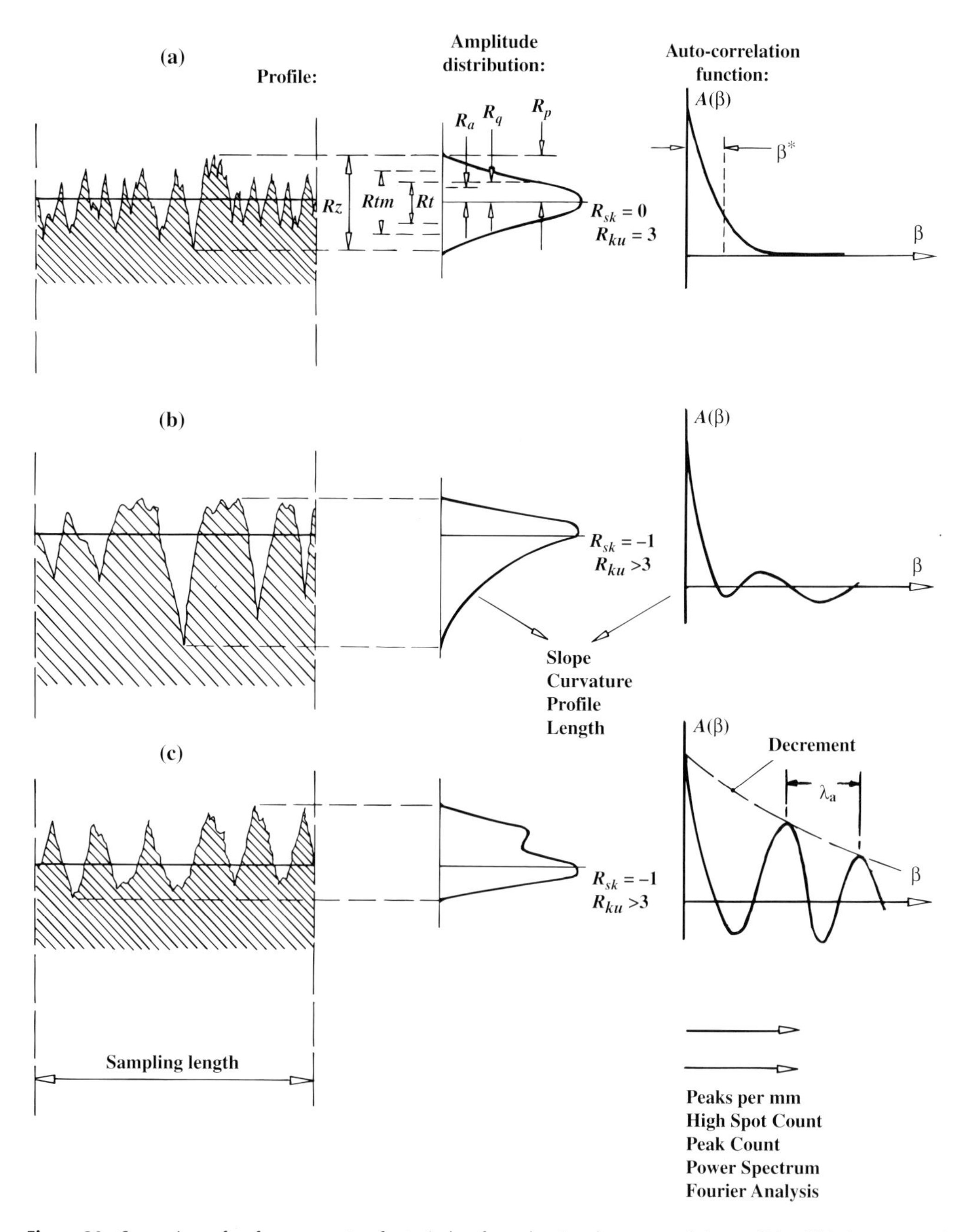

Figure 30. Comparison of surface parameters for typical surfaces showing those currently in use. [After Whitehouse, 1978.]

manner profile repetitions or similarities can be discovered. When the profile of the surface displays either an isotropic or random shape then no portions will be similar, exhibiting low auto-correlation function. On the other hand, if an anisotropic or periodic surface topography occurs – for example as in single-point turning operations – then repetition of the surface profile occurs and auto-correlation will be high.

Figure 30 shows an example of the **ACF** relating to specific surface profile conditions. As mentioned above, the **ACF** contains information relating to the spacings of peaks and valleys in the profile. An alternative to the **ACF** is the equivalent term known as the power spectrum, or power spectral density, in which spacings are replaced by frequencies – one being a Fourier transform of the other. For the **ACF** examples shown in Figure 30, the scale of the horizontal axis can be the mean distance between peaks or the average distance between mean line crossings.

The ACF is derived from the normalised auto-covariance function (ACVF) – this being as follows:

$$ACF(\Delta x) = \frac{ACVF(\Delta x)}{(Rq)^2}$$

where

$$ACVF(\Delta x) = \lim_{lm \to \infty} \int_0^{lm} y(x)\, y(x + \Delta x)\, dx$$

where lm = assessment length, $y(x)$ = profile height at position x and Rq = rms parameter.

The ACF does not contain information about the amplitude of the profile, with its values ranging from plus one – giving perfect correlation – to negative one – this having a correlation of the inverted but shifted profile. Due to the fact that the ACF is independent of the amplitude of the profile it is more popular than the ACVF. In the case of an ACF having an isotropic (random) profile, it quickly decays to zero (Figure 30a), with the profile trace being of an unrepetitive nature. Conversely, the tendency for a truly anisotropic (periodic) generated surface is that it does not decay to zero, with Figure 30(c) illustrating some periodicity in its profile trace – indicating that decay here is significantly less than that for Figure 30(a). Despite the fact that both Figure 30(a) and (c) have similar *Rz* values, their respective ACF are distinctly different and can be used as a means to discriminate between surfaces and indicate whether they will fulfil the desired in-service applications, which might not otherwise be apparent.

Fourier analysis (FFT)

Any surface profile will normally be of some complexity, this being comprised of an array of differing waveforms that are superimposed onto one another. An actual profile's specific composition will be dependent on the shape and size of the waveforms existing within the profile; both their amplitude and frequency, termed "harmonics", can be established by the application of Fourier analysis.

The technique known as a "fast Fourier transform" – FFT for short – enables one to establish a series of sine waves being generated from the surface profile, and when combined together contribute to the original profile. Individual harmonics represent a specific frequency–wavelength, in combination with their associated amplitude.

The surface profile depicted in Figure 31(a) was selected to illustrate a periodic form, which in isolation may mask significant hidden harmonic detail which could play an important role in its later in-service application. Results from the FFT analysis of the profile (Figure31b) reveal the differing sine wave amplitudes necessary to generate the original profile. In this case, the largest sine wave amplitude equals a value of 0.7 mm, this being termed the "dominant wave".

It should be emphasised that the information obtained from the FFT analysis will be exactly the same as that acquired by the auto-correlation method, but obviously it is displayed in a altered form. Due to the isolation of the harmonics into specific bands for a particular surface profile, FFT can impart valuable understanding as to how the production process might have been influenced by variables in processing for that product. For example, the dominant harmonic in the FFT display for a profile might give a valuable insight into the vibrational frequency for a specific machine tool. Knowledge of this vibrational effect on the resulting surface profile may enable the attenuation of this vibrational signature by judicious adjustment of either cutting data, or specific maintenance to the machine tool to improve the subsequent surface profile's condition.

1.11 Appearance of peaks and valleys

For many engineering and statistical graphs no *intrinsic* relationship exists between the plotted measurement units; for example, when plotting the rise in temperature of a heat treatment oven against elapsed time the ordinates would be represented by both degrees Celsius and hours – two quite dissimilar units. As a consequence of these disparate units, acceptance of the shape of the graph is acknowledged, whatever scales are applied for the ordinates. It is quite different when dealing with a surface profile graph, where both the horizontal and vertical ordinates are represented by the same units (i.e., length). At first introduction, it may be difficult to appreciate that due to the difference of scale the graph will not instantaneously give an indication as to the *shape* of the irregularities. Figure 32 illustrates the effect, with this disparity being one of the most difficult factors in surface texture analysis to comprehend.

It is very important to note that although the *visual* profile's shape is distorted due to the application of different horizontal and vertical magnifications, *measurements scaled from the graph are*

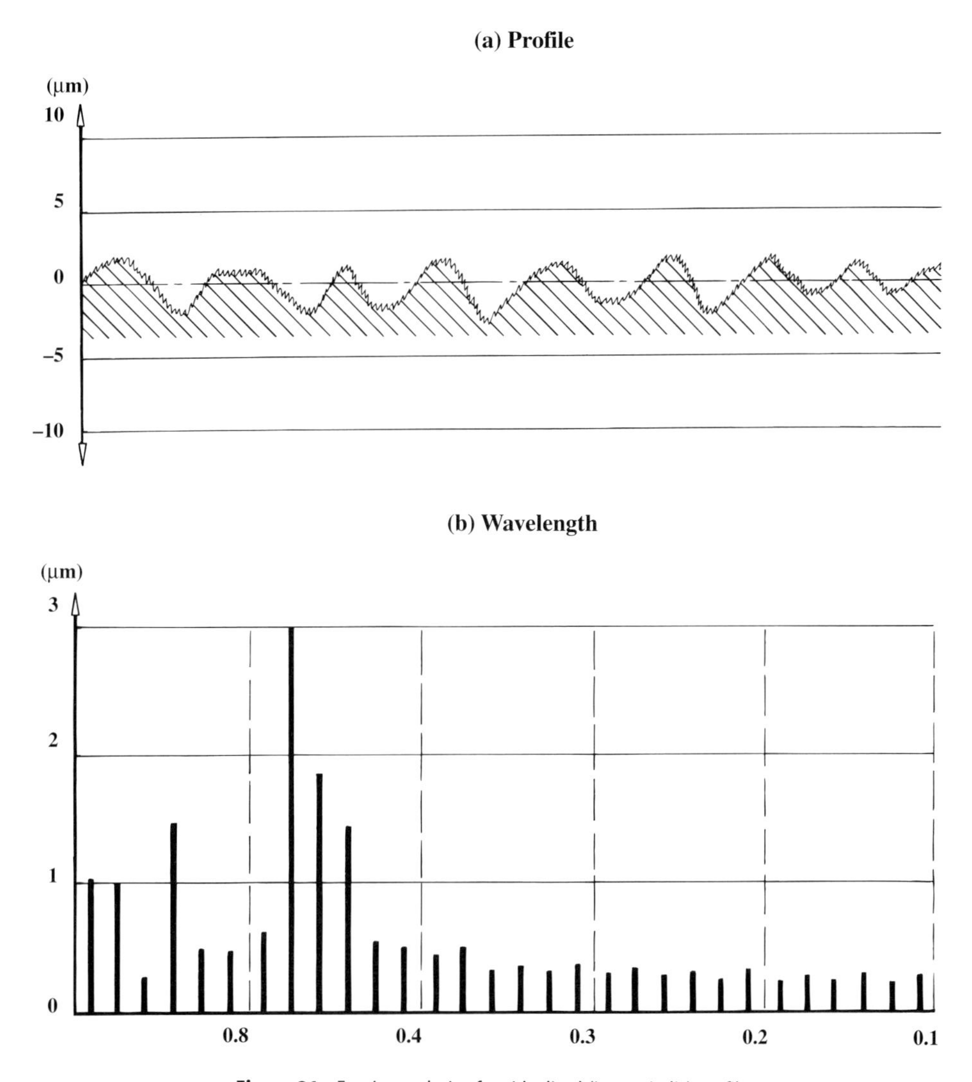

Figure 31. Fourier analysis of an idealised (i.e. periodic) profile.

correct. Such effects can be visually demonstrated with reference to the peak angles of the profile illustrated in Figure 32, indicating how the profile shape varies as horizontal/vertical magnification is altered. Figure 32(III) verifies this fact, as the horizontal magnification of the surface profile is increased (Figure 32II), the length X–X expands to X^1–X^1 and the peaks; A, B, C and D appear flatter. Increasing the horizontal magnification still further – until it equals the vertical magnification (Figure 32I) – expands the sampling length Y–Y to Y^1–Y^1. Once again, with this still higher magnification, the peaks E and G and valleys F and H now have a much flatter appearance. A point worth making is that the actual difference in respective heights of corresponding peaks and valleys in I, II and III are identical.

As a result of the spiky appearance of both the peaks and valleys on the profile graph, this can lead to considerable confusion when visually assessing the actual graph. However, the geometric relationship between angles on the graph and those that occur on the physical profile of the component under test are more easily determined. As an example of this interpretation, Figure 33(a), on the profile graph, depicts a symmetrical peak with an included angle 2α. Figure 33(b) exhibits the corresponding peak, on the actual surface, having an angle 2β. The ratio of the angles is determined by

$$\frac{\tan \beta}{\tan \alpha} = \frac{V_v}{V_h}$$

This ratio is strictly true for symmetrical peaks, although it is only slightly different for asperities of

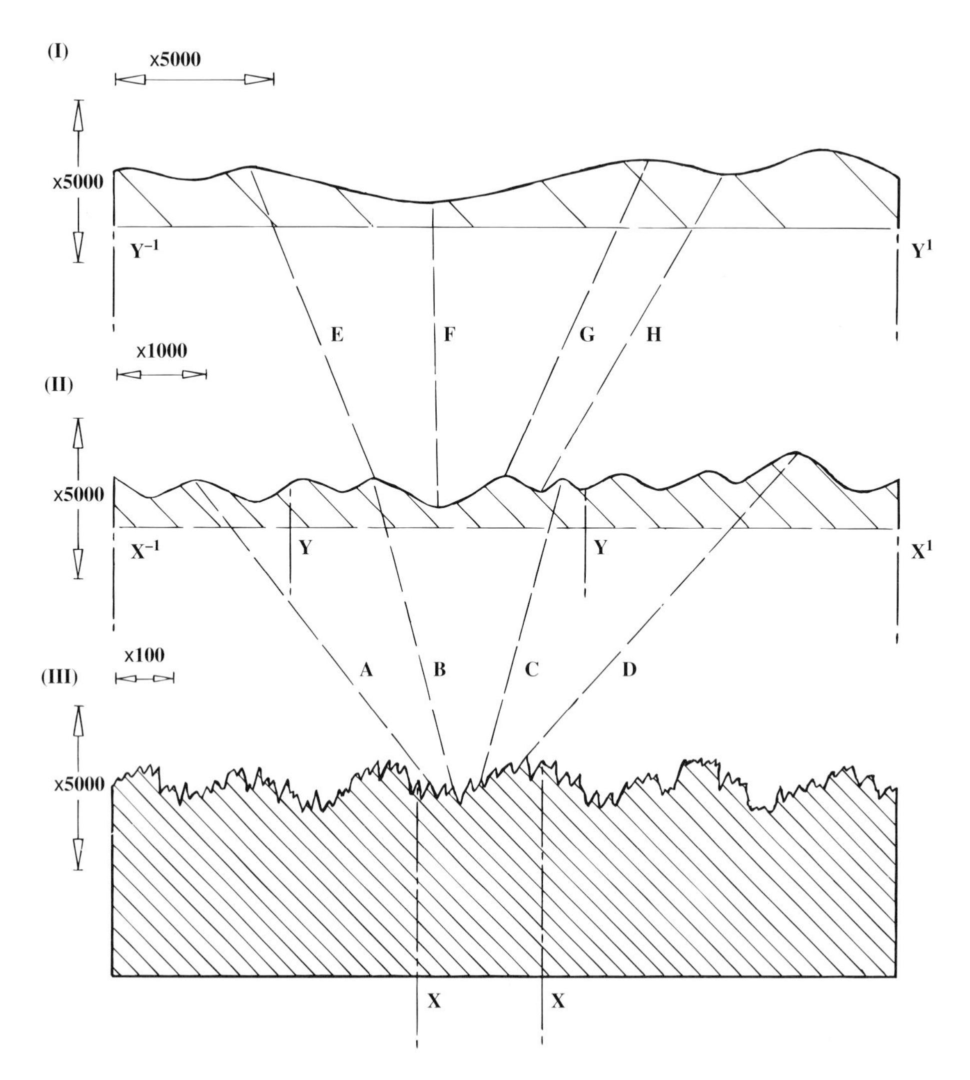

Figure 32. Diagram illustrating how profile shape varies as the horizontal magnification is reduced: (I) surface profile magnified x5000 equally in all directions; (II) profile with A Vv: V_h ratio of 5:1; (III) profile graph recorded with a Vv:V_h ratio of 50:1. (Courtesy of Taylor Hobson.)

an asymmetrical nature. For example, the data obtained from a profile graph in which the vertical and horizontal magnifications and peak geometry might be $V_v = \times 5000$, $V_h = \times 50$ and $2\alpha = 10^\circ$. This will therefore represent

$\alpha = 5^\circ$

$\tan \alpha = 0.0875$

hence

$\tan \beta = 0.0875 \times 5000/50$

$= 8.75$

giving

$\beta = 83.5^\circ$

Thus

$2\beta = 167^\circ$

This example of peak geometry (Figure 33a) represents a fairly sharp peak on the profile graph, but only a gradual slope – rising and falling at 6.5° from the horizontal – on the actual component's surface (Figure 33b). Due to the fact that it becomes difficult to accurately measure the peak angle on the graph, and it is rare that an accurate assessment of the surface angles is required, an approximation of $2\alpha : 2\beta$ is normally sufficient.

Table 3 provides approximate conversions for several $V_v : V_h$ ratios.

(a) Peak on profile graph

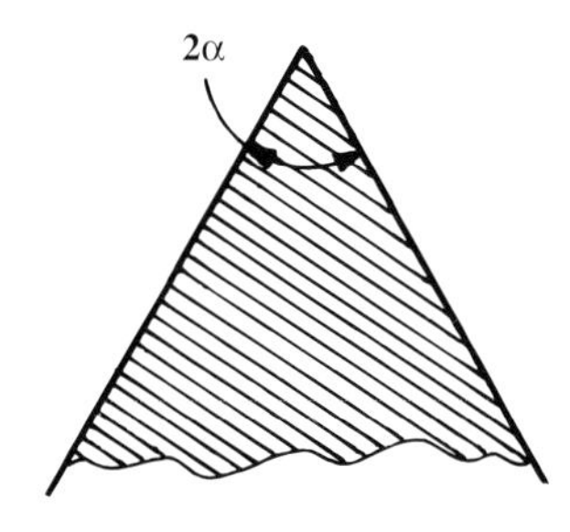

(b) Corresponding peak on the surface

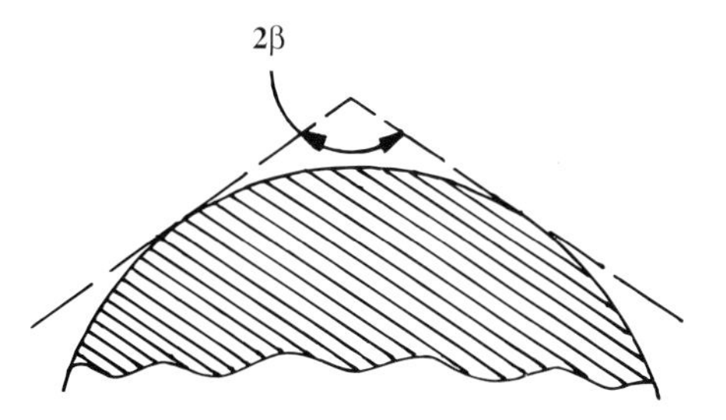

Figure 33. Relationship between peak angles on the graph and the actual surface. (Courtesy of Taylor Hobson.)

It has been previously stated by way of a light-hearted analogy for the representation of two rough contacting surfaces that they resemble "Switzerland on top of Austria", although this simply shows the influence of the visual aspect of the profile graph. A more preferable analogy might be "The Netherlands on top of Iowa", as the peaks on the surface are in general of quite blunt and low aspect.

When considering the profile with its associated peaks and valleys, it has been presumed that the peak height on a profile graph is characterised by the summit of individual peaks traced by contact with the stylus for a surface under test. This may be true for the major ridges for a surface exhibiting marked directional lay (see Figure 5A) being traced at 90° to the direction of the dominant surface pattern lay. Conversely, for a surface displaying random textural influences, such as the "multi-directional" example in Figure 4, it is extremely unlikely that a stylus's path will trace across the summit of every peak. In these circumstances, some peaks and valleys will be displaced from the path of the stylus by varying distances, with the result that the stylus will also traverse across many flanks. Stylus vertical displacement, although somewhat less than the full peak height or valley encountered, will moreover be correct for the traced feature. Assumptions made by many are that every peak on the graph is the summit of a peak on the actual surface, but where random surfaces occur there will be sufficient actual summits encountered to prevent any significant inaccuracy as a result.

1.12 Stylus-based and non-contact systems

Effect of stylus

It should be appreciated by now that the stylus and skid, when employed, are the only parts which are in active contact with the surface under test. Therefore both the shape and dimensions of this pick-up are critical factors and will strongly influence information retrieved from the surface. The geometry of a typical stylus – its shape and size – was discussed in Section 1.4 and illustrated in Figure 11(a); it influences the accuracy obtainable from the surface. Due to the finite size of the stylus some surface imperfections, such as visible scratches, are

Table 3. Peak angles (2β, Figure 33a) on the surface corresponding to peak angles of 2α on the graph

Ratio: V_v/V_h	Peak angle on graph (2α, Figure 33)							
	1°	5°	10°	15°	20°	30°	40°	50°
10	10°	47°	82°	105°	121°	139°	149°	155°
25	25°	95°	131°	146°	154°	162°	167°	170°
100	82°	154°	167°	171°	173°	176°	177°	177°
250	131°	169°	175°	176°	177°	178°	179°	179°
500	154°	175°	177°	178°	178°	179°	–	–
1000	167°	177°	179°	179°	179°	–	–	–
2500	174°	179°	–	–	–	–	–	–
5000	177°	–	–	–	–	–	–	–

Courtesy of Taylor Hobson.

too fine to be completely penetrated, even with the smallest tip available. This is not too much of a hindrance to assessment as the *Ra* is an "averaging parameter", so the omission of the finest textural features will not unduly affect the resulting value. To illustrate this point, many years ago a series of tests were conducted on ground surfaces by Reason et al. (1944), these samples had four increasing levels of roughness. The reduction in measured *Ra* value when utilising either a 1.25 μm or 10 μm stylus was 4.5 to 3.25, 8 to 7.5, 11 to 10 and 26 to 25 μm, respectively. This small variation occured despite the fact that the larger stylus was eight times greater than the smaller.

For undistorted surfaces (periodic), there is a predictable relationship between the stylus radius and the resulting error introduced into the *Ra* reading. Figure 34 charts the approximate values of this error, being applicable to regular profiles with 150° included angle triangular peaks and valleys – typical of surfaces utilised for calibration standards. For surfaces other than of this regular triangulated geometry, the values tend to be only approximate, although they do indicate the order of any error likely to be occasioned. The horizontal ordinates in Figure 34 are the ratio of the stylus tip radius to the actual *Ra* value of the surface, with the vertical ordinates being the amount by which the measured *Ra* value is too low. The chart demonstrates that only when the tip radius is greater than approximately 20 times the *Ra* value is this error of any significance. By traversing a stylus across a standard incorporating narrow grooves of known widths – 20, 10, 5 and 2.5 μm – this is a technique for estimating stylus wear or damage. Obviously, the blunter the tip the less distance it can descend into the groove, this penetration depth being measured from the profile graph. The effective stylus radius can be established from the calibration curve supplied with the artefact. Stylus cone angle can also be measured by this method; because the stylus wears with use, the change in effective tip size can be monitored.

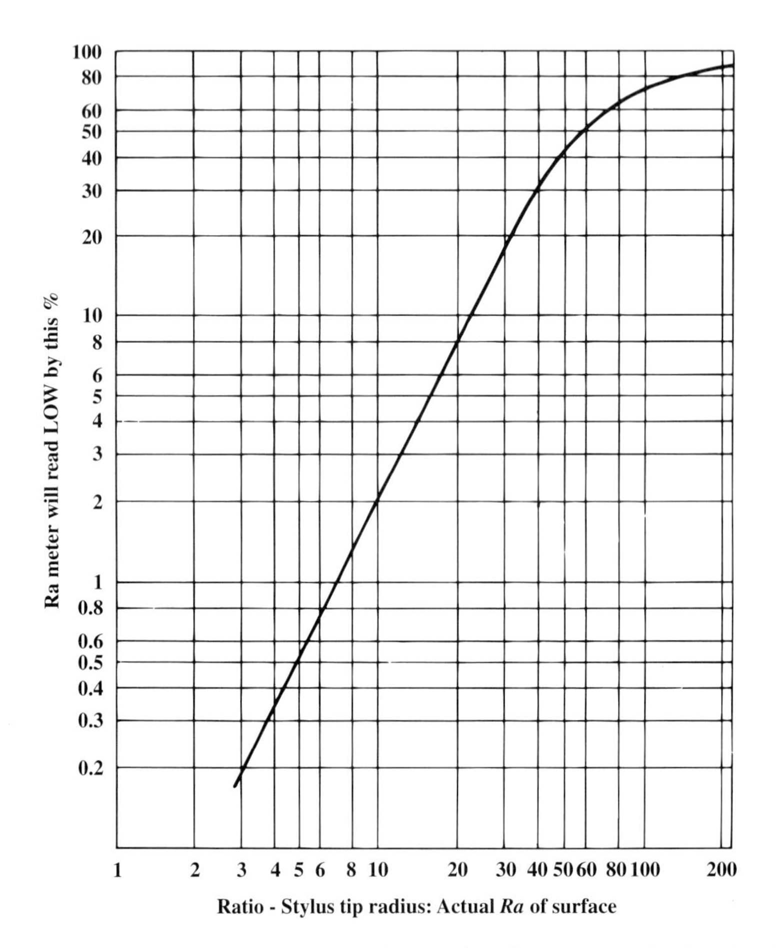

Figure 34. Error due to finite stylus tip radius, for a specific surface. (Courtesy of Taylor Hobson.)

Illustrative example: When measuring a surface of 0.5 μm *Ra* with a stylus tip of 5 μm (10:1 ratio) the display will be approximately 2% low. If the stylus has double this radius (i.e., 10 μm) this error increases to 8%.

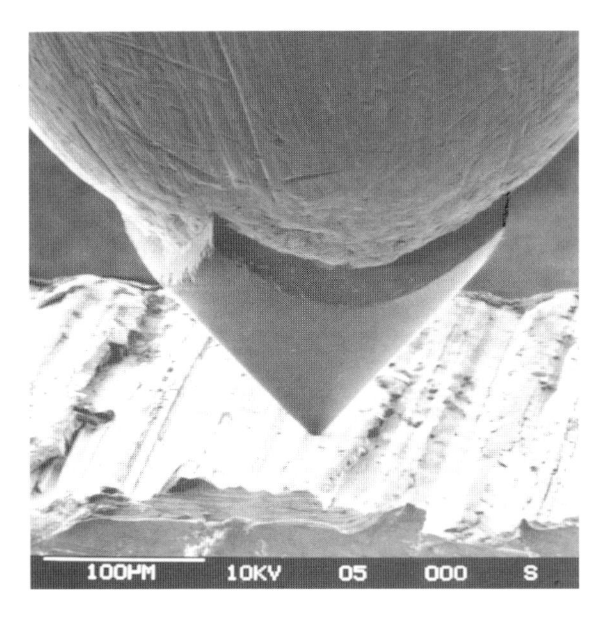

Figure 35. 90° conical stylus tip mechanically tracing across a portion of a honed surface. SEM photomicrograph, X200 magnification. [Courtesy of PTB Braunschweig/Feinpruf Perthen GmbH.]

Stylus force on the surface must not be great enough to deform or scratch the surface as this may promote erroneous results. Conversely, the force imparted by the stylus must be of a sufficient level to ensure that it maintains continuous contact with the surface at the desired traverse speed. This factor is important in the design of the pick-up, with most surface texture instruments having a stylus tip radius of between 1 and 10 μm. However, the greater the radius of the tip, the larger the allowable force that can bear on the surface. If a tip radius is small, say 0.1 μm (Figure 11a), then a very low force must be used, typically around 0.75 mN.

The conical stylus geometry was previously mentioned in Section 1.4 and illustrated in Figure 10, invariably having a 90° tip angle. In action (Figure 35), the conical tip angle of 90° can easily cope with surface features such as that shown, having the anisotropic profile of a honed surface. When surface features become greater than 45° to the horizontal, or the valley is smaller than, say, 5 μm, then the mechanical motion of the tip becomes to some extent distorted, as illustrated in Figure 11(c). Specifically, under these conditions, the tip acts in a similar fashion to a low-pass filter with high-frequency features not being recorded. With a tip angle of 90°, this means that the stylus will foul any peak or valley having an included angle of less than 45° – which may at first seem a considerable limitation, but in reality does not cause too much of a problem.

Chisel-edge styli (see Figure 11a) are a practical variation on the cone-shaped stylus (Figure 35), being particularly successful at entering porous materials such as sintered parts. One problem associated with the chisel-type stylus is the difficulty in maintaining its attitude at right angles to the lay, being more sensitive to error from peaks rather than valleys. Stylus misalignment can be minimised by the application of "smart parameters" that mathematically ignore certain features. Normally, it is recommended to utilise a conventional stylus geometry, unless particular problems are encountered that may cause the standard stylus to impair the validity of the resulting data.

Manufacturers of stylus/skid geometries normally apply identifying marks on the pick-up, or at the very least on its accompanying transportation case. Table 4(a) illustrates the pick-up markings, whereas Table 4(b) highlights the standard configurations.

Table 4(a). Typical (traceable) markings on pick-up

Identification marks:	±100	5	90	0.8	30/6	1/0
Stylus data:						
Measuring range (μm)	X					
Tip Radius (μm)		X				
Tip angle (°)			X			
Measuring force (mN)				X		
Skid data:						
Skid radius, longitudinal/traverse					X	
Distance skid-to-stylus, longitudinal/traverse						X

Table 4(b). Configurations for some standard pick-ups

Stylus			
Tip angle (°)	60° ± 5° or	90° ± 5°	
Tip radius (μm)	2 ± 1	5 ± 2	10 ± 3
Static measuring force (ie at electric zero, mN)	≤ 0.7	≤ 4	≤ 16
Variation in static measuring force (mN/m)	≤ 0.035	≤ 0.2	≤ 0.8
Skid			
Distance from stylus dependent upon pick-up design;			
Radius			
(longitudinal, mm)	0.3 1.3 3 10 30 infinity		
(lateral, mm)	depends upon design of pick-up		
Surface roughness (μm)	≤ 0.1		
Skid force (N) for:			
Hard surface	≤ 0.5		
Soft surface	≤ 0.25		
Linearity of system (%)	± 1		

1.12.1 Pick-up

The function of the transducer or pick-up is to convert the minute vertical movements of the stylus, as it progresses along the surface, into proportionate variations of an electrical signal. The sensitivity demanded of the pick-up is such that it should be able to respond to stylus movements of approximately 0.1 nm or better. This output is minute and of necessity must undergo significant amplification to enable stylus trace movements of between 50,000 and 100,000 times greater than the stylus movement (V_v ×50,000 or ×100,000).

In general, pick-ups can be classified into two groups, depending on their operating principle:

1. *Analogue transducers* (Figure 36)

- *Position-sensitive pick-ups* (Figure 36a) give a signal proportional to displacement, even when the stylus is stationary. Output is independent of the speed at which the stylus is displaced and is only related to the position of the stylus within a permissible range of vertical movement. A big advantage of this type of pick-up is that it enables a true recording of waviness and profile to be obtained.
- *Motion-sensitive pick-ups* (Figure 36b) produce an output only when the stylus is in motion, with the output being related to the speed at which the stylus is displaced, dropping to zero when stationary. If displacement is very slow, perhaps due to widely spaced waviness or to change in form, then the output for practical purposes is almost zero. This low output means that variations of waviness and form are excluded from the profile. These pick-ups tend to be utilised if instruments do not have recording facilities, such as some portable equipment.

The *variable-inductance pick-up*, illustrated in Figure 36(ai), has been widely used for many years. The stylus is situated at one end of a beam pivoted at its centre on knife edges, while at the opposite end an armature is carried which moves between two coils, changing the relative inductance. This pick-up's operating principle is as follows: coils are connected in an AC bridge circuit (Figure 36aii) such

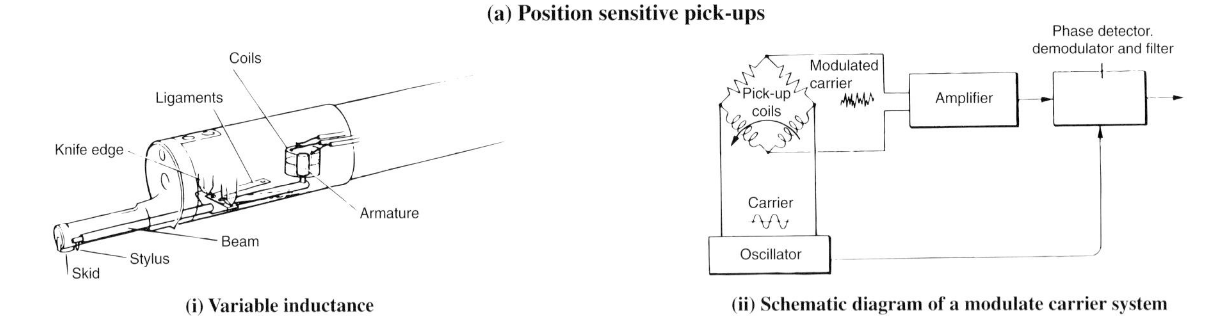

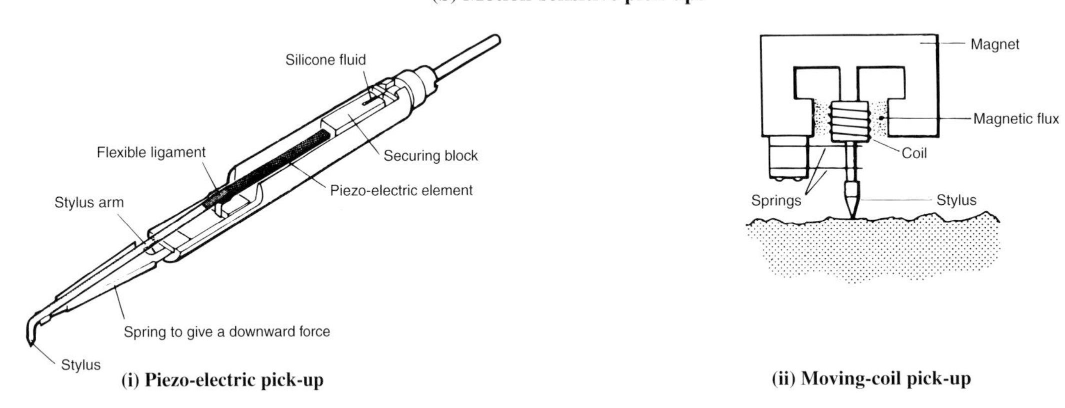

Figure 36. Basic design elements of pick-ups of the analogue transducer configuration. (Courtesy of Taylor Hobson.)

that when the armature is centrally positioned between the bridge it is balanced, giving no output. Movement of the armature unbalances the bridge, providing an output proportional to its displacement; the relative phase of the signal depends on the direction of movement. This signal is amplified and compared with that of the oscillator, to determine in which direction it has moved from the central (zero) position. It is necessary to utilise an oscillator to produce a constant AC output, because the pick-up – unlike a motion-sensitive pick-up (Figure 36b) – does not generate any output; it merely serves to modify the carrier. Simultaneously, the knife edges exert light pressure from a very weak spring acting on the beam, enabling subsequent stylus contact with the surface under test. The ligaments prevent unwanted motion of the beam in the horizontal plane, with the result that stylus movement is only possible normal to the surface being assessed.

This type of pick-up is also fitted to newer versions of measuring instruments, particularly where the range-to-resolution ratio has increased from around 1000 to 64,000. Improvements in electronics for the latest pick-up designs of this type have obviated the need for reliance on a skid, meaning that precise pivot bearings have replaced the knife-edge pivots.

The *piezoelectric pick-up* (Figure 36b) has been widely used in the past, but now it tends to be used for the less sophisticated hand-held measuring instruments. When the stylus deforms the piezoelectric crystal, it has the property of developing a voltage across electrodes, the advantage being that it is virtually instantaneous. As the stylus follows the contours of the surface, the piezoelectric crystal distorts by bending a flexure that causes crystal compression and it becomes charged. The resulting charge is then amplified and electronically integrated, producing signals proportional to the surface profile. A piezoelectric transducer exerts a proportionally larger stylus force to that of an equivalent inductive transducer, with the former pick-up possibly damaging softer and more delicate surfaces.

The operating mechanism of this piezoelectric transducer can be described in the following manner: the flexible ligament interposed between the stylus arm and the piezoelectric element has enough stiffness to transmit normal vertical motion of the stylus to the crystal, but will flex as the stylus is suddenly subjected to instantaneous shock and in this way protection of the somewhat fragile piezoelectric element is achieved. A light spring force provides downward force on the stylus, keeping it in contact with the surface. A small drop of silicone oil is held by capillary force between the securing block and a thin metal blade fastened to the rear of the housing, giving the pick-up suitable dynamic characteristics. The silicone fluid exhibits low stiffness and at lower frequencies; therefore large but slow stylus displacements allow free movement of the far end of the crystal, hence no bending occurs and zero output is generated. At higher frequencies the fluid stiffness increases, effectively preventing the local end of the piezoelectric crystal from moving; therefore the bridge deforms generating an output.

A piezoelectric device is a position-sensitive device. A voltage on the electrodes persists while the element is deformed (providing no current is drawn from it), due to practical considerations such as finite input impedance of the amplifier and cable losses. These considerations ensure that it can only be utilised in the motion-sensitive mode.

The *moving-coil transducer pick-up* (Figure 36bii) operates on the same principle as a DC generator or an electric dynamo. In operation, as the stylus motion occurs the coil moves inside a permanent magnet, inducing a voltage in the coil. This voltage is proportional to the velocity of the coil. This design allows the transducer element (the coil) to be coupled directly to the stylus without the need for an extended arm or hinged beam. Because of its overall size it is somewhat restrictive in use and is not generally popular today.

In a similar fashion to the piezoelectric transducer, the moving-coil pick-up does not measure displacement, but velocity. Stylus velocity has to be integrated in order that the absolute position of the stylus can be determined. As a result of the coil velocity rather than measurement of stylus displacement, integration errors will increase as the signal frequencies decrease. Hence, this type of pick-up is unsuitable for the measurement of either profile or waviness (low characteristic frequencies). Therefore the piezoelectric transducer is only suitable for the assessment of roughness. Yet another disadvantage of the moving coil transducer is its poor linearity. It should only be used in the form of a comparator by comparing similar component surface roughnesses against a known roughness calibration standard.

2. *Digital transducers* (Figure 37)
With stylus motion along the surface under test, pulses occur that correspond to multiples of the transducer resolution which are fed to an up–down electronic counter that displays the gauge displacement, with its range being determined by physical constraints of the gauge. Typical range-to-resolution ratios are of the order of 700,000, hence the advantage is that there is no need for range switching, enabling the maximum resolution to be available over the complete operational pick-up range. The displayed displacement is relative to the stylus position, where the counter is zeroed. This means that every time the electronics are switched off and on

(a) Non-contact laser pick-up. (Courtesy of Mahr Perthen.)

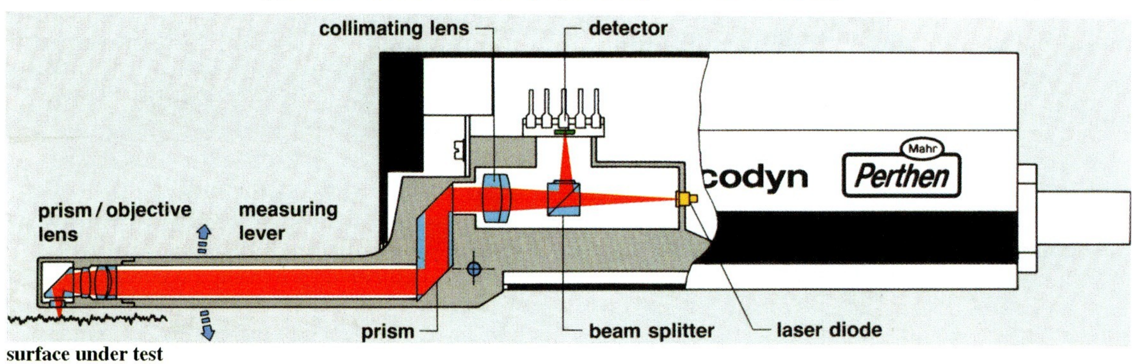

(b) Phase grating interferometric (PGI) pick-up. (Coutesy of Taylor Hobson.)

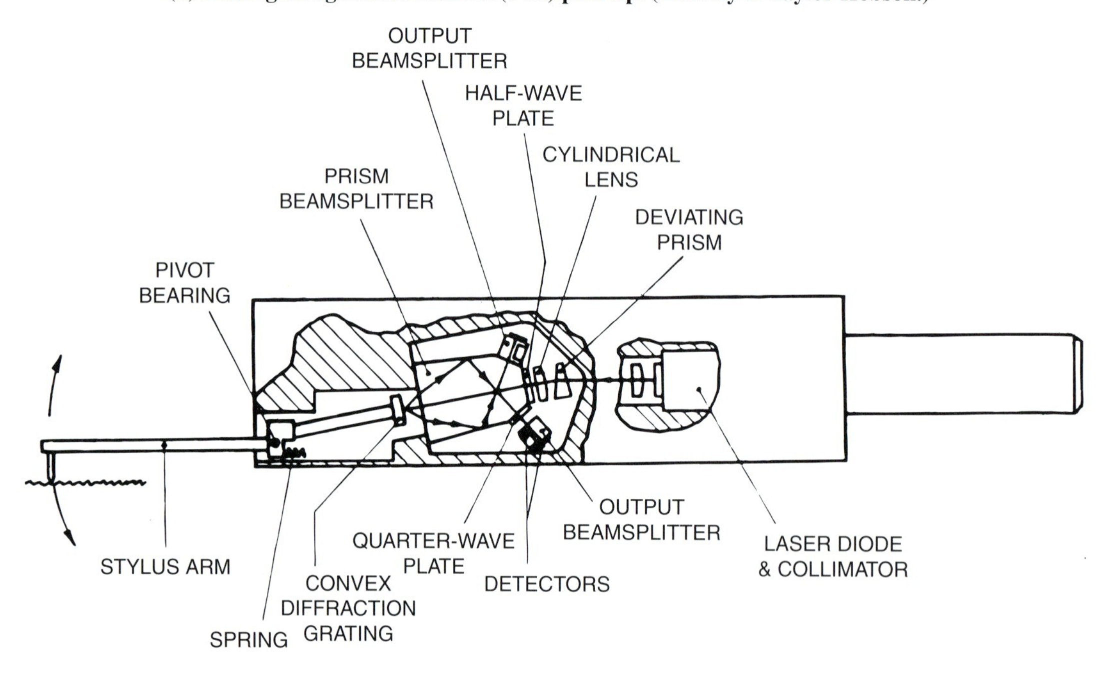

Figure 37. Basic design elements of pick-ups of the digital transducer configuration.

again, the counter must be zeroed at a particular datum position. In a similar manner, if a certain maximum stylus speed has been exceeded the counter may lose count, requiring zeroing again at the datum position.

The following pick-ups are typical of those currently in general use for surface texture instrumentation:

- *Non-contact laser* (Figure 37a): These optical configurations vary depending upon the measuring instrument company, but most systems utilise a miniature Michelson laser interferometer with a laser wavelength of 632.8 nm to provide the measurement reference. Interference patterns are detected by multiple photodiodes, enabling an output signal to be interpolated, typically giving a basic resolution of 10 nm with an operating range of around 6 mm. A focal point of 2 μm is typical, allowing soft or delicate surfaces to be assessed.
- *Phase grating interferometric* (Figure 37b). This pick-up has been developed to complement the laser interferometric type of pick-up as it offers a greater range, with a corresponding reduction in physical size, by employing a laser diode instead of a gas (HeNe) laser source.

The non-contact laser pick-up (Figure 37a) operates on the following principle: infra-red light is emitted by a laser diode; its path through the optical system to the micro-objective is shown in Figure 37(a), with the beam being focused onto the surface to a focal point of 2 μm diameter. Light is reflected back by the workpiece surface, returning along an identical path, later being deflected onto a detector, whereupon an electrical output is generated that corresponds to the distance between the focal point and the surface feature. A powerful linear motor continuously readjusts the hinged measuring lever during the measuring cycle, enabling the focal point to coincide with the surface feature(s) being measured. The focus follows the surface as its motion translates across that portion of workpiece under test, in a similar fashion to a stylus-based pick-up. The vertical movements of the measuring lever are converted into electrical signals by an inductive transducer. Due to the relatively small size of the point of focus (approximately 2 μm), quite minute profile irregularities can be assessed.

Some caution should be expressed in general about non-contact pick-ups, as the surface features being assessed can have a tendency to act as secondary sources of light. When the radius of curvature of a part feature is smaller than 10 μm, diffraction effects appear which can affect the beam's edge. This is not a serious problem for large surface features but becomes crucial for finer surface finishes, illustrating why optical instruments tend to produce higher values for surface texture than their stylus-based equivalents. Other limitations for non-contact pick-ups are that they cannot easily measure small bores, they traverse widely changing surface features, and they cannot "sweep aside" dirt on the surface, unlike stylus instruments.

The phase grating interferometric pick-up (Figure 37b) has positioned on the end of the arm of the pivoted stylus a curved phase grating, which is the moving element in the interferometer. The pitch (grating wavelength) provides the measurement reference. An interference pattern is detected by four photodiodes, enabling the output signal to be measured, giving a basic resolution of 12.8 nm with a range of 10 mm – ensuring that it is a useful pick-up for form measurement.

1.12.2 Skid or pick-up operation

Two types of pick-up operation are normally employed for surface texture assessment:

- with a skid;
- skidless with a reference plane.

The geometry of a skid can vary, from either a large radius (Figure 38b) to a flat (Figure 38c). The skid supports the pick-up, which in turn rests on the surface. The relative motion of the pick-up as it traverses along the workpiece causes the skid to act as a datum, giving rise to a "floating action" of rising and falling of the pick-up (Figure 38b). The macro-irregularities of the surface, namely of waviness and profile, means that the stylus with its significantly smaller radius to that of the skid will follow this surface topography, sinking into valleys that the skid bridges. However, if the crests and valleys in the workpiece surface are too widely spaced, then this distance with respect to the skid motion (Figure 38a) will also cause the skid to rise and fall and thereby lose the datum registration plane (Figure 38a). If the distance between the stylus tip and the skid (Figure 38a) is half the waviness spacing, then these waviness amplitudes are doubled. Conversely, when the distance of the stylus tip (Figure 38a) corresponds exactly to that of the waviness spacing, then the waviness is mechanically eliminated. The skid type of pick-up is only marginally affected by instrument vibration, because the measurement reference is situated close to the point of measurement. Therefore, the mechanical relationship of the pick-up-to-skid system may be an important factor in its design. If different pick-ups are utilised for the same measuring task, then in extreme situations these measured differences could approach that of 100%.

Ideally, the skid and stylus should be coincident, thereby eliminating the "Abbe error"; the Abbe principle states that *the line of measurement and the measuring plane should be coincident.* For practical reasons, the stylus is slightly displaced from the skid and is situated either in front or behind the stylus. Phasing errors can be induced by the skid (Figure 38b), causing distortion of the graphical trace, while its magnitude depends on the waviness crest spacing. In practice, if the waviness crest spacing is identical to that of the skid/stylus spacing, the skid will rise on one crest, with the stylus rising on an adjacent peak. This relative motion completely suppresses the peak. When the spacings differ, the effect illustrated in Figure 38(b) occurs, where a spurious valley is introduced into the graph at a distance from the peak, this being equal to the skid/stylus spacing. If two skids are used on either side of the stylus, then the "phase effect" of the single skid/stylus spacing is diminished. In some pick-ups the line of contact of the skid is laterally offset, which prevents the skid from burnishing the track (mechanically deforming asperities and giving a false reading) during repeated traverses; this is particularly relevant when the instrument is automatically setting the vertical magnification (auto-ranging).

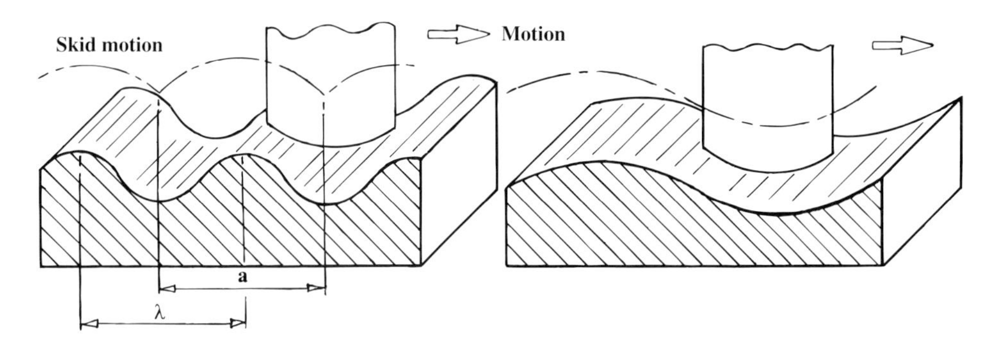
(a) A rounded skid does not provide a serviceable datum,
if the surface irregularities are more widely spaced
Skid motion
Motion
a
λ

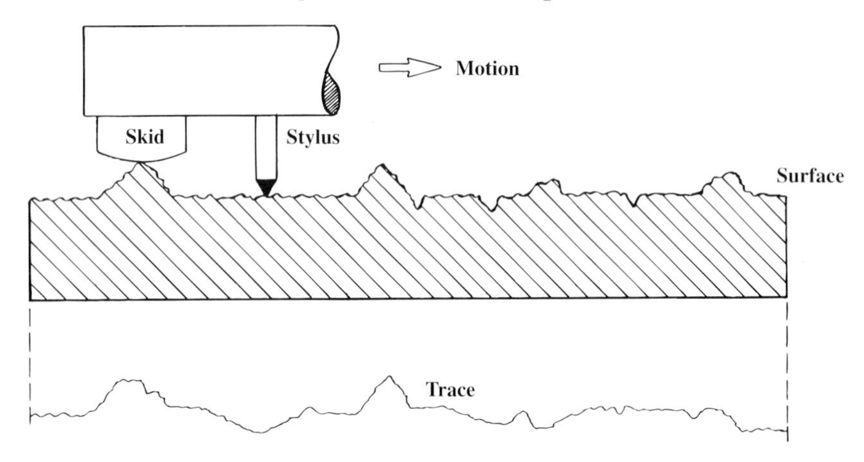
(b) The "phase effects" when using a skid
Motion
Skid
Stylus
Surface
Trace

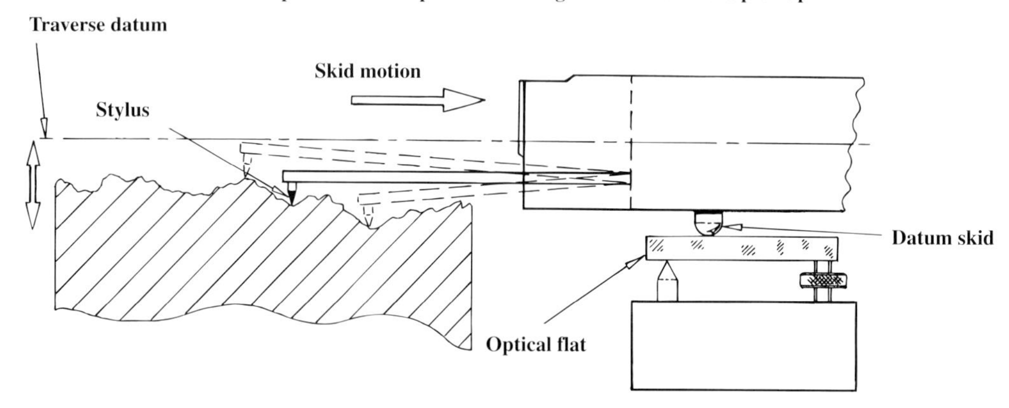
(c) Diagram illustrating an independent datum
An independent datum provides a straight reference for the pick-up
Traverse datum
Skid motion
Stylus
Datum skid
Optical flat

Figure 38. Configuration of surface texture instrument and relative motion of the stylus with respect to an independent datum. (Courtesy of Taylor Hobson.)

The skid acts as a mechanical filter, modifying the profile to a greater or lesser extent. Its radius of curvature, therefore, should be suitably chosen to minimise filtering effects. Generally, if a skid is selected with a radius of curvature of approximately 50 mm, this is normally sufficient for general-purpose applications.

The skidless or reference plane technique can take the form of two distinct types:

1 the pick-up skid travels along a flat surface such as an adjacent optical flat, or similar and not on the surface being measured;
2 a shaft which moves in an accurate straight line, with the pick-up being rigidly fastened to it – usually a glass block providing an independent datum.

The main reason for employing skidless pick-ups are that they do not eliminate the contributions of waviness and profile errors from the measured topography – which the skid obviously does – and often waviness, in particular, is important to the manufacturing process and must be included in the assessment. The skidless technique provides an undistorted assessment of the actual profile, the proviso being that the reference surface has ideal form.

Figure 39. Portable hand-held surface finish instrument "Surtronic Duo" can be used at any angle/attitude. (Courtesy of Taylor Hobson.)

1.12.3 Portable surface texture instruments

As the assessment of surfaces is not confined to those inspected in a metrology laboratory and often entails the measurement to be made by both production and maintenance personnel, then some easily portable yet robust instruments are necessary. One such instrument is that shown in Figure 39. This battery-operated surface texture measuring instrument can be used across a wide range of angles and attitudes, increasing the range of versatility of the equipment. This instrument has a piezoelectric pick-up which is self-calibrating and can be used in remote operation – split mode – through the infrared (IrDA) link, up to a distance of 1 m without the inconvenience of cable supply. This is an important factor when attempting to measure otherwise inaccessible component features such as bores.

When in transit the stylus has protection, as the "park" position fully protects the stylus assembly. Digital circuitry via *surface mount technology* provides both reliability and accuracy and the stylus assembly closes up (in its robust case) to fit either in the pocket (weight 200 g), or on a belt, which is useful for off-site surface evaluation in the field. The large and clear metric or imperial LCD display provides simple-to-read surface-related information and instrumental status data, with the following specification:

- "standard" cut-off (0.8 mm ± 15%);
- accuracy: *Ra* (range: 0.1–40 μm ± 5%); *Rz* (range: 0.1-199 μm ± 5%);
- diamond stylus (5 μm radius);
- traverse length 5 mm;
- minimum bore penetration 65 mm.

Portable and hand-held instruments of this type (within the specification and parameters of the equipment) can establish the surface condition, without the need to break-down assemblies, or the necessity of removing the surface in question to a dedicated and more sophisticated instrument. Such versatility and portability in surface measurement when combined with the confidence in knowing that a representative roughness has been established, albeit for either *Ra* or *Rz*, helps in the interpretation of component topographical information from normally inaccessible positions, when compared to that of conventional desk-based equipment. They are, however, limited in their surface analysis and parameter scope.

1.12.4 Surface form measurement

Curved datums

In the previous discussion, all the surfaces have been nominally considered to be flat with only marginal form present. There are occasions when it is necessary to measure the surface texture of curved surfaces such as balls, rollers, the involute curvature of gear teeth and fillet radii of crankshafts (see Figure 40). In the past a wide range of gauges had to be available with sufficient resolution to assess the surface texture together with form, while setting up the surface under test in such a manner that the geometry of its shape did not influence the actual surface texture measurement.

If the limiting factor in the determination of the surface texture was due to a large radius, then this problem could be minimised by utilising a short cut-off length and a skid fitted to the gauge to reduce the *measurement loop*. The skid will induce problems in obtaining a successful and representative surface texture measurement, if fitted in front or behind the stylus – the former technique is illustrated in Figure 41(a). However, the stylus displacement from the measuring plane can be negated by introducing a skid which surrounds the stylus. This tip projects through the skid's centre (see Figure 41c), thus obeying "Abbe's principle". The solution to curved surface measurement can be addressed by utilising several of the following types of equipment, or techniques:

Figure 40. Form measurement with an inductive contour pick-up, including "customer-engineered" work-holding, for crankshafts. (Courtesy of Taylor Hobson.)

- as mentioned above – instead of using a straight datum (skid) the gauge is guided by a curved datum (Figure 41c);
- the gauge is held stationary and the surface under it is slowly rotated at a peripheral speed sufficient to gather surface data information;
- the gauge is traversed concentrically around the stationary workpiece surface;
- a wide-range gauge is utilised to enable it to allow for the curvature of the workpiece's surface profile as it traverses the part.

The inspection of a curved surface offers a particular challenge, which can be achieved by guidance of the stylus gauge along a curved path having the same radius as the part surface. The curved surface measurement condition is attainable if datum kits are used (allowing curved surface paths to be traced by the gauge – see Figure 41b). These datum kits tend to be time-consuming in setting up, necessitating a degree of technical expertise in both set-up and alignment. The curved surface is rotated within the datum kit's conical seating against the stylus of a gauge, the stylus being incorporated into the conical seat housing (Figure 41b). The advantage of this set-up is that the tip of the stylus is always aligned to the centre of the component, with the likelihood of *tip flanking* on the part being eliminated. Datum kits are normally only practicable when the radius is constant during measurement, but if a series of radii occur then under this situation they are not to be recommended. Today, such datum kits are not normally used due to the necessity of long set-up times.

Form analysis

The introduction of wide-gauge ranges on high-resolution pick-ups and the development of various form options (reference line fitting) has meant that guidance of the pick-up along a curved path has been virtually eliminated. Typical of these surface form-fitting features that are currently available are the following software-based options:

- removal of a least-squares best-fit radius;
- removal of aspheric form from the raw data – form error, surface slope error and tilt comparison with operator-designed data in the form of a polynomial expression (see Figure 42a);
- elimination of elliptical/hyperbolic (conic) geometry – provides major and minor axes values and tilt, with residual surface texture analysis after removal of best-fit elliptical and hyperbolic forms (see Figure 42b);

(a) Stylus displacement solely due to surface curvature (i.e. $d_1 + d_2$)

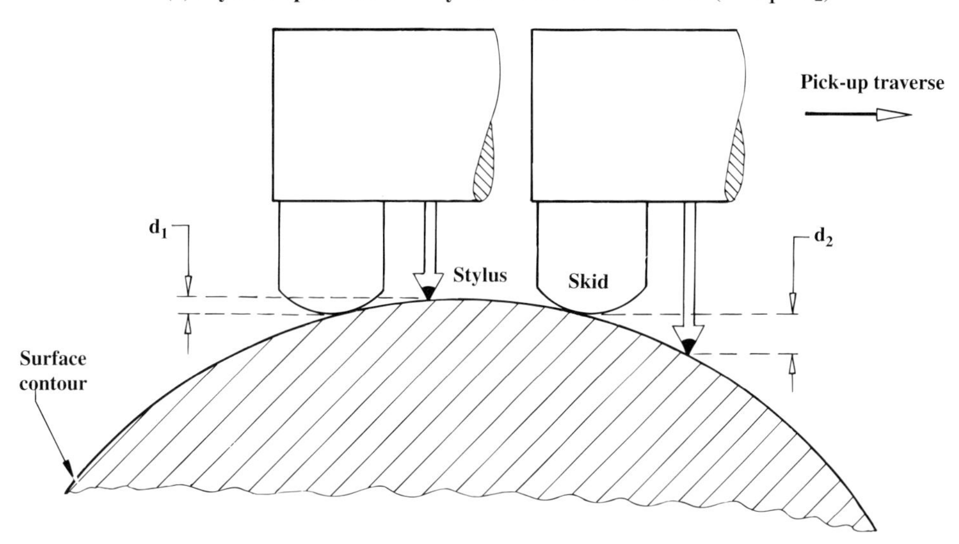

(b) Contour measurement–principle of rotating workpiece instrument

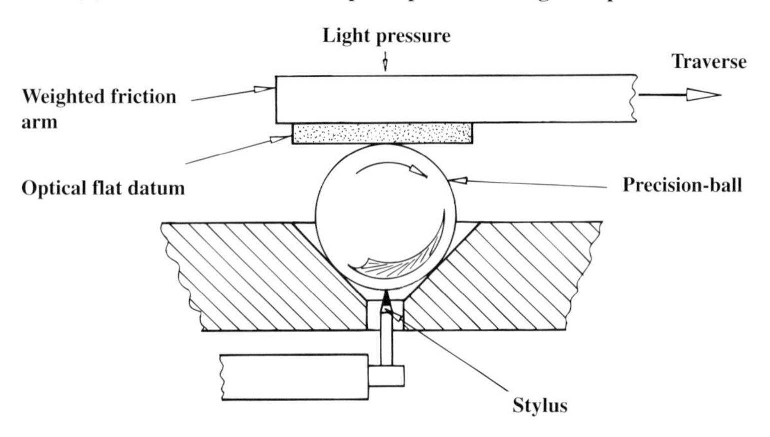

(c) Side skid gauge – the stylus projects through the cylindrical form of the skid

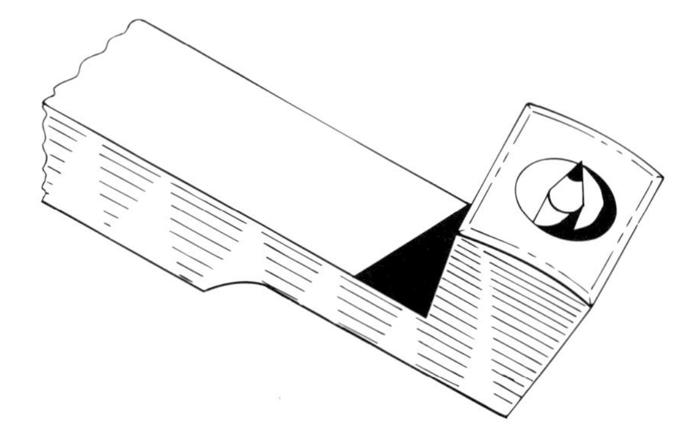

Figure 41. Effect of curved surface measurement. (Courtesy of Taylor Hobson.)

(a) Aspheric form software

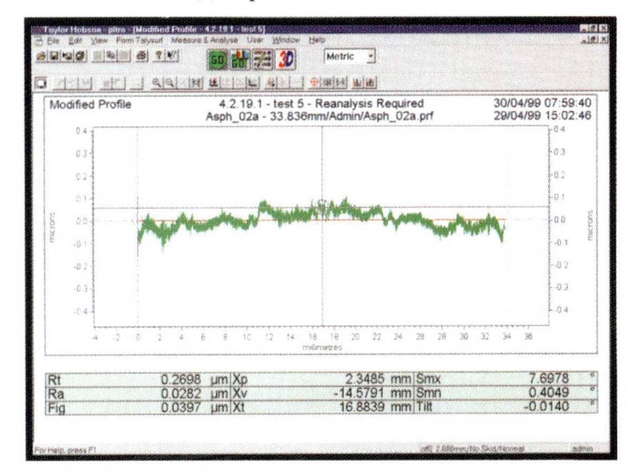

(b) Conical form software

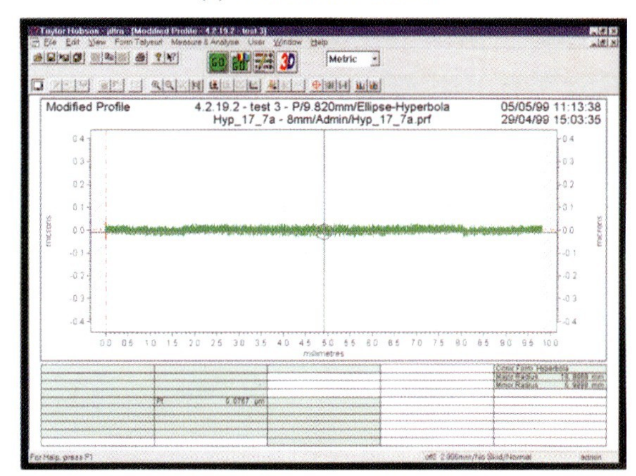

(c) Dual profile software

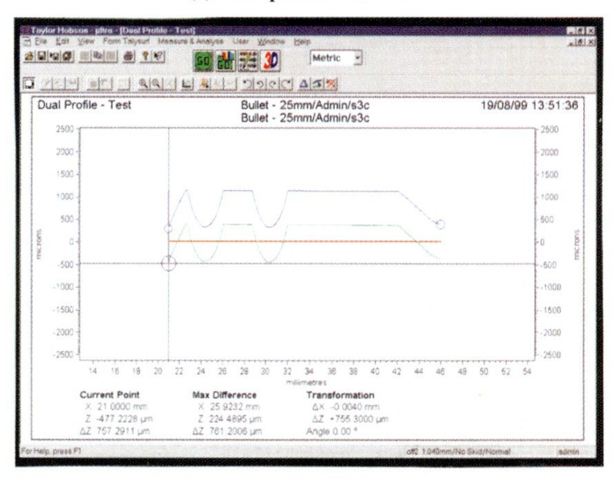

Figure 42. Screen displays of "Windows™-based" software options available for form assessment. (Courtesy of Taylor Hobson.)

- complex symmetrical profile assessment – via an enhanced "Dual Profile" facility, allows the measurement profile of a master shape or prototype component to be saved as a template, enabling future components to be measured and simultaneously displayed along with template data for immediate comparison (see Figure 42c);
- analysis of gothic arch profiles – frequently employed on the ball tracks for recirculating ball-screws;
- three-dimensional contour analysis – a representative 3D visualisation of the part surface: axonometric projection, colour height distribution, wear simulation and form removal, with many additional features available (this topic is a subject on its own and discussed in Chapter 2).

The number of surface texture parameters available to define a surface's condition is immense for form-measuring instruments. Typically, these parameters include 22 profile; 22 waviness; 25 roughness; 12 R+W; 9 aspheric parameters; and this comprehensive list can be enhanced still further. In addition to a full range of surface texture parameters, form analysis software can provide:

- *form error* – calculation with reference to best-fit concave or convex circular arc, straight line measurement, including surface roughness detail. Alternatively, referencing to the minimum zone, this being the minimum separation between two parallel lines containing the data set;
- *radius* – using the least-squares best-fit, concave or convex circular arcs can be automatically calculated from selected data, with the option of being able to exclude any unwanted surface features from the selected data. Conversely, the absolute radius can be set to analyse actual deviations from a design master, with other calculated parameters including its centre coordinate and the pitch;
- *angle* – using a straight edge or minimum zone algorithm, surface tilt can be established and then removed, prior to parameter analysis, with other

calculated values including its intercept and pitch;
- *dimension* – the linear relationship of surface features can be assessed and then compared, owing to the ability to calculate both the true *X*- and *Z*-coordinate positions.

Contours: profile curving, or an irregular-shaped figure

A typical commercial system is currently available for the analysis of contour measurement; assessment of features such as radii, angles, length and height (see Figure 43 for a typical application) can be achieved with:

- *measurement macros* – these can be learnt and edited forming user programs for repeated inspection routines, which minimise repetitive operator input (achieved via a series of definable "fastkeys");
- *individual feature tolerancing* – allowing individual sections of the measurement template to be toleranced with a variety of values (a wide tolerance on a flat area of a surface can be defined, with tighter tolerancing to the radii);

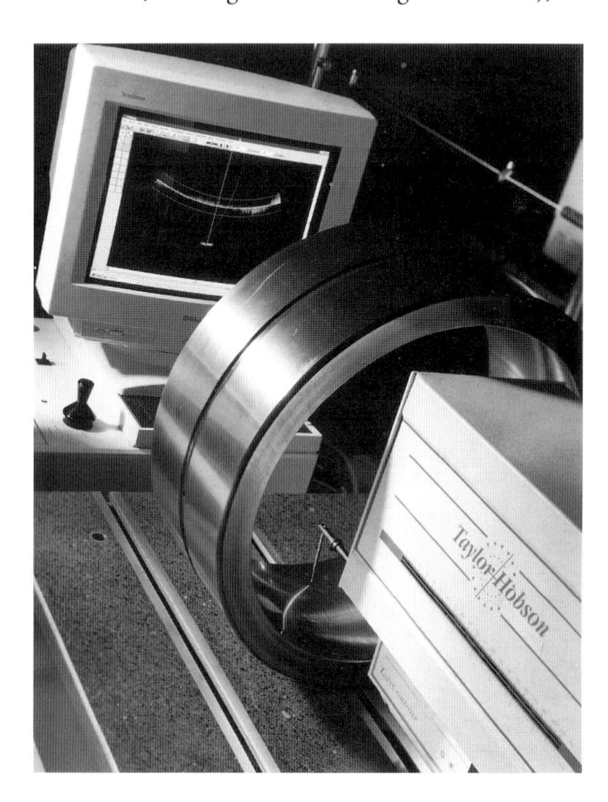

Figure 43. Form and surface measurement instrument "Talycontour" being utilised to assess a spherical bearing outer race. (Courtesy of Taylor Hobson.)

- *comparison of DFX files to the contour* – if measuring a known contour "best-fit" individual geometric elements are calculated with reference to a template and the comparison of each point in the contour is assessed against the nearest element in this template;
- *geometric element fit to an unknown contour* – when measuring components of an unknown contour, the software finds geometric elements that optimise the geometry of the profile determining the size, position and relationship between profile features, plus other functional factors.

The most recent form and surface texture analysis equipment has absolute positional control, offering sophisticated high-technology instruments providing complex analytical abilities in a single traverse of the workpiece. Lately, with the high thermal stability of the pick-up gauge (phase grating interferometer – PGI), it typically has a large range with around nanometric resolution, equating to a range-to-resolution of 780,000:1. One of the current developments in stylus protection employs a "snap-off" connector, allowing the stylus to be ejected from the gauge if the loading stylus force exceeds safe limits – reducing possible damage to the gauge because of operator error. Interchangeable styli can be fitted to such pick-ups, for better access to the component feature being inspected. If the part is of a delicate nature and liable to deform, or could be easily scratched, then non-contact gauge heads can be utilised. These non-contacting types of gauges comprise a focus follower with a programmable controller for fast, simple use, replacing inductive pick-ups. Non-contact gauges are used where the component surface applications consist of:

- soft and touch-sensitive surfaces – printing pastes and coatings;
- wet or dry solder pastes and printed conductive pastes;
- contact and intra-ocular ophthalmic lenses;
- and many other industrial/scientific applications.

Due to variations in the orientation of the workpiece under test to the traverse unit, some of the latest versions of form analysis instrumentation can be operated in inverted attitudes, or right-angle attachments can be fitted for measurements between shoulders on components, such as on crankshafts (see Figure 40).

1.12.5 Non-contact systems

Contact and non-contact operational aspects on surfaces

As has been previously mentioned, in contact systems the stylus contacts the surface utilising a precisely manufactured stylus and, today, it is invariably diamond-tipped. However, owing to the stylus geometry – its shape and size – some styli on certain surfaces will not be able to penetrate into valleys and as a result create a distorted or filtered measure of the surface texture (see Figures 11 and 12). A recent study has indicated that the actual radius of curvature of a stylus can be very different from its nominal value, while the effect of stylus forces can significantly affect the measured results, with too high a force causing surface damage.

The stylus, to enable a representative surface to be assessed, must traverse across the surface following a parallel path with respect to the nominal workpiece surface. The datum in this case would most likely be a mechanical slideway of some sort (see Figure 38). The datum comprises a skid that has a large radius of curvature (spherical or different radii in two orthogonal directions), fixed to the end of a hinged pick-up (see Figures 36 and 37). At the front end of the pick-up body the skid rests on the workpiece surface; alternatively other designs occur, such as a flat shoe being free to pivot so that it can align itself to the surface, or two skids are situated either side of the stylus. It should be reiterated that ISO 3274: 1996 does not allow the use of a skid, but many are still used in metrological applications in industry.

Notwithstanding the long history of stylus-based instrumentation a number of problems exist, associated with either their operation or the interpretation of results. For example, none of these instruments measure the surface alone: if an inhomogeneous workpiece is inspected utilising a mechanical stylus it responds to the topography and changes in the surface mechanical properties, such as its elastic moduli and local hardness. Another important aspect in contact surface texture inspection techniques is the scale – horizontal and vertical – although it is not common for instrument manufacturers to incorporate metrology into the scanning axes of their instruments, with many companies failing to calibrate scanning axes.

Stylus-based surface texture instruments have an infrastructure of many specification standards, unlike optical/non-contact instrumentation. This point regarding standards for stylus instruments only embraces two-dimensional measurements and they have yet to be developed for their three-dimensional counterparts. Yet despite this two-dimensional standards infrastructure, there is a distinct need for both good practice and a clear understanding of which surface texture parameter is most suitable for an industrial application to pass down to the shop floor – relating to standards.

For non-contact instrumentation, if one considers some form of optical interaction with the surface that enables local height variations to be established, then it will sample a different surface from that of stylus-based instruments. As a practical example of this difference, if consideration is given to the optical assessment of metals, their surface homogeneity (different phases) could introduce apparent height changes up to 10 nm due to the phase change on reflection; whereas glasses or ceramics may have local refractive index changes and contaminant films introducing nanometric changes. While phase changes at the conducting surface can introduce variations as a function of incident angle, surface field anomalies that are the result of multiple scattering, or sharp points acting as secondary sources of light, together with various edge and wall effects, will all introduce measurement uncertainty.

It was briefly stated that optical methods do not have a standards infrastructure, unlike that of stylus-based surface texture instrumentation, meaning that there exists no formal techniques for calibrating optical instruments. Therefore, as a result of this lack of calibration methodology, care should be taken when optical assessment of a range of surfaces is to be undertaken, particularly if they are of differing physical surface characteristics. For example, if a glass step-height standard is utilised to calibrate the vertical magnification factor of an instrument, it would be unwise to measure a metal surface set at an identical calibration value. More will be said on the topic of calibration in Chapter 6.

Laser triangulation

The laser triangulation, or chromatic abberration techniques can be utilised to measure surface topography, which are dependent on the sensor model employed. Cut-off filters available range from 0.8 mm to 8 mm with large measuring envelopes of up to 150 mm by 100 mm, enabling large or multiple artefacts to be scanned within the same measurement period. A typical instrument will have a vertical range of 300 μm and resolutions of 10 nm from a spot size of 2 μm in diameter, with scanning capabilities of up to 2000 points per second. The measuring ranges currently available are from 0.3 mm to 30 mm, depending on the sensor model.

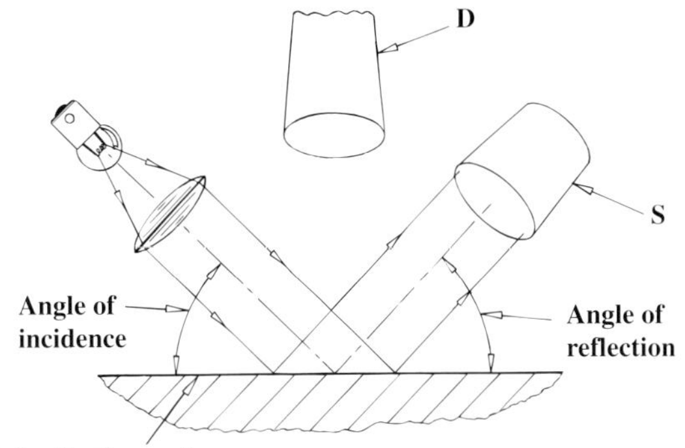

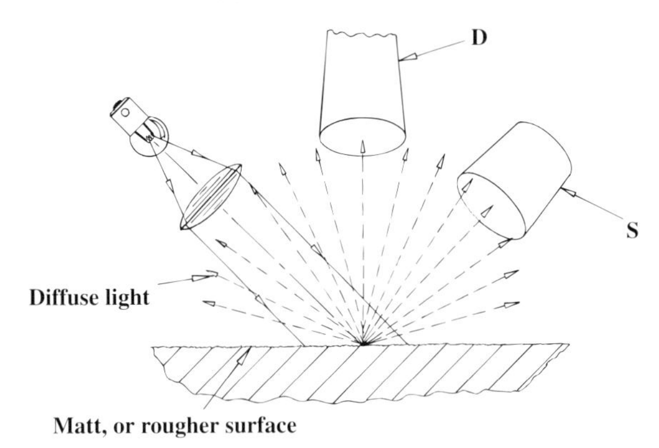

Figure 44. Diagram illustrating the principle of surface assessment by specular and diffuse reflection. (Courtesy of Taylor Hobson.)

One such system has a tested repeatability of 0.03 μm, but this claim applies to a specimen having a 2.90 μm *Ra* roughness, measured with a 2.54 mm cut-off length. Optional equipment can be added to investigate feature location and alignment or vacuum stages can be incorporated to hold awkwardly-shaped artefacts and replication kits can be supplied to measure inaccessible or difficult to assess surface features.

It is claimed that these systems can measure surface colour or texture. These instruments offer fast and multiple scanning facilities that reduce the time taken for each measurement, compared to conventional contact instrumentation. Common to all optical equipment (as described in the following section), light scattering causes measurement accuracy problems in some component surfaces. If a fixed resolution occurs for each sensor model, this would imply that up to six sensors would be required to enable satisfactory coverage of all surface textures, or types of measurement needed for a complete metrological range of applications.

Optical techniques: reflection and scattering

Highly reflective surfaces are normally assumed to be very smooth, so by measuring reflectivity an indirect technique for measuring roughness can be established. Reflectivity testing is particularly beneficial where visual appearance is important (cosmetic items), or on soft surfaces such as paper where non-contact measurement is necessary. Advantages of optical methods include assessing and averaging over an area and high-speed inspection.

Smooth surfaces invariably appear to be visually "glossy" (reflective); conversely, a rough surface has a matt appearance. In the past, the *gloss meter principle* was used to verify how parallel beams of incidence light at an angle on the surface are reflected.

If a perfectly reflective surface is present (Figure 44a), then much of the light will be specularly reflected (angle of reflection equals angle of incidence) and enter the viewing system S, with no light entering the viewing system at D, being positioned perpendicularly above the surface. However, if a perfectly matt surface is inspected (Figure 44b), the light is diffusely reflected (scattered in all directions), causing equal amounts of diffused light to enter both S and D. The ratio of specular-to-diffuse reflection is a measure of the gloss of the surface. The relative proportions of light entering S and D can be measured by either visual means or photoelectric comparison, incorporated into a suitable portable instrument. As it is the *ratio* of the light that is measured, any absorption of light by the surface will not influence the result, and neither will fluctuations in the irradiance distribution of the lamp, or more specifically today, the laser.

Considerable interest has been shown over the years in the use of scattering for surface texture measurement. In essence, three techniques are generally used for scatter-based instrumental design:

1. light scattering theory, when used to construct an instrument giving absolute measurements of surface texture – when a number of conditions are met;
2. general theoretical approach in instrument design – assumptions are made that have general connection with a particular surface texture parameter;
3. application-specific approach, where instrument design is developed to solve an immediate problem.

In just one example, an *integrating sphere* (Figure 45) can be used to measure gauge block surface

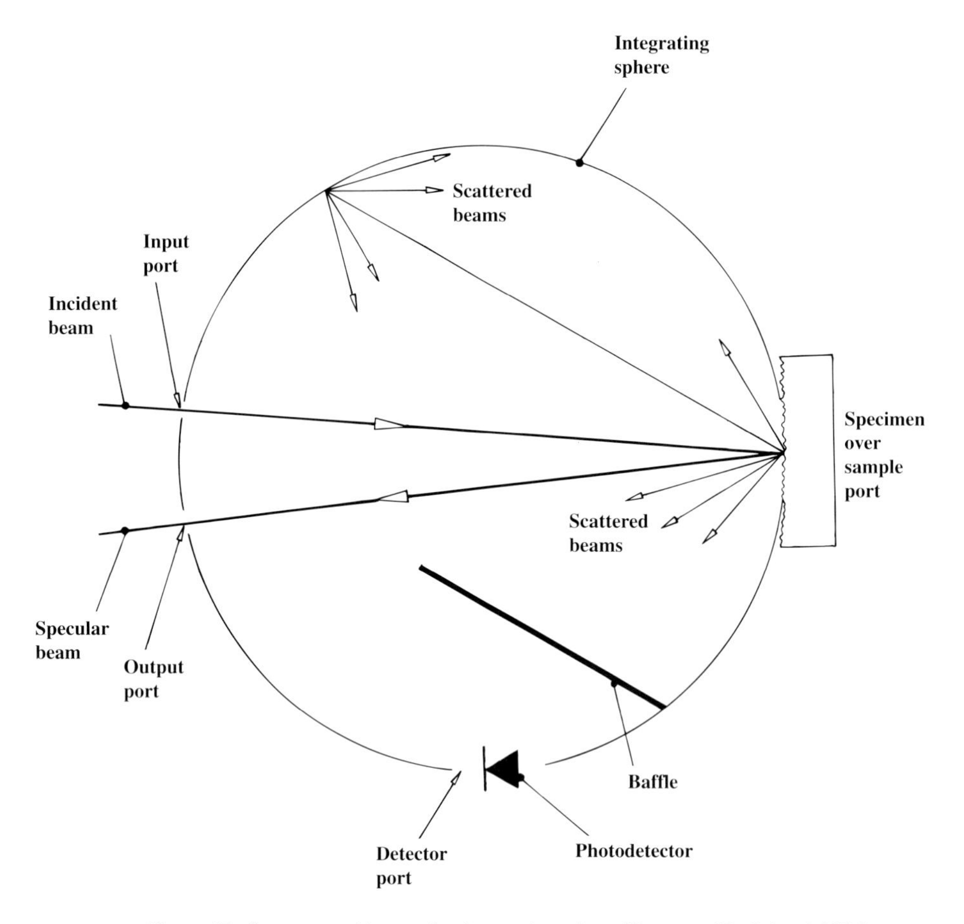

Figure 45. Geometry and layout of an integrating sphere. [Courtesy of Dr R. Leach/NPL.]

texture. The instrumental technique measures the total intensity of light from a gauge surface, then compares it with the light intensity that was diffusely scattered. The ratio of two intensities is termed *total integrated scatter* (TIS) and it is proportional to the square of the RMS surface texture parameter *Rq*. This method relies on a number of assumptions that must be met for the type of surface to be measured. The diffuse component of reflected intensity is measured by collecting light in an integrating sphere. An integrating sphere can be practically represented by a hollow sphere that is coated internally with a highly diffuse reflecting material. Any light reflected from the inner sphere's surface gives rise to a constant flux of electromagnetic power through an element of the sphere. This sphere has a number of ports, allowing radiation both in and out, with the gauge to be irradiated and photodetectors to detect the intensity of integrated light.

Interference instruments

Surface texture interferometry occurs in several instrumental configurations, probably the most common types being interference microscopy, plus full field methods. Becoming popular of late are phase-stepping techniques and swept-/multi-frequency source methods. Today, there are several commercially available phase-shifting interferometric microscopes, giving three-dimensional surface imaging with very short measuring times. The National Physical Laboratory (NPL) has developed a sub-nanometre resolution system that employs a microscope objective with a numerical aperture of 0.5, in conjunction with a birefringent lens providing a focused spot from a laser source and a defocused orthogonally polarised beam 10 μm in diameter. Fourier analysis has been obtained from surfaces, indicating that with this objective the system produces a surface wavelength range from 0.5 to 15 μm. The trace displayed by the system is an interferometrically obtained path difference of the focused probe beam and the defocused beam.

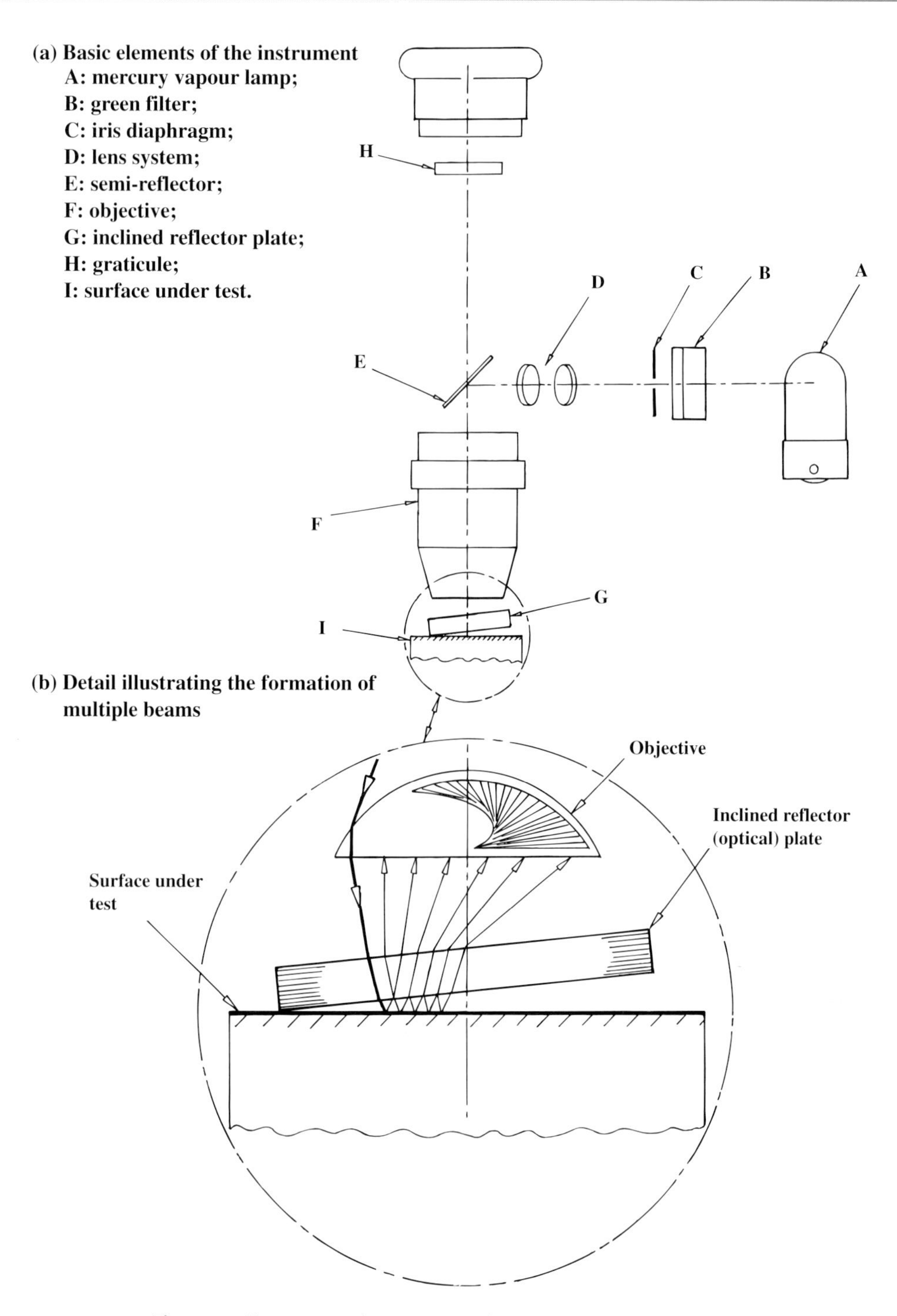

Figure 46. Fizeau-type surface texture interferometer. (Courtesy of Taylor Hobson.)

As has already been mentioned, light interference can be successfully used as a non-contact optical technique to assess surface features. Essentially, the principle of interference of light relates to how light rays are reflected between two surfaces; the differing light path lengths along various points of the surface cause phase changes in the light being reflected back to the observer. The result of this reflection will be to promote alternate dark and light interference fringes. The shape and spacing of these fringes will depend on the regularity of both the surface and the reference reflector. The surface texture irregularities are reproduced as modifications in the interference pattern and, when viewed under specific conditions, displacement of the fringes is a measure of the roughness height.

For some interferometer instruments such as the Fizeau type, interferometric surface assessment relies on the fact that a "wedge of air" occurs on the test surface, with the resulting interference lines being of equal height and their respective height differences given by

$$\text{Fringe spacing} = \lambda/2$$

where λ = wavelength.

Under examination, the surface roughness produces air wedge thickness variations causing deflection of the interference fringe, from which the total height parameter Rt results. Hence the total height becomes

$$Rt = a/A\,(\lambda/2)$$

where A = interference fringe, a = fringe deflection and $\lambda/2$ = fringe spacing.

The height of irregularities that can be resolved range from 0.005 to 1 μm using conventional interferometric methods. The range can be extended by using the technique of *oblique incidence*, which increases fringe spacings from $\lambda/2$ to 2λ, enabling one to measure a surface roughness height four times larger than that measured using a standard interferometric method. Furthermore, if cast replicas of the test surface are made, then immersing them in fluids of a suitable refractive index enables transmission interferometry to assess roughnesses up to 25 μm.

Maximum contrast is important for optical interferometry, and the wavelength of emitted light must remain constant to eliminate variations in the phase difference between interfering beams. Any defects in optical surface quality within the interferometer can promote a reduction in fringe contrast, as can vibrations of optical components, together with parasitic light caused by unwanted reflections. Due to the strongly convergent light utilised in these interferometric optical configurations (Linnick and Mirau interferometers), this can lead to obliquity effects, which can introduce errors in estimation of, say, a scratch depth by as much as 12%. The magnitude of error is dependent on the numerical aperture of the objective and the surface's scratch depths. However, in the case of more commonly available full field techniques this obliquity and hence error is not a problem.

In essence, the Fizeau surface texture interferometer is basically a microscope, with a built-in illumination and special-purpose optical system (see Figure 46a). Between the objective and the surface a semi-reflector G can be positioned, which may be slightly inclined to the work surface I, allowing a wedge-shaped air gap to occur (detailed in Figure 46b). Multiple reflections between the surface and the reflecting plate produce good-contrast, sharp fringes (see Figure 47 for typical interferometric images) which are viewed in the eyepiece. Utilising a spherical semi-reflecting surface, it is possible to examine three-dimensional curved surfaces, such as precision ball surface finishes (Figure 47c). Multiple reflection (Figure 46a) on the semi-silvered surface E and the workpiece I means that all oncoming beams are split into several partial beams, which cause interference. Interference fringes that occur are both higher in contrast and narrower as the reflection coefficient becomes greater. If too great a reflection coefficient occurs, the contrast will degrade. Such conditions allow multiple beam reflection to occur only when the distance to the reflection surface and that of its test surface is quite small.

Large apertures can be utilised by this Fizeau surface interferometer, enabling full exploitation of the light microscope. When employing high apertures, the line standard (graduated scale) is dependent on the selected aperture and, considering the associated aperture angle, the fringe spacing is

$$\text{Fringe spacing} = \lambda/2\ (2/1 + \cos\mu)$$

where λ is the wavelength, μ is a cosine error.

Normally, the largest aperture (A) utilised is 0.65, hence the fringe spacing equation above can be simplified to

$$\text{Fringe spacing} = 1.14\lambda/2$$

Generally, it can be said that applications of this particular Fizeau surface texture interferometer have limited use, although the technique does have some merit:

- it can examine a relatively large specimen surface area;
- it indicates any form error present;
- being a non-contact method, it may be possible to estimate the scratch depths.

An optical and practical limitation of this type of interferometry when applied to precision surfaces occurs due to visual viewing, which tends to be time-consuming and somewhat fatiguing. By using automated image analysis techniques for detection, this limitation could be overcome, but at a considerable cost disadvantage to the user.

Although three-dimensional surface texture analysis will be discussed in Chapter 2, one instrumental technique has been included in this chapter. An automated surface profiling interferometer is

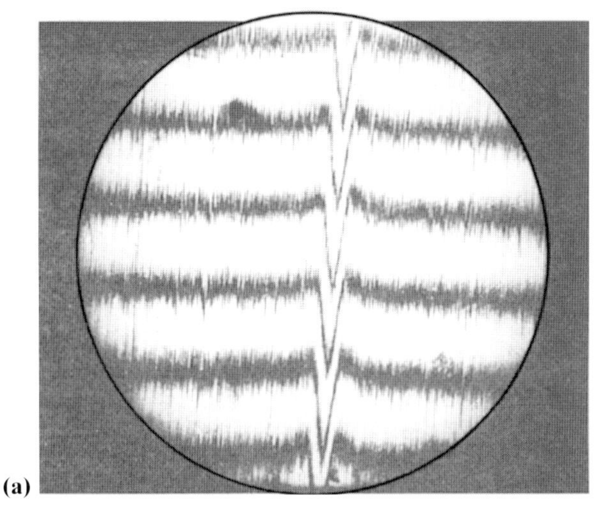

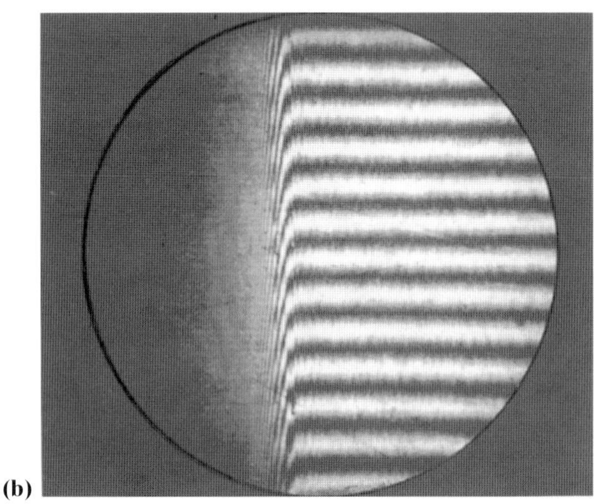

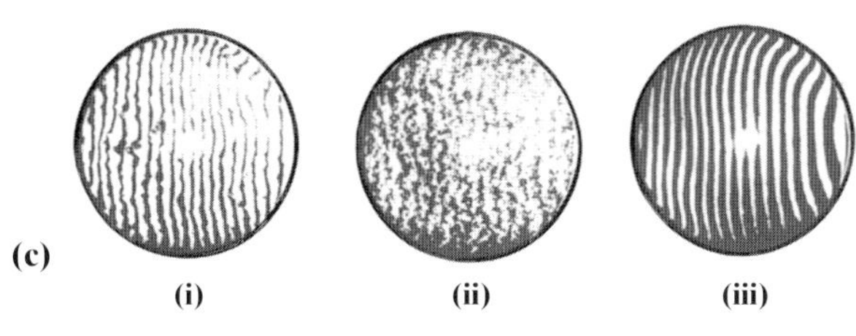

Figure 47. Typical interference fringes from a surface texture interferometer, showing both flat and spherical surfaces (Courtesy of Taylor Hobson).

(a) Lapped surface of a gauge block, showing a scratch. Fringe spacing is 0.25 μm. Due to a scratch extending across adjacent fringes, its depth must also be 0.25 μm. (b) Lapped gauge block edge, indicating edge rounding. (c) Interference patterns on spherical surfaces (i) pitch polished steel ball; (ii) commercial steel ball; (iii) glass sphere.

illustrated in Figure 48 that has a wide range of industrial applications, from research to manufacturing process control. Of particular relevance is its use in semiconductors, disk drives, printing plates and fuel injector seals, together with applications where the surface texture is used to control the product's performance. Until recently, microscopes tended to be utilised purely for attribute sampling, such as imaging surface topographical details, while surface profilers provided accurate measurements to characterise the surface details. One of the latest developments in surface texture (white light) interferometers is illustrated in Figure 48, which combines these two technologies, providing a fast, quantitative surface measurement on many types of surface topographies. The surface details that can be quantified include surface roughness, step heights, critical dimensions and other surface topography features within a matter of seconds. Profile heights that can be assessed range from >1 nm up to 5000 μm, at specimen translation speeds up to 10 μm/s with a 0.1 nm height resolution, which is independent of both the magnification and feature height. This interferometer can resolve sub-micrometre *X*- and *Y*-plane features on profile areas up to 50 mm^2, achieving this by *image stitching*. This stitching technique has been designed to analyse surface details much larger than is possible with a single measurement. Stitching occurs by taking several measurements of the test specimen as it is moved by a motorised stage and then combines – stitches – the multiple data sets into one surface image. In effect, stitching increases the field of view, without compromising lateral or vertical resolution. The scanning actuator is of a closed-loop piezoelectric variety, employing low-noise capacitance sensors to ensure that both accurate and repeatable linear motion occurs over the whole operational range. Other notable modes include both "Phase Stepping Interferometry" (PSI) and "White-Light Scanning Interferometry" (WSI) techniques.

Use of speckle

This technique for surface texture inspection utilises partially coherent light, with the reflected beam from the surface consisting of random patterns of dark and bright regions termed *speckle*. The spatial patterns and contrast of the speckle depend on the optical system configuration utilised for observation and the condition of coherence of the surface texture. Speckle has two important attributes:

- contrast;
- number of spots per unit area.

The contrast can be shown to have a relationship to the surface's correlation length, whereas the speckle density is primarily concerned with the image system resolving capacity. Information on the specimen's surface can be obtained using the contrast of speckle patterns produced in the first instance, near

(a)

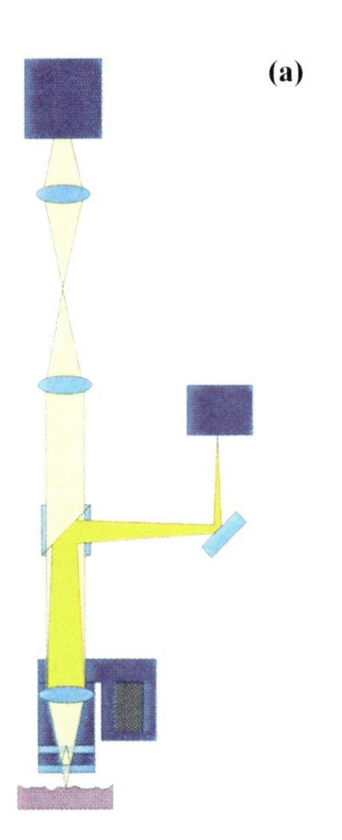

(b)

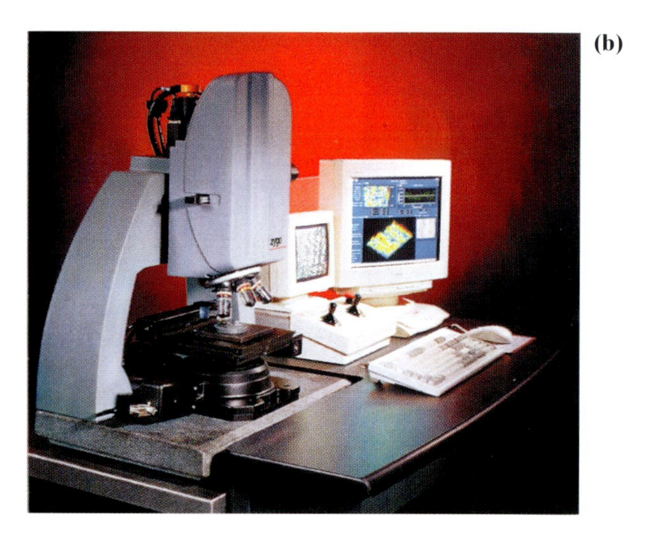

(c)

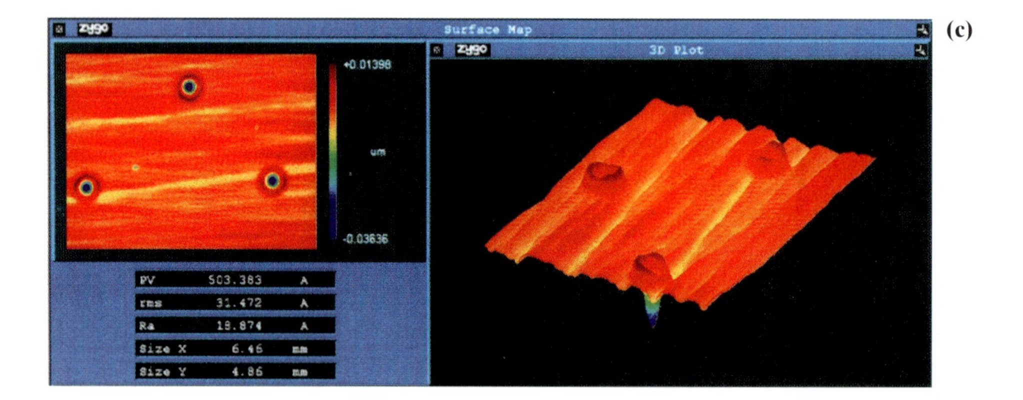

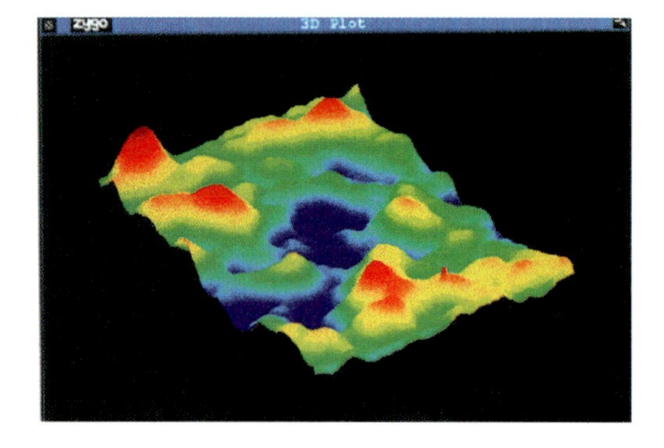

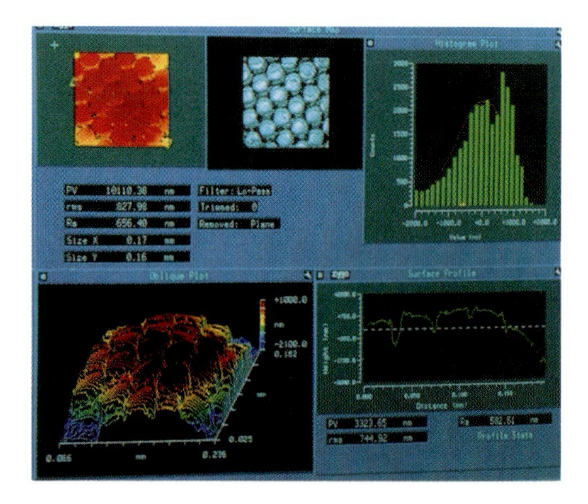

Figure 48. Automated non-contact three-dimensional surface profiling interferometer. (Courtesy of Zygo Corporation.) (a) Optical configuration. (b) View of interferometer. (c) Typical screen displays.

(a) Optical path of Schmaltz microscope

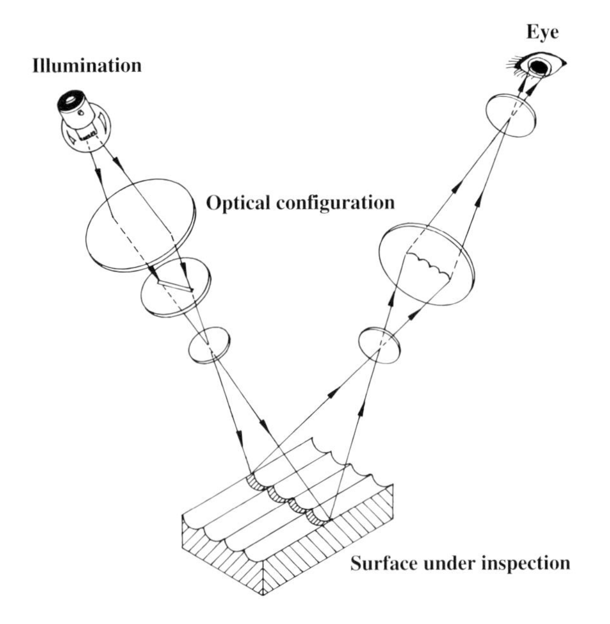

(b) Optical sectioning images, projected onto a screen:

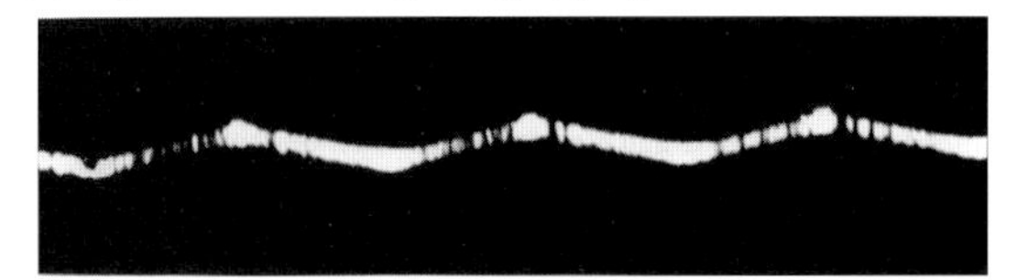

(i) feedmarks from machining, indication cusp heights

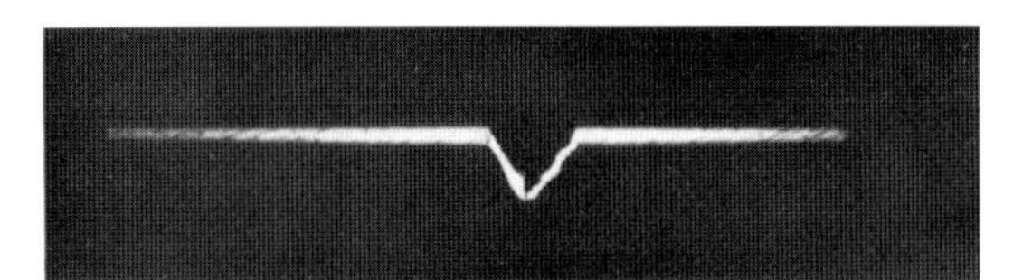

(ii) an engraved line on a flat surface

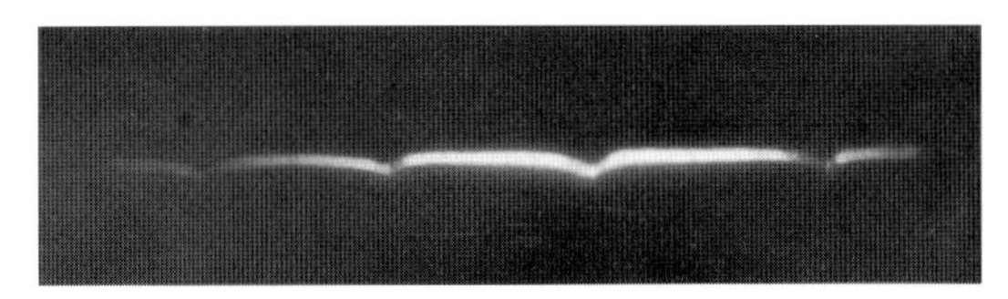

(iii) embossed plastic surface

Figure 49. The "Schmaltz technique" – optical sectioning principle and associated projected images. (Courtesy of Taylor Hobson.)

the image plane, and secondly, close to the diffraction or defocus plane. Yet another technique utilises the polychromatic speckle patterns, while others employ the correlation properties of two speckle patterns. From both of these methods, surface texture values ranging from 0.01 to 25 μm have been estimated.

Optical sectioning

The method of optical sectioning termed the *Schmaltz technique* after its inventor (sometimes termed the *light section microscope*) produces virtually identical results to that of physical sectioning, but in a simpler, non-destructive manner. However, the magnification and data obtainable from the projected image are quite limited, when compared with stylus-based surface texture measuring instruments. The ratio of vertical to horizontal magnification is around unity, meaning that the field of view is small when a large magnification (×400) is used, hence for fine surfaces it is rather limited.

The optical sectioning principle is illustrated in Figure 49(a). The surface to be examined is illuminated by a thin band of light delineating a profile section, which is then viewed at an angle with a microscope. Typically, illumination and viewing angles are 45° to the work surface producing the clearest profiles (see Figure 49b). For example, the projected surface indicated in Figure 49(bi) illustrates an apparent machined profile cusp height of

$$\text{Cusp height} = h \times \sqrt{2}$$

where h is the actual profile height (Rt).

Measurements are normally undertaken using either an eyepiece graticle or an eyepiece incorporating a micrometer for topographical feature assessment.

An alternative technique for optical sectioning can be achieved by utilising an optical projector, as this has the added advantage of an enlarged image being projected onto a screen. The surface's profile height can then be quickly measured with a specially prepared and calibrated template, which compensates for any distortion introduced by the viewing angle.

The optical sectioning technique is suitable for surface topographies having a roughness range of Rt between >1 and 200 μm. Moreover, it is eminently suitable for the inspection of soft and slightly pliable surfaces (Figure 49biii, which could be deformed by a stylus), or for estimation of depths of surface scratches/engraved lines (Figure 49bii).

The reflectance of a surface is a "sensitive function" of its relative roughness, and consequently the wavelength of light is considerably greater than the root mean square value of the surface texture. Hence, the reflectance will depend only on the surface roughness rather than the peak slope of any irregularities. Measuring reflectance at two distinct wavelengths enables one to determine the surface roughness and slope of asperities (peaks).

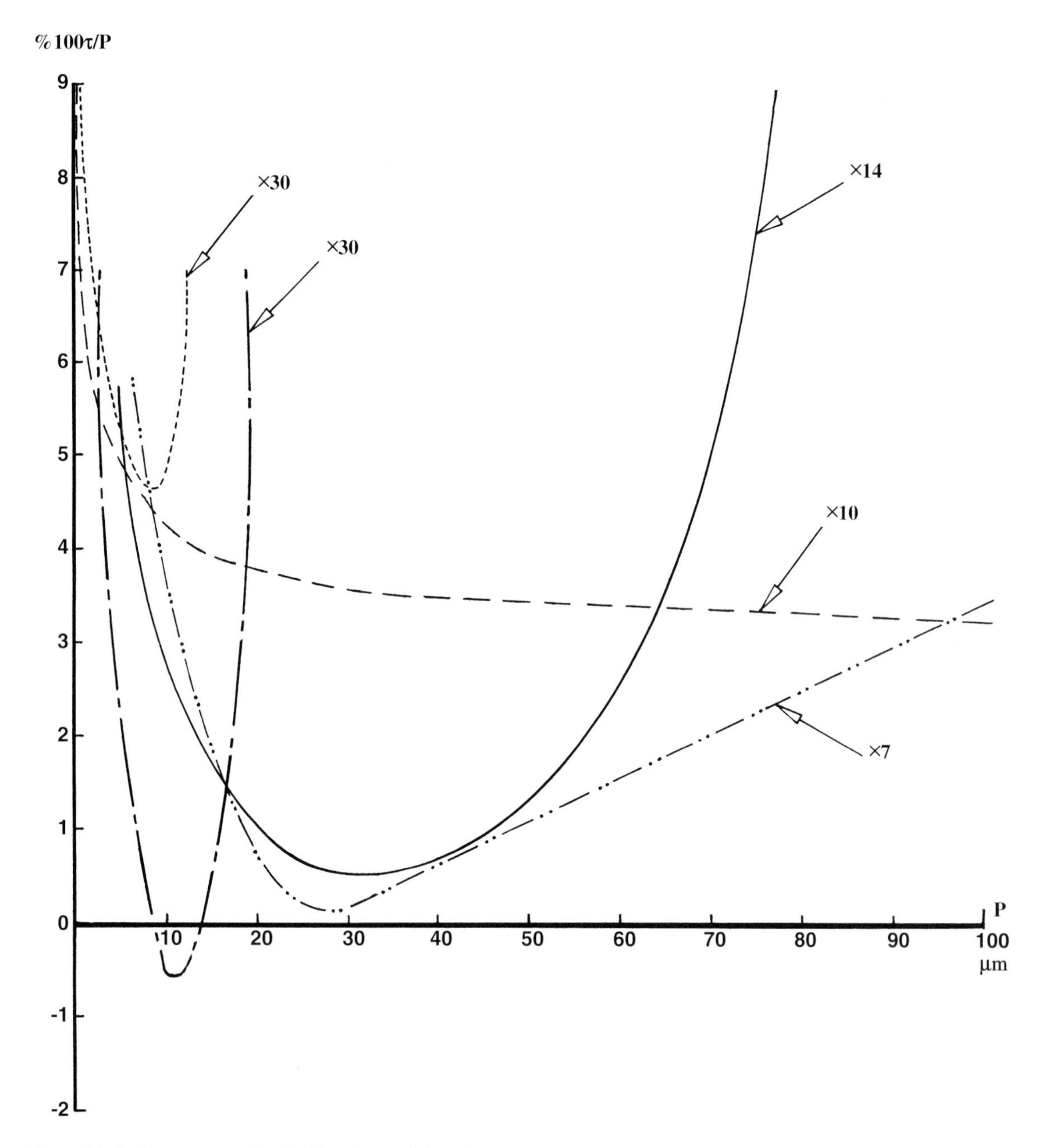

Figure 50. Performance curves for the Schmaltz – optical sectioning – microscope, illustrating the relative error of indication for various microscopes. [After Thomas, 1974.]

Performance tests undertaken for the Schmaltz (optical sectioning) microscope (illustrated in Figure 50) show that for an objective of ×60 magnification the relative error of indication does not exceed 7.7% over the whole application range (*Rt* values between 0.8 and 3.0 μm). Accepting that practical accuracy limits are in the region of 9%, then there is little point in using a ×60 objective, since the necessary accuracy can be achieved with the ×30 objective – with the added advantages of a larger field of view and depth of focus. In the case of the ×30 objective, the relative error of indication does not exceed 8.6%. Now, if one enlarges its measuring range (1.5–10 μm), up to values of *Rt* of 1.5–14.5 μm, a practical accuracy of ±9% may be obtained. The objective having a magnification of ×14 has a relative error of indication within the recommended manufacturer's guidelines, namely, from 3 to 12 μm,

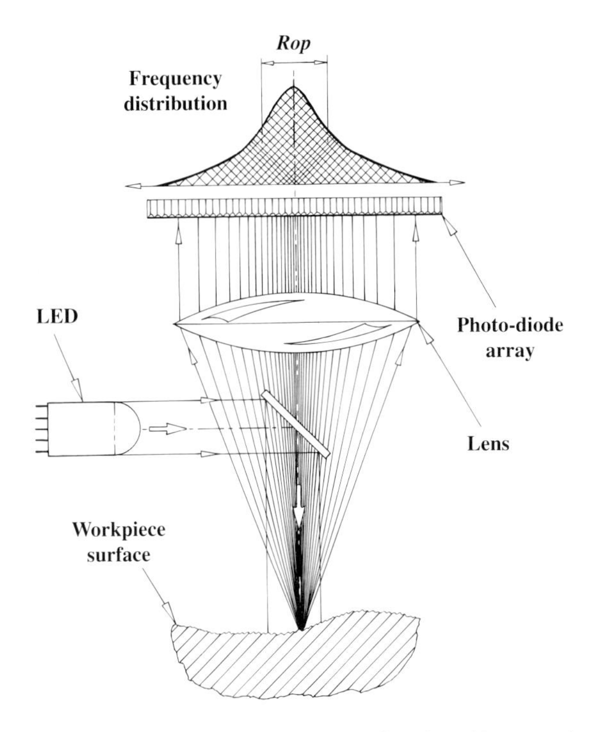

Figure 51. Diagrammatic representation of sensing with scattered light.

varying from 6% to 12%. Moreover, if the measuring range is altered for this ×14 objective to that of *Rt* ranging from 5–65 μm, the value of its relative error of indication is maintained within the limits of ±5%. In a similar manner, for the ×7 magnification objective, by enlarging the measuring range from 8 to 100 μm the relative error of indication will still be within a ±4% limit.

Scattered light: light-emitting diode (LED) source

Utilising "scattered light" is a fast in-process surface roughness technique, which is schematically illustrated in Figure 51. The parallel light is emitted by an infra-red LED and is projected onto the workpiece surface. For this instrument, the size of the source's light spot is approximately 0.8 mm at the focal point on the surface. Any reflected light tends to be scattered or diffuse in nature, resulting from the roughness of the workpiece surface. The lens will collimate the light and project it onto an array of photo-diodes, from where the variance of the distribution of scatter can be established. Any variance is then normalised, which eliminates intensity variations resulting from the reflected light. Normalising the variance compensates for the twin effects of different levels of reflection and that of workpiece materials; moreover, this technique eliminates the influence from variations in intensities of LED. Other techniques have been developed, for example *angle-resolved scatter* (ARS), but the discussion here will be confined to the current scattering technique.

Variance in the scatter distribution is termed *Rop* (the optical roughness parameter). This variability is proportional to the angle of the surface roughness. *Rop* can be utilised to compare the workpiece surface roughness against a calibrated standard, although this parameter can only be validated when measurement is made under similar conditions. If component surfaces are quite smooth, then these are ideal conditions for obtaining the highest sensitivity from the instrument. Typically, if a component's general surface roughness (*Ra*) lies between 0.05 and 0.5 μm, it is appropriate for evaluation by this technique. Any surfaces that might be manufactured by electrical discharge machining (EDM), electro-polishing, or grinding (pseudo-random production operations) are ideal for assessment by the scattered light technique, rather than other systematic manufacturing processes such as turning and milling operations. Only a brief time period is necessary for an object's surface measurement by the scattered light instrument – typically 300 ms – enabling measurement "on the fly" (while the workpiece is in motion).

A practical application of this technique might be to establish the surface roughness of an in situ measurement for a cold-rolled aluminium strip as it progresses beneath the focused beam at maximum strip passing speeds of around 70 km/h. Due to its ability to tolerate relatively fast component motion and quickly estimate surface finish, the scattered light technique can achieve 100% inspection on critical surfaces, enabling almost real-time process control for continuous processing operations.

Optical diffraction

The principle behind this instrument's operation is the technique of using a surface under test as a phase grating, enabling the instrument (Figure 52) to capture the far-field Fraunhoffer diffraction pattern on a linear CCD detector array, thus allowing statistical assessment of the surface condition.

A schematic diagram of the instrument is depicted in Figure 52, with a photograph of the complete assembly illustrated in Figure 53. The system is based upon two laser diodes having operational wavelengths of 670 nm and 780 nm, respectively. These two wavelengths enable diffraction order numbers to be identified, in conjunction with a low birefringence single-mode fibre terminated by an achromatic collimating lens. The purpose of this fibre is twofold:

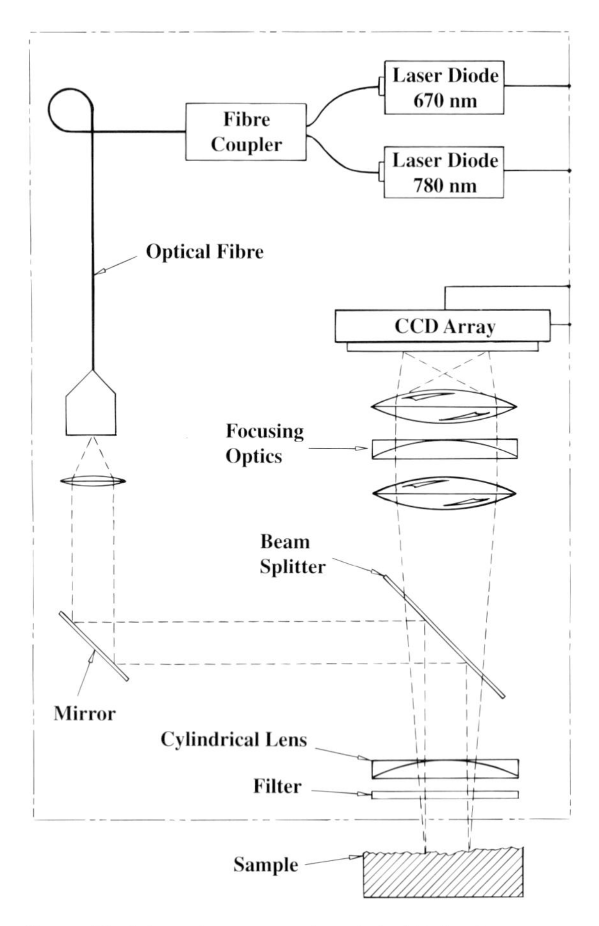

Figure 52. Schematic representation of the Talyfine instrument. (Courtesy of Taylor Hobson.)

- to act as a spatial filter, by selectively transmitting the Gaussian TEM 00 mode;
- providing a convenient method for the production of spatially coincident wavefronts.

The Gaussian beam for both laser wavefronts is directed at normal incidence through a cylindrical lens onto the workpiece surface. It forms a line image at the beam waist, so that a restriction of diffraction occurs only on one plane, enabling a range of metrological parameters to be found such as *Ra*, etc, and the surface power spectrum. Surfaces can be assessed whether they are random or unidirectional in nature.

Diffraction patterns are produced from the phase modulation of the surface and will be focused onto a 2048 CCD detector. Lastly, data processing is undertaken on a PC, enabling the visualisation of diffraction patterns, power spectra and surface parameters.

With this instrument the surfaces under test must be quite smooth, in order to obtain a satisfactory visual performance from the equipment. Typical components that might be inspected with this instrument include diamond-machined brass (Naval non-leaded type), aluminium (grade: 6061 T6) and copper (OFHC) flats, germanium flats or similar precision surfaces.

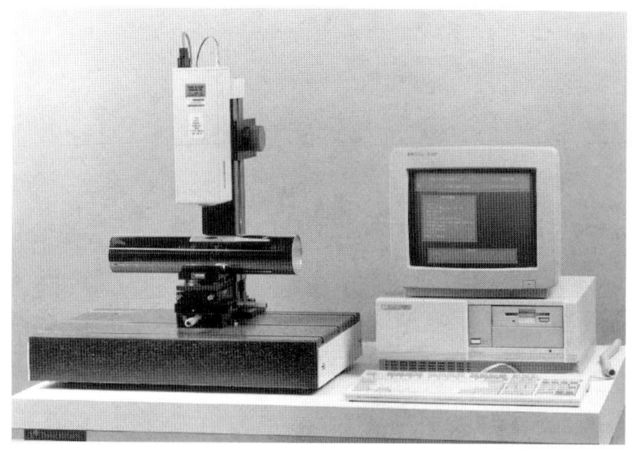

Figure 53. Assessment of surface texture on a diamond-turned component, within the range 5–10 nm *Rq*, utilising a non-contact optical instrument. (Courtesy of Taylor Hobson.)

One of the principal reasons for utilising laser-based diffraction techniques is their ability to characterise low *Rq* surfaces, in particular the advantages obtained from directly filtering the diffraction pattern. Moreover, this instrument has the capability of measuring a substantially periodic surface to an *Rq* of approximately 140 nm. In addition, measurements of *Rq* on a rotating surface can be achieved with no distinction between whether the surface is in motion or static (unlike a stylus-based instrument).

The surface diffraction physics is rather complex and is therefore beyond the scope of the current discussion on the general operational and performance characteristics of this laser-based instrument.

Microscope applications

A cursory examination of a surface with a microscope equipped with a graticle will reveal any prominent spacing irregularities. The height of any dominant peak cannot be estimated, although if suitable illumination is utilised (see Figure 54 for various inspection modes that can be employed) some detail can be visually assessed.

In metallographical inspection of samples, if destructive means can be practised to visually assess surface features at approximately right angles to any interesting irregularities, this form of visual inspection reveals considerable information. Rather than section the material perpendicularly to the surface, a much better and informative strategy is to take a *taper section* of the surface. With taper sectioning,

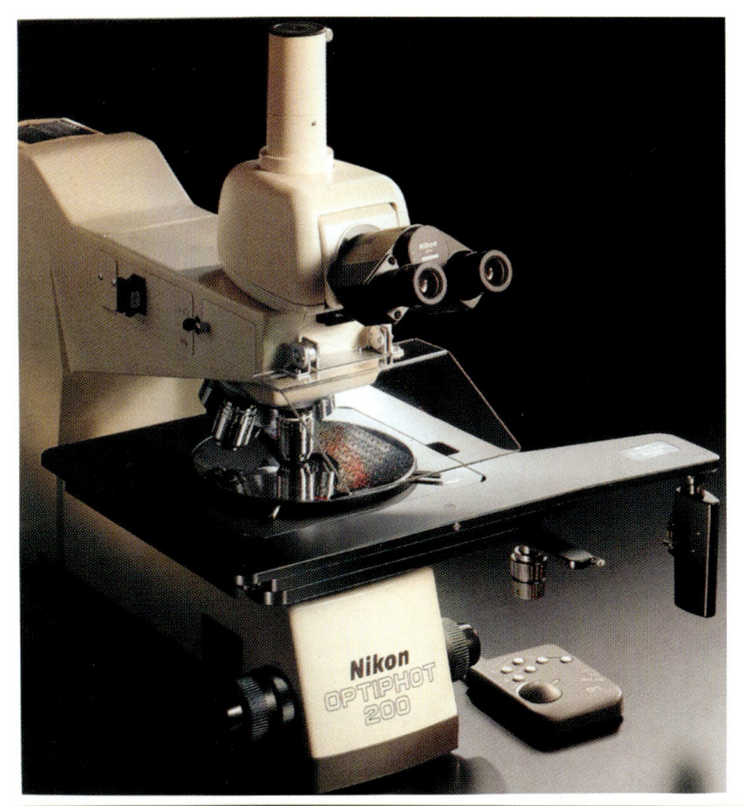

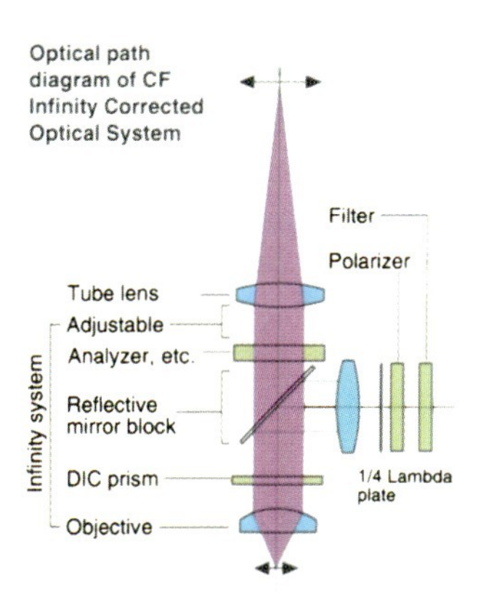

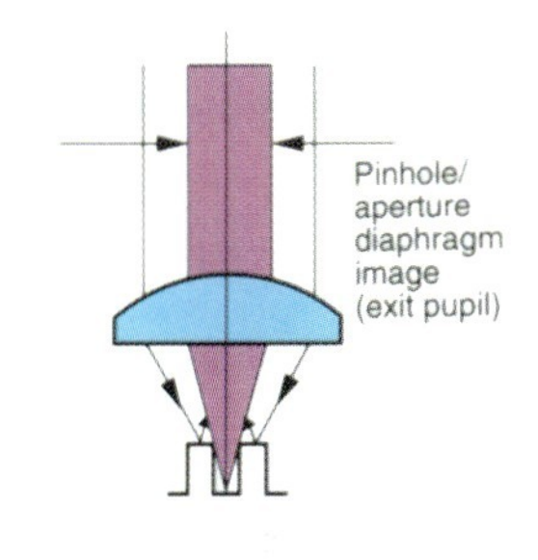

INSPECTION MODES

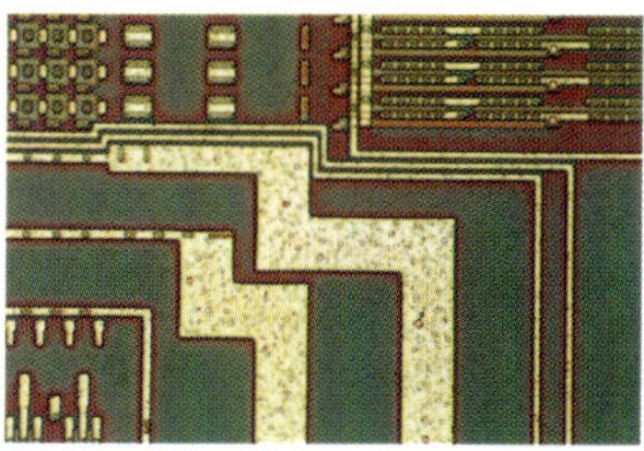

Brightfield Observation
Brightfield microscopy uses differences in reflection to enhance the natural color and form of specimens.

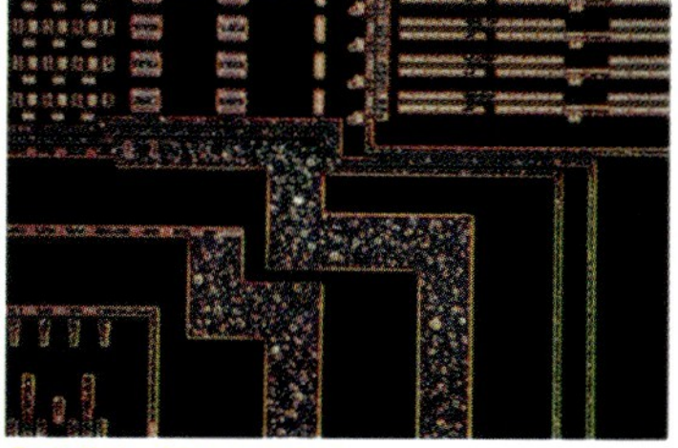

Darkfield Observation
Darkfield microscopy detects tiny flaws, subtle irregularities, impurities and defects on the surface of wafers or masks. It permits highly accurate inspections.

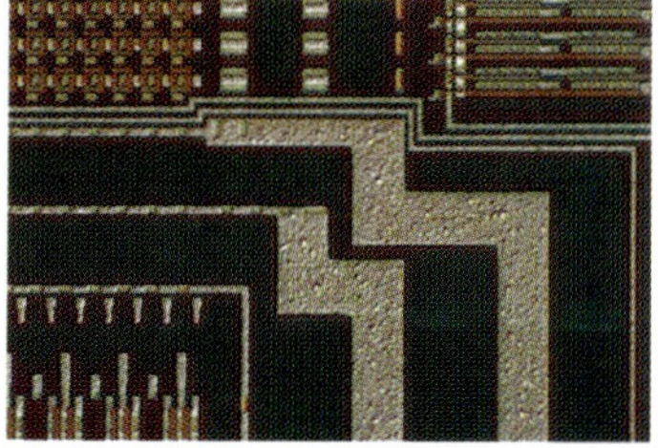

Nomarski DIC Observation
Differential interference contrast microscopy reveals the tiny flaws in wafers and the subtle irregularities of photo resist patterns as an interference color, thereby allowing three-dimensional observation. DIC represents subtle inclination (differential coefficient) with sharp differential contrast, enabling detection of very small differences in height.

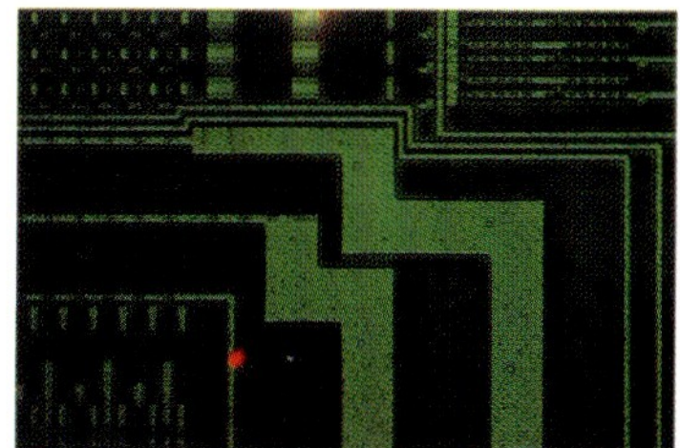

Fluorescence Observation
The epi-fluorescence filter blocks facilitate detection and analysis of residual photo resists in the semiconductor manufacturing process. The filter detects particles as fluorescent images and determines their wavelength. It is ideal for locating photo resist irregularities and the cause of particle generation.

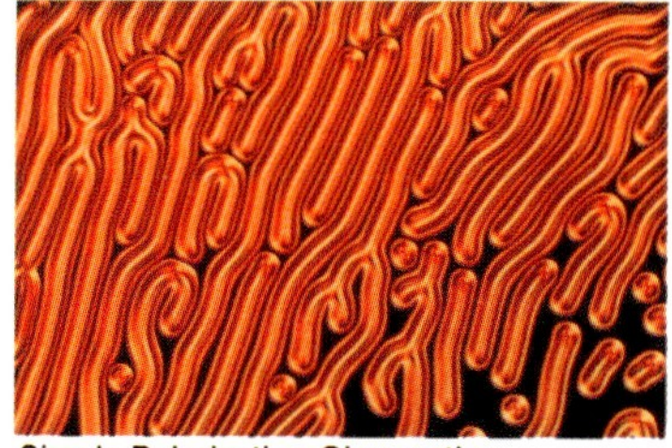

Simple Polarization Observation
This method is applied when studying or analyzing specific optical characteristics—such as optical isotropy and anisotropy—of a given material. An interference color can be applied to allow polarized contrast observations. This method is ideally suited for observing crystal conditions and detecting the stress on a wafer.

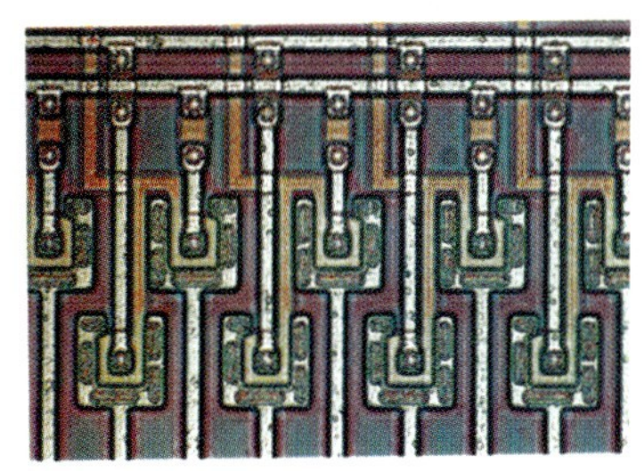

Illumination using pinhole aperture

Figure 54. Microscope applications for surface assessment. (Courtesy of Nikon UK Ltd.)

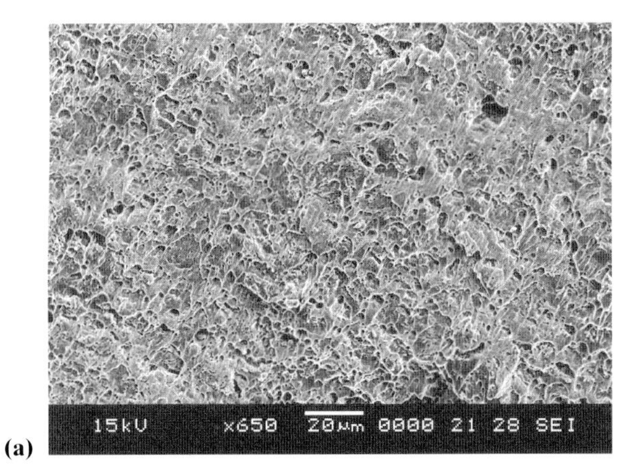

(a)

(b)

Figure 55. Images from a scanning electron microscope (SEM) for optical assessment of fracture and machined surfaces. (Courtesy of Jeol UK Ltd.)

(a) Mild steel ductile fracture – photomicrograph from an SEM. Magnification ×650; acc. V 15 kV; signal SEI; WD 21 mm; SS 28; pressure Pa. (b) Stereo image (left-hand SEM stage tilt – 1.5°) of an austenitic stainless steel (grade 316) by high-speed machining (milling with a 0.5 mm feedrate). Magnification ×50; acc. V 20 kV; signal SEI; WD 20 mm; SS 30; pressure Pa.

the surface is prepared at a shallow angle (typically 11°) perpendicular to the surface (see Figure 161). This slight taper does not unduly modify the shape of the surface irregularities, but increases the apparent magnification of the surface under test, revealing a considerably greater area for visual or micro-hardness assessment (more will be said on this latter topic in Chapter 5). Surface preparation is a difficult task, particularly during the polishing phase prior to suitable etching (normally necessary for metallographical inspection of metals) as surface profile distortion may occur if sufficient care is not taken. Sectioning reveals modifications in sub-surface microstructural details that might not otherwise be apparent and this will be discussed in Chapter 5 on surface integrity; this may offer engineering solutions as to reasons why a surface behaved in a particular manner in service. Engineering surfaces may fail (in service) for a variety of unanticipated reasons; these might include high surface residual hardness, the introduction of unstable metallurgical changes sub-surface grains exposed at the surface occuring under conditions of load or tribological action. Together with a combination of these factors (see Figure 55a), their failure mode might need to be investigated.

A stereoscan electron photomicrograph offers a more informative visual three-dimensional image of the surface area than can be obtained from a conventional optical technique (see Figure 55b), particularly when additional aids of *soft imaging* with topographical height profiles are utilised, as discussed later in Chapter 3 on surface microscopy.

Optical microscopes have wide-ranging capabilities and offer superb yet subtle techniques to enhance surface image quality, including the following inspection modes (see Figure 54):

- *Brightfield* – the most frequently used method of observation, employing differences in reflection to obtain natural colour and shape of surface;
- *Darkfield* – this method becomes of some significance when observing and photomicrographing surface irregularities, minute flaws, differences in height levels and samples with low reflection levels (paper, plastics, composites and fibres);
- *Normarski* – this technique can capture the surface's subtle irregularities and flaws as interference colours, indicating them as three-dimensional forms. Owing to the Normarski's ability to represent very small tilt with sharp differential contrast, this observational technique can detect minute height differences and subtle irregularities in metals, crystalline structures, integrated circuits (IC) and large-scale integrated (LSI) circuits;
- *Fluorescence* – an epi-fluorescence attachment allows for detection of positive and negative photo resists. Selecting from different filter combinations offers an optimum configuration for any application, being ideal when trying to locate resist irregularities and the cause of particle generation;
- *Simple polarisation* – specific optical characteristics can be investigated with this method of polarised light microscopy. The technique is widely utilised for geological research, together with evaluation and examination of minerals, plastics, crystal conditions and stress detection on IC wafers;
- *Pinhole aperture* – allows high-intensity/-resolution images at large depth of focus, which is ideal for surfaces having multi-layer film three-

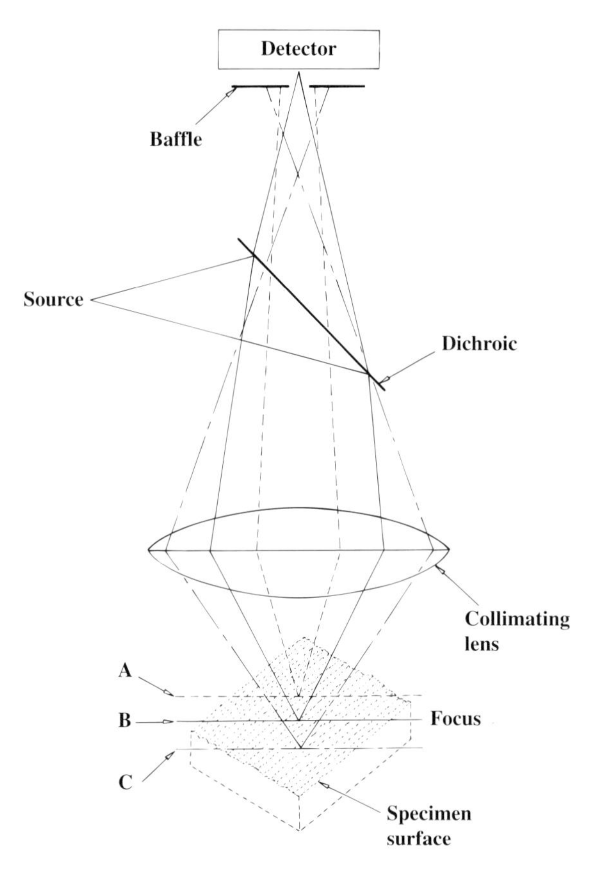

Figure 56. Schematic diagram of an optical profiler. (Courtesy of Dr R. Leach/NPL.)

dimensional structures. Typically, the aperture diaphragm can be "stopped-down" to around 30% of the exit pupil diameter using a ×150 objective, providing large depths of field for surface inspection.

Many of the latest microscope systems can be used for *image capture and processing*, enabling the instrument's software to "grab" live images from the microscope via video cameras, scanner computer files and archive packages. Such image analysis systems enable image enhancement and manipulation (flip and rotate the surface image), with the ability to change contrast/brightness, simple thresholds, sharpening and smoothing, morphology, component separation, retouching and masking, together with scaling and filtering. Many software-enhancing techniques allow operators to take up to 36 geometrical and analytical measurements, together with reference images for comparison, with Windows™-based menus for ease of image and data manipulation. Such systems can be networked for multi-user applications and sophisticated archiving facilities, allowing access by different users.

Confocal microscope

Confocal microscopes generally utilise the principle of depth discrimination. Essentially, the imaging and receiving optics are identical, offering excellent properties. In a similar manner to the so-called "flying spot technique", the optics project a point onto the surface, with the reflected light being picked up by a point detector (see Figure 56). The resulting signal is utilised to modulate the spot brightness on the CRT screen, which is scanned in synchronisation with the object under test. Essentially, the lens focuses a diffraction-limited spot onto the object, with the lens collecting light only from the same vicinity of the surface that was previously illuminated. As a consequence of employing the single-point detector, both the image forming/receiving systems contribute to the signal at the detector. Thus, by rotating the variable aperture over the specimen, which itself in some systems rotates, it will build up an optical image (via software data capture) of the sample's surface. By this means, the "flying spot" covers the complete specimen's surface to the same optical magnification and resolution of a conventional microscope, but with the advantage of a significant improvement of the field of view. These images can be optical slices or sections through the surface, which can be processed providing non-contact three-dimensional information about the surface.

1.13 Nanotopographic instruments

In the mid-1960s, out of the growing requirement to measure very smooth surfaces and thin films, an instrument was developed having a resolution of approximately 0.5 nm combined with a maximum magnification of 1,000,000, a measuring range of 13 μm and an arcuate traverse of ±1 mm. Originally such instruments were developed in order to measure minute step heights, typically around 30 nm as illustrated in Figure 57. From this instrument and others having enhanced design and performance refinements came the current genre of nanotechnology equipment, typified by the "Nanostep" shown in Figures 58 and 59 (pioneered by the NPL). Instruments of this level of accuracy and precision must be virtually free from thermal expansion effects; therefore the majority of the components are manufactured from thermally stable materials, typically glass ceramic "Zerodur™".

The kinematic (translational component members) and operational features of the instrument are such that the workpiece is supported on a levelling

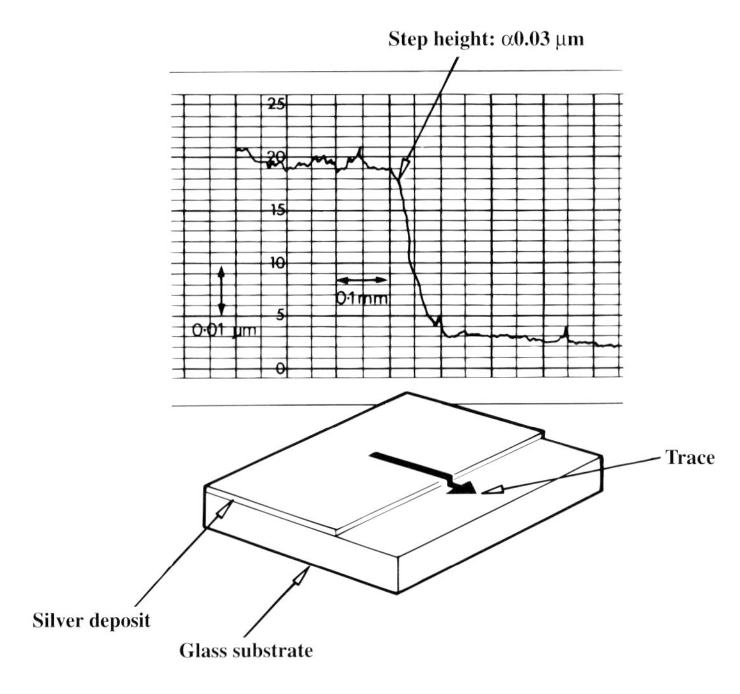

Figure 57. A typical "Talystep" graph of silver deposit on a glass substrate. Thickness of the deposit is approximately 0.03 μm. (Courtesy of Taylor Hobson.)

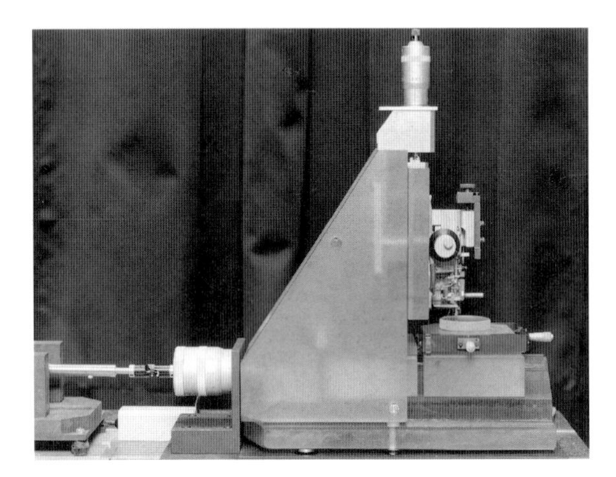

Figure 58. Constructional detail of the "Nanostep" and its associated precision slideway in low coefficient of expansion (i.e., glass ceramic: "Zerodur"). (Courtesy of Taylor Hobson.)

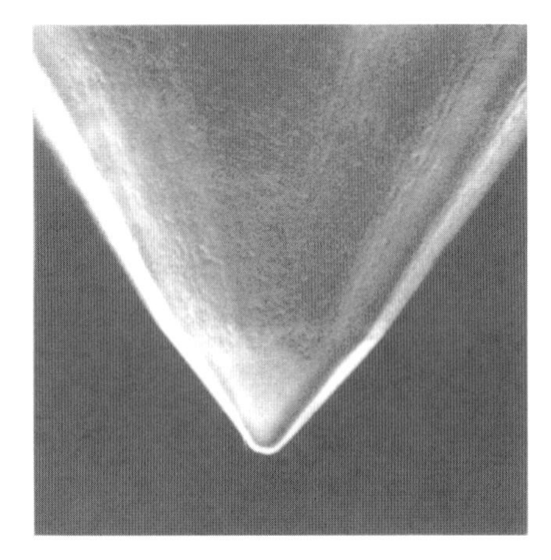

Figure 59. Detail of the interchangeable "Nanostep" stylus. (Courtesy of Taylor Hobson.)

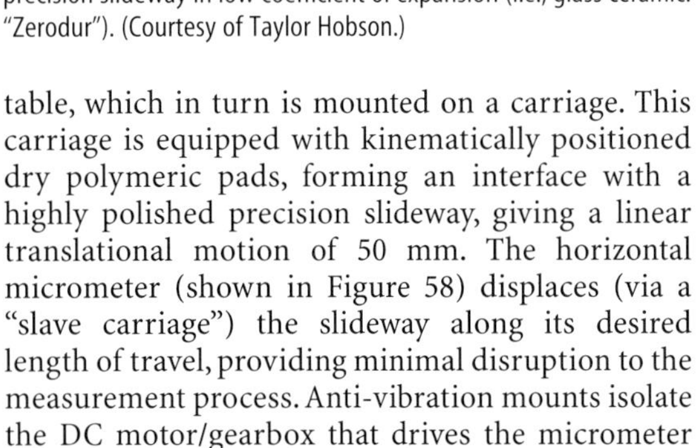

table, which in turn is mounted on a carriage. This carriage is equipped with kinematically positioned dry polymeric pads, forming an interface with a highly polished precision slideway, giving a linear translational motion of 50 mm. The horizontal micrometer (shown in Figure 58) displaces (via a "slave carriage") the slideway along its desired length of travel, providing minimal disruption to the measurement process. Anti-vibration mounts isolate the DC motor/gearbox that drives the micrometer from the main structural elements of the instrument, producing ultra-low instrument noise down to 0.03 nm under optimum conditions. The motor can drive the micrometer at measurement speeds varying from 0.005 to 0.5 mm/s. The stylus/transducer assembly illustrated in Figure 60 can be vertically positioned on the workpiece and in contact with it via the vertical micrometer (depicted in Figure 58), giving a vertical step height range up to 20 μm. Finally, measurement control and signal output are provided by a suitably configured PC.

The software offers various levels of surface

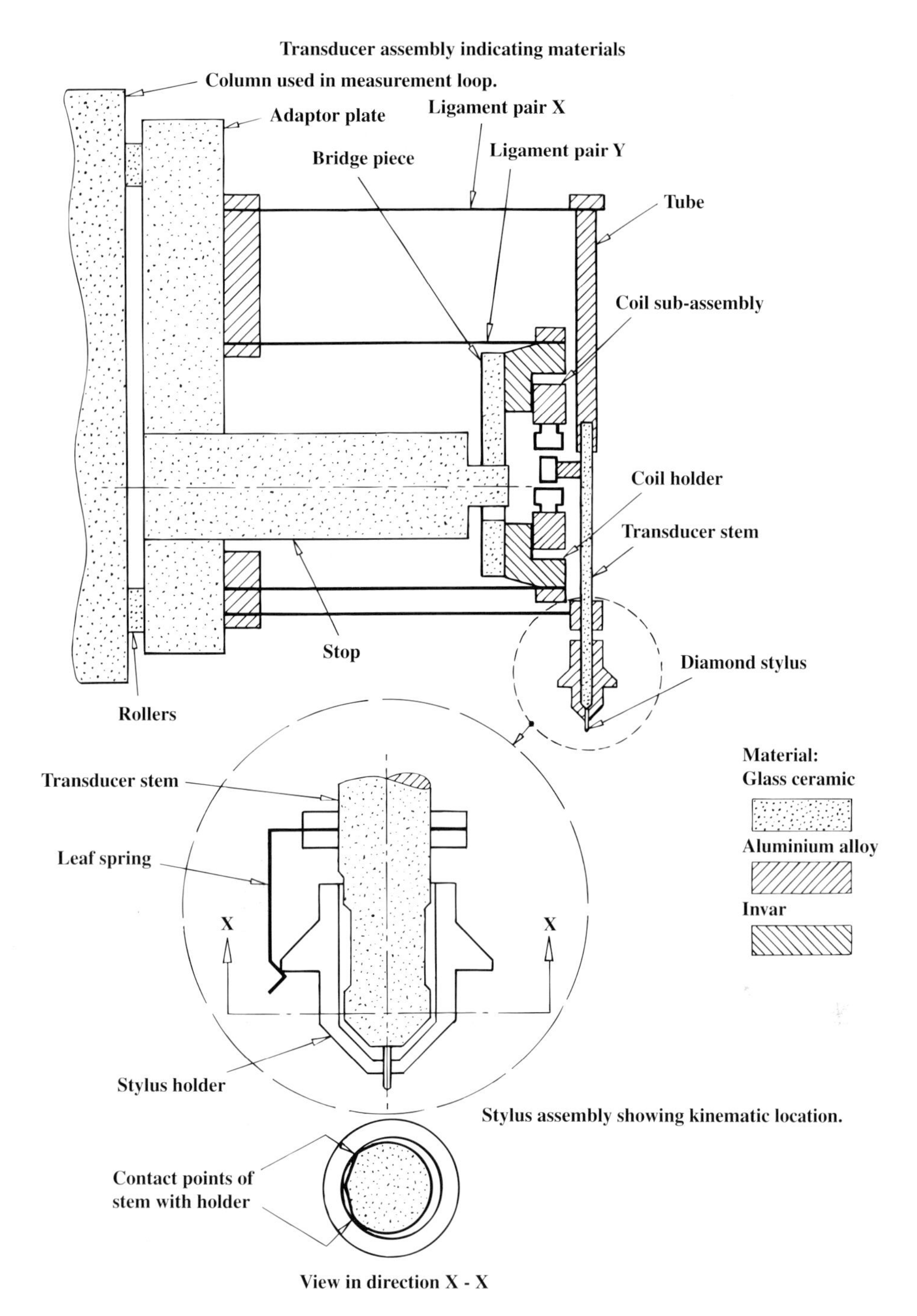

Figure 60. Internal construction of the "Nanostep" instrument. (Courtesy of Taylor Hobson.)

texture analysis, assessed relative to the best-fit reference line, covering a diverse range of international parameters. Additionally, the form software provides for

- *form removal* – best-fit form is calculated (least-squares straight line, minimum-zone straight line or least-squares arc) and then removed, enabling the texture of the surface to be analysed;
- *angle* – providing a facility to remove any compound surface tilt prior to parameter analysis;
- *dimension* – provides for the linear relationship of surface features, which can be assessed and compared.

Software has also been developed for this instrument enabling it to assess an *enhanced dual profile*. This provision enables one to measure the profile of a "master component", which can then be saved as a template; any subsequent components can then be both measured and simultaneously displayed with this template for immediate comparison. A statistical process control software package can be employed, along with other specialised software packages.

A number of interchangeable styli (see Figure 59) can be fitted to the instrument to suit a range of different measuring applications, ranging approximately from 1–10 μm conispherical shape to 0.1 × 0.5 μm pyramidical geometry, the latter being used for high-resolution work. The stylus force can be varied between 10 and 700 μN, allowing it to non-destructively measure delicate or easily scratched components. Once the instrument has been set to the required depth of measurement, an automatic facility provides for the raising and lowering of the stylus when repositioning or changing samples. For the measurement of ultra-fine surfaces, the instrument offers a nominal gauge resolution down to 31 pm (with the 2 μm gauge range). Typical performance of the 10 μm conical stylus with an applied force of 40 μN can produce a surface roughness Rq value of 4.9 nm. Such nano-instrumentation has a component capacity of diameter 100 mm, with a depth of 22 mm, covering a large range of precision components.

Instrument applications are both wide and varied and it might typically be employed for

- ultra-precision bearing and surface defect measurement – surface finish/integrity;
- silicon wafer measurement;
- integrated circuit production – process monitoring;
- magnetic tape measurement;
- diamond-turned components – small diameter;
- calibration standards;
- precision optics;
- mirrors for laser gyroscopes;
- laser and X-ray mirrors.

In this chapter little or no reference has been made to the role of expert systems or neural networks in surface texture analysis. This omission was intentional, as the chapter length would otherwise have simply been too long and less specific in nature. In Chapter 2 a review of three-dimensional surface texture measurements will be made, then later on contour/fractal effects will be discussed, which are now becoming an important technology in describing surfaces.

References

Journal and conference papers

Bell, T. Surface engineering: past, present and future. *Surface Engineering* 6(1), 1990, 31–40.

Bjuggren, M., Krummenacher, L. and Mattsson, L. Noncontact surface roughness measurement of engineering surfaces by total integrated infrared scattering. *Precision Engineering* 20(1), 1997, 33–45.

Bousefield, B. and Bousefield, T. Progress towards a metallography standard. *Metals and Materials* March 1990, 146–148.

Brown A.J.C. and Gilbert, F. Industrial applications for optical profilometry. *Sensor Review* January 1990, 35–37.

Caber, P.J. An interferometric profiler for rough surfaces. *Applied Optics* 32, 1993, 3438–3441.

Downs M.J., Mason, N.M. and Nelson, J.C.C. Measurement of the profiles of super-smooth surfaces using optical interferometry. *Proceedings of SPIE-1009*, 1989, 14–17.

Drews, W.E. Surface measurement: an advanced technology. *Quality Progress* April 1987, 43–46.

Dyson, J. *Interferometry as a Measuring Tool*. Machinery Pub. Co., 1970.

Garratt, J. and Bottomly, S.C. Technology transfer in the development of a nanotopographic instrument. *Nanotechnology* 1, 1990, 38–43.

Garratt J. and Mills, M. Measurement of the roughness of super-smooth surfaces using a stylus instrument. *Nanotechnology* 7, 1996, 13–20.

Gee, M.G. and McCormick, N.J. The application of confocal scanning microscopy to the examination of ceramic wear surfaces. *Journal of Physics D: Applied Physics* 25, 1990, 230–235.

Harvie, A. and Beattie, J.S. Surface texture measurement. *Production Engineer*, 1978, 25–29.

Hongai, Z., Zhixiang, C. and Riyao, C. The control of roughness with expert system controller. *Proceedings of the First International Mach. Mon. and Diagnostics*, Las Vegas, 1989, 813–817.

Kuwamura, S. and Yamaguchi, I. Wavelength scanning profilometry for real-time surface shape measurement. *Applied Optics* 36, 1997, 4473–4482.

Leach, R.K. Measurement of a correction for the phase change on reflection due to surface roughness. *Proceedings of SPIE 3477*, 1998, 138–151.

Leach, R.K. Traceable measurement of surface texture at the national physical laboratory using Nanosurf IV. *Measurement Science and Technology* 11, 2000, 1162–1172.

Leach, R.K. Traceable measurement of surface texture in the optics industry. *Large Lenses and Prisms Conference*, University College London, 27–30 March 2001.

Leach, R.K. and Hart, A.. Investigation into the shape of diamond styli used for surface texture measurement. *NPL Report CBTLM10*, April 2001, 1–12.

Mansfield, D. Surface characterisation via optical diffraction. *SPIE (International Society for Optical Engineering) Vol. 1573: Commercial Applications of Precision Manufacture at the Sub-Micron Level*, 1991, 163–169.

Prostrednik, D. and Osanna, P.H. The Abbott curve: well known in metrology but not on technical drawings. *International Journal of Machine Tools Manufacture* 38(5–6), 1998, 741–745.

Schaffer, G.H. The many faces of surface texture. *American Machinist*, June 1988, 61– 68.

Schneider, U., Steckroth, A. and Hubner, G. An approach to the evaluation of surface profiles by separating them into functionally different parts. *Surface Topography* 1, 1988, 71–83.

Stout, K. How smooth is smooth? Surface measurements and their relevance in manufacturing. *Production Engineer* May 1980, 17–22.

Thomas, T.R. Trends in surface roughness. *International Journal of Machine Tools Manufacture* 38(5–6), 1998, 405–411.

Trumpold, H. and Heldt, E. Why filtering surface profiles. *International Journal of Machine Tools Manufacture* 38 (5–6), 1998, 639–646.

Ulbricht, R. Die Bestimmung der mittleren räumlichen Lichtintensität durch nur eine. *Messung Electrotechnische Zeitschrift* 29, 1900, 595–601.

Vorburger, T.V. and Teague, E.C. Optical techniques for online measurement of surface topography. *Precision Engineering* 3, 1981, 63–83.

Westburg, J. Opportunities and problems when standardising and implementing surface structure parameters in industry. *International Journal of Machine Tools and Manufacture* 38(5–6), 1998, 413–416.

Whitehouse, D.J. Some ultimate limits on the measurement of surfaces using stylus techniques. *Measurement and Control* 8, 1975, 147–151.

Whitehouse, D.J. Beta functions for surface topology? *Annals of the CIRP* 27, 1978, 491–497.

Whitehouse, D.J. Surfaces: a link between manufacture and function. *Proceedings of IMechE* 192, 1978, 179–187.

Whitehouse, D.J. Conditioning of the manufacturing process using surface finish. *Proceedings of the Third Lamdamap Conference, Computational Mechanics* July 1997, 3–20.

Whitehouse, D.J. Some theoretical aspects of surface peak parameters. *Precision Engineering* 23, 1999, 94–102.

Whitehouse, D.J. Surface measurement fidelity. *Proceedings of the Fourth Lamdamap Conference*, University of Northumbria, WIT Press, July 1999, 267–276.

Whitehouse, D.J. Characterizing the machined surface condition by appropriate parameters. *Proceedings of the Third Industrial Tooling Conference*, Southampton Institute, Molyneux Press, September 1999, 8–31.

Wyant, J.C. Computerized interferometric measurement of surface microstructure. *Proceedings of SPIE-2576*, 1995, 122–130.

Zahwi, S. and Mekawi, A.M. Some effects of stylus force on scratching surfaces. *Eighth International Conference on Metrology and Properties of Engineering Surfaces*, University of Huddersfield, 26–28 April 2000.

Books, booklets and guides

Bennett, J.M. and Mattsson, L. *Introduction to Surface Roughness and Scattering*. Washington, DC: Optical Society of America, 1989.

Busch, T. and Wilkie Brothers Foundation. *Fundamentals of Dimensional Metrology*. Delmar, 1989.

Dagnall, M.A. *Exploring Surface Texture*. Taylor Hobson Precision, 1997.

Dyson, J. *Interferometry as a Measuring Tool*. Machinery Pub. Co., 1970.

Gayler, J.F.W. and Shotbolt, C.R. *Metrology for Engineers*. Cassell, 1990.

Haycocks, J.A. *Novel Probes for Surface Texture Metrology*. NPL Report MOM105, July 1991.

Hume, K.J. *Engineering Metrology*. Macdonald, 1970.

Hume, K.J. *A History of Engineering Metrology*. Mechanical Engineering Pub., 1980.

Leach, R.K. NPL Good Practice Guide No. 37, Measurement of Surface Texture using Stylus Instruments. NPL, 2001

Leach, R.K. and Hart, A. Investigation into the Shape of Diamond Styli used for Surface Texture Assessment. NPL Report CBTLM 10, April 2001.

Mainsah, E., Greenwood, J.A. and Chetwynd, D.G. *Metrology and Properties of Engineering Surfaces*. Kluwer Academic Pub., 2001.

Mummery, L. *Surface Texture Analysis: The Handbook*. Hommelwerke GmbH, 1990.

Nicolls, M.O. *The Measurement of Surface Finish*. DeBeers Technical Service Centre, DeBeers Industrial Diamond Division, Charters, UK, 1980.

Reason, R.E. *The Measurement of Surface Texture*. Cleaver-Hume Press, 1960.

Sander, M. *A Practical Guide to the Assessment of Surface Texture*. Feinpruf GmbH, 1991.

Stover, J.C. *Optical Scattering*. McGraw-Hill, 1990.

Thomas, G.G. *Engineering Metrology*. Butterworths, 1974.

Thomas, T.R. *Rough Surfaces*. Imperial College Press, 1999.

Whitehouse, D.J. and Reason, R.E. *The Equation of the Mean Line of Surface Texture Found by an Electric Wave Filter*. Rank Taylor Hobson Pub., 1965.

Whitehouse, D.J. *Handbook of Surface Metrology*. Institute of Physics, Bristol and Philadelphia, 1994.

Whitehouse, D.J. *Surfaces and their Measurement*. Hermes Penton Science, 1991.

Williams, D.C. *Optical Methods in Engineering Metrology*. Chapman & Hall, 1993.

Surface texture: three-dimensional

"Brevis esse laboro,
Obscurus fio."

Translation
"It is when I struggle to be brief
that I become obscure"
(*Ars Poetica*, 25, Horace, 65–8 BC)

Outline on three-dimensional surfaces

Invariably the ability of a two-dimensional trace to satisfactorily characterise a surface has been somewhat limited, because in reality surfaces are three-dimensional in nature rather than two. Three-dimensional techniques have been developed to answer this criticism, but at present there are only proposed standards of assessment - by surface research groups - that are now being discussed, with many of these potential parameters mirroring their two-dimensional counterparts. The areal nature of three-dimensional surfaces relates to their texture, directionality and lay, and this can create problems in surface interpretation. For example, if one were to inspect a highly polished copper component, then both our visual and tactile senses would establish that it was shiny and smooth, respectively. Conversely, a shot-peened steel surface will appear to have a matt or dull optical appearance and feel somewhat rough to the touch. These three-dimensional visual and textural attributes create a problem of surface interpretation, but an even more fundamentally challenging question is: "What parameters does one select to determine a representative surface in three dimensions?"

The anisotropic and isotropic nature of surfaces could, in the main, be disregarded for two-dimensional surface characterisation, but they become of fundamental importance for surfaces that are three-dimensional in nature. Not only is considerably greater computer power necessary to process the significantly greater number of data points, but this will tend to slow the whole data-processing time down, although this latter point is becoming of less importance with the recent improvements in a PC's computing power.

Despite the objections raised above to the adoption of a three-dimensional surface texture inspection strategy, the surface characterisation technique has some real merit, in that:

- the practice enables the user to gain an improved visualisation of manufactured surfaces;
- it enables engineers to classify and subsequently control the properties of three-dimensional surfaces, as these features affect its in-service performance;
- three-dimensional surfaces overcome the current problem of inadequately describing a surface by two-dimensions.

This chapter is predominantly concerned with some of the three-dimensional instrumentation and software that is currently available. Moreover, the factors relating to three-dimensional parameters and their characterisation are discussed, together with some of the more esoteric and contentious areas of the subject, namely, the research based on fractal and topological feature recognition. With these latter techniques, one is looking into means of quantifying three-dimensional surfaces by their geometry and form, of similarities to those found in the natural world. As a result, they can be considered as experimental approaches to the surface topologies and still need to be refined if they are to be utilised in an industrial context. The same reasoning could be said to be true for the discussion on neural networks decision-making abilities of surface characterisations. These somewhat controversial issues have been included to show current thinking in surface developments.

2.1 Introduction

The ability to undertake surface texture measurements has been refined and developed continuously for many years; the basic principles were mentioned in Chapter 1. Simplistically, the object is to move the stylus, or probe in a straight line and at a fixed speed, with measurement of the "Z altitude" in relation to the "X axis", this being the basis of two-dimensional surface texture assessment (see Figure 18). In the past there was not a choice as to whether two- or three-dimensional measurements were to be taken, as only the former existed. A criticism levelled at two-dimensional surface texture assessment was its inability to represent a surface's functional characteristics, thereby causing a proliferation and uncontrolled expansion in two-dimensional parameters. Many of these newly developed parameters proved to be inadequate in providing essential information relating to a surface's functional aspects. More will be said on this topic in Section 2.1.3.

Previously, interest in three-dimensional surface texture was always the objective when attempting a full understanding of topographical surface details, but instrumentation limitations combined with inadequate data processing restricted its development. Today, three-dimensional measurements can be carried out, consisting of measuring many parallel and regularly spaced profiles in order to reconstruct the surface topography from a rectangular area of the surface. In many circumstances three-dimensional surface measurement has become an indispensable tool to today's metrologist seeking a clearer representation of the surface's topographical details.

An anisotropic surface, typified by Figure 10, results in radically different properties dependent on the direction from which the two-dimensional measurements are obtained. Clearly, if readings are taken on topographical surface features such as machined cusps or ridges, then the roughness measurement obtained would be significantly different due to "lay effects" (as illustrated in Figure 4), depending on whether the trace is parallel or perpendicular to the ridges. It is only when three-dimensional measurements and analysis are undertaken that a true conceptual understanding of the properties of a surface are apparent.

2.1.1 Stylus speed and dynamics

Conventional stylus contact instruments would typically have operating traverse speeds of 0.5 mm/s. If such speeds were used for just a small surface area, then it could take from 20 minutes to several hours to completely map the surface, depending on the number of data points and track width distances required to obtain a representative 3-D surface. The problem cannot be simply remedied by increasing the stylus speed; in particular, at high traverse speeds a conventional stylus loses contact with the workpiece surface as it moves over sharp surface features. This aspect of stylus performance is often termed "trackability", being dependent on the following:

- inertia of the pivoting system;
- static stylus force on the workpiece surface;
- stylus traverse speed;
- amplitude and wavelength of surface features.

To portray this "trackability effect", Figure 61(a) illustrates a schematic representation of a simple stylus system. The equations of motion are

(i) $I(Z/l) = T_s - T_r$

where I = moment of inertia of the system; Z = vertical position of tip; l = length of the stylus from pivot to tip; T_s = net sum of static torques acting about stylus point (i.e., due to mass of stylus and any external applied force); T_r = torque due to reaction of stylus on surface (i.e., stylus loses contact with surface when $T_r = 0$).

By way of example, consider a stylus motion across a simple sinusoidally varying surface, such that the stylus tip's position (when in contact with the surface) can be expressed as

(ii) $Z(t) = A \sin(\omega t)$

where A = amplitude of sinusoidal surface; ω = angular frequency of oscillation of tip as it moves across surface.

The frequency of tip oscillation is related to the spatial wavelength λ of the sinusoidal surface and traverse speed ν by

(iii) $\omega = 2\pi\nu/\lambda$

By substituting equations (ii) and (iii) into equation (i), it can be shown that the stylus will lift off the surface when

(iv) $(2\pi\nu/\lambda)^2 A = T_s l/I$

Namely, for a sinusoidal feature of given amplitude A and spatial period λ, there will be a maximum speed with which the stylus can be traversed across any feature yet still remaining in contact with the workpiece surface. At higher speeds, the stylus will be prone to lift off, thereby no longer representing with any accuracy the surface profile.

So that the dynamic performance of the stylus can be optimised for high-speed operation, the right-hand side of equation (iv) should be maximised. This maximisation can be accomplished by applying a static torque to the stylus via an external transducer (in this case by an electromagnetic actuator) – see Figure 61(c) – where it is attached to part A. Increases in traverse speed are limited by the square root of the static pressure, requiring large increases in actual static stylus pressure applied to the surface, if higher speeds are to be employed. However, this increase in the stylus pressure is somewhat undesirable and generally should be avoided, if surface damage promoted by it is to be minimised as it traverses across the part.

In order to rectify the unwanted stylus's static pressure mentioned above and simplify the model, a small lightweight spring can be utilised to distribute the mass evenly between the pivot point and the tip, which provides static force on the surface. This factor can now be expressed by modifying equation (iv) as follows:

(v) $\omega^2 A = 3/m(T_{sp}/l) + 3/2g$

where T_{sp}/l = static force of stylus on surface due to the lightweight spring; m = mass of stylus; g = acceleration due to gravity; ω = frequency oscillation of tip, in motion on surface.

From equation (v) it can be gleaned that this particular stylus can have its dynamic response optimised by reducing the stylus's mass, while maintaining an identical force on the surface. An advantage accrues from this modification, in that the dynamic response of the system can be optimised.

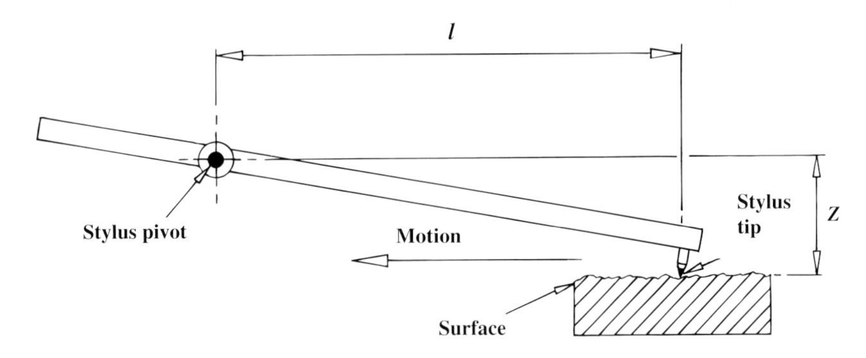

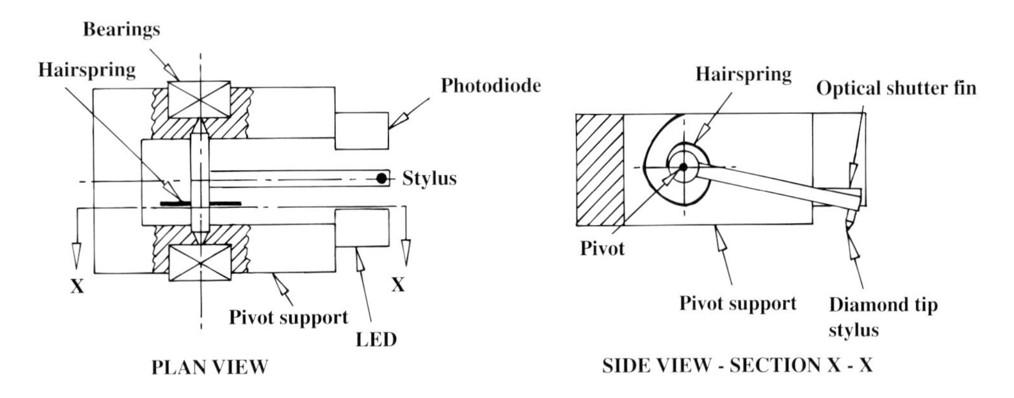

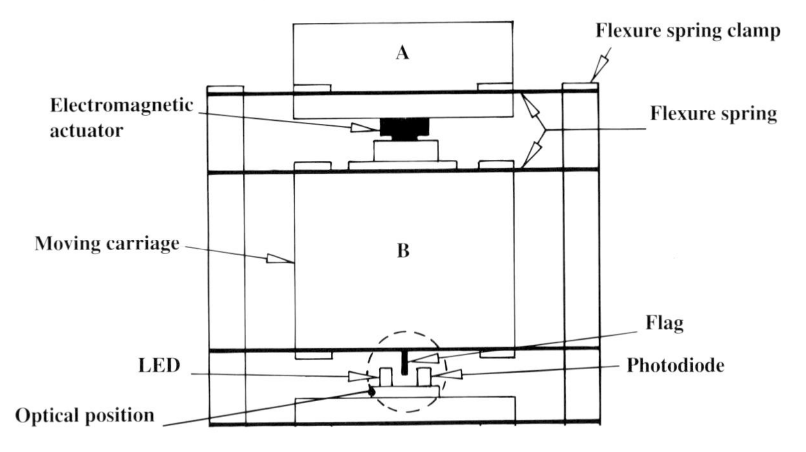

Figure 61. Conventional stylus, with a comparison to the high-speed variety general arrangement fitted to a "Talystep" instrument. (Courtesy of Taylor Hobson.)

As a consequence, this now allows a shorter, lightweight stylus to be fitted, which ensures that any mechanical resonances in the stylus arm are at a high frequency and as a result do not unduly compromise its performance.

The arrangement of the scanner, which is based on a precision compound flexure spring (Figure 61c) allows the stylus gauge to be rapidly and precisely moved across a surface in two orthogonal directions. Being of compact design, this has the benefit of ensuring that the measuring head is both robust and portable, being ideally suited for industrial applications. Furthermore, the instrument's higher sampling rates ensure that three-dimensional spatial resolution can be maintained. Hence the requirement was to ensure that the stylus gauge had sufficient bandwidth to cope with the increased data rate, together with minimal mechanical resonance in the stylus system at the desired bandwidth, which might otherwise lead to distortion of the measured profile. Recently, such instruments (typified by the "Talystep") have allowed stylus traverse speeds to be increased 10-fold by this careful redesign of the stylus gauge. This enables stylus speeds of approximately 5 mm/sec to be used without losses in either tracking or signal fidelity as it measures areas up to 500×500 μm, with a sampling rate of 1 μm, illustrating that the operation can be completed in less than a minute.

2.1.2 Envelope and mean systems

In the mid-1960s it was proposed that the "envelope system" (E-system) might offer advantages in assessment of both two- and three-dimensional surfaces. This E-system approach was at that time somewhat impracticable, as the hardware and software had not been developed to exploit this technique. Although the E-system procedure provides a rigorous and sound approach to such data computation, it could not be developed further at the time. With the introduction of fast and inexpensive PCs to complement the hardware developments in surface texture instrumentation, this has virtually eliminated the requirement for hardware execution of surface filtering and characterisation. This E-system is a specific example of the application of a morphological filter: this being a dilation of the surface/profile by the application of a spherical/circular element. In two-dimensional E-system filtering (Figure 62a) it can be assumed that instead of a stylus being used a hypothetical circle is rolled over the profile under test. As the circle rolls along the profile the locus of its circle's centre approximates to that of the profile, with its path dependent upon the circle radius relative to the separation of the average peaks. The formation of the artificial two-dimensional envelope curve is obtained by shifting down the loci of circle centres by a linear distance equal to its radius (illustrated in Figure 62a). With the latest development of this E-system technique, both the computational time and ensuing processing speed have been considerably enhanced by previously identifying prominent peaks.

The three-dimensional envelope system is simply an extension of that for two-dimensions (see Figure 62b), with the envelope surface resulting from shifting down the radial distance of the locus of the sphere's centre as it rolls over the three-dimensional data points. A "mean surface" can be obtained, as in the "M-system", by shifting the envelope surface down below the highest contact point (peak) in order that an equal volume of material to air occurs. This new "M-system" envelope approximates the previous envelope surface of the original rolling sphere radius. In this manner a form of three-dimensional filtering can be attained, with the filter being controlled by the radius of the rolling sphere. In practice, the envelope surface consists of just portions of these rolling spheres, with each having a radius equal to that of the rolling radius. Complex mathematical definitions for the envelope system would be extremely difficult to derive; therefore data is stored numerically and processed by suitable algorithms to speed up processing time.

2.1.3 Three-dimensional characterisation

The goal in any three-dimensional characterisation of surface topography is to integrate the surface features in a representative manner as accurately as possible. Many methods have been utilised to obtain a degree of surface visual characterisation, with notably the best technique at present being to describe the surface condition by a predefined series of parameters which can be quantifiably measured then related to practical operational performance. In recent years, many practitioners have developed new parameters and even some that are inappropriate – which has led to the much-quoted term "parameter rash" – to describe the unchecked evolution of particular topographical conditions, or specific functional applications. Moreover, up to the present, nearly all measurements related to surface topography are based on two-dimensional techniques, although a certain number of three-dimensional parameters have been developed and at this time have not been formally adopted into specification standards. As a result of this uncertainty in specifying three-dimensional parameters, they offer only marginal correlation, with the important in-service requirements of properties typified by:

- tribological factors – frictional and wear characteristics;
- fluid retention – lubrication performance;
- resistance to galling – minimising surface delamination (i.e., tearing) through wear.

Two-dimensional roughness parameters are in general denoted by the lead letter *R*, as discussed in Chapter 1, whereas their three-dimensional counterparts are designated by the logical choice of the letter *S*. This *S* prefix was assigned to indicate the "mean surface" of the profile, which has been derived from a previously EU-funded project co-ordinated by the University of Birmingham. This work developed a range of three-dimensional surface descriptors and has since been known as the *Birmingham 14 – Primary Set*. This "set" has since been expanded and is shown in Table 5. Other parameters have also been developed for specific production requirements, most notably in the steel-

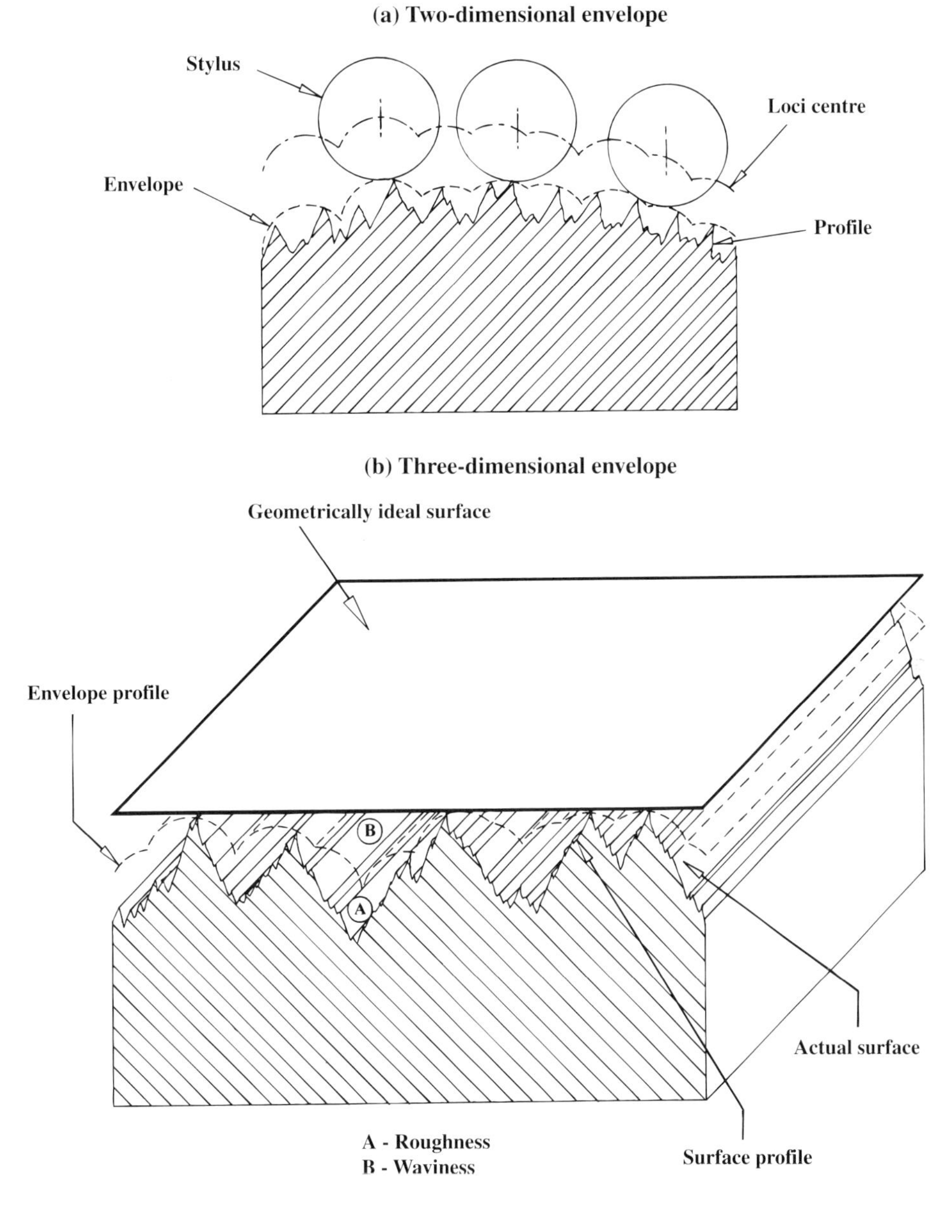

Figure 62. The two- and three-dimensional evaluation of a surface, according to the "E-system". (After Haesing, 1964, Tholath and Radhakrishnan, 1999.)

Table 5. Three-dimensional parameter characterisations

Amplitude parameters	Spatial parameters	Functional parameters	Hybrid parameters
"The Original Birmingham 14 Data Set" (After Stout et al., 1993)			
RMS deviation *Sq*	Density of summit *Sds*	Surface bearing index *Sbi*	RMS slope $S\Delta q$
Ten-point height *Sz*	Texture aspect ratio *Str*	Core fluid retention index *Sci*	Mean summit curvature *SSc*
Skewness *Ssk*	Fastest decay autocorrelation length *Sal*	Valley fluid retention index *Svi*	Developed area ratio *Sdr*
Kurtosis *Sku*	Texture direction *Std*		
Taylor Hobson "Talymap"			
Additional parameters include:			
Sa St Sp Sv	*Spc*	*STP SHTp*	
		Smvr Smmr > Volume parameters	

Derived from ISO 13565–2:
Sk Spk Svk
SR1 SR2

rolling/texturing industries, particularly for fluid retention applications, but more will be said on this topic later in the chapter. However, some confusion exists in either parameter classification into appropriate subgroupings as well as their associated designatory letters and prefixes, because the continuity of groupings has as yet not been fully established. Generally, three-dimensional parameters can be classified by the following groups, which are assigned by their mathematical derivation:

- amplitude characterisation;
- spatial characterisation;
- functional characterisation;
- hybrid characterisation;
- fractal characterisation.

Amplitude characterisation

The characterisation of surfaces by statistical methods has been widely exploited by industry and academia. Because surfaces tend to be represented by random data, it seems natural that statistical techniques can be applied to them for appropriate analysis. In two-dimensional parameter characterisation, ISO 4287: 1996 describes surface texture and the techniques for defining terms and parameters, while ISO 4288: 1996 deals with the rules and procedures for surface texture assessment. It is anticipated that the equivalent three-dimensional parameter characterisation will also be developed in due course. A range of amplitude or height distribution parameters can be selected for the examination of a three-dimensional surface and include the following:

- *Dispersion* (*Sq*) – this is the *root mean square deviation* of the surface which relates to surface departures within the sampling area, being statistically termed the standard deviation of the height distribution:

$$Sq = \sqrt{\frac{1}{A}\int_A z^2(x, y)\,\mathrm{d}x\,\mathrm{d}y}$$

  where domain of integration is the measurement area $A = \mathrm{d}x\,\mathrm{d}y$.
- *Extremes* (*Sz*) – this is the *ten point height of the surface*, being an extreme parameter which can be defined as the average value of the absolute heights of the five highest peaks and deepest valleys – by the "eight neighbours method" – within the sampling area.
- *Asymmetry* (*Ssk*) – this is the *skewness of topography – height distribution*, which measures the asymmetry of surface deviations above the mean plane. This parameter describes the shape of the topography height distribution. For example, for a "Gaussian surface" – having symmetrical shape to its surface height distribution – this would equate to zero skewness, whereas an asymmetric

distribution of surface heights may produce a negative skewness, when it has a longer tail on the lower portion of its mean plane. This skewness parameter provides an estimation of the existence of "spiky topographical features":

$$Ssk = \frac{1}{mn} Sq \sum_{i=1}^{m} \sum_{j=1}^{n} z_{i,j}^{3}$$

- *Sharpness* (*Sku*) – this is the *kurtosis of topography – height distribution*, which measures the sharpness of the distribution, characterising its spread. For example, a "Gaussian surface" equates to a kurtosis value of 3; conversely, a surface topography that appears centrally distributed would produce a kurtosis value in excess of 3, while a more evenly spread topography would have a value less than 3. By combining both the *Ssk* and *Sku* parameters, this may allow identification of surface topographies having reasonably flat tops with deep valleys.

Spatial characterisation

Parameters relating to the a surface's spatial properties offer some difficulty in their characterisation, due to their general wavelength randomness combined with multi-wavelength variations, which in turn are coupled to their high sensitivity to the sampling interval. Techniques utilised for spatial characterisation include the following:

- *Auto-correlation* – a version of this technique in surface texture characterisation is that of areal auto-correlation function (AACF), where the dependence of one position of data values is dependent upon another. Typical parameters that originate from the AACF, might be:
- *Sds* – this is the *density of peaks*, being the number of peaks of a unit sampling area, by the "eight neighbours method":

$$Sds = \left(\frac{1}{6\pi\sqrt{3}}\right)\left(\frac{m_4}{m_2}\right)$$

where

$$m_2 = \frac{2[R(0) - R(\Delta x)]}{(\Delta x)^4}$$

$$m_4 = \frac{2[3R(0) - 4R(\Delta x) + R(2\Delta x)]}{(\Delta x)^4}$$

- *Str* – this is the *isotropy index* or *texture aspect of the surface*, defined as its ratio of fastest-to-slowest decay (0.2) with respect to correlation length and is used to determine aspects relating to a surface texture's uniformity via the auto-correlation function. Principally, the texture aspect ratios vary between 0 and 1, with values <0.5 indicating that no defined lay occurs and ones of >0.5 exhibiting a pronounced lay. In some instances with a finite size of the sampling area, the possibility exists that for certain anisotropic surfaces the AACF's slowest decay will not reach 0.2. Under such conditions, the surface's slowest decay direction should be employed;
- *Sal* – this is the *fastest decay auto-correlation length*, defined as the horizontal distance of the AACF having the fastest decay to 0.2. This value of 0.2 is the shortest auto-correlation length that the AACF decays to in any direction. For example, with an anisotropic surface, *Sal* occurs perpendicular to the lay surface and a large value of *Sal* denotes that the surface is dominated by long-wavelength components. Conversely, a small value indicates short wavelengths;
- *Std* – this is the *texture direction of the surface*, which determines the most conspicuous direction of surface texture (lay) with respect to the *Y*-axis within the frequency domain (Fourier).

Functional characterisation

The classification of three-dimensional surfaces by its "bearing area" implies that a set of techniques are engaged relating to specific functional properties of a surface. Unfortunately, due to the diverse range of functional aspects demanded by industrial applications, no single functional characteristic can adequately cover all of these techniques. However, it is possible to utilise the following parameters to establish the potential in-service performance of a surface:

- *Sbi* – this is the *surface bearing index*, which relates to the *Sq* parameter over the surface's height at the value of 5% bearing area, with a greater surface bearing index indicating superior bearing properties;
- The RMS deviation *Sq* is defined by analogy with its two-dimensional form;
- *Sci* – this is the *core fluid retention index*, being the ratio of void volume of the unit sampling area at the core zone (i.e., 5–80% bearing area) over parameter *Sq*. Good fluid retention occurs with greater values of *Sci* and for a "Gaussian surface" the index is approximately 1.56:

$$Sci = \frac{1}{A} Sq \left[1 - \int_{z^{*}0.8}^{z^{*}0.05} p(z^{*})\,dz^{*}\right]$$

where $z^* = z/Sq$ and any suffices refer to bearing area fractions.

- *Svi* – this is the *valley fluid retention index*, which is the ratio of the void volume of the sampling area within the valley zone (i.e., 80–100%) over the *Sq* parameter. For example, if a larger *Svi* occurs, this indicates excellent fluid retention in the zone of the valleys.

$$Svi = \frac{1}{A} Sq \left[1 - \int_{z^*1}^{z^*0.08} p(z^*)\, dz^* \right]$$

where $z^* = z/Sq$ and any suffices refer to bearing area fractions.

Hybrid characterisation

The result of combining two or more of the previously discussed three-dimensional classification techniques is termed a "hybrid". By combining such parameters, it is possible to acquire many hybrid parameters, with the most significant being:

- *SΔq* – this is the *root mean square slope* of the three-dimensional surface texture, within the sampling area:

$$S\Delta q = \sqrt{\frac{1}{A} \iint_A z'(x, y)\, dx\, dy}$$

where

$$z'(x, y) = \sqrt{\left(\frac{\delta z}{\delta x}\right)^2 + \left(\frac{\delta z}{\delta y}\right)^2}$$

- *Ssc* – this is the *(arithmetic) mean summit curvature* of the surface and is defined as the average of the principal curvatures of the summits within the sampling area. Namely, the sum of curvatures of a surface at a point along any two orthogonal directions equals the sum of predominate curvatures;
- *Sdr* – this is the *developed (interfacial) area ratio*, which is the ratio of interfacial area of a surface over its sampling area. The parameter describes the hybrid nature of surfaces, with a large numerical value equating to either significant amplitude, spacing or both surface conditions.

Fractal characterisation

The assumption that industrially-developed surfaces might "mirror" natural phenomena (fractal geometry; see Figure 1) that are not particularly influenced by the laws of Euclidean geometry, namely defined planes and analytical functions, is of relatively recent origin. The reasons for this situation are not at present particularly clear, but presumably occur from the operation of process hierarchies that chaotically dissipate energy across many dimensional scales. Initial work in this field was principally concerned with discovering the existence of fractal behaviour over a large variety of distinct circumstances. Later, fractal research was influenced by attempting to understand and uncover the causes and relevance of this behaviour and, in particular, trying to find functional dependencies and relationships that correlate fractal dimensions of engineered surfaces to specific in-service performance.

A considerable body of research-based publications has been written on the measurement of fractal dimensions, with some techniques being inappropriate for surface assessment, namely of a general self-affine nature, as opposed to those being self-similar, while other work has proved unsuitable for anisotropic surface characterisation. For most of this fractal research work the techniques employed have been used to estimate the bounds of the fractal dimension with varying degrees of numerical precision. Generally, the lack of specific physical meanings in fractal characterisation has meant that work in this field has been restricted in the main to academia.

In-service performance

Depending upon its potential end-use, a surface can be "engineered" to meet the anticipated in-service demands and now, with three-dimensional characterisation, quantitative measurements can be made to ensure that the desired conditions have been met. For example, in many sheet metal applications the retention of a lubricant at specifically defined sites is considered an important factor and parameters have been suggested that define lubricant trapping; Figure 63 indicates just such a site. This figure can be defined by intersecting lines at different positions along the profile heights to produce the percentage length. The three-dimensional version is an extension of its two-dimensional counterpart, although currently there is not a specification for such filtering. The technique is generally more suitable for characterising surface conditions where the profile has either a flat top or when porosity/grooving is present. In fact, the ability to estimate and determine both isolated valleys and the possibility of valley connectivity have been investigated through a "Maxwellian/Motif landscaping

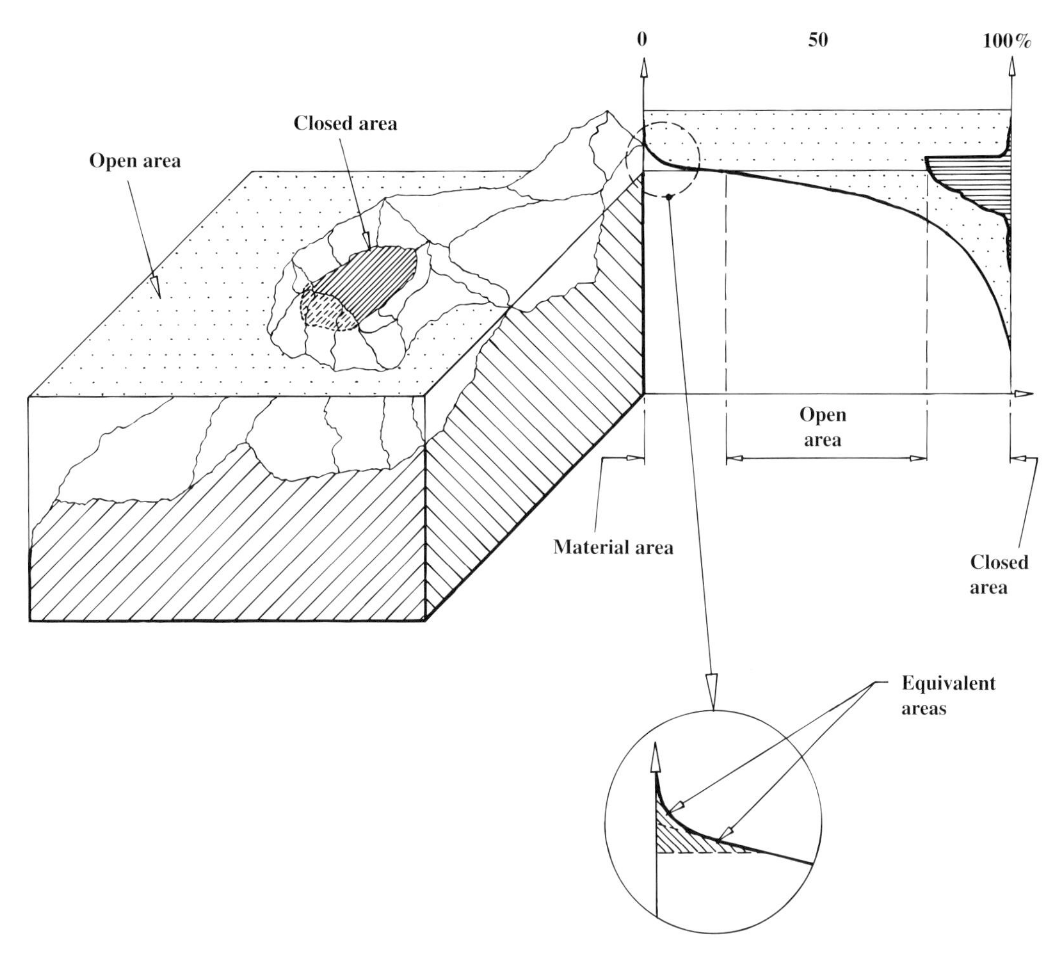

Figure 63. Abbott–Firestone curve with closed areas. (After Sacerdotti et al., 2000.)

technique" or "change tree approach" – but more will be said on this topic later in the chapter under Section 2.4 (Fractal techniques).

In Table 6 are indicated the functional characteristics for a range of conditions, showing the three-dimensional surface texture parameters and their expected correlation on a matrix for a series of characterisations. The table cannot be an exhaustive list of functional performances but indicates just some of the current applications where they may, or indeed may not, be utilised. Such tables indicating functional performance are important, but exact classification cannot be quite as clear-cut as indicated in this chart. It has been argued that three-dimensional measurements are significant because, in practical terms, surfaces are three-dimensional and not two-dimensional in nature. In Section 2.2.1 is a discussion on aspects of three-dimensional functional performance of surfaces.

2.2 Three-dimensional analysis software

In the last few years three-dimensional measurement technology has been introduced by many of the larger surface instrument manufacturers, with the specific objective of being able to overcome deficiencies resulting from two-dimensional assessment. In general, industry has been somewhat reluctant to invest in three-dimensional equipment, apart from academia (see Table 7). The success of any three-dimensional surface texture application is derived through the abilities of its software surface performance and characterisation. This section will briefly review just some of the visual images generated from the three-dimensional characterisation of surfaces; it is not meant to be an exhaustive account of the subject but simply a "visual tour" of 3-D surface potentialities.

Table 6. Functional performance correlations

Function	Characterisation			
	Amplitude	Spacing	Functional	Hybrid
Forming and drawing	***	****	****	**
Painting and plating	***	****	****	***
Friction	****	****	****	****
Galling	***		****	****
Wear	****	****	****	****
Joint stiffness	****	**	****	**
Slideways	****	****	****	**
Electro-contracts	****	****	****	****
Bonding and adhesion	****		****	**
Fatigue	***		****	
Stress and fracture	**		****	**
Reflectivity	**	**	****	****
Hygiene	***		****	**
Bearings	****	**	****	****
Seals	****	**	****	****
Related 3-D Parameters	*Sa, Sq, Sz, Ssk, Sku*	*Sds, Str, Std, Sal*	*Spk, Svk, Sk, Sr1, Sr2, Svi*	*SΔa, Ssc, Sdr*

Key: **** high correlation; * very little correlation.
After Griffiths (1988).

Table 7. Two- and three-dimensional stylus profilometer usage

Industry applications:	Surface texture usage	
	Two-dimensional	Three-dimensional
Aerospace	10%	<2%
Automotive	75%	2%
Electrical	10%	<1%
Paper	10%	0%
Cosmetic	10%	<1%
Academic	5%	94%

After Griffiths (1999).

Many of the three-dimensional software systems employ a Windows(-based software interface, giving the user a highly intuitive visual capability with logical, menu-driven functions having comprehensive help facilities for on-line user support. Typical of such fast operational industrial-based software is that illustrated in the following self-explanatory examples, giving not only clear visual and easily interpretative imagery but also considerable direct surface-related 3-D information.

A representative example of such comprehensive 3-D surface software is described below. This software enables the user, once the 3-D profilometry raw data has been captured, to manipulate a measured surface in numerous ways by selecting appropriate colour palettes or individual colours – for clarity – together with zoom function ability, giving:

- colour representation – for example, an axonometric surface characterisation – either of the wired or meshed type (Figure 64);
- colour altitude coding – indicating relative heights of peaks and valleys (Figure 64 again);
- autocorrelation imaging – analyses directions (isotrophy) and periodicity of the 3-D relief (not shown);
- photo-simulation with lighting effects – which enables the captured image to be rotated or inverted to gain a complete visual impression of the measured surface's terrain (Figure 65);
- volume studies – defining, say, a hole's volume, perimeter, area, mean depth and maximum depth of an irregular cavity in the surface (Figure 66);
- surface filtering – employed to separate out any waviness, roughness and form data into discrete entities; this occurs by FFT filtering to separate out components of the 3-D relief, frequency by frequency and direction by direction (Figures 67 and 68);
- contour diagram studies – enabling points having identical heights to be joined (Figure 69);
- pattern recognition – highlighting the points with specific topological properties (Figure 70);
- threshold operator utilisation – enables excessive peaks to be removed or extreme holes to be filled in (Figure 71);
- a re-sampling operator – this technique clarifies the image after successive zooming operations have occurred and insufficient data points exist

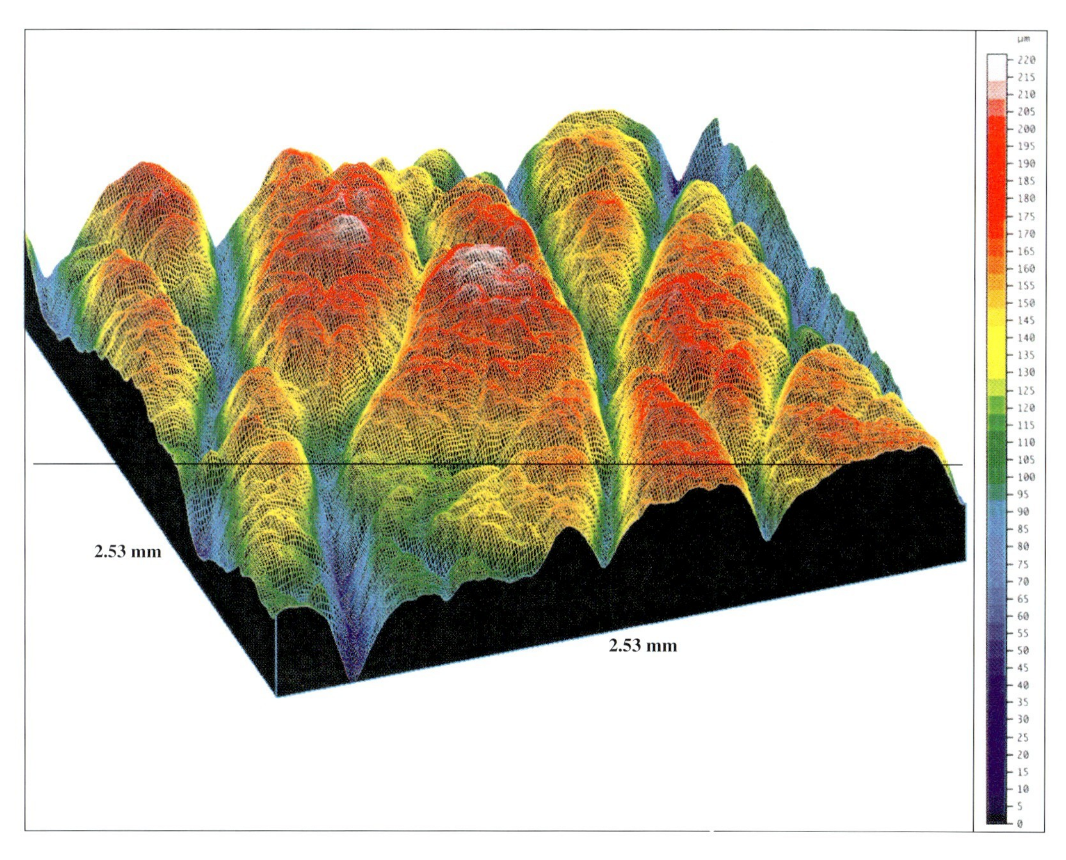

Figure 64. 3-D representation of an axonometric surface produced from a stylus-based instrument. (Courtesy of Taylor Hobson.)

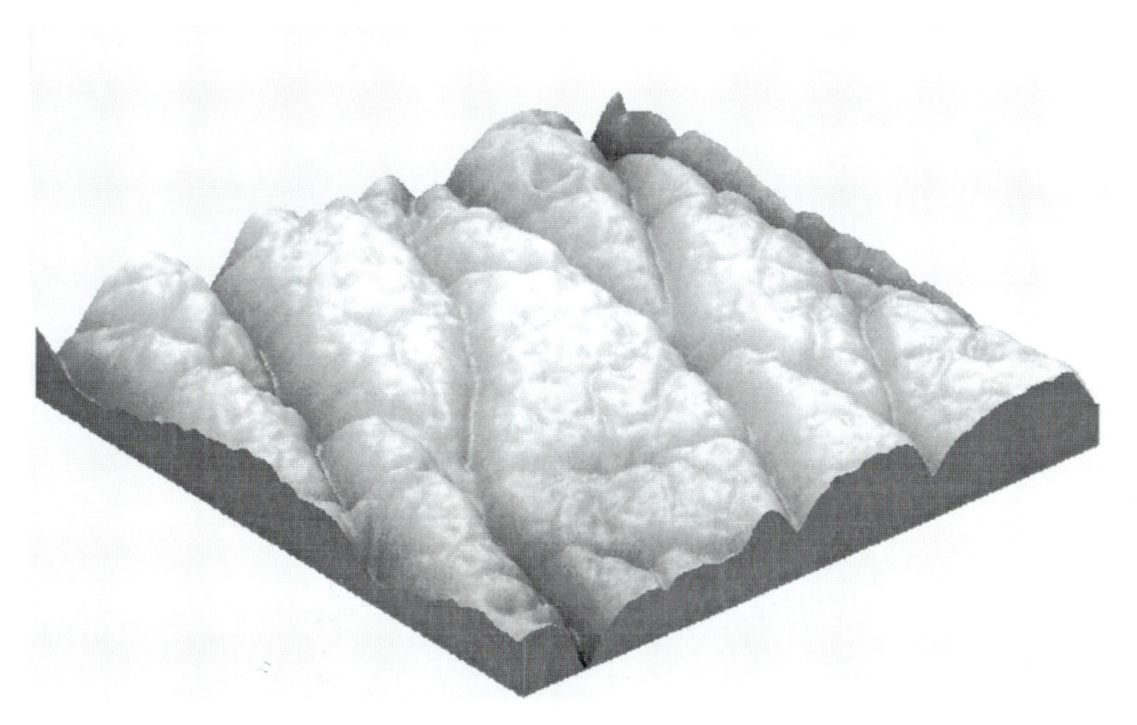

Figure 65. A continuous 3-D axonometric representation of a surface reproduced from a stylus-based instrument. (Courtesy of Taylor Hobson.)

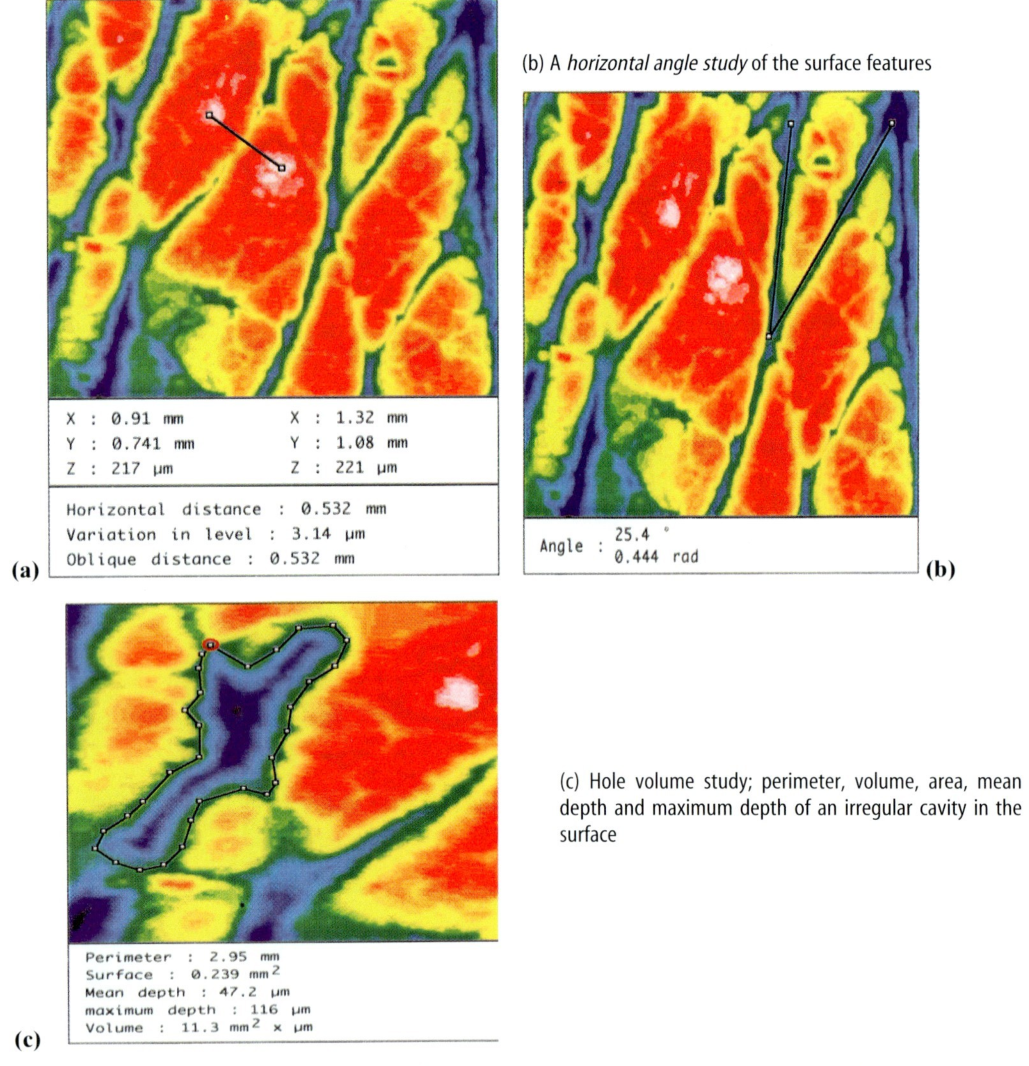

(d) A horizontal slices study illustrates which part of the surface is below, or above a given height. NB the area of the "highest slice" is known as the *bearing ratio*

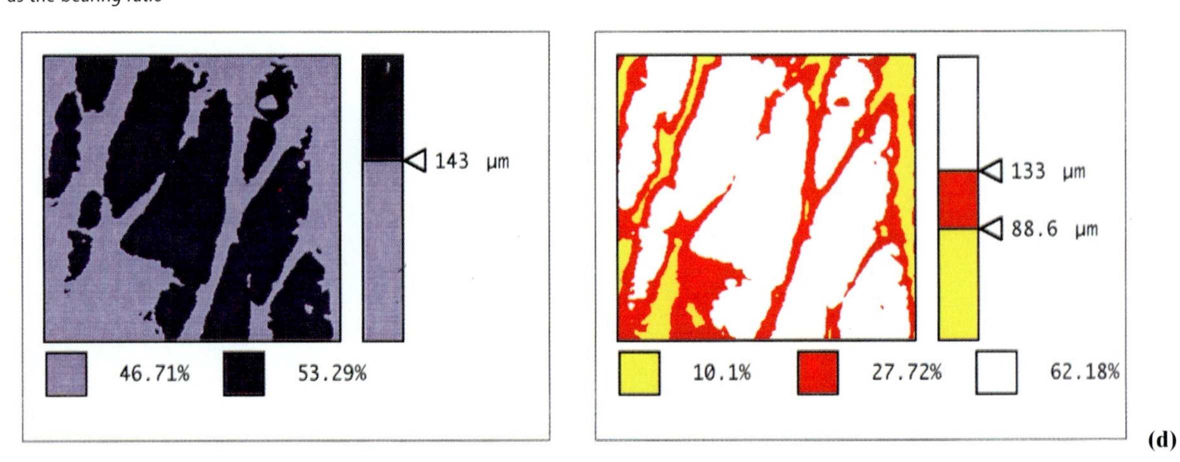

Figure 66. 3-D software manipulation used to indicate important features on the surface topography. (Courtesy of Taylor Hobson.)

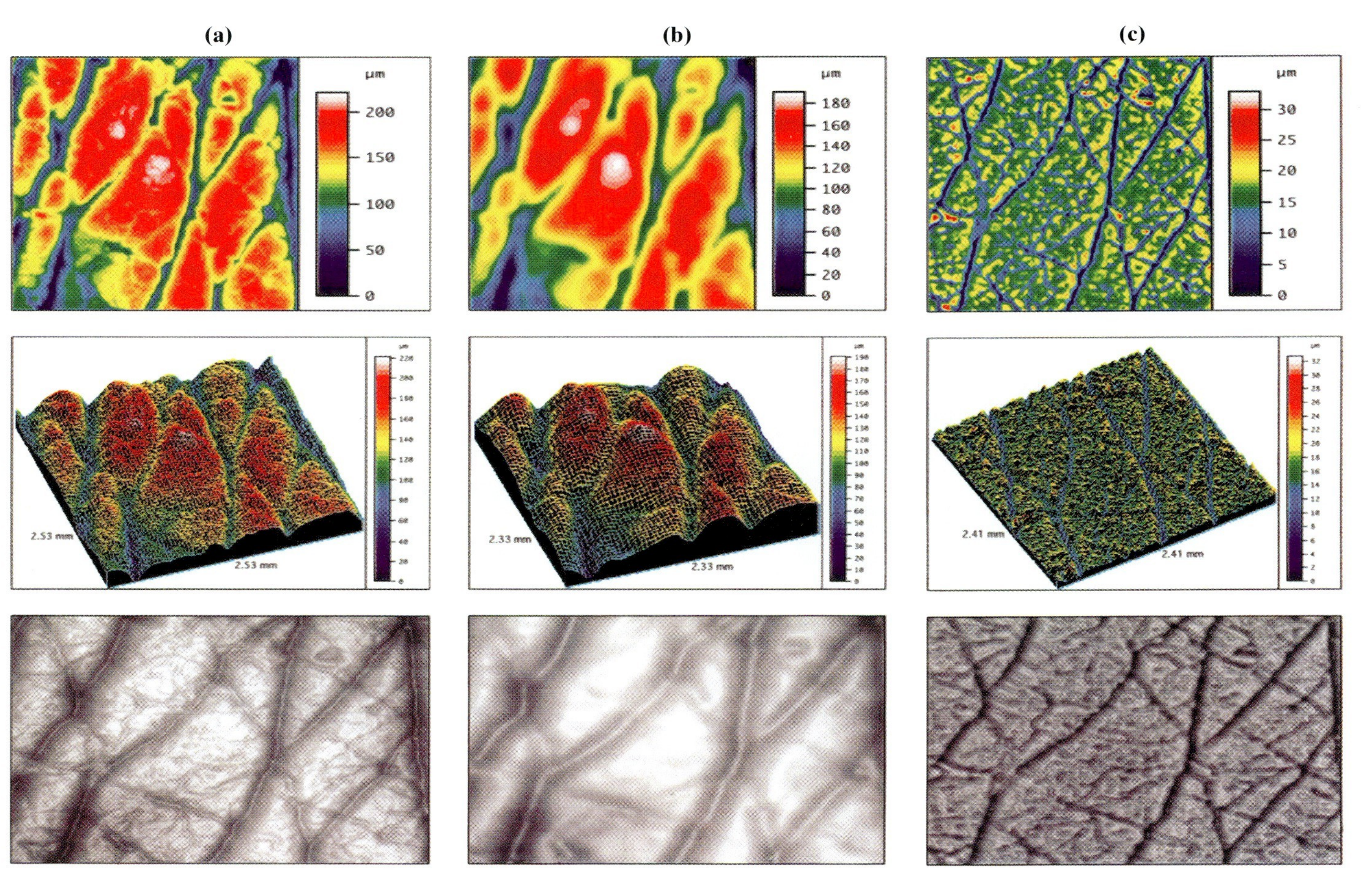

Figure 67. The *filter operator* can be used to separate out waviness and roughness. (a) The unfiltered surface. (b) The waviness of the surface. (c) The roughness of the surface. (Courtesy of Taylor Hobson.)

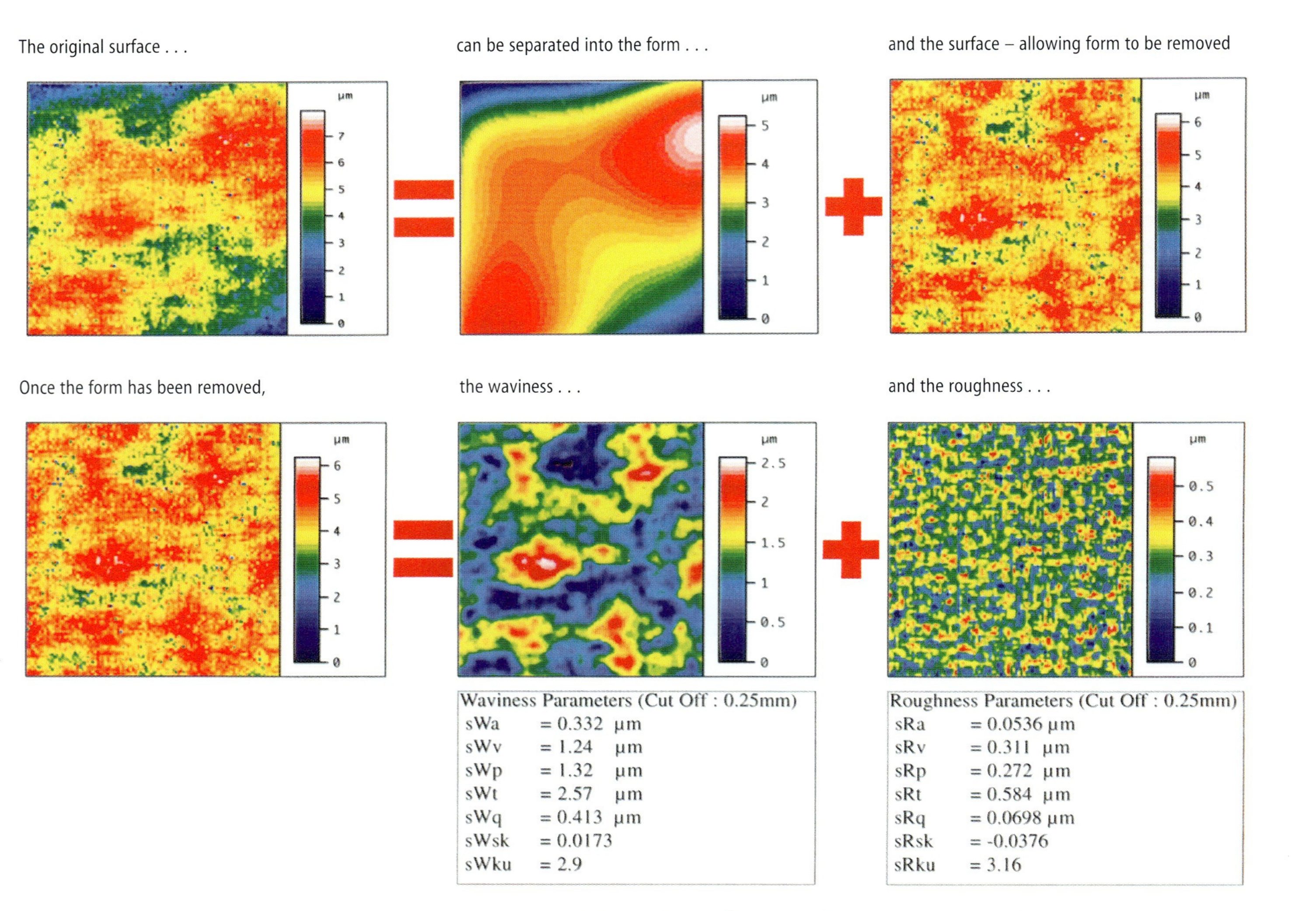

Figure 68. It is possible to remove *form* from the surface utilising the *filter operator.* (Courtesy of Taylor Hobson.)

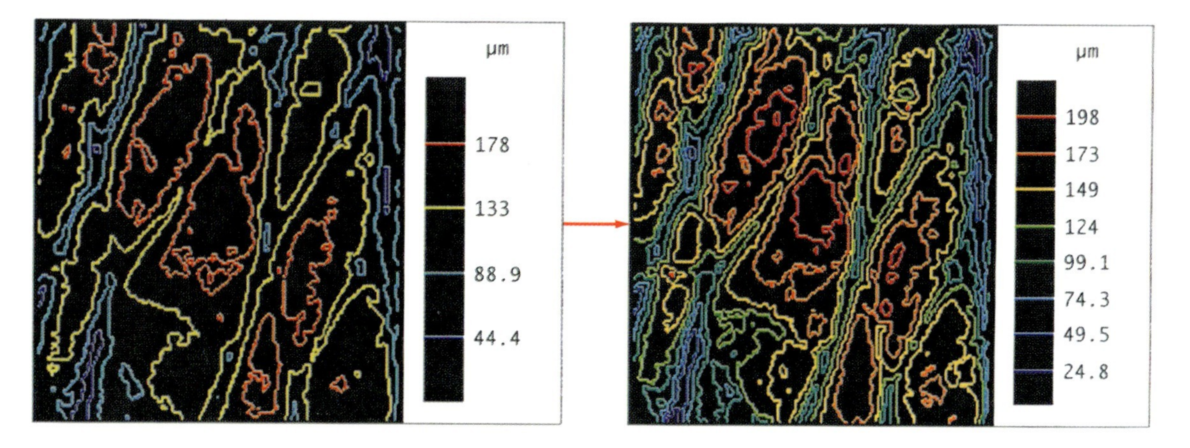

Figure 69. A 3-D *contour diagram study* joins points which have the same height. (Courtesy of Taylor Hobson.)

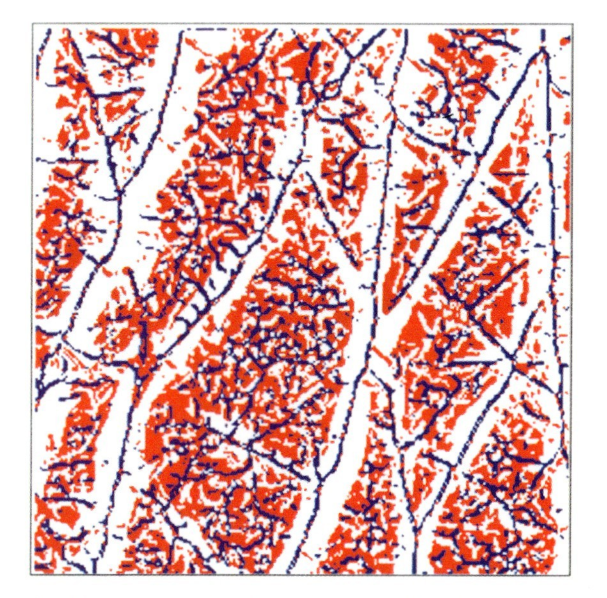

The plateaus (i.e., in red) and the valleys (i.e., in blue) of a human skin surface

Figure 70. *Pattern recognition study* highlights the points with specific topological properties. (Courtesy of Taylor Hobson.)

to enhance the image; this can be achieved by increasing the number of pixels, which includes optimising the "Z resolution" (Figure 72);

- profile extraction – this enables the 3-D topography to be examined and at particular positions of interest on the surface the user can pre-select a line – straight, or manipulated around and across specific surface features – to be drawn and a two-dimensional trace at this "cut surface" can be illustrated (Figure 73).

2.2.1 Functional 3-D performance

The surface topography characteristics of an engineered component may have important implications in the fields of:

- *tribology* – wear studies;
- *lubrication* – oil retention in surface valleys;
- *adhesion* – for paint and plating applications;
- *absorbency* – measuring paper surface volume to ascertain the amount of ink needed in the printing process;
- *plateau honing* – determining honed cylinder liner characteristics;
- *reflectivity* – analysis of pattern and direction of car headlamp glass covers, ensuring optimum lighting in foggy conditions;
- *ageing* – comparative analysis of components before and after accelerated weathering/ageing processes (typically to predict material life for painted and canvas surfaces).

In many industrial applications speed of data manipulation is paramount and if identical analytical processing is necessary then "macros" allow operations, calculations, studies and comments added to the image produced – to be efficiently executed in a few seconds. In many industries there is a fundamental requirement for the measurement of surface holes and depressions, this being necessary, when:

- "surface engineering" is the requirement where cavities or holes in the surface topography are designed for fluid retention;
- texturised surface features – depressions – need to be confirmed by 3-D measurement;
- cavities present in the surface need to be measured that might affect in-service performance.

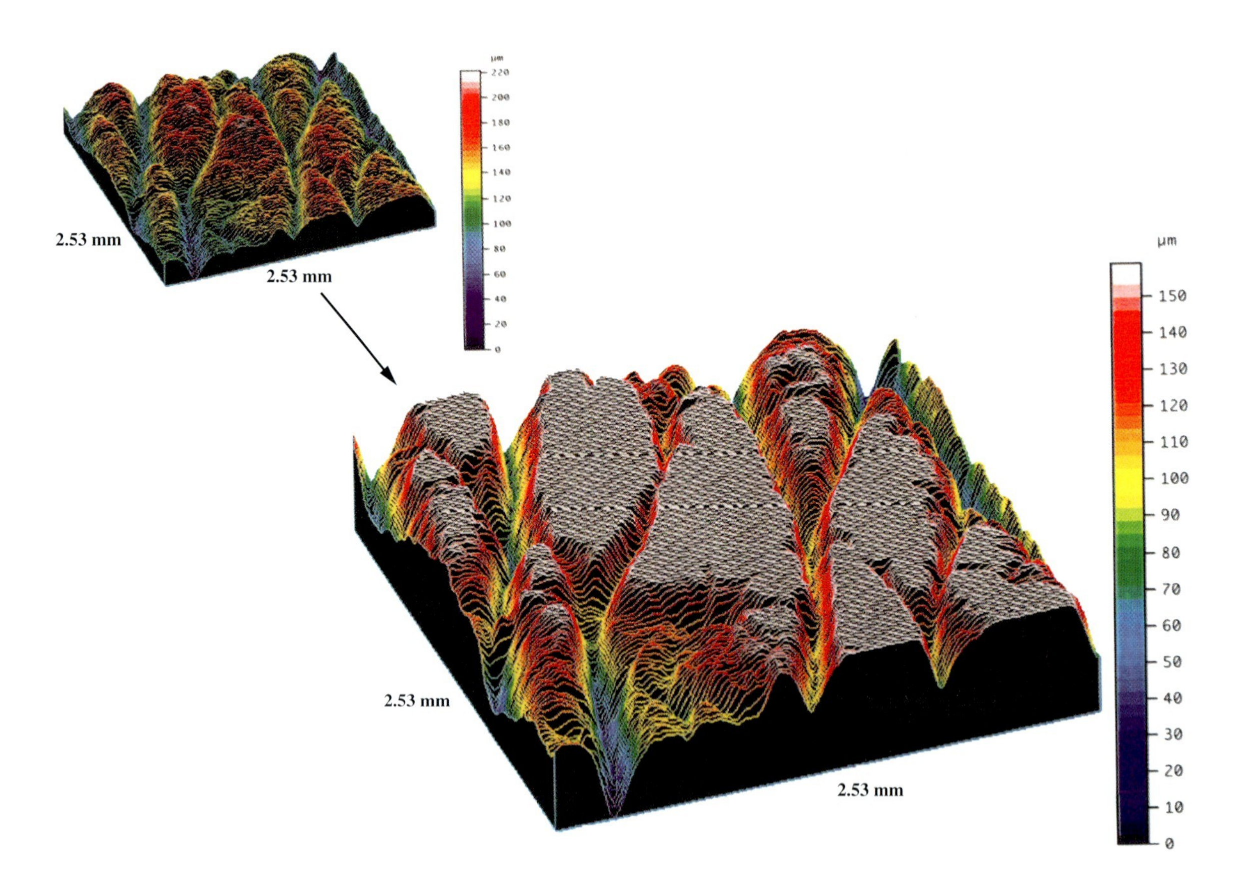

Figure 71. By utilising the *threshold operator* excessive peaks can be removed. This also applies to the filling in of excessive holes. [Courtesy of Taylor Hobson.

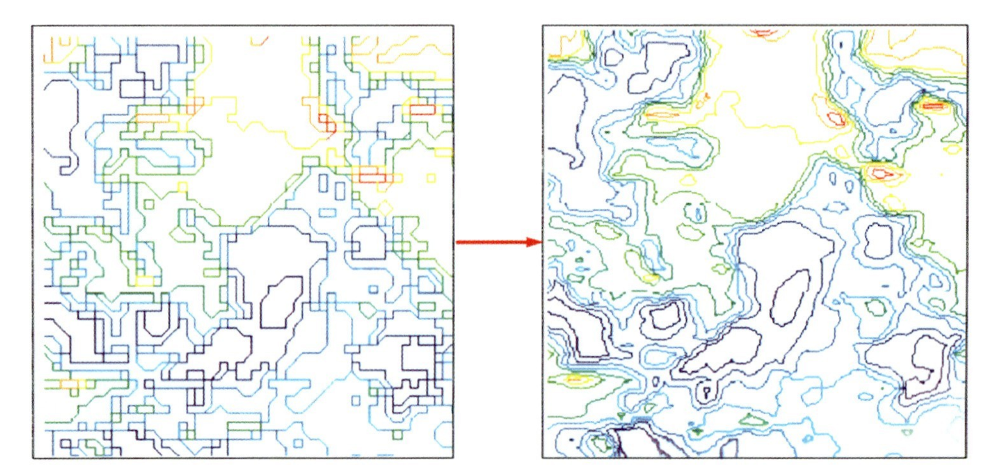

Re-sampling enhances the representations when there are no longer enough data points (e.g., after successive zoom operations).

Figure 72. The *re-sampling operator* allows one to change the data sampling space. (Courtesy of Taylor Hobson.)

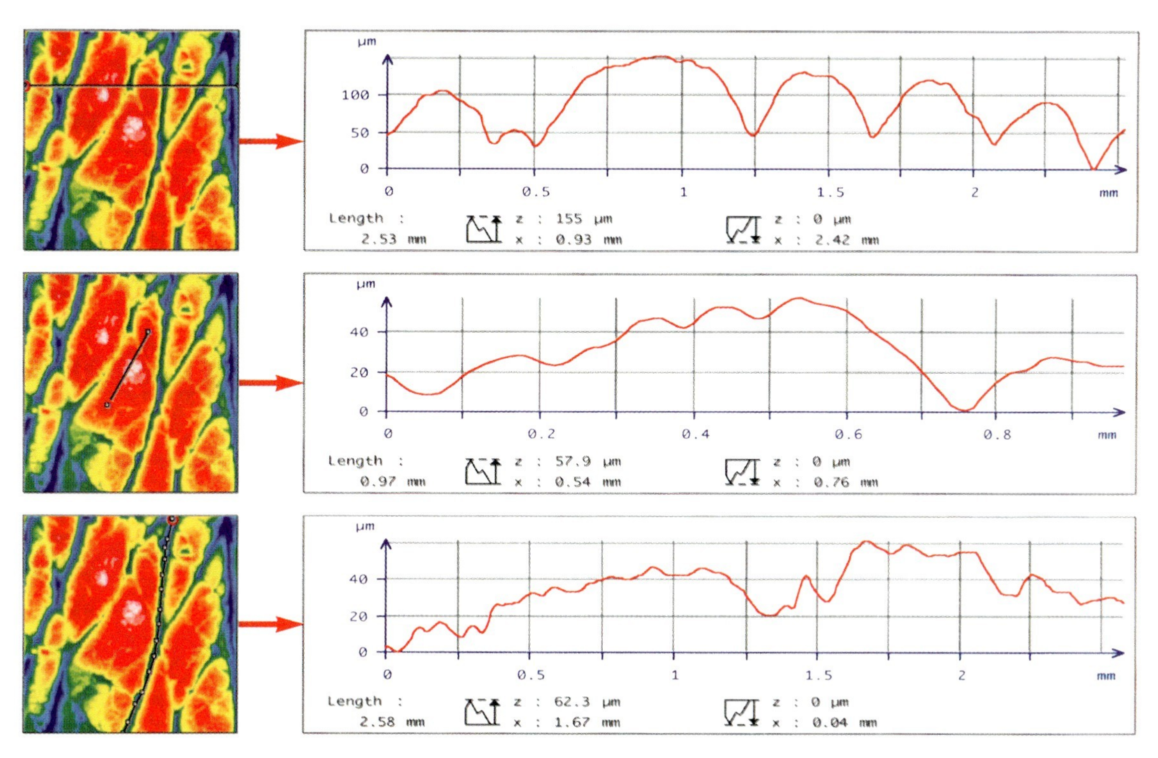

NB: the path followed along the surface can be a straight line, the bottom of a valley, a circular line, or any other user-defined contour.

Figure 73. The *profile extraction operator* allows one to extract one trace out of the surface. (Courtesy of Taylor Hobson.)

Hole calculation

In Figure 66(c) a "hole volume study" of a cavity was shown being examined. It might be prudent to ask the question: "Why calculate the hole's volume?" As has been described above, many industries have specific requirements to measure the volumes of holes in high-grade fine surfaces, in order to quantify:

- abrasive effectiveness – grinding, honing, lapping and super-finishing cutting efficiencies;
- textural condition of laminating rollers;
- wear and corrosion efficiency;
- predictions of oil reserves on mechanical components;
- porosity of an interior (open aspect) surface, such as those found on powder metallurgy components;
- impact of graining on a surface;
- efficiency of a punch;
- effectiveness of cosmetics;
- volume of an embossed bump (after height inversion, for example, solder on a printed circuit board).

Importance must be given to differentiating methods for calculating the volume of a single "blind hole", whose periphery can be delineated by either tracing around it with a mouse or moving the cursors. Such an outlining technique allows for calculations of either the average volume of a cavity or, conversely, for material that might be situated between two altitudes on a 3-D trace (see Figure 63). Some practical examples and solutions to this irregular hole measurement dilemma are given below:

- *Measurement of a hole on an uneven trace* – this represents some degree of difficulty if suitable software is not available. By way of illustration: if there is a requirement to measure the healing capacity on a leg wound, the wound's surface volume is uneven – due to the anatomy of muscle and bone – and tends to be represented by a cylindrical rather than a planar profile. If a silicon replica of the wound is taken this will be three-dimensional and follow the wound's contours, but when laid flat much of the cylindrical contouring effect will have been lost. However, the constraints imposed when taking an imprint – replica – deform a zone around the inspected hole. Such deformations cannot be ignored, since they have an amplitude akin to that of the hollow to be measured. Suppression of the deformation of the replica can be achieved by applying a *polynomial form removal* to the contour of the hole by separating out both the local effects at the inside and outside of the hole, this being delineated by using the mouse;

- *Measuring an irregular hole volume, when closely surrounded by other holes* – in order to calculate the main hole's volume, it is necessary to carefully define the *reference plane*. For example, on a hemispherical (dished) surface its volume depends upon its relative depth from the top surface. By using one of the software functions, typified by a *zoom operator*, it is an extremely difficult task to isolate the hole, since this creates a "squared portion" of a surface – thereby containing the adjacent closely packed holes. However, the *erase defect operator* enables other holes present in this zoomed area to be erased, giving an acceptable hole contour which acts as a reference plane for subsequent calculation of the hole's volume;
- *Zone surrounding a hole has a complicated relief* – rather than attempting to exactly define the reference plane and then calculate the hole's volume between the bottom of the hole and this plane (as in the previous case), one can vertically *cut the image into elementary profiles*. For each of these separate profiles a calculation can be made for the surface between the hole's bottom and a line defined by the edge of the hole. For each hollow its "top" will differ with every vertical cut of the image, since the software recalculates every respective hollow. The advantage is that a complex shape to the hollow's contour can be assessed, but the disadvantage is that small local variations in the contour's relief can introduce significant variations in the measured volume. Furthermore, this technique has a tendency to consider possible lips adjacent to the hole, typically crater edges and material displaced from the hole, whereas the *reference plane* method ignores the presence of disturbances promoted by lips when calculating the height of the hole;
- *Average volume, or volume of a single hole* – certain studies require the calculation of the average hole volume, while others just need the volume of a single hole. An example might be when it is necessary to calculate the quantity of adhesive needed to bond two surfaces together, requiring a calculation of the volume of air between these two surfaces. The software best suited to this problem is one that can produce a *volume study*, giving the total cavity volume between the highest and lowest points of the surface, or between altitudes that can be readily calculated.

Form removal

The topography of a complex relief surface can be measured by profilometry, by separating out its component parts depending on their respective wavelengths and amplitudes. The smaller the wavelength, the greater will be the variations in the altitude for identical horizontal lengths. By moving from a larger to a smaller wavelength, the four following components can be distinguished:

1. *Form* – this is a component of topography having a wavelength almost equating to the object's length. Its measured form is an amalgamation of the *theoretical form* defined at conception (sphere, cylinder, etc) and the *form variation*, which is the difference between the manufactured and theoretical forms. It should be noted that *form* cannot be measured using a skid in situ with its stylus, as the skid attempts to follow the intrinsic shape of the object; therefore the altitude signal will only contain variations in the relief (see Figure 38);
2. *Waviness* – this is a component of the surface that gradually varies with respect to its horizontal position. For example, waviness might be defined by a wavelength of between 0.5 and 2.5 mm; moreover, on a production part such waviness could be the result of low-frequency vibrations of the component in relation to the machine tool;
3. *Roughness* – this is a component of the surface that rapidly varies in relation to the horizontal position. Roughness could be defined by wavelengths between 20 and 500 μm; with a manufactured part this roughness might be due to a number of different but discrete production processes, such as shot-/sand-blasting, grinding particles (i.e., abrasives) and tool deflection/vibration;
4. *Micro-roughness* – this, as its name suggests, is the finest component on the relief, with wavelengths of, say, less than 20 μm. Micro-roughness is generally associated with the structure of the measured material, rather than being the result of a particular production process. Prior to calculating the roughness parameters, micro-roughness can be eliminated by suitable filtering.

The analytical study of a relief requires that the components described above, which relate to different properties in the overall relief, are normally separated out before close inspection occurs. As previously mentioned, the separating out of waviness, roughness and micro-roughness can be achieved by filtering. Generally, two main filters tend to be used:

- *Gaussian filter* (convolution of the profile to a "Gaussian function");
- *RC2 filter* (mathematical simulation of a "second-order electric capacitor–resistor filter").

The delineation between waviness and roughness is termed "cut-off" (see Section 1.6) and can take a certain number of normalised values. This technique is utilised for both two-dimensional (profile) and three-dimensional (surface) measurements.

Form separation is not normally carried out by filtering and probably the main reason for this is that filtering, as defined by standards, may relinquish an important section of the surface when employed. When utilising surface texture instruments, the aim is to suppress the form component so that an isolated study of waviness and roughness can be attained. The form can be suppressed by using a *polynomial approximation*, while imposing the minimum of constraints. The method consists of seeking the polynomial which best adjusts to the surface or profile being measured. The coefficients of the polynomial are automatically calculated using the *least-squares* technique. The only constraint imposed is the *degree "n" of the polynomial*, which defines the complexity of the form to be subtracted. Therefore, the higher n becomes, the more the polynomial "mimics" the surface relief, hence:

- subtracting the *first-order* polynomial is the equivalent to subtracting the least-squares plane;
- subtracting the *second-order* polynomial allows uniform curvature to be eliminated from the surface;
- the higher the *degree* of polynomial, the more the form described by the polynomial becomes complex and the less relief remains inside the waviness. The choice of *degree* will therefore depend on the limit that one defines between waviness and form.

2.3 Portable three-dimensional measuring instruments

This instrument for the assessment of three-dimensional surfaces (Figure 74) is used in conjunction with 3-D analysis software. It can be readily calibrated automatically and features a relatively low cost but high-speed measurement facility in conjunction with a portable scanning head. The design enables the measuring instrument to be placed directly onto the surface and provides an in-

Figure 74. Portable 3-D surface profiling instrument allowing surface topography measurement and Windows scanning software to create visualisation of the 3-D surface. (Courtesy of Taylor Hobson.)

depth analysis of the topographical features at high resolution and accuracy.

The fast operational stylus pick-up is scanned across the surface under examination using two traverse mechanisms. The mechanism controlling the Y-position of the pick-up simply steps the stylus sideways in synchronisation with the X-position controller. These two systems result in a series of profiles being obtained from the measurement head, which are built-up into a 3-D presentation of the surface under examination. The measurement process can be enhanced for small components by a positioning table having: X-, Y- and Z-positional control over a 10 mm volumetric envelope.

The main features of the measurement head's specification, include traverse area of 1 mm^2, measurement time <60 s for a 0.5 mm^2 area, gauge range of 200 μm, gauge resolution 3 nm, gauge range/resolution 4096:1; traverse speed ranges from 10 mm/s (fast) to 2.5 mm/s (slow), with a head weight of 0.85 kg.

The software mapping capabilities are comprehensive for such a relatively inexpensive instrument, and include the following:

- *Visual functions*
  - 3-D plots with heights coded by colour (Figure 64);
  - 3-D continuous plots (Figure 65);
  - surface smoothing abilities;
  - a contour diagrammatic representation of the surface (Figure 67);
  - height viewing in pseudo colour for an alternative visual interpretation;
  - photo simulation.

- *Analysis functions*
  - 3-D surface parameters – previously mentioned in Section 2.1.3;

- limited 2-D surface texture parameters – for comparison;
- zooming abilities, for specific topographical feature investigation/analysis;
- both Gaussian and 2CR filtering capabilities;
- form removal – if this either influences the surface topography or is visually distorting the parameters, such as a scratch (Figure 68);
- hole volume computation, for regular/irregular peripheries (Figure 66c);
- distance measurement between two topographical features (Figure 66b);
- rotation of any angle/mirror, enabling a clearer visual interpretation of the captured image to be assessed;
- inversion of heights, employed to measure surface replicas;
- Abbott-Firestone curve and amplitude distibutions to be generated.

- *Desk-top publishing functions*
  - macros for rapid analysis and report generation.

The complete system requires connection to a suitable PC, which ideally requires a screen resolution of at least 1024 × 768 pixels having 65,000 colour codes, if the system is to perform in an adequate manner for visual interpretation and assessment.

2.4 Fractal techniques

Introduction

It has been recently shown in published work that many types of engineered surfaces exhibit a fractal geometry. Under certain conditions modelling the surface generation predicts this effect and provides correlation between dimensions, historical development of the surface and its properties. Observations have indicated that many surfaces are non-Euclidean in nature (are not defined by planes and analytical functions), but apparently have a fractal geometry – mirroring the natural world (see Figure 1 background). The reasons why fractals occur cannot as yet be readily determined; presumably they arise from the operation of hierarchical processes that dissipate energy in a chaotic fashion across many dimensional scales. Some of the initial papers were primarily concerned with identifying the existence of fractal behaviour in a wide variety of differing circumstances. Later work attempted to investigate and understand the causes and importance of this fractal behaviour and to detect functional dependencies and relationships that correlate the surface's fractal dimension with specific history and performance.

Numerous procedures are now available for the determination of fractal dimensions. Some of these procedures are unsuitable for textural surfaces, which can be considered as generally self-affine, as opposed to self-similar, while other procedures are inappropriate for anisotropic surfaces. Many of the fractal techniques estimate bounds of the dimension and these methods have varying levels of numerical precision. However, a fractal model – via its surface height distribution – can predict or give a reasonable estimate of the fractal dimension for a rough anisotropic surface and can even simulate such a surface. The accuracy and robustness of these current "direct" and "Monte Carlo" simulation techniques for fractal models still require further research to give higher degrees of correlation to actual situations.

The following technique indicates a change of emphasis in determining embossed (convex) or concave features on three-dimensional engineering surfaces, by employing an "areal motif" technique, which has its roots in the discipline of cartographic surveying.

2.4.1 Topological characterisation

In recent years a new development in attempting to determine the "connectability" of three-dimensional surface features via "areal motifs" – previously developed by the French automotive industry – using cartographic characterisation has been proposed (see Standards in References). This areal motif method is based on the cartography procedure, whose discipline was established for measuring surfaces over thousands of years, with an extensive array of techniques to characterise features on a surface, albeit in the macro-environment.

A singular motif consists of that portion of a profile between two peaks having two adjacent peaks together with a dominant valley between them being used to characterise individual motifs; namely, these stationary points (critical points) are employed to characterise the individual motifs. The initial motifs can be defined between neighbouring peaks on the profile and rules are then applied to combine adjacent motifs, in pairs, to form even larger new motifs. The concept here is to eliminate any "insignificant peaks" while retaining "significant peaks". By this method the combined motifs enable a series of parameters to be calculated which characterise the surface texture. In order to achieve satisfactory areal motif analysis by this technique, two basic concepts are needed:

- a definition of the areal motif itself – the areal equivalent of the profile motif;
- defining areal combination rules.

(a) Contour map illustrating critical points and lines, with peaks "P", pits "V" and saddle points "S".

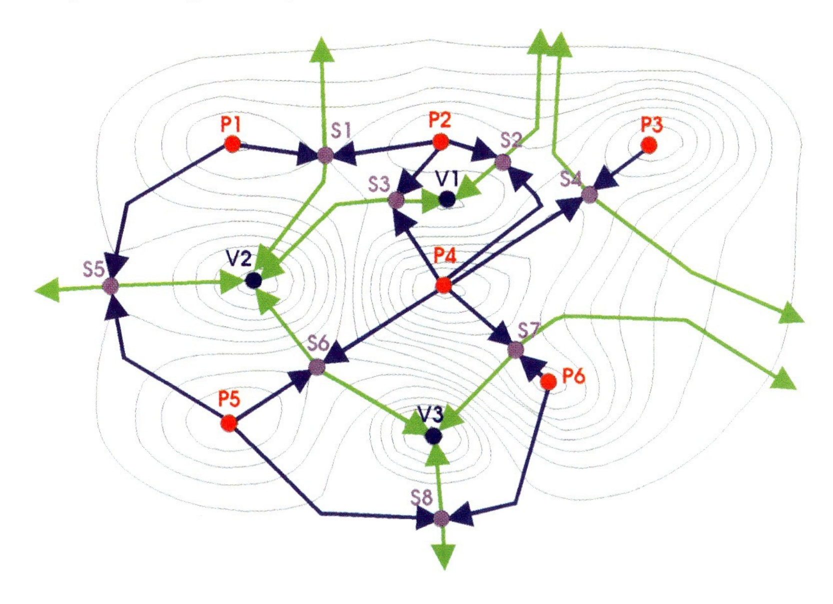

(b) "Change tree" chart derived from the surface contour lines.

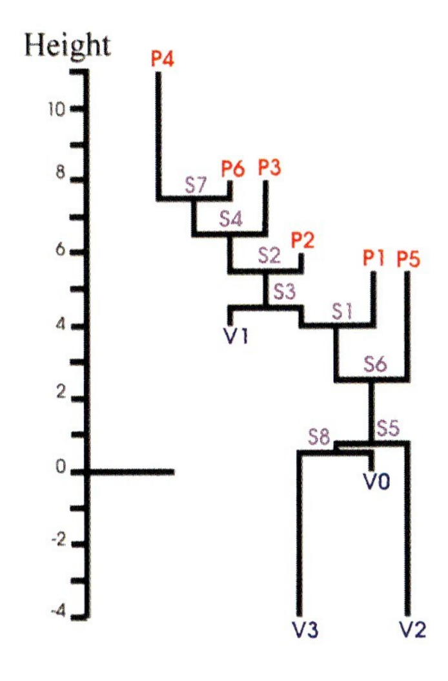

Figure 75. Topological surface texture characterisation by fractal analysis. (Courtesy of Taylor Hobson.)

Cartographic characterisation

Many years ago, Maxwell (1870) proposed dividing a landscape to be assessed into those regions that consisted of hills and those that consisted of dales. In principle, a Maxwellian hill is an area by which maximum uphill paths lead to a particular peak; conversely, a Maxwellian dale is an area from which maximum downhill paths lead to a specific pit. By definition, the boundaries between hills are termed course lines (watercourses) and those boundaries between dales are ridge lines (watersheds). Maxwell was able to demonstrate that ridge and course lines are respectively maximum uphill and downhill paths emanating from saddle points, which either terminate in peaks or pits (see Figure 75a).

Using these Maxwellian concepts and rules, Scott (1998) has proposed an extension of Maxwell's definitions to define a dale as consisting of a single dominant pit surrounded by a ring of ridge lines connecting both peaks and saddle points; a hill, he argued, would consist of a single dominant peak which is surrounded by a ring of course lines that connected the pits and saddle points together (Figure 75a). Moreover, within the dale or peak in question, there may be other pits/peaks, but they are irrelevant compared to dominant pit/peak features.

In this example a dale is considered to be the "areal equivalent" of the profile motif, while the hill can be thought of as a useful complementary concept. Similarly as for the profile motif method, several types of surface-specific lines and points can characterise dales and hills, respectively. These features include the critical points (peaks, pits and saddle points) and critical lines (ridge and course lines; see Figure 75a).

Change tree charts

The topographical "change tree" chart (Figure 75b) is ideal for organising the connectability relationships between critical points in hills and dales, while still retaining relevant information. The change tree represents the relationships between contour lines from the surface, with the vertical direction representing height. At a predetermined set height all individual contour lines can be represented by a point which is part of a line representing that contour line continuously varying with height. Saddle points occur by the merging of two or more of these lines into one. Peaks and pits are indicated by the termination of a line.

For example, consider a dale that gradually fills with water. The point where the water eventually flows over the edge of the dale is a saddle point. The pit in the dale is connected to this saddle point, as indicated in the change tree. By continuing to fill the new lake, the next important occurrence where water flows out of the lake is again another saddle point. As before, the line on the change tree represents the contour of the lake on the shore line, being connected to this saddle point in the change tree. This process can be continued, establishing connections between the pits and saddle points in the change tree. If the landscape is inverted, so that now the peaks become pits and so on, then a similar process establishes connections between peaks and saddle points and the change tree. In general, at least three types of change tree can be utilised:

- *dale change tree* – which represents the relationships between pits and saddle points;
- *hill change tree* – representing relationships between peaks and saddle points;
- *full change tree* – representing relationships between critical points in hills and dales (Figure 75b), from which dale and hill change trees can be calculated.

In all of the previous discussion on these change trees it was assumed that the landscape did not have edges, which unfortunately in practice is not the case. Four types of edge critical points can occur: these are edge peaks, edge pits, saddle peaks and saddle pits.

Areal combination

Due to noise and other factors, in practice change trees tend to be dominated by quite short contour lines that impede interpretation. Therefore in order to reduce factors such as noise, a mechanism is needed to "prune" the change tree while still retaining relevant information. By utilising the technique of areal combination, this simplifies the change tree and any relevant information is still kept. For profile motifs, they can combine with adjacent features, while in the areal case the problem is somewhat more complicated. Eleven different types of possible combination on a change tree exist for areal combinations (see Table 8); Scott (1998) gives a full descriptive account of such combinations.

The logical steps developed below highlight how the areal combination algorithm has been developed for a full change tree. This algorithm can be modified for hill and dale combinations, although for dale combinations (see Table 8) six occur, but in practical terms types 3 and 11 are not usually applied, leaving only four basic types of dale combinations.

The following rules apply when attempting to develop an outline for the areal combination algorithm for a full change tree, but they are not a comprehensive account of all the potential hill and dale combinations, as they can be modified, as experience dictates:

Table 8. The application of areal combination types

Type	1	2	3	4	5	6	7	8	9	10	11
Dale combination	✘	✔	✔	✔	✘	✘	✔	✘	✘	✔	✔
Hill combination	✔	✘	✔	✘	✔	✔	✘	✘	✔	✘	✔
Full change tree	✔	✔	✔	✔	✔	✔	✔	✔	✔	✔	✔

After Scott (1998)

Step 1 – find all Maxwellian hills and dales, then generate the full change tree.
Step 2 – classify all peaks, pits, edge peaks and edge pits which are significant, or not, according to the function of the surface.
Step 3 – combine non-significant peaks and pits using combination types 1, 2, 3, 6 and 7.
Step 4 – combine non-significant edge peaks and edge pits that do not involve a fourth-order saddle, using combination types 4, 5, 9, 10 and 11.
Step 5 – combine possible type 8 combinations, if the fourth-order saddle has at least one of the following: a non-significant edge peak and edge pit.
Step 6 – if no further combinations occur, cease, or otherwise go back to *Step 4*.

The resulting change tree indicates the significant peaks, pits, edge peaks and edge pits, together with the relationships between them. This "pruned" change tree has reduced noise, yet it still retains relevant topographical information concerning the three-dimensional surface.

2.5 Textured metal sheets

Textured sheet metal is produced in rolling mills (using rolls that have previously been suitably textured) which enable a textured surface to be reproduced on the sheet as it passes through the roll stand. The techniques that have been used to texture rolls with a three-dimensional deterministic structure are laser beam and electron beam texturing of the roll's peripheral surface. As a result of the deterministic nature of the resulting topographies, they have completely differing features, requiring appropriate techniques to characterise them. Such suitable parameters can be derived from a so-called "mechanical–rheological model" (see Figure 76). The basic parameters of this method are:

- the material area ratio, α_{ma};
- the open void area ratio, α_{op};
- the closed void area ratio, α_{cl}.

Other methods can be employed to texture the rolls and hence the sheet, but here the discussion will centre on the nature of deterministic surface sheet structures. One of the functions of a generated sheet surface is to influence the tribological properties in the forming process, enabling adherence and retention of surface conditions such as its painted appearance. Typically, the stochastic surface structures resulting from processing by shot-blast texturing do not always meet the demands of production for the final sheet product. Both laser and electron beam texturing provide surface topographies of a deterministic nature, which are fundamentally different from shot-blast rolling production processes. Therefore, the more commonly described two-dimensional parameter characterisations will not be pertinent here for textured sheet surface assessment. Adequate characterisation for deterministically textured metal sheets necessitates suitable 3-D parameters (mentioned above) to describe features of the sheet topography.

In this model, in the case of mixed lubrication the load on the surface can be transmitted by three distinct bearing ratios, these being the solid contact and lubricant contact area, both the static and dynamic lubricant of the pockets. In such situations, the surface ratio of the solid contact relates to the relative quantity of the real contact area, this being equivalent to the asperity area of contact. Dynamic lubricant pockets are the zones of lubrication having a connection to the boundary of the loaded area. Hence in the process of sheet forming this lubricant is squeezed out. Under such conditions the transmitted load is the result of hydrodynamic pressure, which in turn depends upon the rheological parameters of either the lubricant or the surface's structure. By way of contrast, the pockets of static lubrication have no connection to the boundary of the loaded area, thereby trapping the lubricant, resulting in hydrostatic build-up of pressure. For actual sheet surfaces, these pockets of static lubricant will only transpire when a flattened surface topography condition occurs. Normal stresses in the sheet can be transmitted by the three contact mechanisms, while shear stresses may only result from solid contact. From the ratio of solid contact associated with dynamic and static pockets of lubrication, it is apparent that the bearing ratios play a significant part in influencing tribological properties in forming processes. Thus, from these bearing ratios of the mechanical–rheological model, the three-dimensional parameters are derived.

From Figure 76, the relative portion of the solid contact represents the material area ratio. In brief, the three-dimensional parameters for either the dynamic or static lubricant pockets are the open void and closed void area ratios, respectively.

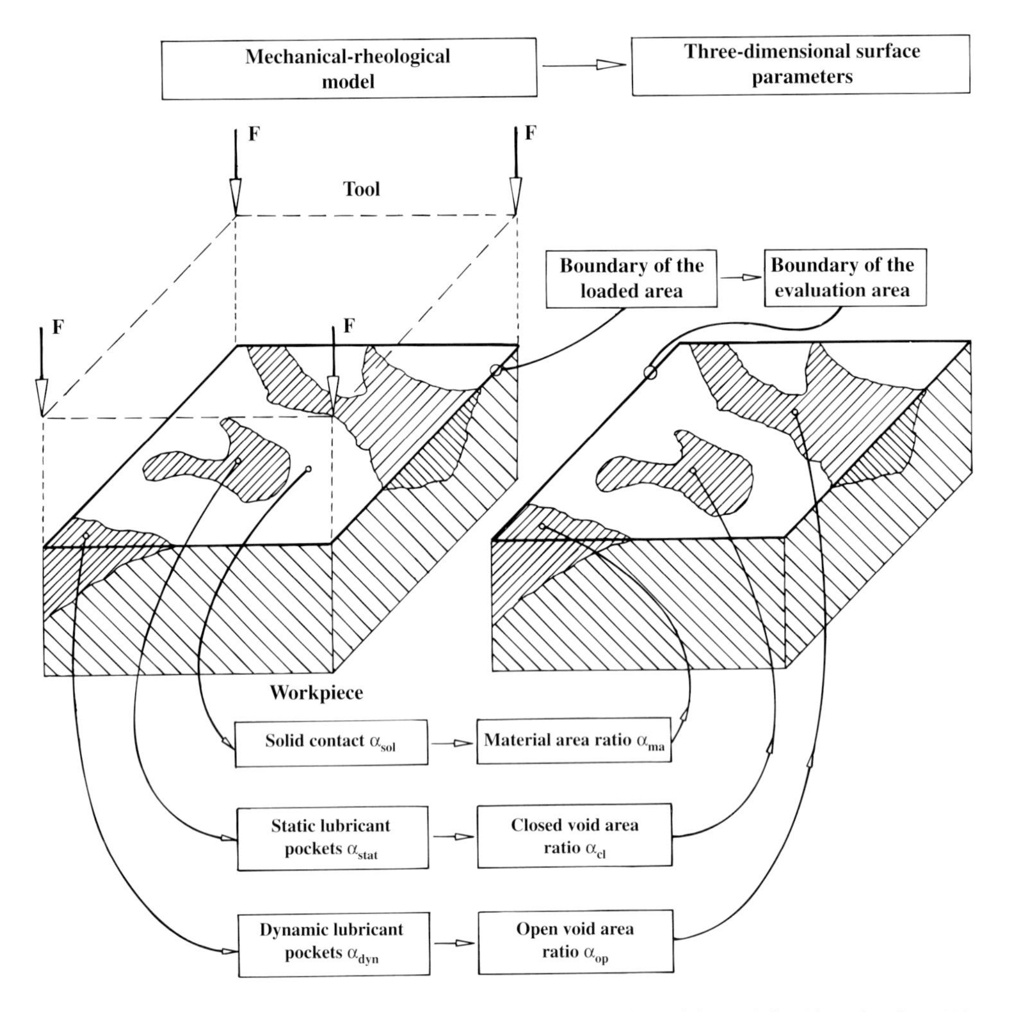

Figure 76. Definition of three-dimensional surface parameters – textured metal sheets. (After Pfestorf et al., 1998.)

2.6 Surface topography characterisation by neural networks

Introduction

Prior to mentioning in detail the subject of neural network techniques, these have as yet to be truly utilised in an industrial context, particularly from a decision-making outcome of the anticipated surface characterisation. However, such techniques have been included, as they may see further development in the future, as some of the research literature is currently indicating.

Artificial neural networks (ANNs) mimic the networking of the functional behaviour of biological neurons, and can be used to integrate and fuse information data streams from a multiple-sensor source. The advantages of ANNs are that they can be employed to integrate and fuse data, then adapt to instructed environments, coupled to their robustness to noise, fault tolerance, simultaneous processing and feasibility of on-line realisation, via hardware implementation. These inherent capabilities of ANNs are composed of many simple processing nodes that operate simultaneously. The functional behaviour of the overall system is primarily determined by the pattern of connectivity of the nodes. The ANN algorithms can be many and varied, with a typical simple architecture shown in Figure 77, based on a "feed-forward multi-layered perceptron".

Operationally, this ANN has input layer neurons that send out signals across the hidden layer(s) in the network to the output layer, with the desired and actual outputs being compared to evaluate the system error – usually Euclidean error. This system

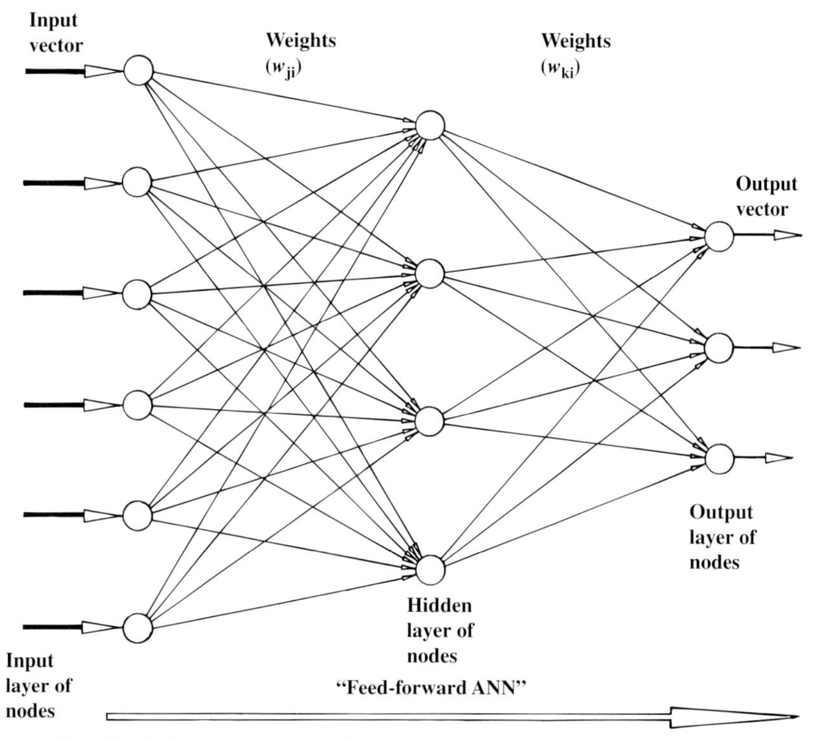

Figure 77. Schematic diagram of a feed-forward "artificial neural network" (ANN) architecture.

error is applied to adjust the node strengths of connectivity amongst the network's neurons, via a feed-forward propagation in the case of Figure 77. Hidden layer nodal capabilities are developed during network training, so that the extracted features are optimised for the classification task. Within the ANN's algorithm, it can be configured around the number of input-to-hidden-to-output nodes required for a given application, which may necessitate some initial trials in order to optimise the correct architecture. The greater the layers and nodal structure the better the discrimination, but at the expense of longer "training" of the ANN. Hence a "trade-off" occurs between a highly accurate and robust system, termed "dimensionality", to one of slightly less discriminatory ability and with faster training.

Many ANNs can perform computational tasks by learning from examples, then generalising them to new conditions. Practically, these examples of training – sometimes referred to as training data, patterns or vectors – are normally observations associated with modelling the physical process. ANNs can still extract the underlying structure from training data that might be corrupted by noise or are fuzzy in nature, providing a virtual representation of the modelled physical process. As a result of this compliance, surface topography data fits well to analysis by the ANN approach. The learning algorithm constrains the arrangement of pattern connectivities, although the concept of these learning algorithms can be adapted to any architecture.

From the training data, the fundamental ANN learning problem can be resolved in the following manner, by determining a set of network weights that can operate in the required manner. This assignment of weights can be achieved by initially assigning minute values of randomness to these weights, then progressively adjusting them according to the following formula:

Adjustment of modal weights: $$w_{jk}^{\text{new}} = w_{jk}^{\text{old}} + \Delta w_{jk}$$

where w_{jk} is the weight from unit k to unit j; and Δw_{jk} can be calculated in a manner of ways, according to different algorithms.

Two major paradigms of learning can be employed:

- *supervised learning* – where a "teacher" is necessary to facilitate the network's response during training;
- *unsupervised learning* – where no "teacher" is present and the network must discover for itself the underlying structure while training.

Surface topography utilising ANNs

In any application of 3-D surface topography assessment utilising ANNs, reduction of dimensionality is vital, as training becomes lengthy and tedious, being exacerbated by considerable noise. Dimensionality reduction can considerably restrict the number of degrees of freedom of the ANN, typically from 16,384 (128^2 data points), down to 1024 (32^2 data points); the latter might be employed to suitably train the network from a moderate amount of data samples.

For example, to perform an intra-surface characterisation – to discriminate between worn and unworn surfaces for a particular manufacturing process – it is desirable to identify surface topography features, whose presence/absence might uniquely characterise the surface. When attempting to discriminate between, say, the worn and unworn surface conditions that might be apparent from an isotropically manufactured surface, such as honed bores in automobile engines, then a demarcation between them by some geometric or statistical parameter provides a degree of objectivity in the system. An obvious choice for this plateau-honed surface in the selection of three-dimensional parameters would be one to measure amplitude, such as that of dispersion *Sq* ("root mean square deviation" – relating to surface departures within the sampling area). Shot-blasting is another production process that can benefit from utilising the *Sq* parameter. Conversely, for the more anisotropic a surface topography, such as that resulting from face turning, the greater would probably be the discrimination by using another 3-D parameter, such as the spacing parameter *Sds* ("density of peaks" in the unit sampling area). The major strength of the ANN approach to 3-D surface topography assessment lies in the fact that, once trained, little human intervention is required – eliminating the tedium associated with the conventional approach – and consistency occurs, provided the basic data integrity rules have not been corrupted.

Systems can be built that are independent of the methodology of data capture and a practical system might link the ANN to an automated data capture station – such as a vision system – to investigate, for example, on-line the effects of process variables for a turned surface finish. Such a system would be able to highlight any changes within the production process, e.g. the tool's flank wear, which might unduly affect the surface finish. The complete system could then be incorporated into a feedback loop to modify some or all of the cutting data, such as the feedrate, cutting speed, depth of cut or coolant, as necessary to maintain consistent turned products. This artificial intelligence (AI) technique can be considered as a form of adaptive control optimisation (ACO), which has been the goal for many years in unmanned manufacturing environments.

2.7 Non-contact measurement

Non-contact measurement provides a means of assessing not only rigid components but, more importantly, fragile and delicate parts which would either distort under the measuring pressure – causing induced measurement errors – or, more significantly, degradation to a delicate and unsubstantial surface under test. In fact, for the latter case, it is the only real means available to inspect soft and sometimes pliable part surfaces with any degree of confidence in a speedy and efficient manner.

Many of the optical systems previously mentioned in Chapter 1, can operate in 3D, especially the interferometers, confocal microscopes and triagulation systems.

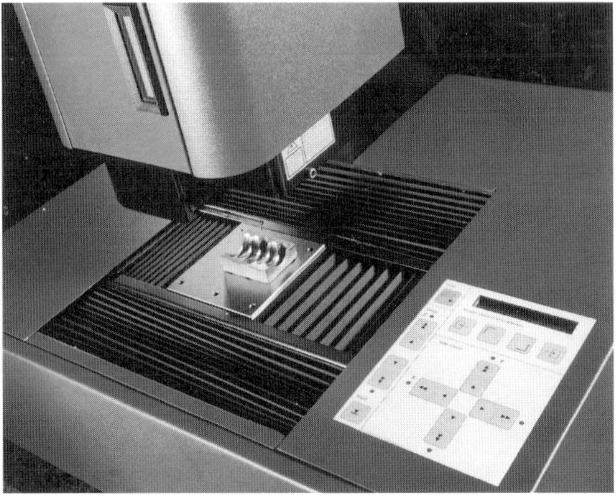

Figure 78. Multi-sensing scanning instrument for contact/non-contact 3-D measurement. (Courtesy of Taylor Hobson.)

Table 9. Technical data for 3-D multi-scanning instrument

Slides			
Traverse length	97 mm × 97 mm (*X*, *Y*)		
Motorised column	83mm (*Z*)		
Straightness:			
Profile (measured in *X* direction)	**10 mm traverse**	**50 mm traverse**	**95 mm traverse**
Peak-to-valley	0.25 μm	0.6 μm	0.9 μm
PHT_p (0.1 to 99.9%)			
Surface (measured in *X*-*Y* direction)	**10mm²**	**50mm²**	**95mm²**
Peak-to-valley	0.7 μm	1.6 μm	3.5 μm
SHT_p (0.1 to 99.9%)			
Minimum spacing			*X* = 0.5 μm; *Y* = 5 μm
Maximum scanning speed			10500 μm/s
Maximum sample load			5 kg
	Inductive gauge	**Laser gauge**	
Vertical resolution	0.06 μm (60 nm)	1 μm	
Range	2.5mm	Vertical accuracy	1 μm (on diffusing surfaces homogeneous optical properties)
Principle	FTSS inductive probe		
	Range	**10 mm**	
	Principle illumination	Laser beam triangulation	
	Maximum angle of measurement surface	90°	

Courtesy of Taylor Hobson.

Non-contact three-dimensional surface texture measurement offers considerable flexibility for a wide range of component assessment. With the instrument illustrated in Figure 78, scanned surface areas are quite large, typically 10 mm², which can be completed by a laser probe in approximately 5 minutes. This particular instrument type is very versatile, as it can be used with either an inductive laser or dual-gauge probe. The non-contact probe utilises laser triangulation technology, so that speed and large surface areas can be mapped, whereas the contact probe is of the inductive gauge variety that has been extended to a range of 2.5 mm. Switching from a non-contact to a contact operation can be completed in 5 s. Both single and multiple areas on any component shape are easily selected, ranging from a minimum size of 50 μm² up to a maximum of 97 mm².

The instrument has Windows™ "Talymap 3-D" analysis software, offering a range of functions, including view rotation, axis directions, automatic levelling, resampling, defect removal, multiple profiles and cylindrical/spherical/polynomial form removal, together with multiple output formats and custom colour palettes.

Technically the instrument, as illustrated in Table 9, has a comprehensive and versatile specification, which can be employed in a host of industrial applications, ranging from soft material analysis – cosmetics, biomedical and paper technologies – through to precision engineering, advanced materials, semiconductor and automotive/aerospace technologies. However, they do have some limitations when compared to two-dimensional surface capture techniques – see Chapter 1.

Holographic interferometry

Holographic interferometry utilising a non-contact 3-D measuring instrument is shown in Figure 79, the instrument being connected to a PC. This particular instrument operates solely in a non-contact mode, with the capability of accurately measuring component features over an area of 300 mm². The system has been designed with no moving parts, ensuring reliable and repeatable measurements. The measurement is undertaken by simply positioning the component to be inspected in the cabinet, then flood

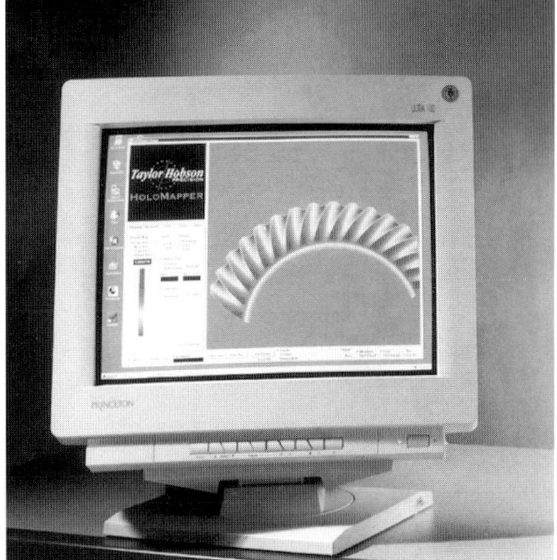

Figure 79. Non-contact holographic imaging instrument for 3-D measurement. (Courtesy of Taylor Hobson.)

illumination is applied across the complete field of view, prior to extensive data being obtained during inspection. This measuring cycle is achieved in approximately 2 minutes, with a Z-resolution of 0.1 μm; the system can also cater for batches of parts to further increase throughput. The instrument is equipped with three field-of-view options – 50 mm^2, 175 mm^2 and 300 mm^2 – with each capable of achieving >1,000,000 data points. The measurement is achieved by a 3-D holographic interferometer, utilising a diode laser, with precision optics and a diamond-turned parabolic mirror, coupled to a high-resolution CCD camera having proprietary data-processing software algorithms. The depth of focus is 100 mm (fixed window), with a Z-axis display resolution of 100 nm. All the advantages relating to precision measurement of hard/solid parts are still valid for non-contact inspection; in addition, the enhanced ability to inspect delicate/pliable parts, as previously mentioned, is relevant with this technique.

The software features are enhanced for both visual and surface measurement output, providing a comprehensive 3-D reporting procedure, and this allows the following advantages:

- measurement of complex and discontinuous surfaces for flatness;
- automatic levelling of the workpiece measurement planes;
- incorporation of advanced functionality, the system being simple to operate.

Holographic interferometry is at the pinnacle of development for three-dimensional capabilities of today's instrumentation, offering new means of visually representing 3-D surfaces, giving unique insights into surface inspection assessment.

References

Journal and conference papers

Ahmose. *The Rhind Mathematical Papyrus*. Year 33 of the Reign of King Auserre (Apophis): Fifteenth Dynasty, Second Intermediate Period (Ancient Egypt), mid-1600s BC.

Bay, N. and Wanheim, T. Contact phenomena under bulk plastic deformation conditions. *Advanced Technology of Plasticity* 4, 1677–1691.

Blackmore, D. and Zhou, G. A new fractal model for anisotropic surfaces. *International Journal of Machine Tools and Manufacture* 38(5–6), 1998, 551–557.

Burrows, J.M. and Griffiths, B.J. A vector modelling technique for the representation of 3-dimensional surface topography. International Journal of Machine Tools and Manufacture 38(5–6), 1998, 537–542.

De Chiffre, L., Christiansen, S. and Skade, S. Advantages and industrial application of three-dimensional surface roughness analysis. *Annals of the CIRP* 43(1), 1994, 473–478.

Dong, W.P., Sullivan, P.J. and Stout, K.J. A comprehensive study of parameters for characterising three-dimensional surface topography. III: Parameters for characterising amplitude and some functional properties. *Wear* 178, 1994, 29–43.

E000.14.015.N. *Etats Geometriques de Surface Calcul des Criteres de Profil*, August 1983. CNOMO.

Griffiths, B.J. Manufacturing surface design and monitoring for performance. *Surface Topography* 1(1), 1988, 61–69.

Hall, C. and Griffiths, B.J. A comparison of two and three-dimensional stylus surface measurement. *Quality Today* July 1999, S30–S34.

Haesing, J. Nestimmung der Glaettungstiefe rauher Flaechen. *PTB-Mitteilingen*, 4, 1964, 339–340.

ISO 12 085. *Geometrical Product Specifications – Surface Texture – Profile Method – Motif Method for Measurement of Surface Roughness and Waviness.*

Kiran, M.B., Ramamoorthy, B. and Radhakrishnan, V. Evaluation of surface roughness by vision system. *International Journal of Machine Tools and Manufacture* 38(5–6), 1998, 685–690.

Kweon, I.S. and Kanade, T. Extracting topographic terrain features from elevation maps. *CVGIP: Image Understanding* 59(2), 1994, 171–182.

Li, C., Dong, S. and Zhang, G. Evaluation of the root-mean-square of 3D surface topography. *International Journal of Machine Tools and Manufacture* 40, 2000, 445–454.

Littlefair, G. et al. On-line tool condition monitoring using artificial neural networks. *Insight* 38(5), May 1996, 351–354.

Littlefair, G., Javed, M.A. and Smith, G.T. Fusion of integrated multi-sensor data for tool wear. *Proc of IEEE Conference*, Perth, Australia, November 1995.

Long, F., Fractals, engineering surfaces and tribology. *Wear* 136, 1990, 141–156.

Lopez, J. Hansali, G., Le Bosse, J. and Mathia, T. Caractérisation fractale de la rugosité tridimensionelle d'une surface. *Journal de Physique III* 4, 1994, 2501–2519.

Mainsah, E. and Ndumu, D.T. Neural network applications in surface topography. *International Journal of Machine Tools and Manufacture* 38(5–6), 1998, 591–598.

Mainsah, E., Sullivan, P.J. and Stout, K.J. Effects of quantisation on 3-D characters. *Measurement Science and Technology* 5(2), 1994, 172–181.

Majumber, A. and Tien, C. Fractal characterization and simulation of rough surfaces. *Wear* 136, 1990, 313–327.

Mandelbrot, B. Self-affine fractals and fractal dimension. *Physica Scripta* 32, 1985, 257–260.

Maxwell, J.C. On hills and dales. *London, Edinburgh and Dublin Philosophical Magazine and Journal of Science* 40(4), 1870, 421–425.

Morrison, E. The development of a prototype high-speed stylus profilometer and its application to rapid 3d surface measurement. *Nanotechnology* 7, 1996, 37–42.

Pfestorf, M. Engel U. and Geiger, M. 3D-surface parameters and their application on deterministic textured metal sheets. *International Journal of Machine Tools Manufacture* 38(5–6), 1998, 607–614.

Radhakishnan, V. and von Weingraber, H. Die Analyse Digitalisierter Oberflaechenprofil nach dem E-system. *Fachberichte f(r Oberflaechentechnik* (11/12) 1969, 215–223.

Russ, J.R. Fractal dimension measurement of engineering surfaces. *International Journal of Machine Tools and Manufacture* 38(5–6), 1998, 567–571.

Sacerdotti, F., Griffiths, B.J., Butler, C. and Benati, F. Surface topography in autobody manufacture: the state of the art. To be published.

Scott, P.J. Foundations of topological characterization of surface texture. *International Journal of Machine Tools and Manufacture* 38(5–6), 1998, 559–566.

Sherrington, I. and Howarth, G.W. Approximate numerical models of 3-D surface topography generated using sparse frequency domain descriptors. *International Journal of Machine Tools and Manufacture* 38(5–6), 1998, 599–606.

Smith, G.T. Developments in 3D characterisation of surfaces, *Metalworking Production* August 1998, 59–62.

Sobis, T., Engel, U. and Geiger, A. A theoretical study of wear simulation in metal forming processes. *Journal of Materials Processing Technology* 34, 1994, 233–240.

Tholath, J. and Radhakrishnan, V. Three-dimensional filtering of engineering surfaces using envelope system. *Precision Engineering* 23, 1999, 221–228.

Whitehouse, D.J., Parameter rash: is there a cure? *Wear* 83, 1982, 75–78.

Williams, J. Surface data puts users in the picture. *Machinery and Production Engineering* December 1989, 1–15.

Books, booklets and guides

Bishop, C.M. *Neural Networks for Pattern Recognition*. Clarendon Press, Oxford, 1995.

Blunt, L. et al. *Development of Methods for Characterisation of Roughness in Three Dimensions*. Penton Press, 2000.

Blunt, L. and Stout, K. *Three Dimensional Surface Roughness.* Hermes Penton Science, 2002.

Stout, L. and Blunt, L. *Surface topography*. Hermes Penton Science, 2002.

Falconer, K. *Techniques in Fractal Geometry*. Wiley, New York, 1996.

Hertz, J., Krogh, A and Palmer, R. *An Introduction to the Theory of Neural Computation*. Addison-Wesley, Reading, MA, 1991.

ISO Technical Report 16610 Parts 5 and 6. *Geometrical Product Specifications (GPS): Data Extraction Techniques by Sampling, Part 5: Concepts of Envelope Filters, Part 6: Envelope Morphological Techniques ISO.*

Mehaute, A. le. *Fractal Geometries: Theory and Applications.* Penton Press, 1991.

Robertson, W.S. *Lubrication in Practice.* Macmillan, New York, 1987.

Russ, J.C. *Fractal Surfaces*. Plenum Press, New York, 1994.

Stout, K.J. et al. *The Developments of Methods for the Characterisation of Roughness in Three Dimensions.* Pub. no. EUR 15178EN, Commission of the European Communities, 1993.

Stout, K. and Blunt, L. *Surface Topography*. Hermes Penton Science, 2002.

Thomas, T.R. *Rough Surfaces* (2nd edn). Imperial College Press, 1999.

Vicsek, T. *Fractal Growth Phenomena*. World Scientific, Singapore, 1992.

Whitehouse, D.J. *Handbook of Surface Metrology.* Institute of Physics, Bristol, 1994.

Surface microscopy

"With affection beaming out of one eye, and calculation shining out of the other."

(*Martin Chuzzlewit*, Charles Dickens, 1812–1870)

Summary of surface microscopy

What was said about the advantages to be gained in visual recognition of surfaces for three as opposed to two dimensions is equally the case for surface microscopy. Here, the optical definition is significantly greater and so will be the depth of field in focus, particularly for imaging with scanning electron microscopes. Measurements taken on the surface and particularly those for two-dimensional surface topographical characterisation are quantitative in nature. In the case of three-dimensional imaging, if this is compared to the surface geometry information obtained from two-dimensional surfaces, then the surface characterisation data in the main will tend to be possibly more qualitative. Therefore, any attempt at comparisons between the different systems becomes somewhat superfluous and, at best, would be misleading. This takes nothing away from the fact that surface microscopy has superb three-dimensional surface visualisation and as an analytical tool offers considerable and valid topographical details that can be extracted from the surface or subsurface, with certain configurations and types of equipment.

Unlike the previous surface characterisation instruments described in Chapters 1 and 2, much of the equipment to be discussed in this present chapter can interrogate the surface for a wide range of elemental and topographical surface detail analysis. Moreover, in many instances, the specimen under test can be manipulated about a series of rotational axes, to obtain the best possible angle for surface assessment. Magnifications can be increased or decreased quite readily, enabling minute surface features to be visualised, then quantified by an incredible array of analytical techniques. This enables the operator to perform sophisticated analysis in situ, without having to change operational set-ups to any significant degree.

This chapter cannot attempt to cover in anything but superficial detail the information appertaining to a component's surface, nor the complexity and range of equipment currently available to the user. Each surface microscopy instrument covered here in this narrow selection could easily be expanded to fill a complete book in their own right. However, it was thought reasonable to include a brief resumé of some of the current equipment available and to attempt to show where, when and how such instruments complement our understanding of surfaces, which would otherwise be lacking if only two- and three-dimensional surface characterisation had been undertaken.

3.1 Introduction

Surface microscopy covers a wide range of metallographical/morphological and instrumental techniques used either directly, in the case of an optical microscope, and indirectly, via scanning electron, transmission electron and atomic force microscopes as a means of visually depicting surfaces. These and other techniques for viewing the surface topography are complemented by spectrographical analysis instrumentation, typically Auger electron and X-ray photoelectron microscopic methods. This chapter will be principally concerned with electron microscopic techniques, most notably:

- scanning electron microscopes (SEM);
- transmission electron microscopes (TEM);
- atomic force microscopes (AFM);
- X-ray photoelectron spectroscopy (XPS).

In the past and indeed currently, many inspection, quality assurance and research and development facilities employ conventional optical microscopy as the main means of surface assessment. However, such instruments are not without their limitations, with the best optics giving magnifications of around ×2000, which can be somewhat limiting for minute surface detail inspection, or when a larger "depth of field" (i.e., the vertical height which remains in focus) is necessary. Even at ×10 magnification, the depth of field is quite small, requiring sample preparation as thin sections, "tapered sections", or polished and etched surfaces to yield meaningful information at the instrument's higher magnifications. Yet another frustrating occurrence is that although it is possible to see what might be causing the problem, it is not possible to determine by visual investigation what it is, therefore no explanation or estimate for a cure to the problem can be given and it remains one of pure speculation. By utilising electron microscopic techniques, these instruments overcome the drawbacks found with purely optical microscopes, as they can exploit spectrometric techniques, among others, for detailed analysis of surface chemistry.

NB Tapered sections at around 11° increase the apparent field of view, with minimal feature distortion. Etched surfaces are normally necessary on metallic surfaces to preferentially etch the surface and to reveal otherwise hidden surface detail.

In Figure 80(a) can be seen the application areas for a range of complementary electron microscopic instrumentation, indicating their vertical and lateral resolutions. Such equipment can be exploited at a range from the picometre (i.e., 10^{-12} m) area for an AFM, to an area well over 1 mm^2 using an SEM. In

(a) Range of lateral and vertical resolutions of instruments

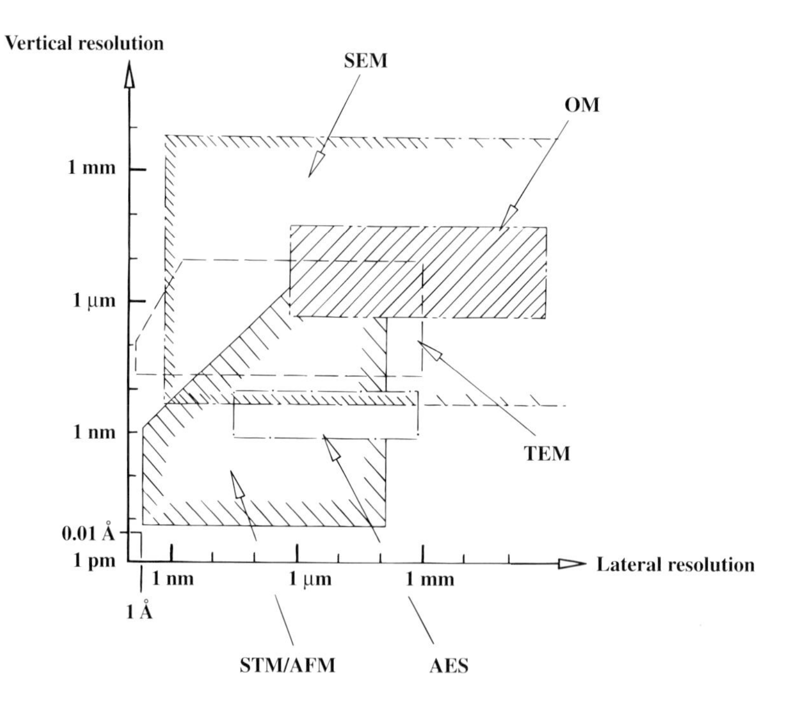

(b) Resolution, excitation method and observations of instruments

STM/AFM : atomic force microscope;
SEM : scanning electron microscope;
TEM : transmission electron microscope;
AES : Auger electron spectroscopy;
XPS : X-ray photoelectron spectroscopy
OM : Optical microscope

Analytical instrument	Resolution		Excitation method	Object of observation	Observation environment
	Vertical	Horizontal			
STM/AFM	0.001 nm	0.1 nm		Tunnelling electrons/atomic force	Atmosphere, gas, vacuum, liquid
SEM	8 nm	5 nm	Electrons	Secondary electrons	vacuum
TEM	0.08 nm		Electrons	Transmitted electrons	vacuum
AES	3 nm	10 nm	Electrons	Auger electrons	Ultrahigh vacuum
XPS	2 nm	100 m	Characteristic X-rays	Photoelectrons	Ultrahigh vacuum

Figure 80. Range and resolution of a selection of surface microscopy instruments. (Courtesy of Jeol (UK) Ltd.)

Figure 80(b) a tabulated chart of the resolutions, excitations and observations available from this group of analytical instruments is shown.

3.2 Scanning electron microscope

SEMs can overcome the limitations of optical microscopy by offering depths of field of several millimetres at low magnifications (×20–200) and their resolution is similar up to ×2000, whereupon the SEM significantly extends its range, at ×100,000, or greater, over conventional optical techniques. Instruments of the SEM variety are not a new development: were first designed in Britain about 50 years ago, but then they were big, complicated and somewhat temperamental instruments requiring great skill, patience (and some luck!) to produce generally mediocre results. The modern SEM is a totally different instrument, controlled by PC technology with Windows™-based software, having a minimum of knobs and buttons to manipulate the specimen and necessitating minimal training for an operator to become proficient. Instrument size has been significantly reduced by utilising PC-based technology and the complete equipment is not much larger than an office desk in terms of its working area (see Figure 81).

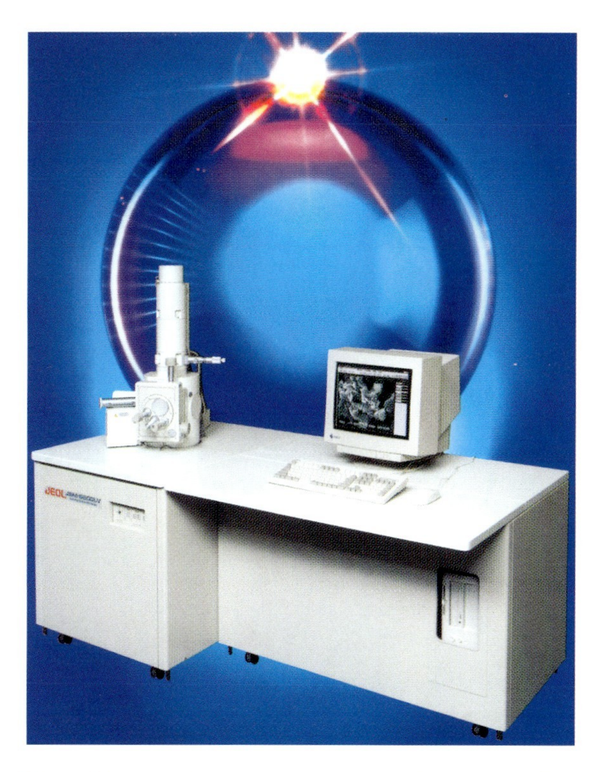

Figure 81. A typical "basic" scanning electron microscope, based on proven PC technology and Windows™-based software, which can be equipped with a range of analytical tools for diagnostic investigation. [Courtesy of Jeol (UK) Ltd.]

Many of the latest instruments available can examine objects up to 200 mm in diameter, weighing up to 3 kg, while still retaining micrometre precision movement of the sample (10^{-6} m positional resolution). The SEM process of examination is non-destructive in nature, thereby retaining the value of an expensive sample being investigated. As a quality control tool the SEM can quickly examine either random or sequentially produced samples, which if necessary can then be reintroduced into the production line, if they prove to be acceptable. After many years of development, the SEM has become a "workplace-hardened instrument" that can be utilised in reasonably controlled production environments, necessitating minimal component preparation without damaging the parts in any way.

Basic SEM operation

The principle of operation of the SEM is quite simple. At the top of the column (see Figure 82) an electron gun is situated, consisting of a tungsten filament in a strong electrical field. As a result of the electrical field the electron gun emits electrons (negatively charged atomic particles), which are then accelerated to high speeds. These high-speed electrons travel down the column, being influenced by lenses lower in the column which squeeze them together, forming an electron beam of very small diameter. This electron beam is then focussed so that it collides with the specimen in the microscope specimen chamber as a diminutive spot. This minute spot is scanned from left to right and up and down over the surface, in an identical manner to that of a normal TV as it scans the screen. By way of an analogy for its visual operation, the SEM can thought of as a TV manipulated so that its screen faces the floor. Under these conditions, an SEM "tube" is redefined and can be thought of as a column, whereas the screen becomes the specimen being examined. Although in the case of an SEM a considerably smaller "screen" is scanned with an appropriately smaller spot. In the same manner that a normal TV tube is operated under a vacuum, the SEM column and sample chamber are also evacuated.

Once the electron beam strikes the specimen surface many different processes occur, but for a basic understanding of the mechanism it is only necessary to state that secondary electrons are generated. These secondary electrons can be thought of as electrons on the specimen's surface which will

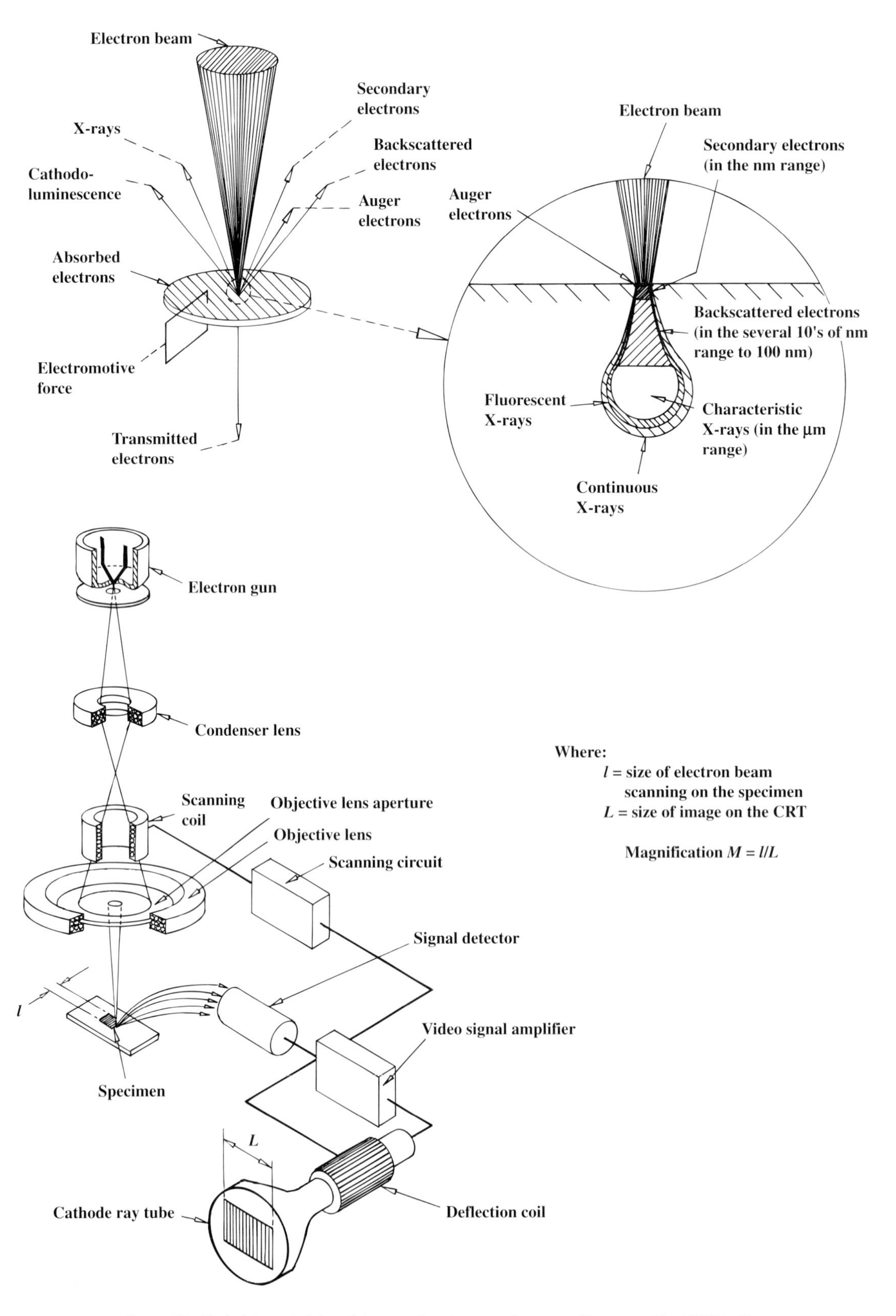

Figure 82. Underlying principles of the scanning electron microscope. [Courtesy of Jeol (UK) Ltd.]

have been displaced by the high-speed electrons that originated from the electron gun. Any secondary electrons that are emitted from the specimen surface as the beam strikes are collected and counted. The total number of electrons counted for a particular point on the specimen is converted to a spot of light on the SEM display screen. The intensity of the spot depends on the total secondary electron count. An identical operation is performed for each point on the specimen being scanned and, an image is built-up which represents the secondary electron yield over the surface. Fortunately, one of the main factors influencing secondary electron yield is the shape of the specimen's surface, hence if the SEM display (viewing screen – CRT) is scanned at the same rate as the electron beam over the surface, an identical image occurs of the specimen's surface. To ensure ease of use, the SEM display and electron beam are synchronised at the TV frame speed, enabling a TV image of the specimen surface to be viewed on the SEM screen.

The spot on the CRT is shown in real time, with the information emitted from the specimen surface being displayed as an image. The magnification of the displayed image is defined as the ratio of the size of the image on the CRT to that of the electron beam scanning on the specimen surface. The type of information obtained can be changed by switching the signal. In this way, the specific desired characteristics of the specimen surface can be viewed on the CRT in a magnified scale.

An SEM can be regarded as a miniature laboratory, with the specimen chamber being used for dynamic experiments such as heating, cooling, tensile, compression and hardness testing, which can be examined in real time and at an appropriate magnification. Not only are secondary electrons collected for imaging, but also many other emissions occur and can be employed to yield additional specimen information. Of these additional emissions, perhaps the most useful one is the generation of characteristic X-rays. When an electron beam strikes a specimen X-rays are emitted (see Figure 82), the energies of which are directly related to the chemical constituents of the specimen at the focal point of the beam. This is the principle of *microprobe analysis*, where a finely focused beam in the SEM is aimed at a particular surface feature that has been identified by normal imaging, the object here being to determine its local chemical composition. Unlike analytical techniques based on "wet chemical" methods, X-ray analysis is very fast, typically 100 s being the standard acquisition time. As opposed to normal optical microscopy where only the image can be seen, using an SEM the chemical composition can be determined. Hence, an SEM enables one to establish where problems occur and the X-ray analysis indicates what causes them; given this information it is possible to deduce why the problem arose. Once this level of understanding of the problem has been gained, preventive measures can be defined and implemented.

Apart from telling the investigator what a specimen's chemical constituent elements are, microprobe analysis can determine respective quantities of each element. Such instruments can provide analytical accuracy to better than several parts per million on a routine basis. X-ray detectors can also be "tuned" to detect only specific substances and then monitor their concentration while the electron beam scans, thereby "mapping" the distribution over the sample surface. Using the SEM in conjunction with X-ray analysis and mapping, provides all the information needed to characterise the problem, thus:

- imaging locates potential problem areas;
- normal X-rays determine what causes them;
- X-ray mapping deduces where else problems occur in the specimen.

Yet another vital facility available to the SEM user is *backscatter electron (BE) detection*. Here, some of the electrons that strike the specimen do not "interact" with it (i.e., producing secondary electrons or X-rays), but simply collide with atoms in the specimen and bounce immediately back – termed *backscatter*. The closer together the atoms are within the specimen, the more potential collisions, and hence the greater the certainty of backscattered electrons. This backscattering of electrons can be detected, giving an *atomic number contrast*. Denser materials exhibit tightly packed atoms, resulting in increased collisions and causing more backscatter than their lighter material counterparts, which have looser packing densities. The resultant image in the CRT will indicate the well-known saying: "Dense is bright and light is dark!" The technique is often utilised to determine irregularities, since they usually appear as bright or dark areas on an otherwise uniform (normal) background. In many circumstances, the backscatter electron detector is utilised in conjunction with X-ray analysis, because the backscattered electron image is normally less complicated than the secondary electron image, and therefore it is simpler to select areas for analysis.

X-ray analysis and backscattered electron detection are essentially the most widely used supplementary techniques, although many other characteristics may be determined from the results of the interaction of an electron beam with the specimen. Other supplementary techniques include crystallographic and magnetic information, surface topography and composition, semi-conductive or electrical characteristics; however, it should be stressed that some of these methods require quite complex analytical stages to obtain valid information.

Principle of magnification

If a specimen surface is scanned with a finely focused electron beam of only a few nanometres (see Figure 82), then information will be emitted from each point of the scanning operation. This emitted information is converted into an electrical signal, amplified, then fed into the CRT for subsequent observation. On the CRT, the information is used to control the brightness of the corresponding spot. The spot on the CRT is shown in real time with the electron beam scanning on the specimen surface. Hence, information emitted from the specimen surface is displayed on the CRT as an image. In Figure 82, the magnification of the displayed image is defined as the ratio of the size of the image on the CRT (L) to that of the size of the electron beam scanning the surface (l). The type of information that could be obtained can be changed by switching the signal; in this manner a specific desired characteristic of the specimen can be visually displayed on the CRT in a magnified scale.

SEM applications

As previously described in this section, various kinds of signals can be obtained by an SEM. These signals carry distinct types of information and, as such, are employed for different purposes, as listed in Table 10.

Topographical observation: secondary electron imaging

This secondary electron image (SEI) technique is probably the most utilised method of general applications for an SEM by industry, particularly so now that image-processing procedures have been developed, to further expand the visual impact to the observer – more will be mentioned about this shortly. The number of secondary electrons emitted from a specimen surface greatly depends on the incident angle of the electron beam to that surface. In other words, the secondary electron signal depends upon undulations of the specimen's surface. Furthermore, since the energy of secondary electrons is very low, they are only emitted from a thin layer on the specimen surface. Thus, secondary electron signals are considered to be the most appropriate signal for observing a specimen's topography.

The following describes just some of the fields of science and engineering that might exploit SEM observation techniques:

- *materials science* – utilised in most advanced materials science fields, typically for superconducting materials, or for advanced composites, etc.;
- *biology* – particularly biotechnology applications or for the examination of entomological taxonomy (insects, bacteria, viruses and animal tissues);
- *electronics* – utilises the power of such instrumentation in the R&D and quality control fields, typified by the semiconductor/electronics applications for failure analysis techniques, etc.;
- *mechanical/industrial engineering* – for investigation of process technologies, machinability research, component failure modes and metallurgical/metallographical investigation.

SEM: images

In Figures 83–86 are depicted a series of photomicrographs produced across a diverse range of metallic – ferrous and non-ferrous – non-metallic and biomechanical structures. It is only not only at the high-resolution range of magnification that

Table 10. Typical scanning electron microscope applications

Signal	Mode of operation	Purpose of SEM – information carried
Secondary electrons	SEI	Topographical observation of surface
Backscattered electrons	BEI	Compositional observation of surface
X-rays	X-ray	Elemental analysis of specimen
Transmitted electrons	TEI	Internal structure observation
Cathodoluminescence	CL	Internal characteristics observation
Electromotive force	EBIC	Internal characteristics observation
Secondary electrons or	ECP	Crystalline structure
Backscattered electrons	MDI	Magnetic domain observation

Courtesy of Jeol (UK) Ltd.

(a) Cleavage planes on natural diamond surface: 15 kV x200.

(b) Paper surface: 15 kV ×100.

(c) Semiconductor memory surface: 15 kV ×1500.

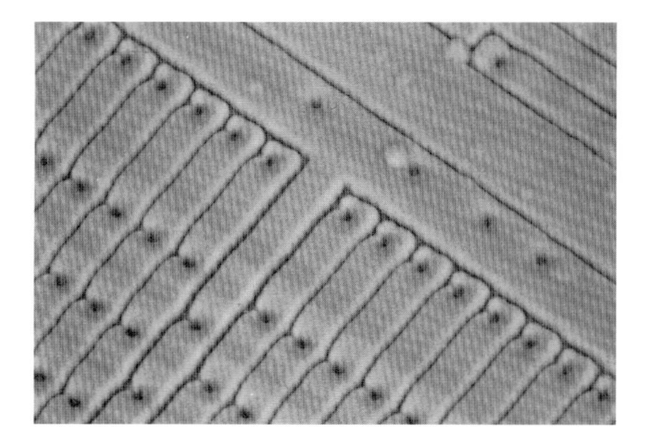

Figure 83. Low-vacuum scanning electron micrographs. (Courtesy of Jeol (UK) Ltd.)

useful analysis can be undertaken, but also at lower magnifications, where these instruments' large depths of field are exploited. In Figure 83(a) can be seen a fractured natural diamond surface, illustrating the cleavage planes which can be preferentially worked to achieve the inclined geometric facets so admired in cosmetic jewellery. Conversely, Figure 83(b) shows a more mundane, but nevertheless important surface – that of paper – which must absorb the ink or colour pigments from printed media. With such surfaces good adherence on the surface is essential; therefore the quality and texture of this paper surface are vital if it is to respond to, say, the ink and not blot/spread, thereby losing line delineation and definition. Figure 83c illustrates a typical semiconductor memory surface, where surface structure and separation of discrete memory elements are vital for efficient operational performance of these IC chips.

The two SEM photomicrographs displayed in Figure 84(a and b) are notable not from the images themselves, which are markedly different in texture and composition, but because of the information provided overlaying the actual images. Of particular relevance is the "error bar" or linear scaled bar which automatically changes its dimensional magnitude as magnifications are either increased or decreased. Such an auto-scaling facility can be employed as a form of "indirect measurement" of notable dimensional/surface features, well beyond the range of most conventional dimensional techniques, certainly up to the nanotechnology level (10^{-9} m) and even approaching that of picotechnological values (10^{-12} m) in size. Each SEM image can be labelled with an identifying number for subsequent later reference and the mode of operation can also be displayed (for example, Figure 84a: SEI = secondary electron image).

In Figure 85 is shown an SEM "soft scanned" image of an "ultra-high-speed" face milled austenitic-grade stainless steel surface machined at 3500 m/min. The SEM processed data in this instance was magnification ×100, accelerated voltage 20 kV, working distance 20 mm, pressure Pa 200 μm, with secondary electron imaging. Of particular note was that the original photomicrograph was in fact a left-hand stereo image of the surface. Both a left- and right-hand stereo image can be superimposed/merged into one image. If the original of such a "merged pair of images" had been viewed through 3-D glasses (green and red – as worn in the early days of 3-D cinematography), then this combined image would take on a three-dimensional aspect. This visual enhancement is often useful when attempting to interpret whether peaks or valleys are being viewed, this being a particular optical distortion that is visually difficult to interpret in plan view, as shown in Figure 85. Moreover, this face-milled image has a topographical height profile superimposed onto the surface, showing peak and valley features, giving greater dimensional and visual interpretation of the machined surface topography.

(a) A processed photo-cell surface.

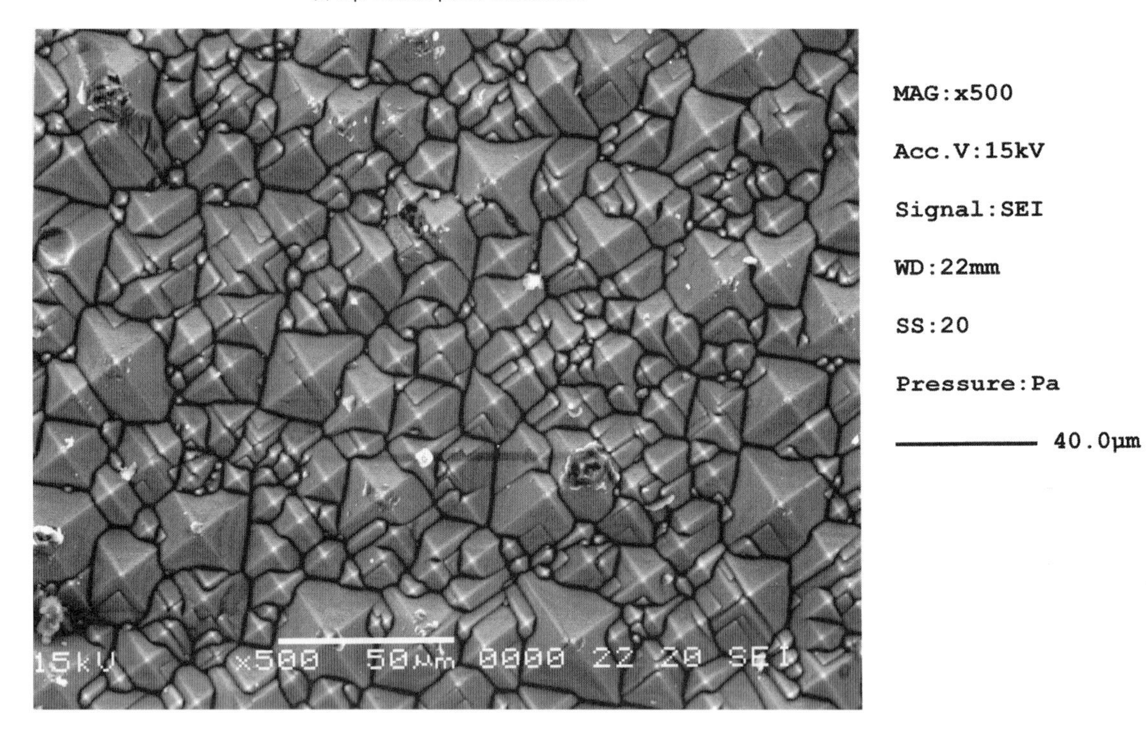

(b) High-speed face-milled surface: 316 grade stainless steel, peripheral speed 2500 m/min.

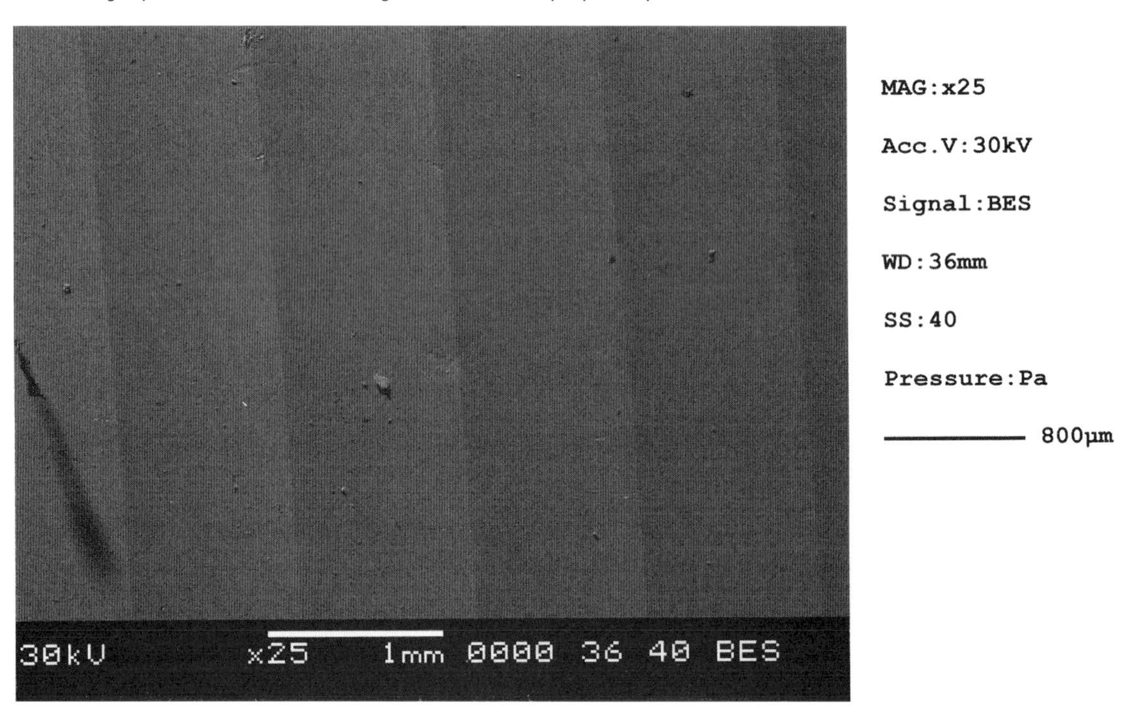

Figure 84. Scanning electron microscope images. [Courtesy of Jeol (UK) Ltd.]

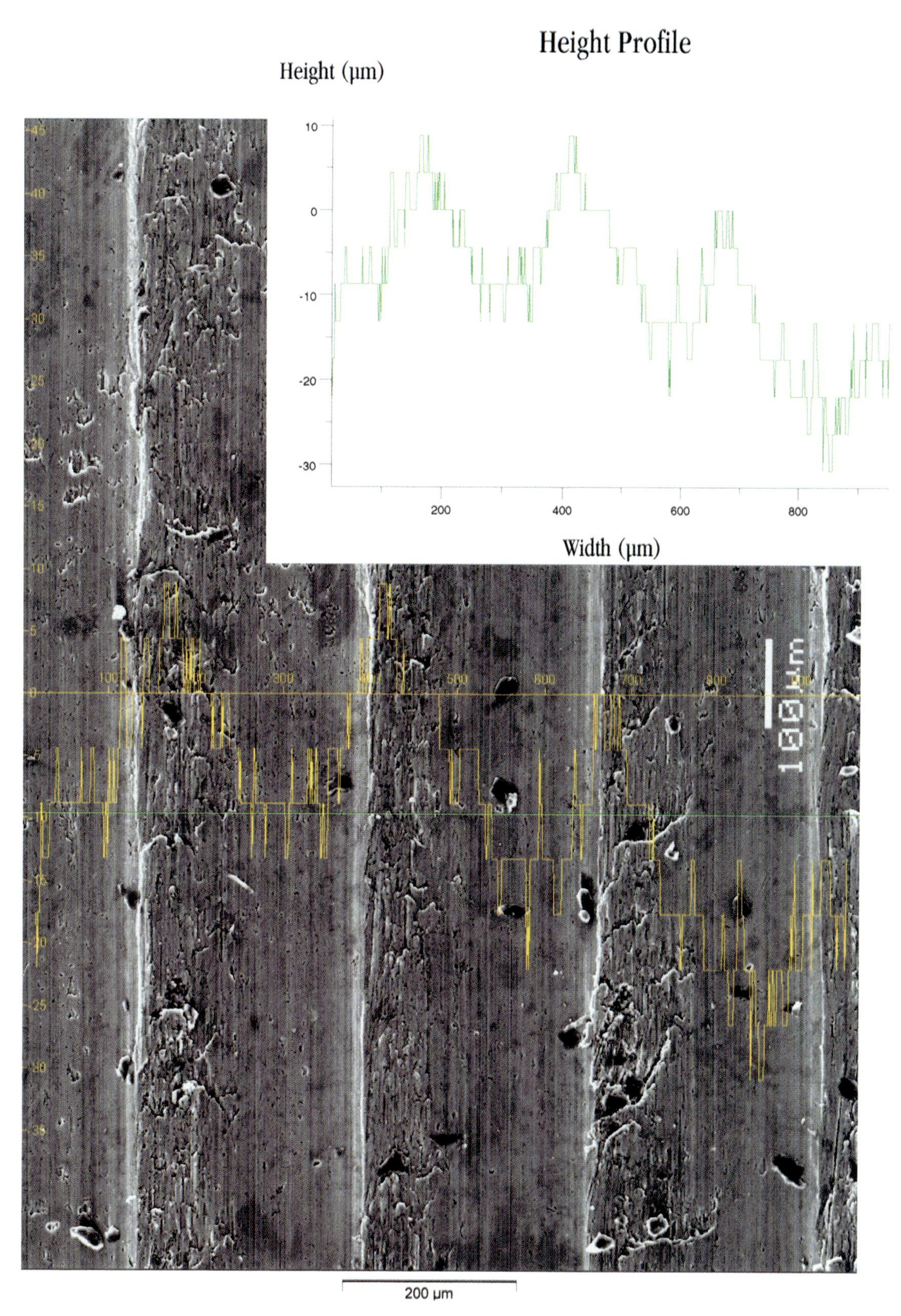

Figure 85. Scanning electron microscope photomicrograph with "soft imaging" and topographical height profiles. [Courtesy of Jeol (UK) Ltd.]

SEM: image processing

Digital images can be subjected to a variety of post-processing functions, with processed images being separately stored from the original images. Therefore, when writing reports the image data can be converted into a suitable form for the content of the report. Windows™-based software enables a host of image processing techniques and enhancements to be carried out; some these are as follows:

- *On-screen measurement* – this is a cursor measurement function allowing measurements between two cursors at 90° to each other (Figure 86i), or a multi-point measurement function which measures many points in any direction (Figure 86ii);
- *Look-up table (LUT) and pseudo-colour display* – when a stored image is to be downloaded, the LUT allows the image to be optimised for visual impact, through adjustment of the contrast and brightness of the original image (Figure 86iii). Furthermore, the image can be visually enhanced by the application a pseudo-colour display (Figure 86iv);
- *Dual and quad displays* – either two or four images can be combined into an overall new image (see Figure 86v and vi, respectively);
- *Dual magnification and digital zoom displays* – the digital zoom function provides an enlarged portion of the stored image, which allows the images to be positioned adjacent to each other (Figure 86vii). The dual image shown in the previous Figure 86(vii) can be increased in size to any desired magnification to observe notable features, for further investigation, or analysis (Figure 86viii).

The following two sections relating to SEM enhancement attempts to describe and illustrate several types of the large array of instrumental equipment that can be fitted to an SEM. Such investigative items considerably increase the versatility and the scope of the SEM as an analytical tool.

3.2.1 Energy-dispersive X-ray spectrometer

The characteristic X-rays emitted from the specimen under test have energies that represent individual elements. These energies can be detected with an energy-dispersive spectrometer (EDS) fitted at a specific location port on the SEM. Its detector orientation and angular relationship to the axis of the X-ray beam is conveniently located in one of the SEM's ports. Angled detectors are preferred for practical reasons, because they are less likely to be damaged by movement of the specimen within the working chamber of the instrument. The X-rays are discovered by employing either a thick- or thin-frame window (typically an Si (Li) detector), for heavier or lighter elements, respectively. This semiconductor detector is positioned at the tip of the detector (see Figure 87). The height of current pulses that are generated by the X-ray illumination is proportional to the energy of incident X-rays. By calibrating the multi-channel analyser prior to performing testing using a standard specimen, characteristic X-rays from a previously unknown specimen can be measured for element identification. The sophistication of these EDS systems enables the investigator to determine by an analysis of the specimen, its percentage weight of elements in the vicinity of the sampling area, spectra for specific elements that might be located in the test zone, or a general spectrographical search for a totally unknown specimen, when attempting to identify its overall elemental composition.

When an electron beam irradiates a specimen, characteristic X-rays are emitted (Figure 87). By detecting and analysing these X-rays, identification of elements contained within them can be undertaken (qualitative analysis). More specifically, it is possible to determine weight concentrations of the contained elements (quantitative analysis). Electron beams are finely focused, so by utilising the "spot mode" on the instrument an elemental analysis of a very small area on the specimen can be investigated, or the more general averaged element concentration, mentioned above. Furthermore, both a "line analysis" and "areal analysis" can be performed as well as X-ray image observation, the procedure being:

1. observing the specimen surface with a secondary electron image and/or a backscattered electron image;
2. designating the point to be analysed with the cross-point of the cursor lines displayed on the image;
3. positioning the electron beam at the designated point on the specimen, then performing a qualitative analysis;
4. observing the distribution of an element, by electing a specified characteristic X-ray.

Such procedures as those just described for EDS analysis enable considerable elemental data to be derived, then interpreted for appropriate future actions, based on sound and factual judgements.

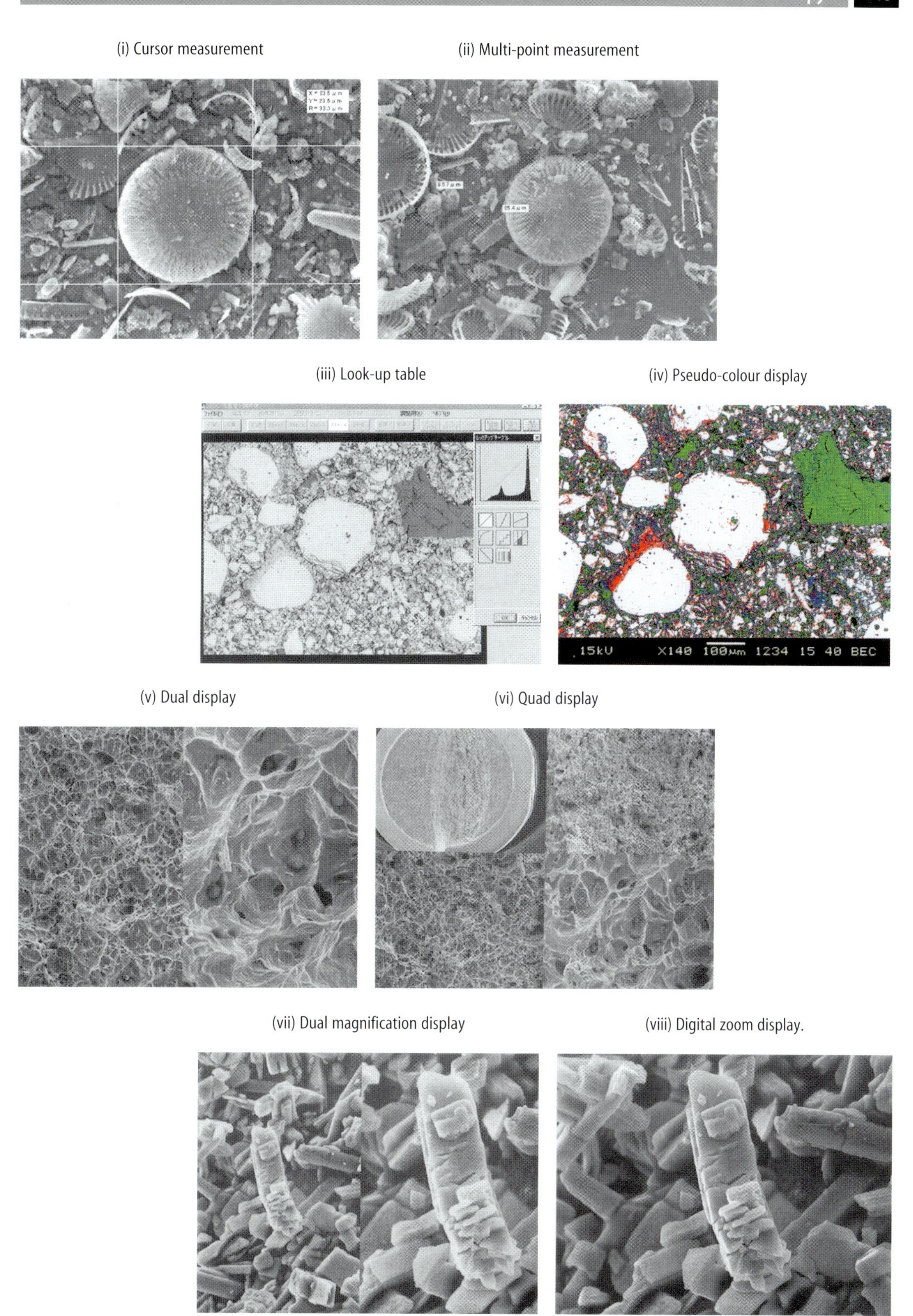

Figure 86. SEM image-processing techniques. [Courtesy of Jeol (UK) Ltd.]

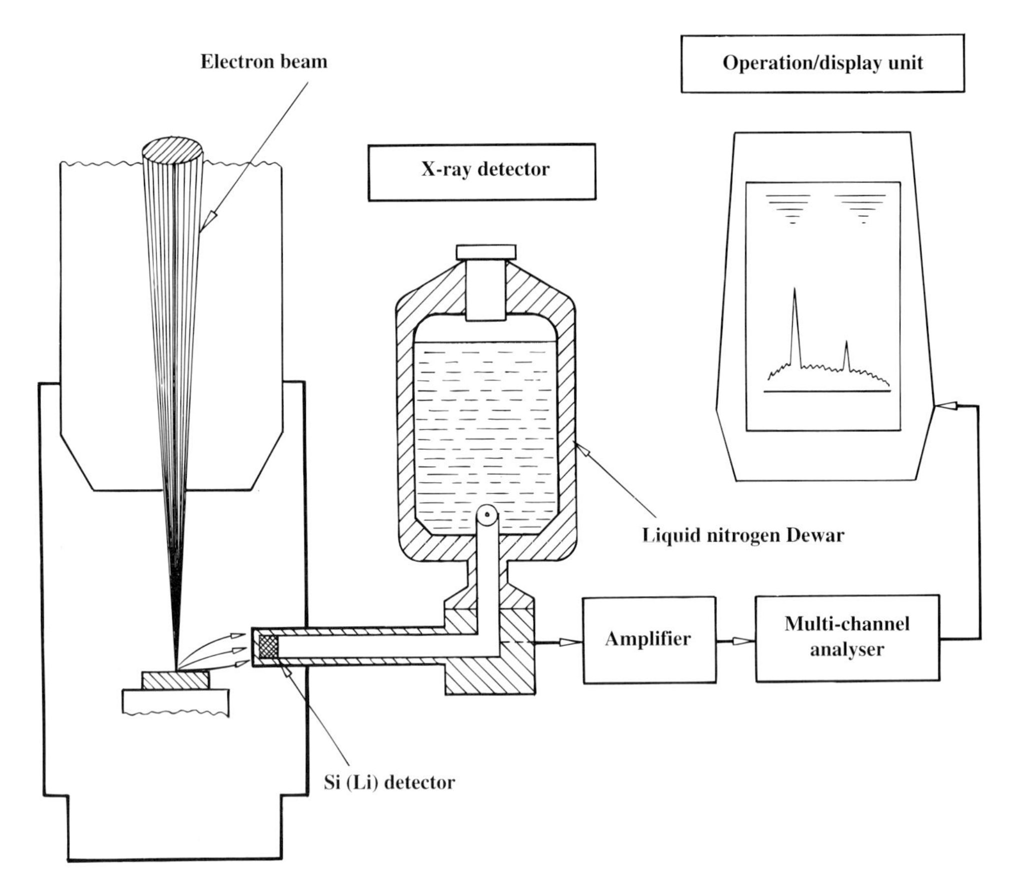

Figure 87. The energy-dispersive X-ray spectrometer (EDS) – block diagram. [Courtesy of Jeol (UK) Ltd.]

3.2.2 Transmitted electron image

Prior to a brief discussion of the operational procedure of this instrument that can be fitted to the SEM, it is should be mentioned that the operation can be performed in its own right on a purpose-built instrument – see the following section. If a specimen is thin enough for the irradiated beam to penetrate it, electrons will be scattered during its subsequent penetration (see Figure 88).

The degree of electron scattering depends on the product of the density of the specimen's surface (ρ) and its thickness (t). Namely, any of the electrons penetrating through a less dense area (having a small "$\rho \times t$" value) will not be as widely scattered as those from a more dense area (large "$\rho \times t$" value). Thus, a larger signal is detected through the less dense area. When a thin section is cut from a bulk specimen and observed with transmitted electrons, information concerning the inner structure of the thin section can be obtained, this being the basis for the transmitted electron image.

3.3 Transmission electron microscope (TEM)

With the recent developments in the fields of high-temperature superconductors and the introduction of new materials, more sophisticated instrumentation has been created. Research at the atomic level has become quite commonplace, being essential for a realistic and an interpretative understanding of the behaviour and properties of such materials. The age of research into subatomic operational activities on both nano-structures and nano-fabrication can be said to be in existence today. The transmission electron microscope (TEM) is a field-emission, high-resolution analytical instrument, which has many operational applications and functions, but most notably it has been widely employed for the investigation into new material developments. A typical instrument is illustrated in Figure 89 and can be configured with, say, a 200 kV thermal field emission gun having approximately 100 times brighter image

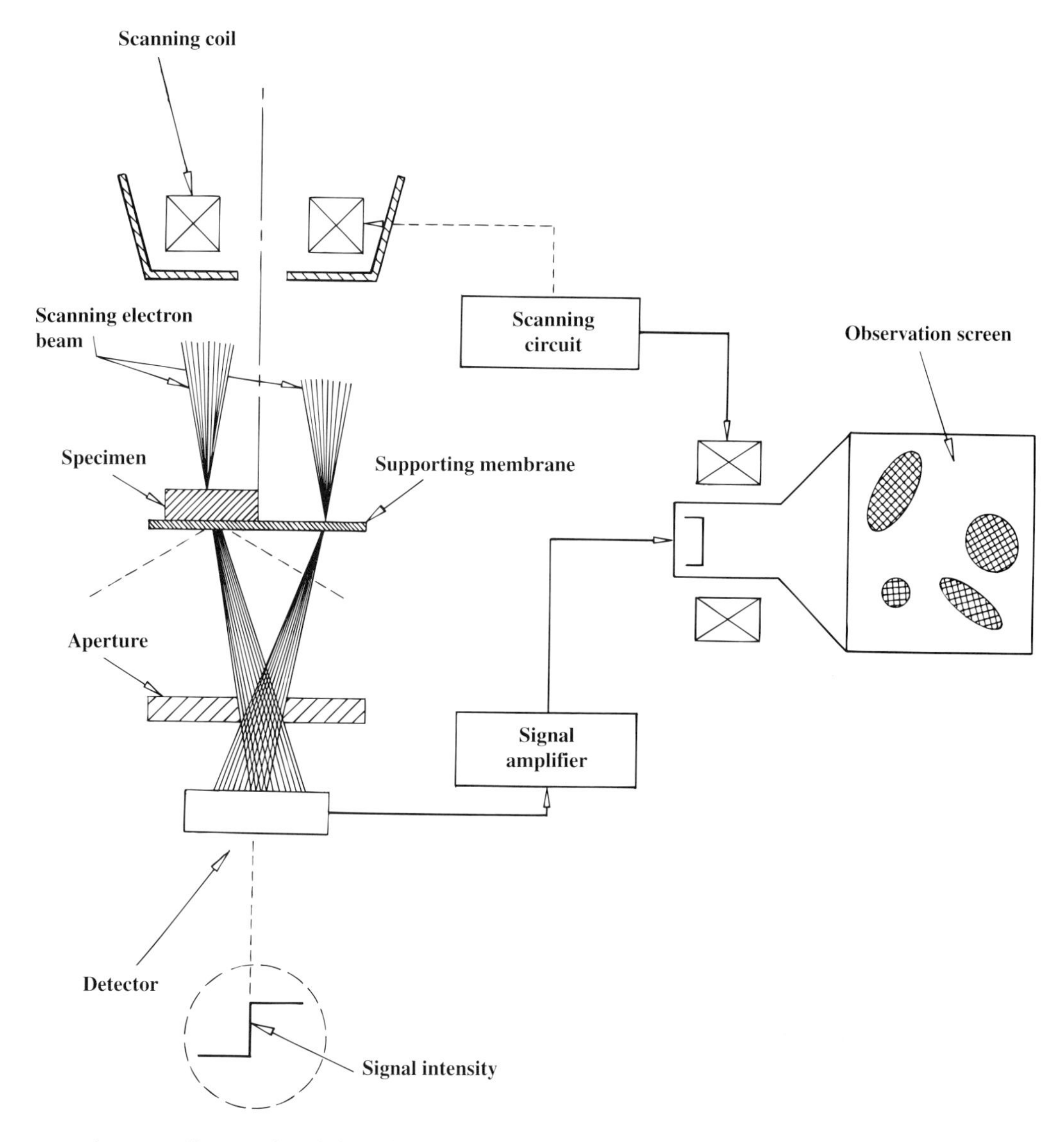

Figure 88. The principle underlying the SEM transmitted electron image observation. [Courtesy of Jeol (UK) Ltd.]

than that found on many earlier instruments, without sacrificing long-term emitter stability. One of the advantages of high-grade versions of this instrument is its ability to instantaneously select the analysis point, while observing a high-resolution image, with certainty that both the observed and analysis points coincide. Other applications for TEMs might include electron holography and coherent convergent beam electron diffraction.

Operational characteristics of the TEM

A typical field emission gun might use either a ZrO/W (100) emitter, or a W (100) emitter. A very basic diagrammatic representation of a TEM is shown in Figure 90. Recent progress in analytical microscopy has enabled improved theoretical resolution, this being calculated from the spherical aberration coefficient, with yet further enhancements in analytical capability through improvements in the illumination system. Such improvements include micro-area diffraction, convergent-beam electron diffraction and high-sensitivity analysis employing finely focused probes.

The illumination lens system might consist of a series of lenses (four), in conjunction with an imaging lens system (i.e., six lenses) – see Figure 90 (many of the lenses are not shown, for simplicity). This imaging system provides various illumination

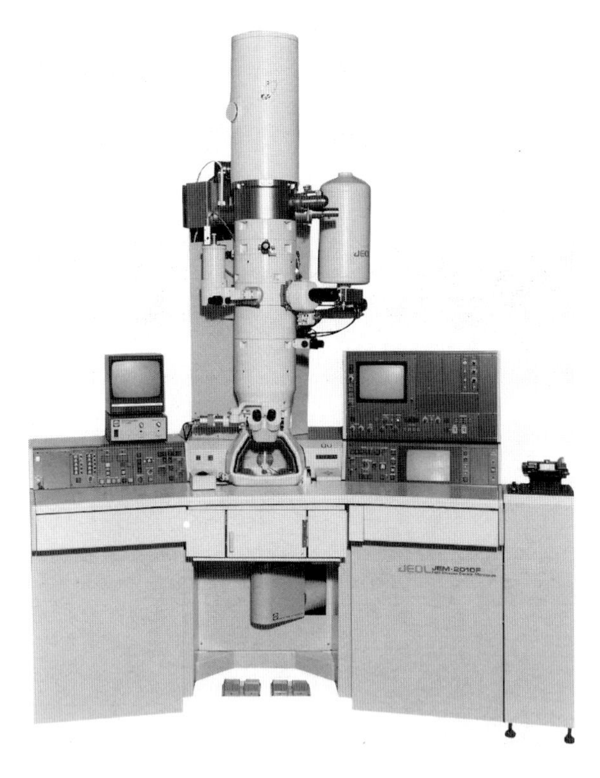

Figure 89. Transmission electron microscope, equipped with:

- an energy dispersive X-ray spectrometer;
- a parallel detection electron beam loss spectrometer (PEELS);
- A scanning image observation device (ASID); and
- TV units.

[Courtesy of Jeol (UK) Ltd.]

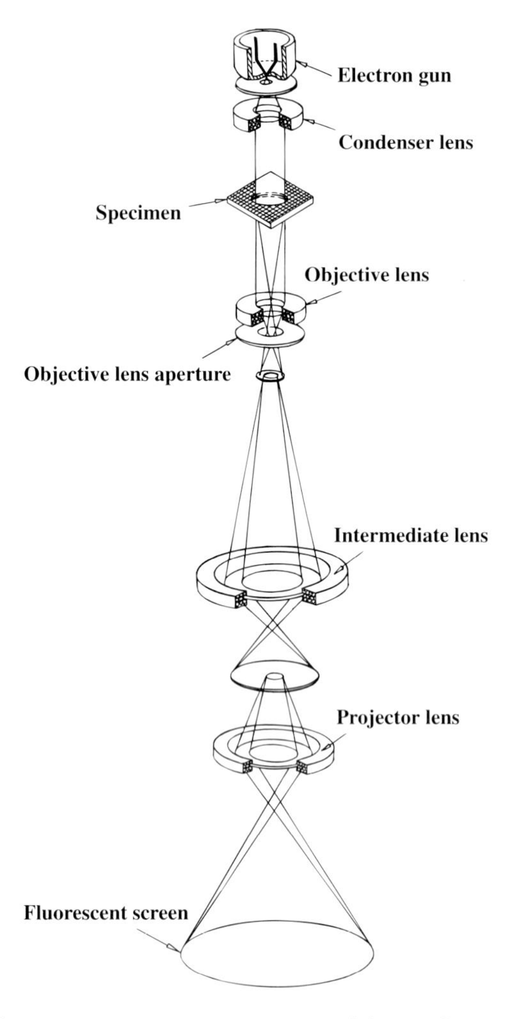

Figure 90. Diagrammatic representation of the TEM. [Courtesy of Jeol (UK) Ltd.]

modes to accomplish high resolution and diffraction, together with analytical requirements for the microscopist, while the imaging lenses provide high magnification, rotation-free imaging and orientation correspondence between the diffraction pattern and its associated image. Typical values for the acceleration tube evacuation are of a high order, being around 3×10^{-8} Pa. The specimen chamber vacuum pressure range is normally up to 3×10^{-5} Pa.

In general, a field emission gun is available as two distinct types:

- cold field emission (CFE) variety – which uses the emitter at ambient temperatures;
- thermal field emission (TFE) variety – which heats the emitter to approximately 1600K.

The CFE normally features a small electron energy spread, but has the disadvantage of both long- and short-term emission fluctuation, due to residual gas contamination of the emitter surface. Conversely, the TFE, being constantly heated, has no gas adsorbed onto the emitter surface, resulting in extremely high stability of the emission current.

Additionally, the TFE has a high probe current, ensuring that it is an ideal electron source for analytical microscopy.

The advantages of the field emission gun can be summarised as follows:

1. the field emission has approximately two orders of magnitude higher brightness than a LaB_6 filament and provides a considerably greater probe current – in a probe of less than several nanometres;
2. the specimen illumination angle can be configured to be exceedingly small, due to the gun brightness; this, together with a small electron energy spread, enables the information limit to be improved;
3. this small energy spread of the emitted electrons improves the energy resolution of the electron energy loss spectrometer (EELS);

(i) Ultra-high resolution.

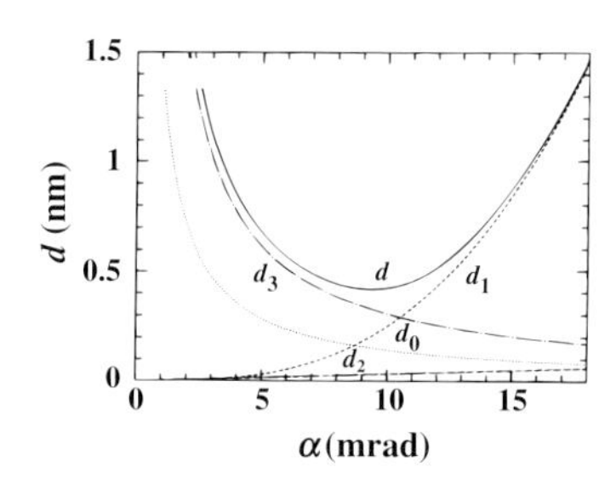

(ii) Specimen high-tilt configuration.

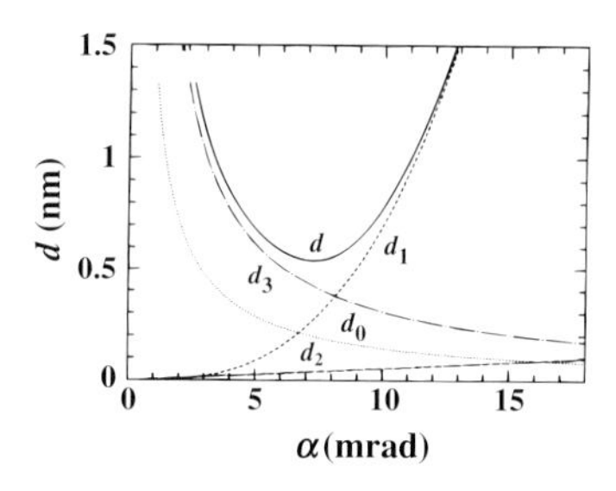

Figure 91. Relationship between illumination angle and probe diameter. [Courtesy of Jeol (UK) Ltd.]

4 the minute effective source size and the high coherence of the beam are effective for magnetic domain observation and are essential for holography.

Functions and ray diagrams of illumination lens system

In Figure 91 are shown the calculated values of the specimen illumination angle (α) versus the probe diameter (d), in both the ultra-high-resolution configuration (i) and the specimen high-tilt configuration (ii) for a typical TEM. From these graphs it can be seen that a probe diameter of 0.5 nm is obtained at $\alpha = 10$ mrad with the ultra-high-resolution configuration (i) and 0.5 nm at $\alpha = 7$ mrad with the high-tilt configuration (ii).

The probe diameter on the specimen is given by the following equations:

$$d = \sqrt{d_0^2 + d_1^2 + d_2^2 + d_3^2}$$

where

$$d_0 = \frac{1}{\pi\alpha}\sqrt{\frac{Ip}{B}}$$

(Gaussian image with no aberrations),

$$d_1 = \frac{1}{2} Cs\, a^3$$

(minimum disc due to spherical aberration),

$$d_2 = Cc\, \alpha \left(\frac{\Delta V}{V}\right)$$

Minimum disc due to chromatic aberration,

$$d_3 = 1.22 \frac{\lambda}{\alpha}$$

Blur due to diffraction (Airy disc).

Here, B, Ip, Cs and Cc are, respectively, the electron gun brightness, probe current, spherical aberration coefficient (Cs of objective lens pre-field), and the chromatic aberration coefficient. With the field emission electron gun, the probe diameter is determined by d_0, d_1 and d_2, owing to the realisation of a sufficiently small and stable light source.

Operating functions and TEM ray illumination lens system configurations

The TEM can resolve down to the atomic spacing of materials, as depicted in Figure 92(a), where the probe has been focused on an Au particle, its profile being taken through the line of measurement as indicated in Figure 92(b). As mentioned above, the TEM can utilise a number of operating modes, some of which are as follows:

- *Nanometre beam diffraction* (NBD) – an NBD pattern can occur with high angular resolution when illuminating the specimen with a nanometre-sized probe having a small illumination angle;
- *Energy-dispersive X-ray spectrometry* (EDS) – a wide range of elements can be investigated, from light to heavy, which can be found by analysing the energies of characteristic X-rays generated from a specimen. The optimum arrangement of the detector and collimator allows for a good P/B ratio and high detection efficiency;
- *Electron energy-loss spectrometry* (EELS) – the electrons that have passed through the specimen are dispersed by a sector magnet to perform energy analysis. This technique features high-energy resolution and the acquisition of information on inner shell excitation.

A variety of ray diagrams can be produced using a TEM, most notably:

(a) An electron probe focused on a specimen and an Au particle, which were photographed on the imaging plate (IP).

(b) Profile on the line in the image at left.

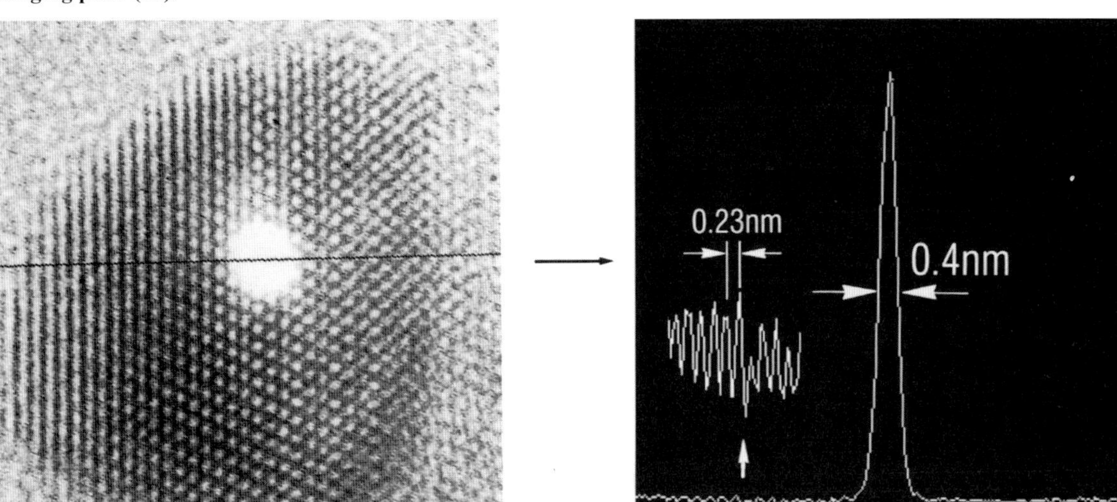

(c) An image of carbon graphite, observed with a scanning image observation device. A lattice image of 0.34 nm is clearly observed, proving that the probe was this order of diameter.

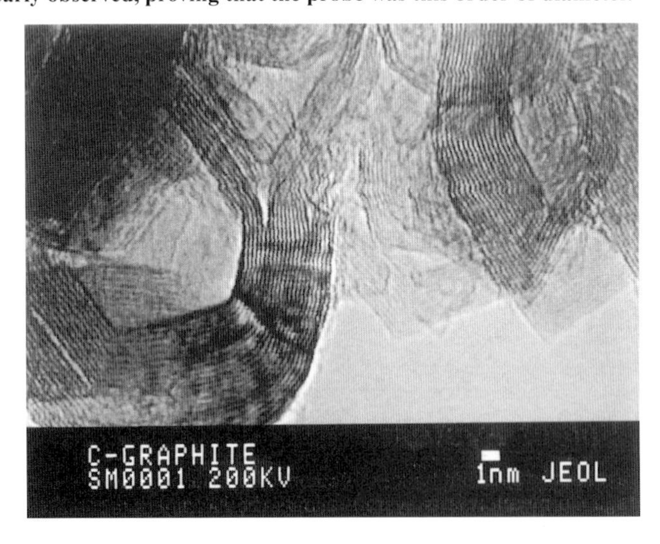

Figure 92. Measured probe diameter and image of carbon graphite taken on a TEM. [Courtesy of Jeol (UK) Ltd.]

1 atomic lattice images – with probes configured to approximately 1 nm diameter. Such diagrams are visual representations of the physical associated and geometrical positioning of atomic lattice structures within materials;
2 diffraction patterns – diffraction can occur whenever the *Bragg law* is satisfied, varying for each material and its orientation, this being related in the following manner:

$\lambda = 2d \cdot \sin\theta$ (Bragg law)

where X-rays of known wavelength (λ) and measuring angle (θ) can determine the atomic spacing (d) for various planes/orientations in a crystal;
3 convergent-beam diffraction – simplistically speaking, the atomic structure here closely resembles that observed in elementary "bubble-raft patterns", which are used to simulate simple grain boundaries and dislocations in crystalline materials.

3.3.1 Transmission electron microscopy: general application

In X-ray diffraction, X-rays incident on a crystal are diffracted by parallel sets of atom planes, in a similar manner to that of light reflection from a mirror. The position of the diffracted beam will be in accordance with the "Bragg reflection conditions", described in the previous section, and they are similar to the way an electron beam is scattered by a crystal. Any dislocations and stacking faults in the surface region can be revealed by a technique which examines the interference between transmitted and diffracted beams, resulting from a beam of electrons incident on a crystal foil of between 100 and 500 nm in thickness. The technique for forming images of these diffracted elements has previously been mentioned. In operation, a parallel beam of electrons – typically accelerated by a potential of 100 kV – is transmitted through a thin foil and is diffracted in a variety of directions by the crystal under test. The diffracted beams are subsequently focused forming a diffraction pattern that can be magnified for investigation, as necessary. A diffraction pattern (see Figure 93) normally consists of a two-dimensional array of spots, with each spot representing a specific series of reflecting planes. From this diffraction pattern, the foil orientation can be determined.

Individual diffraction pattern spots are the product of the incident beam's diffraction from a reasonably large vicinity on the test specimen (typically 10^4 atom spacings in diameter). A single electron beam can produce a corresponding image by inserting an objective aperture into its path.

Two main optical conditions can arise:

- *brightfield image* – resulting from the aperture being centred on the main transmitted beam;
- *darkfield image* – produced by situating the aperture over a diffracted beam.

Generally, the predominant amount of analysis occurs in the brightfield mode of illumination, although some critical applications required the darkfield illumination technique.

Hypothetically, if a test specimen was completely flat and without defects, under these conditions a completely homogeneous image would result when the objective lens was centred over the main transmitted beam. Subsequent changes in image brightness are the result of any effect that changes the diffracted beam's path, so that it enters the objective aperture and interferes with the transmitted beam. If very thin test specimen foils are buckled, the orientation of the specimen's surface with respect to the electron beam will be slightly varied for different zones on the specimen. These variations for local surface reflection conditions are governed by the "Bragg law", so they also must vary. The interference between both the transmitted and diffracted beams will produce intensity variations of the brightfield image – termed *extinction contours*. Such an effect appears as dark bands across the image, which indicates that a set of planes near these bands has a significant diffracting orientation. Therefore, when a foil bend occurs in the TEM during its observation, "extinction contours" seem to appear to move across the test specimen. From a simplified analysis technique, this approach can be employed to account for the contrasting effect resulting from surface regional dislocations. Moreover, "stacking faults" in the specimen will also produce a characteristic diffraction image. Hence, the study of specimen defects by these various diffraction techniques is applicable to all materials that can be produced as adequately thin sections.

In order to show the potential of the TEM's abilities as an analytical tool, Figure 93 illustrates a $TiAl/Ti_3Al$ two-phase alloy in lamellar form. This particular material has shown promise as a new class of lightweight high-temperature structural material. Three different types of $TiAl/Ti_3Al$ intervariant lamellar boundaries, in such a TiAl-based alloy containing Sn are observed by high-resolution electron microscopy. Of the three types of boundaries shown in Figure 93, the ternary element Sn, which is added to improve the mechanical properties, is observed to preferentially segregate onto two types of high-energy boundaries. In Figure 93 the visual image and diffraction patterns, together with graphical displays, show just some of the investigative/analytical abilities of the TEM.

3.4 Atomic force microscope

The atomic force microscope is just one variant of a wide variety of scanning probe microscopes and accompanying techniques that were developed in the early 1980s and include the following:

- *atomic force microscope* (AFM) – contact; AC; also visco-elasticity operational modes;
- *scanning tunnelling microscope* (STM);
- *scanning tunnelling spectroscopy* (STS);
- *friction force microscope* (FFM) – static and lateral operational modes;
- *magnetic force microscope* (MFM);
- *scanning near-field optics microscope* (SNOM).

This section on scanning probe microscopes, will almost exclusively discuss an overview of the AFM

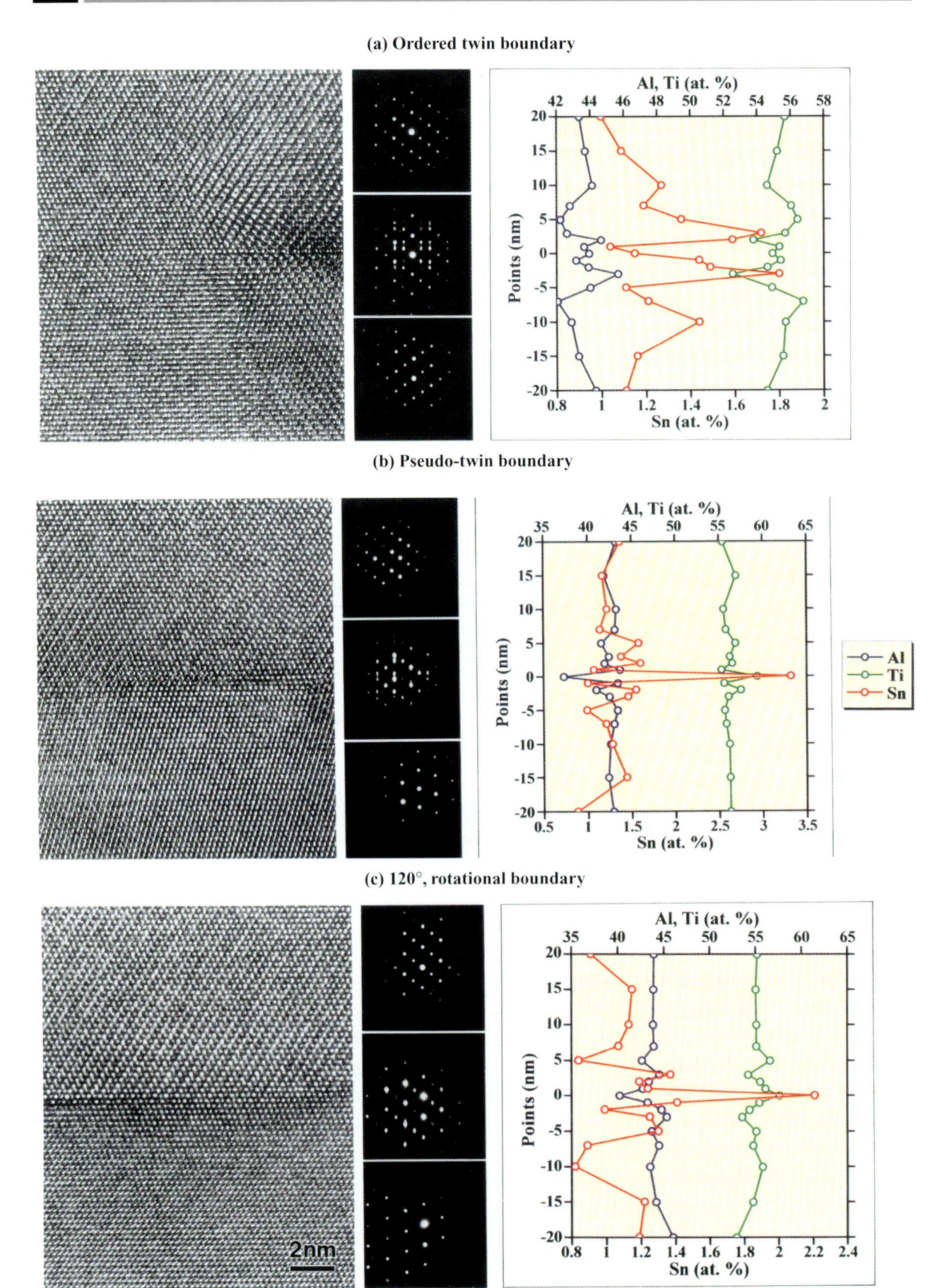

Figure 93. TEM sub-nanometric elemental analysis (i.e., point analysis). (Courtesy of Jeol (UK) Ltd/Dr H. Inui and Prof. M. Yamaguchi, Kyoto University, Japan.)

Table 11. Information obtained from various analytical techniques

Analysis instrument	Morphological observation	Elemental analysis	Element mapping	Crystal structure	Bonding state
STM/AFM	★				
AES	★	★	★		
EPMA	★	★	★		
FTIR		☆			★
SEM	★				
SIMS		★			
TEM	★			★	
XPS		★			☆

AES: Auger electron spectroscopy; EPMA: electron probe micro-analysis; FTIR: Fourier transform infra-red absorption spectroscopy; SEM: scanning electron microscope; SIMS: secondary ion mass spectroscopy; TEM: transmission electron microscope; XPS: X-ray photoelectron spectroscopy.

Courtesy of Jeol (UK) Ltd.

instrument and just some of its potential applications. Today, with the diverse range of scanning probe microscopes and observation methods that exist, they can offer a wide range of analysis techniques, as indicated in Table 11.

Recently, scanning probe microscopes have been utilised across a diverse variety of disciplines, including fundamental surface science and routine surface roughness analysis, together with three-dimensional imaging. The scanning probe microscope is an imaging tool with a large dynamic range, encompassing the realms of optical and electron microscope applications. These instruments can be employed as a three-dimensional profiling instrument – with superb image resolution abilities – that in certain cases can quantify physical properties such as surface conductivity, static charge distribution, localised friction, magnetic fields and elastic moduli. The instrumental applications for scanning probe microscopes can be shown on a chart relating to their lateral and vertical resolutions, as previously indicated in Figure 80.

3.4.1 Criteria for using scanning probe microscopes

As can be seen in Table 11, the AFM can be utilised for morphological analysis, with the visual display having high-resolution capability for extremely flat samples, whose topographical features cannot be readily distinguished even with an SEM. The AFM (see Figure 94) is particularly applicable when

- *topographical features* lie within an observation area of $<1\ \mu m^2$;
- *non-conductive specimens* use either an STM, or more notably an AFM – in accordance with the sample's conductivity;
- *specimens having no conductivity* require an AFM used in either contact or AC mode, depending on the sample's surface roughness and its softness;
- *specimens with conductivity* – both the AFM and STM may be utilised.

Any scanning probe microscope allows specimen observation in various environments, from normal atmospheric conditions, to liquids, also with an ultra-high vacuum. Conversely, some electron microscopic applications necessitate situating the specimen in a vacuum and in certain instances the samples need to be either stained or coated with a conductive film. Certain types of specimens in their original form may be destroyed by such sample

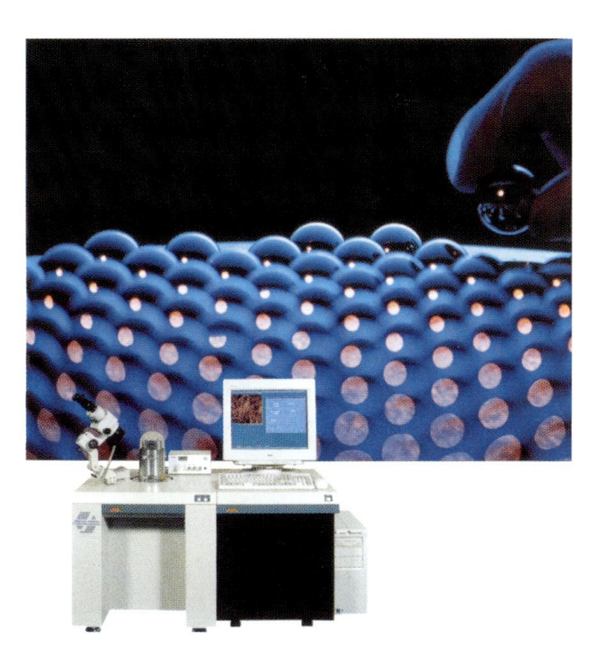

Figure 94. Scanning probe/atomic force microscope. [Courtesy of Jeol (UK) Ltd.]

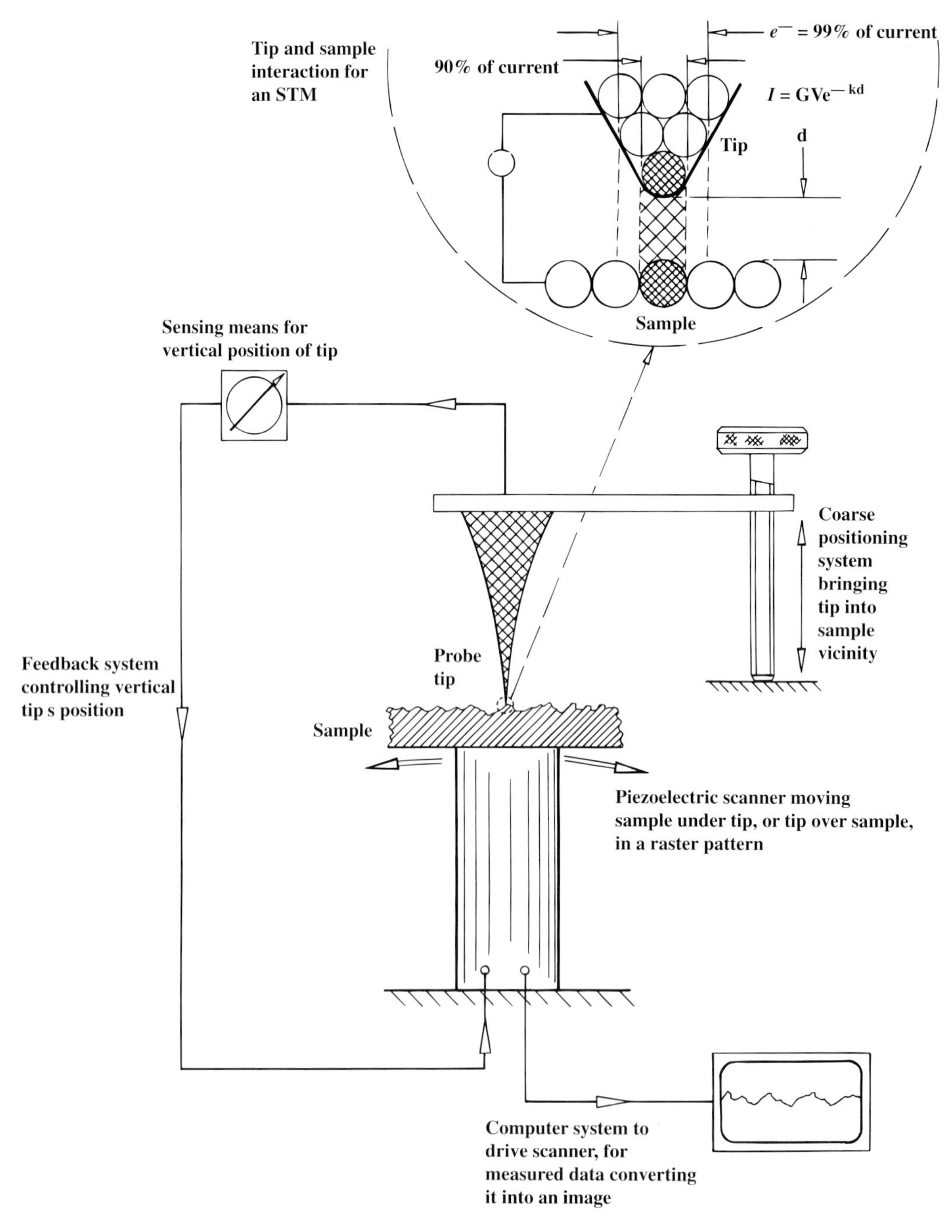

Figure 95. Schematic representation of the generalised operation of a scanning probe microscopes (i.e., STM/AFM). [Courtesy of Nikon/Park Scientific Instruments.]

treatments. If observation occurs in an atmosphere, gas or liquid, the scanning probe microscope can be effectively used, but careful interpretation of images is necessary, due to the potential of adsorbed impurities. These AFM/STM microscopes can be used as an effective means for measuring specific physical properties, such as surface magnetism, friction force, viscosity, elasticity, surface potential and other important surface-related factors, but care is needed when quantitative measurements are required.

The scanning tunnelling microscope (STM) can be considered to be the forerunner of all derivative types of scanning probe microscopes. It was invented by Gerd Binning and Heinrich Rohrer at IBM (Zurich) in 1981, who later received a Nobel prize (Physics, 1986) for this invention. This STM instrument was the first of its type to generate real-space surface images at atomic resolution. STMs use a sharp conducting tip (see Figure 95, detail), with a bias voltage being applied between the tip and sample. When the tip approaches the sample to within 10 Å ("d") electrons from the sample will "tunnel" through this gap into the tip or vice versa, depending on the bias voltage sign. The resulting tunnelling current is variable, this being related to the tip-to-sample spacing which, in turn, is used in the creation of the STM image. In the case of the STM, both the tip and the sample must ideally be conductors, or at least semiconductors, which is not the case for an AFM that can image insulating materials.

From Figure 95 it can be seen that the tunnelling current is an exponential function of the tip-to-sample distance relationship. If this separation distance changes by 10% (i.e., equating to 1 Å), then the tunnelling current will change by an order of magnitude. Such an exponential dependence enables an STM to have a notable sensitivity, allowing surface imaging to sub-Ångström vertical precision, with lateral atomic resolution. STMs are designed to surface scan in either of two modes:

1 *constant height mode* – where the tip's travel is in a plane horizontal to and above that of sample and variation of the tunnelling current is related to the local surface topography and the electronic properties of the sample. This tunnelling current is measured at each sample surface feature, which constitutes a data set for the topographic image;
2 *constant current mode* – the instrument uses feedback to maintain the tunnelling current, by constantly adjusting the height of the tip at each measured topographical feature. If the system detects an increase in tunnelling current, the voltage is adjusted to the piezoelectric scanner to increase the tip-to-sample distance.

In constant current mode the scanner's motion constitutes the topographic data set, such that the system maintains a constant tunnelling current to within several per cent. This control equates to a tip-to-sample distance constancy of less than several hundredths of an Ångström.

The main consideration in the following text is principally reserved for a discussion related to the AFM type of scanning probe microscope, its design, operating principle and potential surface-related applications.

3.4.2 Atomic force microscope: operating principle

The AFM's tip will probe the sample surface with a sharp tip having dimensions of several micrometres long and approximately 100 Å in diameter. The tip is located at the free end of a cantilever which is between 100 and 200 μm long (see Figure 96). Applied forces between the tip and the sample's surface cause the cantilever to either bend or deflect. Cantilever deflections are measured as tip scanning occurs over the sample or, conversely, with the sample being scanned under the tip. These deflections of the cantilever allow the production of a computer-generated surface topography map.

The AFM cantilever can be deflected by the contribution from several microscopic surface-related forces. The principal one of these minute forces is the inter-atomic force termed *van der Waals force*. The *van der Waals theory* states that *atomic attraction increases as atoms are progressively brought together, until their electron clouds begin to electrostatically repel each other*. However, the electrostatic repulsion progressively weakens this attractive force, as the inter-atomic separation continuously decreases. Eventually the van der Waals force reaches zero, when the separation distance between the atoms reaches several Ångströms – which is approximately the length of a chemical bond. When the atoms are in contact, then the total van der Waals force becomes positive (i.e., repulsive). As a result of this repulsive van der Waals force, any attempt to force the atoms into a closer atomic bond are negated.

In practice for the AFM, this results in the cantilever pushing the probe's tip against the sample, and under this condition the cantilever bends rather than forcing the tip and sample atoms closer.

The potential energy that acts between two neutral atoms can be expressed from the formula derived by J.E. Lennard-Jones, as follows:

$$U(d) = 4\varepsilon\left[\left(\frac{\sigma}{d}\right)^{12} - \left(\frac{\sigma}{d}\right)^{6}\right]$$

where, by calculating the potential energy of say, Xe atoms, using $\varepsilon = 0.02$ eV, $\sigma = 0.4$ nm; it is possible to derive a "distance–potential energy graph" (not shown). Moreover, using the relationship for the total potential energy of two adjacent atoms this can be obtained and the force acting on these two atoms may be found, after differentiation, to be in the region of 0.2 nN.

Additionally to the repulsive van der Waals force, two other forces often occur during the AFM-contact operation:

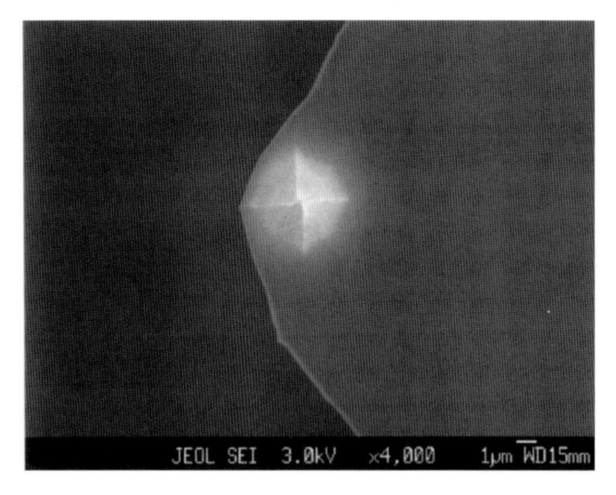

(a) SEM image of an atomic force microscope (AFM) probe, indicating the depth of field of an SEM (specimen tilt 0°).

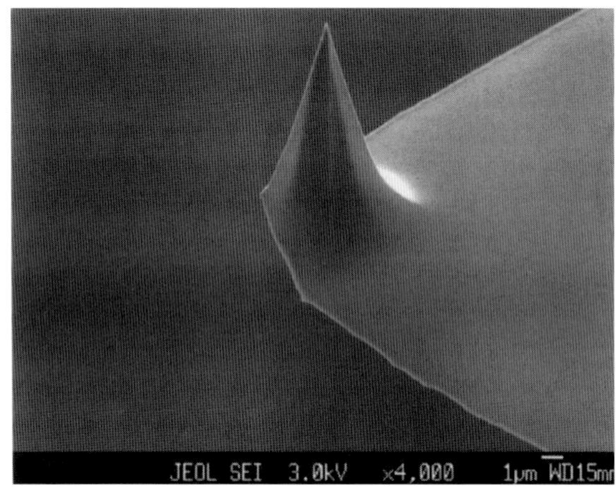

(b) SEM image of AFM probe geometry (specimen tilt 60°).

(c) Indentations in soft material and profile height distribution, as indicated.

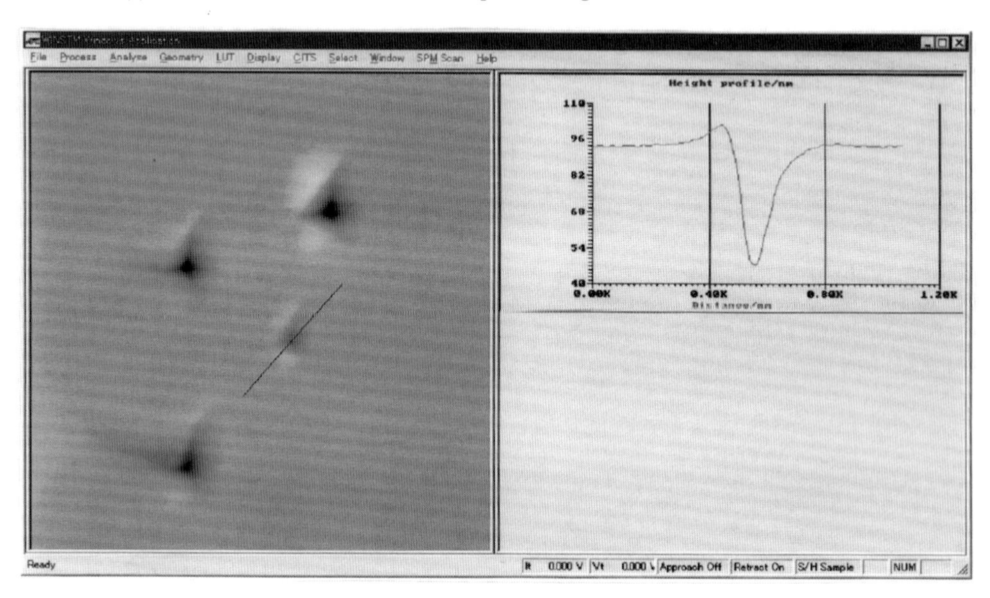

Figure 96. Atomic force microscope probe geometry. [Courtesy of Jeol (UK) Ltd.]

1 *cantilever force* – exertion applied by this mechanical device;
2 *capillary force* – exerted by a thin layer more often inspected in an ambient environment.

The cantilever force tends to act in a compressive spring-type manner, with the magnitude and its associated sign being either attractive or repulsive in nature, depending upon the cantilever deflection and its spring constant. The spring constant k can be found by the following relationship:

$$k = \frac{Ead^3}{4l^3}$$

where E = Young's modulus for cantilever material, a = cantilever width, d = cantilever thickness, l = cantilever length.

To obtain the atomic resolution with an AFM, the spring constant needs to be 0.01 nm and, to detect 10^{-9} to 10^{-12} N between the atoms the constant needs to be

$$k = 1 \times 10^{-9} \sim \frac{1 \times 10^{-12}\,\text{N}}{1 \times 10^{-11}\,\text{m}}$$

Therefore

$$k = 0.1 \sim 10\ [\text{n/m}]$$

The capillary force arises when water "wicks" its way around the tip, enveloping it and applying a strong force attraction (10^{-8} N) holding the tip in contact with the sample's surface, with the magnitude of the capillary force being dependent upon the separation distance of the tip-to-sample. So long as the probe's tip maintains contact with the sample surface (in contact mode of operation), then the capillary force would normally be constant due to the distance between tip and surface being virtually incompressible. An assumption is made that the retained water layer is relatively homogeneous, with the variable force in the contact mode being the force exerted by the cantilever. Hence, the total force exerted by the probe's tip on the surface of the sample, being the sum of the capillary and cantilever forces, must be balanced by the repulsive van der Waals force for the contact AFM. The total force magnitude exerted on the sample will vary around 10^{-8} N – with the cantilever being repulsed as hard as the water provides attraction to the tip – to the general operating range of typically 10^{-6} to 10^{-7} N.

The cantilever arm's measurement positioning must be accurately known and optical techniques are the most common form of sensing media. Typically, the cantilever bounces back a laser beam onto a position-sensitive photodiode (PSPD). The cantilever bends as contact with the surface occurs, resulting in the laser beam's position shifting in the detector. The PSPD is capable of discriminating light measurement displacement down to 10Å, with the ratio of optical path length between the cantilever and its detector to that of the actual cantilever giving a mechanical amplification. This further mechanical amplification enables the overall system to detect sub-Ångström vertical motion of the cantilever tip. There are other methods of detecting cantilever deflection, such as optical interference, or simply by manufacturing the complete cantilever arm from a piezoresistive material and detecting its deflection electrically – due to the fact that any mechanical deformation causes changes in the arm's material resistivity. Once the AFM instrument's electronics have detected the deflection of the cantilever, it can compute and then generate a topographical data set, by operating in two distinct modes – as previously mentioned, these are the *constant height* and *constant force* modes.

3.4.3 Atomic force/scanning probe microscope: applications

The diagrams schematically illustrated in Figure 97 are just some of the typical applications for either an AFM or some form of STM, most notably:

- *Static AFM (contact AFM, Figure 97i)* – this mode obtains sample surface structure by scanning the surface, while keeping a constant repulsive force acting between the tip and sample. The tip is normally in contact with the sample during this mode of imaging. Historically, the contact mode was the first probe microscope mode of operation that enabled the imaging of non-conductive samples.
- *STM (scanning tunnelling microscope, Figure 97ii)* – this mode of operation applies a bias voltage between a sharp conductive probe and sample exhibiting conductivity. When the probe and sample are in close proximity, some of the electrons can "tunnel" between the tip and sample without ohmic contact. Scanning the surface while using a feedback loop to maintain a constant tunnelling current reveals the structure of the sample surface.
- *LFM (lateral force microscope, Figure 97iii)* – this technique is sometimes termed *friction force microscopy*. It measures the friction between the tip and sample by measuring the torsional bending of the cantilever. LFM is both a static and contact mode technique which simultaneously acquires an AFM topography image.

- *Force modulation/phase detection (Figure 97iv)* – the technique adds a modulation signal in the *Z*-direction, while observing the sample with static AFM. Qualitative elasticity of the sample is obtained by measuring the displacement of the cantilever with respect to the modulation signal. The "phase detection" technique measures the phase lag of the cantilever relative to the driving signal. In the case of "discrete contact" mode of operation, qualitative visco-elasticity images of a sample can be obtained by measuring the phase lead or lag of the cantilever relative to the modulation signal.
- *Dynamic AFM (non-contact and discrete contact AFM, Figure 97v)* – dynamic AFM is accomplished by oscillating the cantilever near its reso-

(i) Static AFM (i.e., contact AFM).

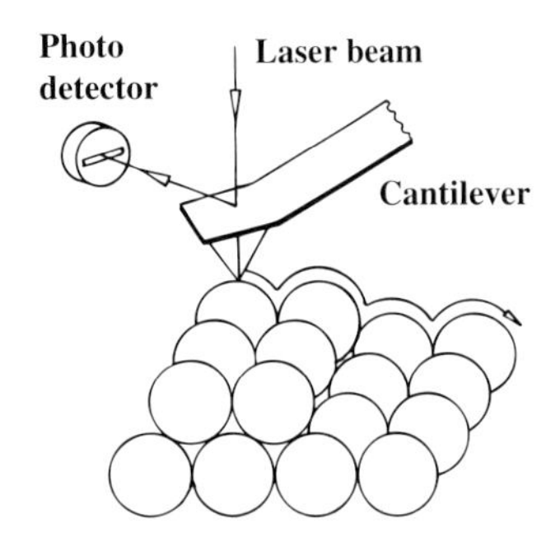

(ii) STM (i.e., scanning tunnelling microscope).

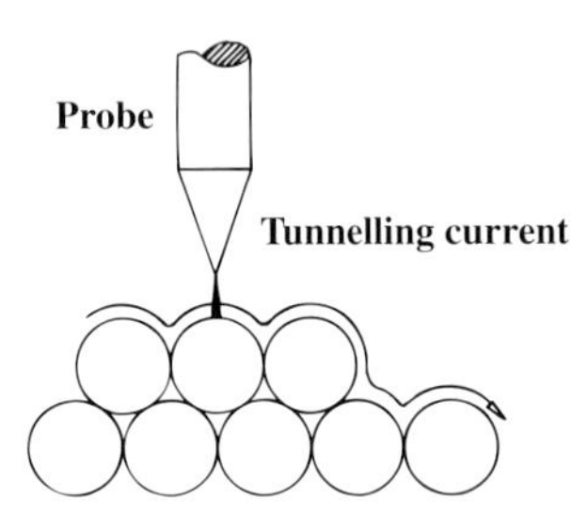

(iii) LTF (i.e., lateral force microscope).

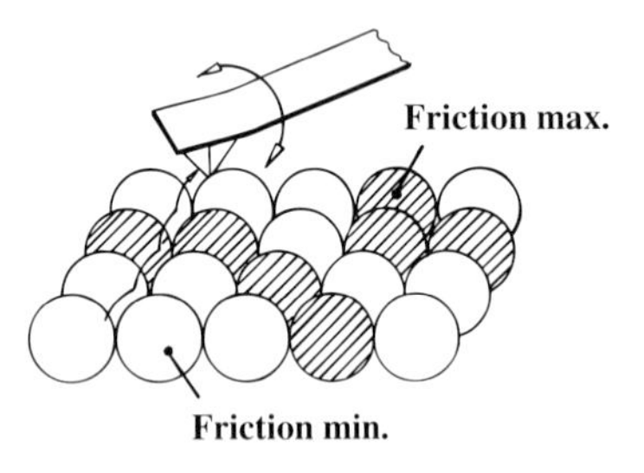

(iv) Force modulation/phase detection.

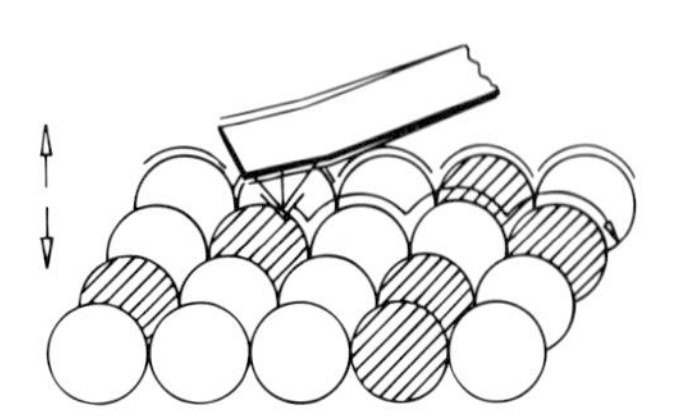

(v) Dynamic AFM (i.e., non-contact, or discrete-contact AFM).

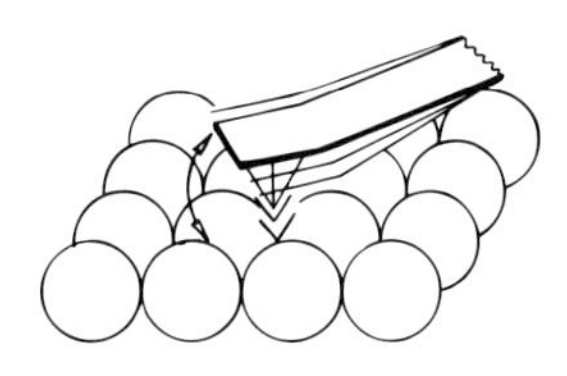

(vi) STS (i.e., scanning tunnelling spectroscopy), or CITS (i.e. current imaging tunnelling spectroscopy).

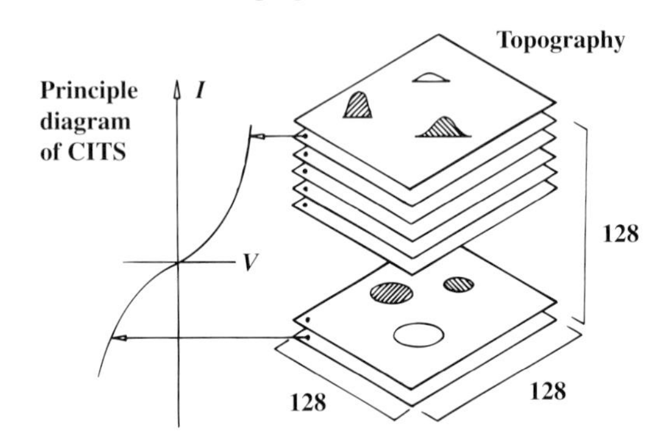

(vii) LTF (i.e., lateral modulation).

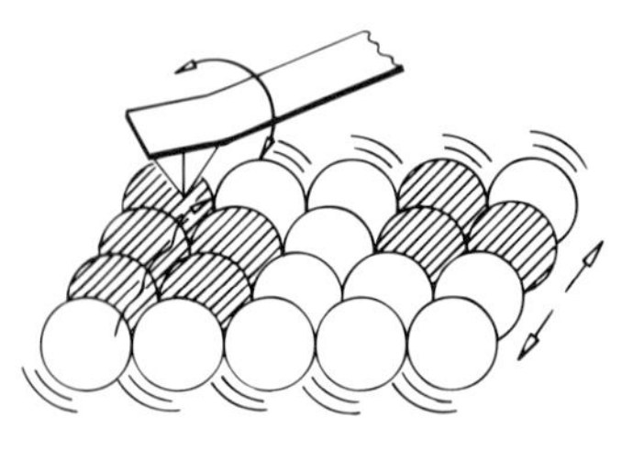

(viii) MFM (i.e. magnetic force microscope).

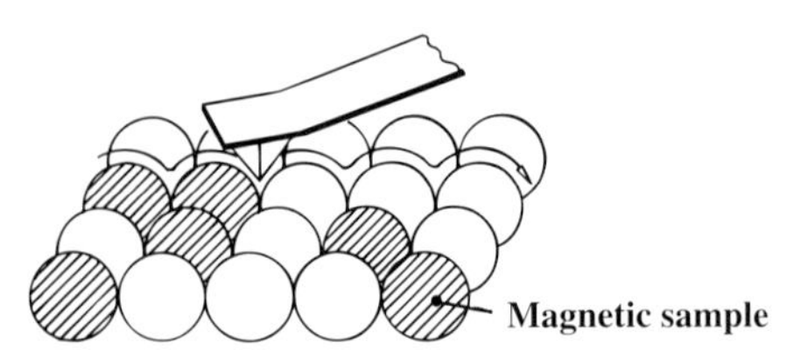

Figure 97. Atomic force/scanning probe microscope applications. [Courtesy of Jeol (UK) Ltd.]

nant frequency and using forces at or near the sample surface to damp the motion of the cantilever. Non-contact mode utilises attractive van der Waals forces to damp the cantilever's motion, allowing imaging to be undertaken without the tip contacting the sample surface. Discrete mode uses tip–surface contact to damp the motion of the cantilever. Discrete contact mode of operation is sometimes referred to as "tapping". Additionally, it is possible to use an "FM detection" method.

- *STS (scanning tunnelling spectroscopy) and CITS (current imaging tunnelling spectroscopy, Figure 97vi)* – STS is a measurement of current versus voltage (*I–V* curves). These *I–V* curves can provide an insight into the characteristics of a sample; typically these *I–V* curves are taken at a point. However, the CITS technique acquires an *I–V* curve at each point in the topographical image.
- *Lateral modulation LFM (Figure, 97vii)* – in this technique a lateral modulation signal is applied in the scanning direction while taking an LFM image. Compared to the non-modulated LFM method, this technique removes most of the "artefacts" associated with the scanning direction and sample morphology.
- *MFM (magnetic force microscope, Figure 97viii)* – this is a variation on the NC-AFM that uses a cantilever coated with a magnetic material. The magnetic coating makes the tip sensitive to magnetic domains on the sample surface. Further, the magnetic domain distribution can be simultaneously obtained with a topographical image.

Figure 98 illustrates an AFM image resulting from the face milling of an austenitic (316-grade) stainless steel specimen being machined in the ultra-high speed region – 3000 m/min. The image area is 80 μm^2, which shows the passage of the milling inserts and the periodicity of the feed motion at a rate of approximately 30 μm/cutter revolution. Such three-dimensional images can impart valuable information on the surface topography, the shearing mechanism and the influence of cutting geometry across the surface, which might otherwise remain unseen.

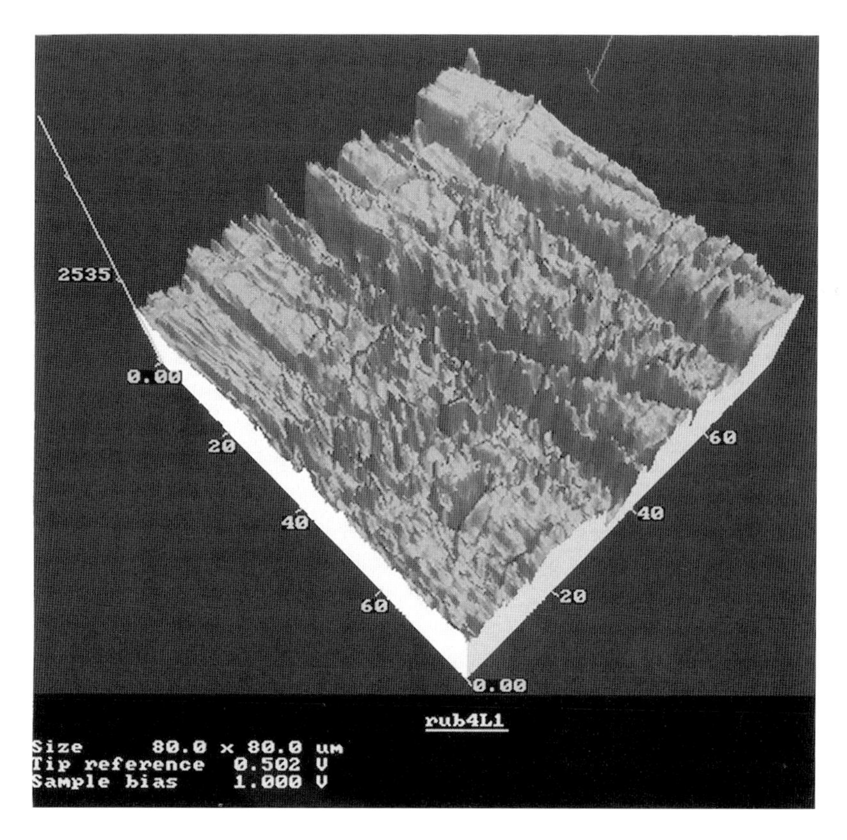

Figure 98. An atomic force microscope (AFM) 3-D image of an ultra-high-speed face-milled austenitic stainless steel surface. [Courtesy of Jeol (UK) Ltd.]

3.4.4 Ultrasonic force microscope: developments

The relatively recent development of an ultrasonic force microscope (UFM) amalgamates acoustic microscopy, with its sensitivity to elastic properties, with atomic force microscopy, having nanoscale resolution imaging for materials. In the UFM instrumental technique, a high-frequency ultrasonic vibration with a frequency of several megahertz is applied to a sample that forces it to elastically "indent" itself against a static AFM tip. This tip is positioned at the end of a force-sensitive cantilever, which has very high stiffness at megahertz frequencies (see Figure 99a). The contact between the AFM tip and the sample, being only a few nanometres in size, serves as a mechanical diode, detecting the amplitude-modulated ultrasonic vibrations in the kilohertz frequency range. The manner in which the system operates is analogous to that of a crystal radio (see Figure 99b). The UFM instrumental technique takes advantage of the intrinsic properties of the AFM cantilever, being of high stiffness at the indentation frequency (i.e., 10^2 to 10^4 times greater stiffness than at the lower modulation frequency) and has significantly better force sensitivity at this modulation frequency.

Owing to the advantages found when using the UFM, the elastic properties of many materials can be readily determined and this indicates the remarkable contrast between them, from soft polymers to very hard ceramics. For example, the AFM topographic image shown in Figure 100(a) indicates an interface on a surface as a jagged trench that divides a silicon carbide fibre (i.e., positions 1 and 2) and a mullite matrix (Al_2O_3–SiO_x) at position 4. By utilising the "elasticity image" in Figure 100(b), much greater definition – and hence detail – of the interface structure can be seen. In Figure 100(b) a relatively soft intermediate concentric carbon-rich layer (position 2) and a softer reaction layer (position 4) separate the higher-stiffness regions of both the SiC and mullite. Other images (Figure 100c and d) are also depicted, comparing topographical images that were simultaneously acquired to the FMM image (different region; Figure 100c) with an FMM image of the area shown in "c" (Figure 100d). When applied to semiconductor low-dimensional structures such as that found in quantum wells and dots, this material-dependent elastic contrast enables one to differentiate between areas of disparate material compositions – for example, image the percentage of Ge in the Si_xGe_{1-x} compound – together with highlighting the growth defects, or to detect residual polishing damage. Defect-free areas such as nanostructures can be employed to test the lateral resolution of the UFM to local elasticity (i.e., finer than 4 nm), with sensitivity differences in the elastic moduli to better than 0.5%.

UFMs have the ability to detect cracks and delaminations, based on the principle that cracks interrupt the propagation of acoustic waves – a widely used technique in ultrasonic non-destructive testing (NDT) and acoustic microscopy. The UFM has the capability of inspecting either surface or subsurface cracks to nanoscale resolution. Although the UFM instrument offers considerable scope in assessment of surface and subsurface conditions, its physics have yet to be fully understood. By way of example, the thin layers of water that cover virtually every surface in an ambient environment act at ultrasonic frequencies as a tough sticky film, modifying the UFM signal. Fortunately for the aspect of UFM imaging, the images can be acquired at different "normal" forces and this enables differentiation of the elastic and adhesive contributions, resulting from the influence of minute levels of water. This approach to discrimination of contributing effects suggests a way towards quantitative measurements of nanoscale elastic and surface adhesive properties, while another manner of controlling the surface adhesion problem is to use the UFM in a liquid environment. This development of an "underwater UFM" not only reduces potential damage to delicate material, but also enables a direct study of nanoscale elastic properties of materials such as biopolymers and medical materials in their natural environment.

One of the latest discoveries is that ultrasonic waves can be directly propagated from the base of the minute AFM cantilever, travelling down to its end, in a similar manner to light propagating along an optical fibre. These ultrasonic waves can vibrate the AFM tip at megahertz and even sub-gigahertz frequencies. The cantilever simultaneously acts as a stiff and high-quality conductor of high-frequency ultrasonic vibrations, while also acting as a gentle, sensitive detector of non-linear ultrasonic force at the sample frequency of several kilohertz. This dual ability is of significant benefit, enabling bulk material samples of materials to be studied by the UFM, as access is only necessary from one side of the sample.

An intriguing further development in the UFM is the combination of ultrasonic vibrations from both the sample and cantilever. When such vibrations occur at adjacent frequencies, the AFM tip detects the oscillating force at the difference frequency, in a similar manner to that of a heterodyne radio receiver, prompting the name for this new instrument: heterodyne force microscope (HFM). Just as modern heterodyne radios are far superior to crystal detectors, the HFM has superior sensitivity to small vibrations than that of a UFM. Even more significant

(a)

Additional piezotransducer allows the launch of an ultrasonic wave down the cantilever in hetrodyne force microscope (HFM)

Microfabricated cantilever

Laser

Position-sensitive photodetector monitors average force on the tip

AFM tip

Studied sample is attached to the piezotransducer and placed on the AFM scanning stage

Piezotransducer-vibrated sample is attached to the piezotransducer and placed on the AFM scanning stage

(b)

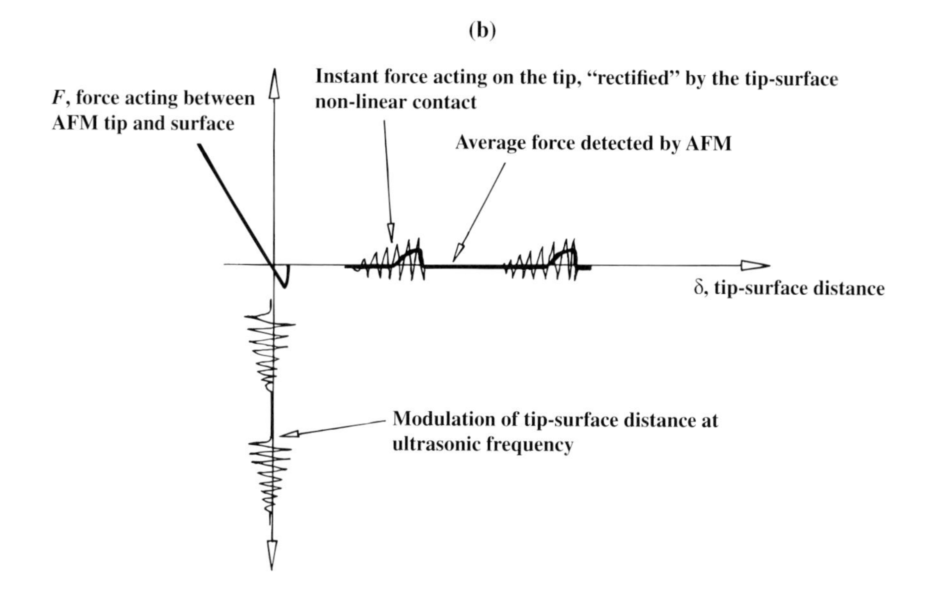

Figure 99. (a) Design of the ultrasonic force microscope. (b) Principle of non-linear detection of ultrasonic vibration in UFM. [Courtesy of Dr O. Kolosov, Isis Innovation and University of Oxford.]

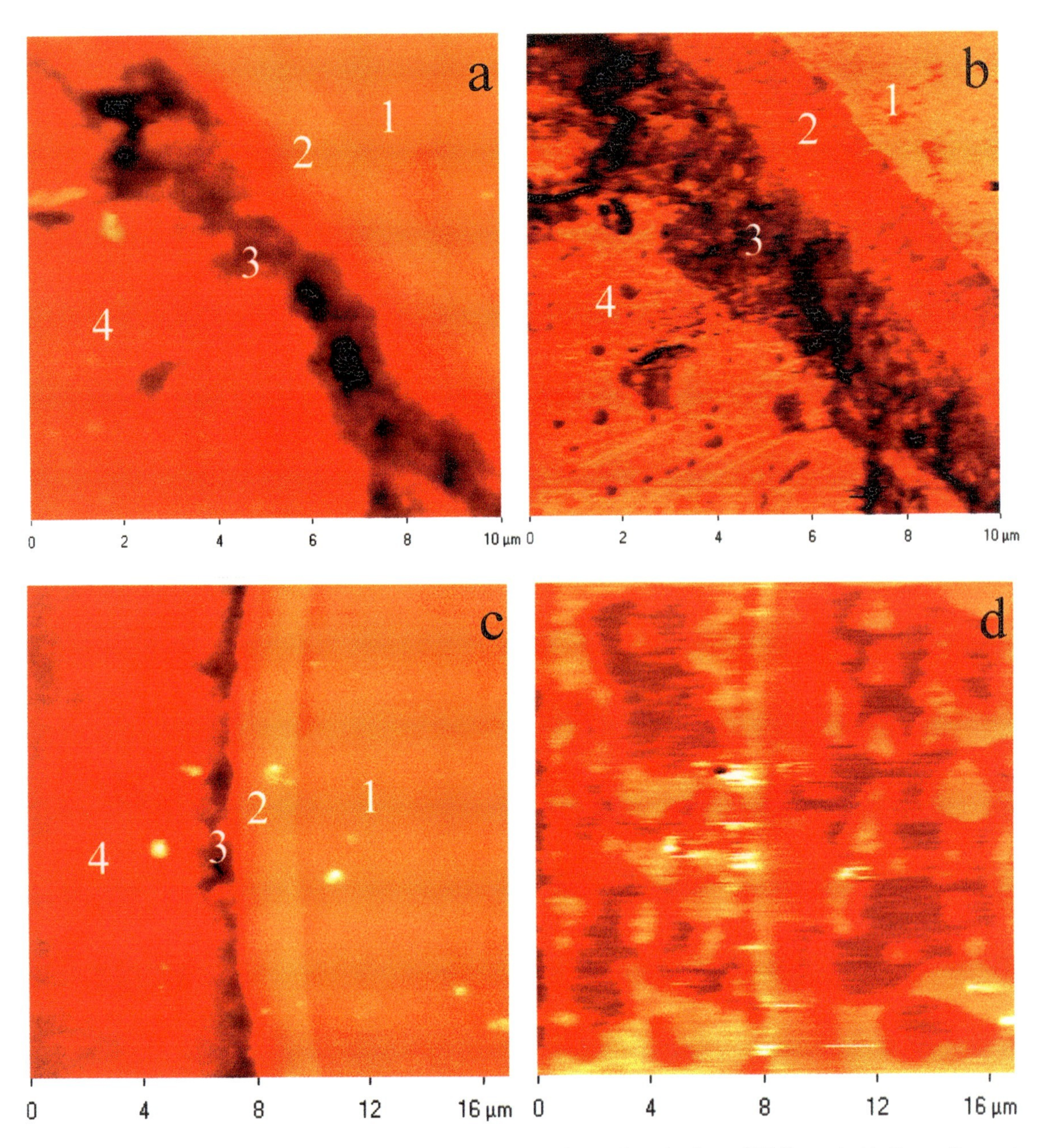

Figure 100. Topography (a) and UFM (b), topography and FMM (d) images of a sample made of coated SiC fibre embedded in a mullite matrix ($A1_2O_3$-SiO_2), image width 10 μm. [Courtesy of Dr O. Kolosov, Isis Innovation and University of Oxford.]

is the fact that the HFM picks up information about the phase of high-frequency vibrations, which are directly related to the time-dependent processes of tip–surface force interaction. This will enable observations of the mechanical structure of materials with both nanometre lateral resolution and nanosecond time sensitivity.

3.5 X-ray photoelectron spectroscopy (XPS)

For most technologies today, crucially important information is required on surfaces and associated chemical composition. This surface detail might include corrosion, lubrication, adhesion, micro-electronics features, catalysis and surface coatings. For

(a) Photoelectron spectrum of lead superimposed on a schematic diagram of the electronic structure of the element

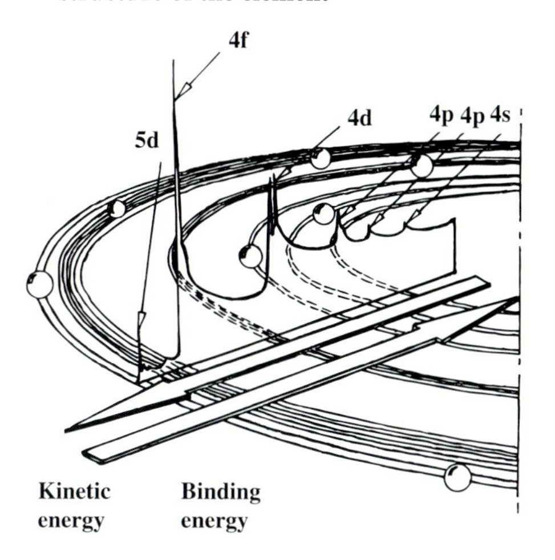

(b) Sigma Probe is an XPS instrument and is optimised for analysis of small areas.

(c) The important features of a typical XPS instrument.

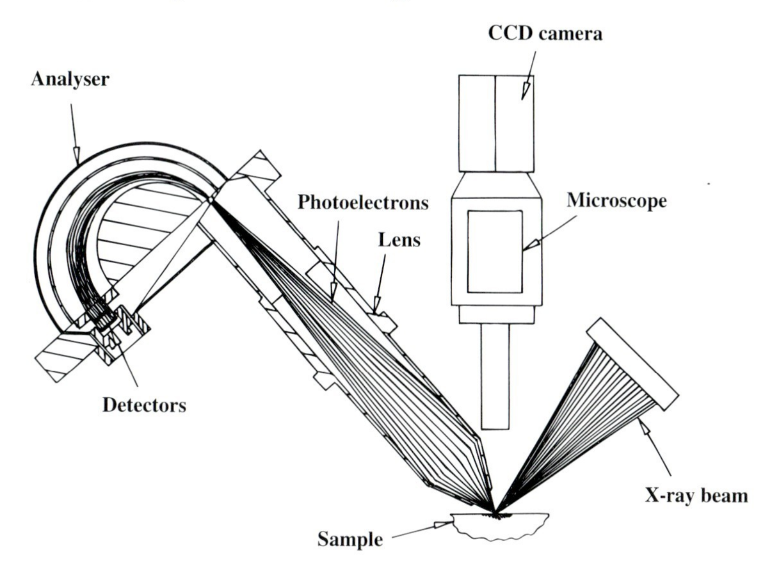

Figure 101. The important elements of an X-ray photoelectron spectroscopic (XPS) instrument. [Courtesy of VG Scientific.]

example, even factors such as the sub-monolayers of material can affect bonding properties. Hence, gaining an understanding of these surfaces requires sophisticated and powerful analysis techniques, typified by that of X-ray photoelectron spectroscopy (XPS).

The XPS instrumental technique and associated equipment are shown in Figure 101; the technique is sometimes referred to as "electron spectroscopy" (for chemical analysis) and has the following characteristics:

- *surface sensitivity* – typically, the information depth is less than 10 nm; by comparison, an EDX information depth might be several micrometres;
- *elemental detection* – all elements with the exception of H and He can be detected;
- *chemical state identification* – chemical shifts in

(a) Spectrum of barium acquired in 10 seconds using a monochromated X-ray source

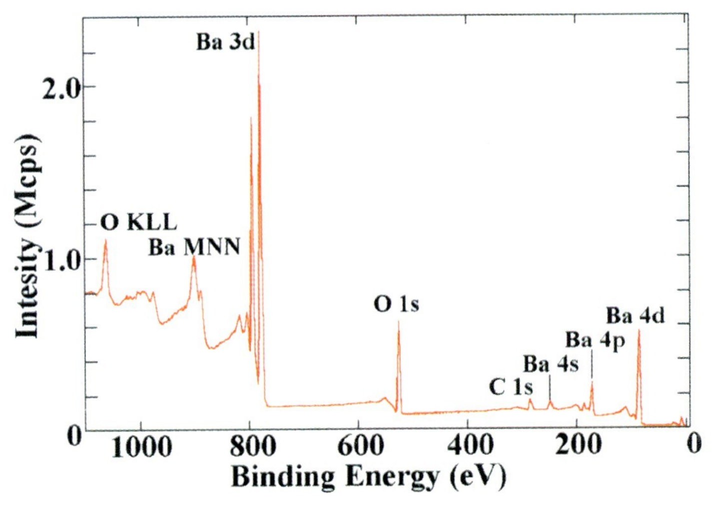

(b) An overlay of two silicon images.

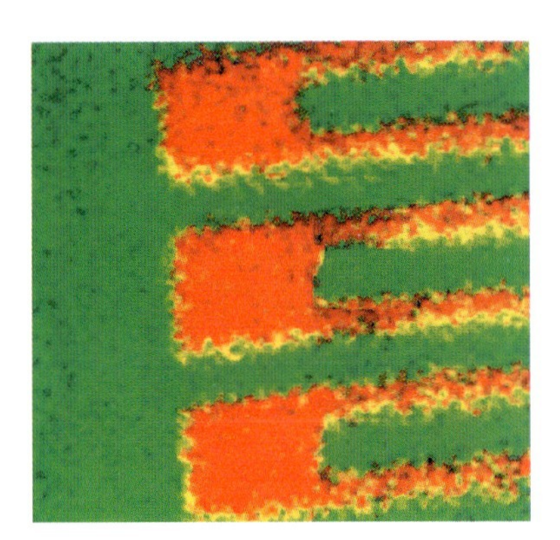

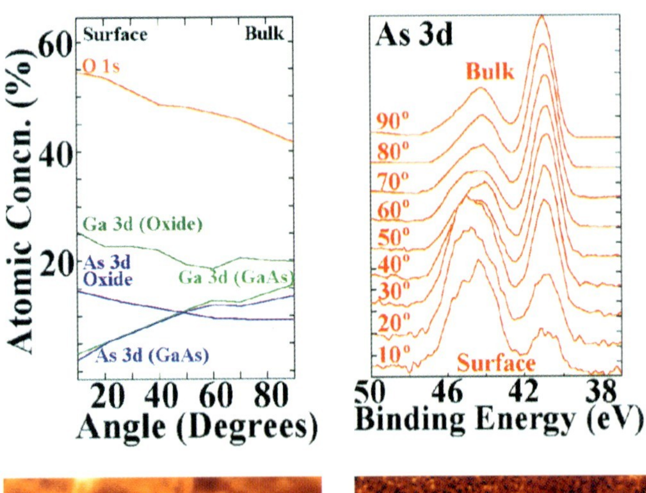

(c) ARXPS measurements from a GaAs surface, showing the presence of an oxidised region at the surface. Right: a montage of
As 3-D spectra taken at a series of angles, showing the change in relative intensity of the As in the oxide and GaAs states

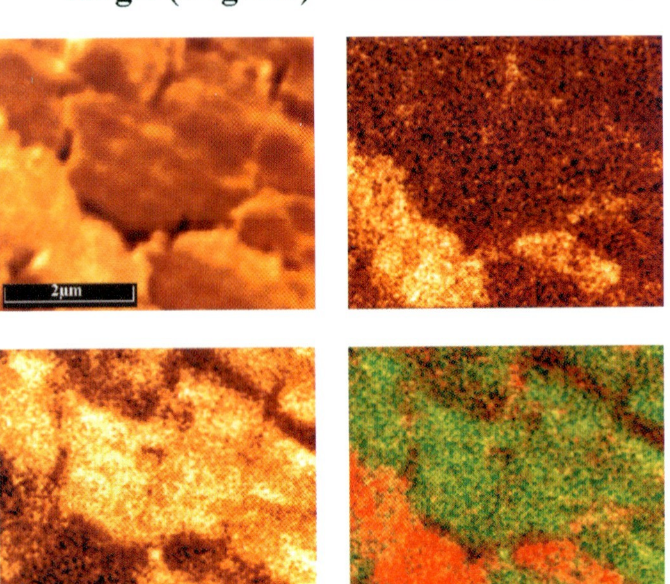

(d) Images from a 5 μm field of view using 2.7 nA of beam current at 5 eV. The spatial resolution in these images is 100 nm. (Top left: SEM; top right: copper; bottom left; titanium; bottom right: copper/titanium overlay)

Figure 102. Just some of the X-ray photoelectron spectroscopic (XPS) graphs and images available from quantitative surface assessment. [Courtesy of VG Scientific.]

the spectrum indicate the chemical state of the element;

- *quantitative analysis* – the sensitivity of XPS to an element is independent of the matrix in which it is found. Sensitivity factors can be used to quantify the spectra produced.

The XPS method relies upon the *photoelectric effect*, where the material to be analysed absorbs the X-rays and electrons are rejected; these are then detected and their kinetic energy analysed. The kinetic energy of a photoelectron depends upon

1. the element from which it was ejected;
2. the orbital in the atom from which the electron originated;
3. the chemical state of that element.

As an example of this XPS spectrum, that of lead is shown in Figure 101(a), superimposed onto a diagram showing its electronic structure. This diagram indicates where the electrons that form each of the peaks originate.

All XPS instruments have three major components: the X-ray source, the energy analyser and the detector (see Figure 101c). The X-ray source can be either monochromated, or non-monochromated. It is accepted that monochromators provide an improved energy resolution over their non-monochromated counterparts, providing a focused spot size of the X-ray beam to smaller than 15 μm – which can be the case. The energy analyser is usually of the spherical sector type. A lens is normally fitted to the energy analyser to maximise the number of photoelectrons entering the analyser and to define the area from which the electrons are collected. The detector is usually an array of channel electron multipliers or a channel plate. The complete instrument must be housed in an ultra-high-vacuum (UHV) system, so that surfaces remain uncontaminated and the photoelectrons have a sufficiently large mean free path to reach the detector. A typical variant of this type of instrument is shown in Figure 101(b), where it has a spatial resolution that is less than 15 μm.

An XPS experimental procedure would usually commence with the acquisition of a survey spectrum, to identify the elements present in the specimen. This acquisition tends to be rapidly taken over a wide range of binding energies at low resolution (see Figure 102a). After acquiring the survey spectrum, a series of high-resolution spectra are normally acquired for each element of interest, providing quantitative information about the surface region. For many technological disciplines, it is essential to be able to analyse very small surface features. These surface features may be such problems as conductive tracks on semiconductor materials, areas of impurity on ceramics, or likewise for polymers impurities. Apart from taking spectra from small areas, some XPS instruments produce chemical images or maps from the surface of the specimen. The illustration shown in Figure 102(b) depicts an overlay of two XPS images from a semiconductor device. Both images use the Si 2p peak, but one (red) is of elemental silicon, while the other (green) is from silicon in the form of its oxide. These are representations of chemical state images and have a spatial resolution of approximately 1 μm.

X-rays generate photoelectrons from deep within the matrix, with those in the top several nanometres being able to escape from the solid without undergoing inelastic collisions. If the electrons are collected along the surface normal, then considerable information can come from this region, near the surface. However, if electrons are collected at a grazing emission, then the surface sensitivity is further increased. The escape depth varies with the sine of the take-off angle, relative to the surface. In angle-resolved XPS (ARXPS), this phenomenon is exploited to study the variation of composition of the specimen with depth. An ARXPS assessment is usually performed by setting the angular acceptance of the spectrometer to a small value (typically 1–10°) and measuring the XPS spectra at a range of sample angles. Figure 102(c) shows ARXPS data from GaAs, which has a thin oxide layer at its surface. Measurements were made of the Ga, As and O peaks as a function of angle and for the Ga and As both the elemental and oxide peaks were measured. The data observed at small angles is dominated by the surface composition, while the data at larger angles has a greater influence from the bulk material. The decrease in the oxygen and oxide peaks will therefore indicate the presence of a thin surface oxide.

Many more analytical activities can be undertaken with an XPS instrument, such as surface topography (Figure 102d), depth profiles or sputter profiling, to identify impurities at interfaces and so on, as it is a quantitative chemically specific analysis instrument for both conducting and non-conducting surfaces. A wide range of XPS-based instruments are available to suit specific and general analysis needs, often being the principal surface analytical instrumentation purchased for surface science laboratory and industrial studies.

References

Journal and conference papers

Dinelli, F. et al. Elastic mapping of heterogeneous nano-structures with ultrasonic force microscope. *Proceedings of Surface and Interface Analysis*, Basel, Switzerland, 1999.

Griffith, J.E. et al. Characterization of scanning probe microscope tips for linewidth measurement. *Journal of Vacuum Science and Technology* B9 (6), 1991, 3586–3589.

Hill-King, M. Measurement in the detail. *Quality Today* March 1999, 22–26.

Hibino, H. et al. High-temperature scanning-tunneling-microscopy observation of phase transitions and reconstruction of a vicinal Si (111) surface. *Physical Review* B47, 1993, 13027–13030.

Hutter, J.L. and Bechhoefer, J. Calibration of atomic force microscope tips. *Reviews of Scientific Instruments* 64, 1993, 1868–1873.

Ichimiya, T. et al. Formation of (21 ((21 structure by gold deposition on Si (111) (3 ((3 Ag surface and the wavering behavior. *Surface Review and Letters* 1, 1994.

Kolosov, O. and Yamanaka, K. Nonlinear detection of ultrasonic vibrations in an atomic force microscope. *Japanese Journal of Applied Physics* 32(Part 2, No. 8a), 1993, L1095–L1098.

Kolosov, O. Imaging the elastic nanostructure of Ge islands by ultrasonic force microscopy. *Physical Review Letters* 81(5), 1998, 1046–1049.

Kolosov, O. UFM shakes out the details at the nanoscopic scale. *Materials World* December 1998, 753–754.

Kuk, Y. and Silverman, P.J. Scanning tunneling microscope instrumentation. *Review of Scientific Instrumentation* 60(2), 1989, 165–181.

Markoff, J. A novel microscope probes the ultra small. *New York Times* 23 February 1993, Ci, C8.

Nechay, B.A. et al. Applications of an atomic force microscope voltage probe with ultrafast time resolution. *Journal of Vacuum Science and Technology* May/June 1995, 1369–1374.

Sato, T. et al. Dynamic observation of Ag deposition process on Si (111) surface by high-temperature scanning tunneling microscopy. *Japanese Journal of Applied Physics* 32, 1993, 2923–2928.

Schöfer, J. and Santer, E. Quantitative wear analysis using atomic force microscopy. *Wear* 222(2), 1998, 74–83.

Stahl, U. et al. Atomic force microscope using piezoresistive cantilevers with a scanning electron microscope. *Applied Physics Letters* 65, 1994, 2878–2880.

Uchida, H. et al. Single-atom manipulation on the Si (111) 7 × 7 surface by scanning tunneling microscopy (STM). *Surface Science* 287/288, 1993, 1056–1061.

Warren, P.D. et al. Characterisation of surface damage via contact probes. *Nanotechnology* 7, 1996, 288–294.

Wolstenholme, J. XPS reveals all about the state of the surface. *Materials World* 7(7), 1999, 412–414.

Yoshimura, M. et al. Low-coverage, low-temperature phase of Al overlays on the Si (111) (-7 (7 structure by scanning tunneling microscopy. Physical Review B47, 1993, 13930–13932.

Yuan, C.W. et al. Low temperature magnetic force microscope utilizing a piezoresistive cantilever. *Applied Physics Letters* 65, 1994, 1308–1310.

Books, booklets and guides

Barrett, C. and Massalski, T.B. *Structure of Metals* (3rd edn). Pergamon Press, Oxford, 1980.

Briggs, D. and Seah, M.P. *Practical Surface Analysis*, Vol. 1. Wiley, New York.

Cullity, B.D. *Elements of X-ray Diffraction*. Addison-Wesley, Reading, MA.

Gonzales, R.C. and Woods, R.E. *Digital Image Processing*. Addison-Wesley, Reading, MA, 1992.

Hull, D. *Introduction to Dislocations* (2nd edn). Pergamon Press, Oxford, 1979.

Jeol (UK) Ltd. *Invitation to the SEM World.*

Howland, R. and Benatar, L. *A Practical Guide to Scanning Probe Microscopy*. Nikon Corp./Park Scientific Instruments, 1996.

Kuzmany, H. *Solid-State Spectroscopy*. Springer, Berlin, 1998.

Vickerman, J.C. *Surface Analysis: The Principal Techniques*. Wiley, New York, 2000.

Walls, J.M. *Methods of Surface Analysis*. Cambridge University Press.

Watts, J.F. *An Introduction to Surface Analysis by Electron Spectroscopy*. Oxford Science.

Wiesendanger, R. *Scanning Probe Microscopy*. Springer, Berlin, 1998.

Yates Jr, J.T. *Experimental Innovations in Surface Science*. Springer, Berlin, 1997.

Desjonquères, M.-C. and Spanjaard, D. *Concepts in Surface Physics*. Springer, Berlin, 1996.

Roundness and cylindricity

"Round and round the rugged rock
The ragged rascal ran."

(Nursery rhyme)

Roundness – setting the scene

In the past, the simplistic technique for the assessment of roundness was measuring two or three diameters on a component to determine diametral variations, with symmetrical part features being assessed for variations in radius about an axis of rotation. These simple roundness measurement techniques give information about geometric and harmonic errors about an axis of rotation, therefore could only be employed as a very superficial guide to a part's potential in-service performance. If there exists from within a selected cross-section a point from which all subsequent points on the periphery of a component are equidistant, then geometrically at this plane – position where measurement occurs – the component can be said to be round. In reality, the radius of, say, a nominally round component tends to deviate – from a true circle – around the periphery of the part, making these variations subject to subjective interpretation of results. Although it should be emphasised that variations in a component's radius are not always detrimental to its potential performance, certain parts which can exhibit periodic radius fluctuation may operate as though they have a constant diameter – that is may be lobed in nature (see Figures 107 and 109). There might be a whole host of reasons for a component to vary in its radius for a stated cross-sectional plane, some of which might be:

- the production process inducing rotational imperfections – from the machine/tool/workpiece system;
- the release of strain, or that induced in the material – the former case may be the result of releasing the part from clamping pressure, while the latter may be the result of plastic deformation that promotes localised surface residual hoop stresses;
- induced radial vibration – potentially resulting from tool displacement;
- circumferential surface texture – residual effect resulting from the production process.

Roundness is a condition of a *surface of revolution*, which can take the form of a cylinder, cone or sphere, where all the peripheral data points intersect. The question that can be asked is: "Why is it so important to be able to establish close control over departures from ideal roundness?" Firstly, a circular cross-section is the most fundamental shape utilised by industry and is generated or formed by the manufacturing process. Secondly, due to the symmetry of a round component, it may exhibit uniform mechanical strength in all directions that are symmetrical to the axis. The dimensional and geometric characteristics of the part will have a significant effect on the in-service performance of the component in any precision work, etc, where either the male/female fitment is vital and/or its rotational assembly must be strictly controlled. The "pressure tightness" in component sealing applications will also be heavily influenced by the manufactured accuracy and precision of assemblies, due to departures from roundness factors. Ball and roller bearings must be round and produced to tight tolerances and have good surface texture. These bearings would otherwise potentially have a reduced bearing life and reliability. Furthermore, such bearings will have increased noise and vibration tendencies, while providing some variation in torque-related assembles. To give a practical example of just one type of potential operational condition where roundness is important, the following is worth consideration. In a rotating ball bearing application, if a minute roundness error occurred in the balls of not greater than 0.25 μm, this error would result in an increase in operational noise by of approximately 15 dBA, which greatly increases the sound level that would otherwise be caused by either track wobble or lack of shoulder squareness. Therefore it can be seen that a part's roundness can have far-reaching effects well beyond that of the simple expedient of its precision manufacture.

4.1 Introduction

Many devices and instruments critically depend upon rotation, coupled in many cases to controlled linear motion. Therefore, an automobile and machine tool could not operate without such kinematics, nor would the smallest watch or the largest power station. Many factories engaged in engineering activities such as precision manufacture have equipment that produces components: bearings, shafts, bushes – the list continues – but all have one significant feature: they have radial symmetry and are *round*. The question is: "How round?"

Departure from roundness assessment, or a more convenient term, *roundness*, has been historically practised by inspectors or by craft-based engineers. Practitioners without access to sophisticated metrology equipment assess the roundness of a component by carefully rotating it in a vee-block, or between bench centres, then measuring the run-out with a dial gauge (Figure 103). This technique, (BS 3730: Part 3) although useful for elementary roundness inspection, has its limitations and may result in misleading results – somewhat negating the simplicity of the operation, but this topic will be discussed further shortly.

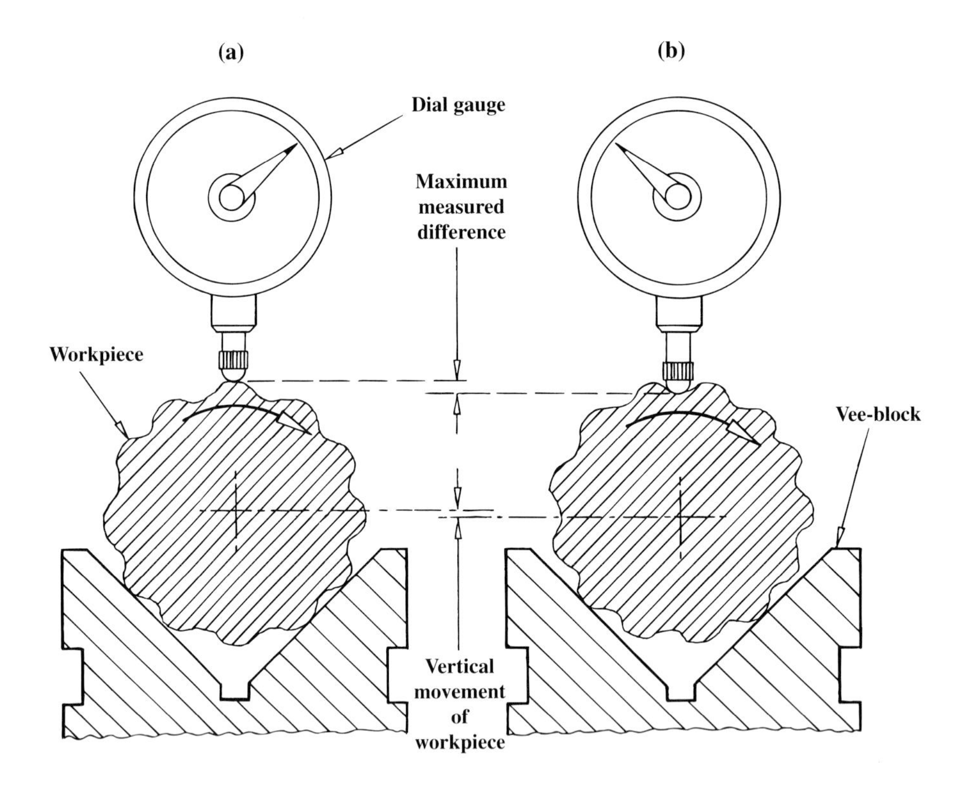

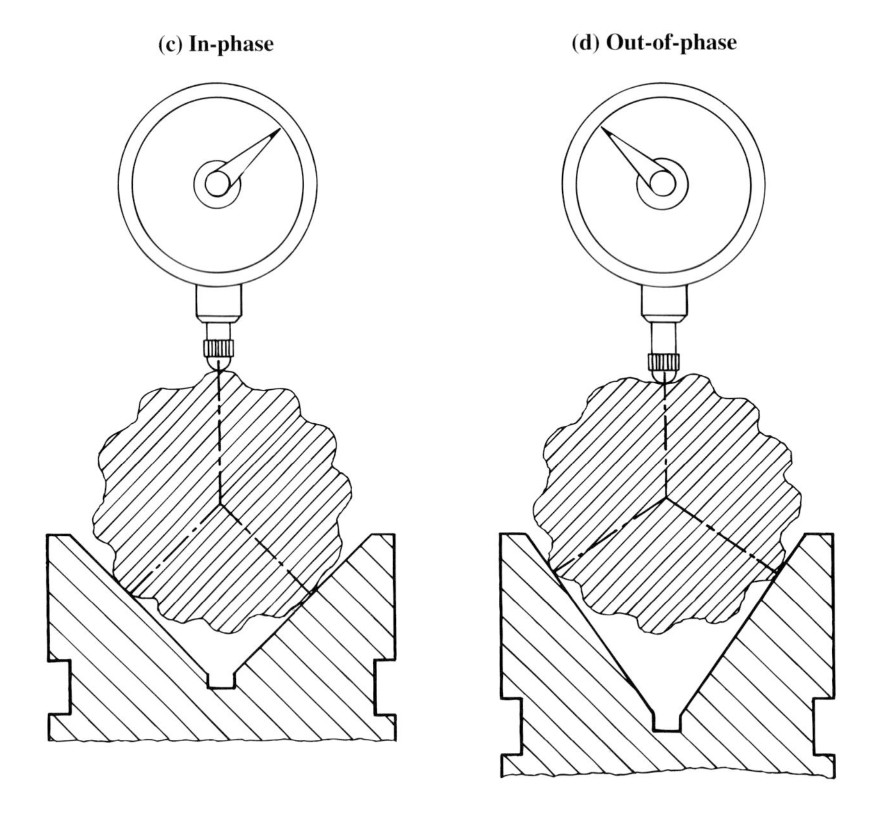

Figure 103. Assessment of roundness in a Vee-block. (Courtesy of Taylor Hobson.)

With industrial demands being firmly focused on precision manufacture, necessitating the requirement for even closer tolerances and greater accuracy in roundness measurement, this has meant that an even greater emphasis is necessary on the control of geometric form. During the last sixty years instruments have been developed with greater levels of sophistication, for measuring a component's roundness and any associated parameters, many of which are still in common use today. It is important that modern-day engineers, designers, machine tool operators and inspectors have a basic understanding of the working principles of these instruments, together with their capabilities and limitations and how to interpret subsequent roundness results. Roundness results may be used to either accept or reject round components. Moreover, results from roundness inspection can be employed to monitor a range of process-related component manufacturing capabilities, typically the performance of a machine tool, tool wear detection, being the resulting effects of poor operating procedures or similar in-service production problems.

Why is roundness important?

A plain bearing and its mating shaft are depicted in Figure 104. Provided that both the bearing and shaft are round and the clearance fit is neither too tight nor too loose, then the shaft will run smoothly, particularly if full-film hydrodynamic lubrication is present. The questions relating to the bearing fitment that might be asked are:

- Is this fit sufficient?
- Will a combination of shaft and bearing continue to offer reasonable in-service performance when heavily loaded and after many years of operation?
- Does the lubricant have ideal conditions to work effectively?

The answer to these questions must depend upon what we consider the word *round* to mean. Superficially, the bearing and shaft fitment illustrated in Figure 104 may appear to be round when visually assessed; it might also have a constant diameter when measured with a micrometer. However, if this same bearing/shaft relationship is considerably magnified, then a different picture arises, as shown by the magnified inset image in Figure 104. From this image it is apparent that departures from the "true" circular profile exist and that these lobes at points "A" will carry most of the load of the shaft as it runs in the plain bearing. The lubricating film thickness must ideally be kept within specific limits if the bearing assembly is to perform as the designer intended, although in this current situation the thickness of the oil film at positions "B" will be considerably greater than at points "A". In a similar manner, the bearing's bore may not be circular: it could be either slightly oval or lobed (having a constant diameter, but not geometrically round – see Figure 109), creating variable lubricating effects to the complete assembly which were not as the designer intended.

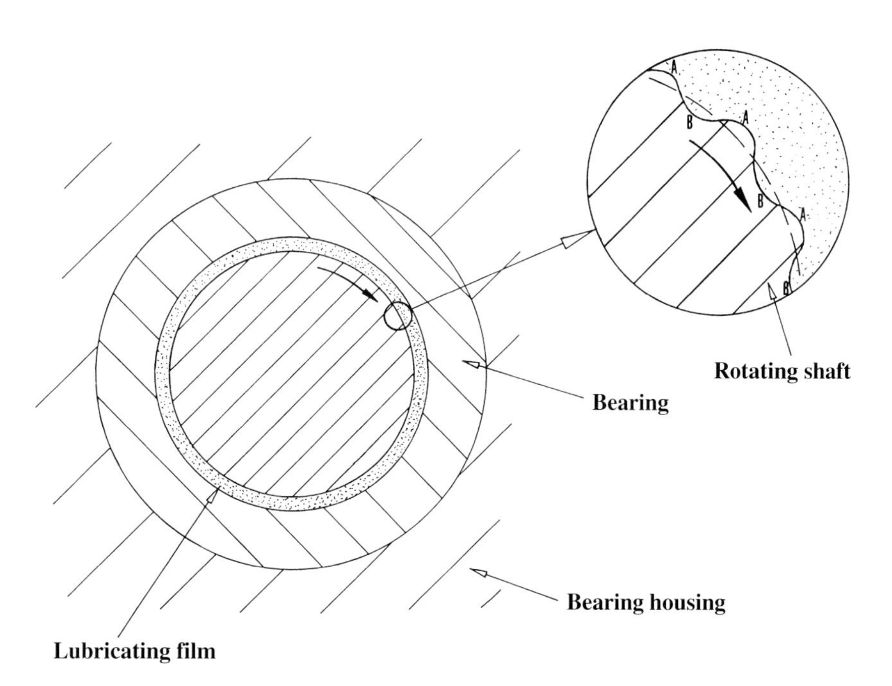

Figure 104. Roundness of a bearing will affect its performance.

Further aspects of the measurement of cylindrical components

The discussion so far has only been related to that of shaft and bearing assemblies, with roundness only described from a single cross-sectional viewpoint. However, other information needs to be known which is connected to circular features in engineering components. For example, in the component errors depicted in Figure 105, it might be necessary to investigate whether

(a) two different diameter portions are concentric with one another;
(b) the bore of the thick-walled tube is coaxial with the outside diameter;
(c) the bore is straight;
(d) the horizontal face is square to that of the axis;
(e) bore is tapered, straight and coaxial with the outside, also that the diameters are concentric and the recessed undercut is square to the axis.

In these examples (Figure 105), the components can be considered to have comparatively simple geometric features; conversely, this is not the case for a crankshaft for an automotive application (Figure 106). This crankshaft diagram indicates that the relative geometry of bearing surfaces becomes critical to its desired operating efficiency. Inspecting this type of component, having a series of complex and interrelated geometric elemental features, is very much within the scope of roundness measuring instruments.

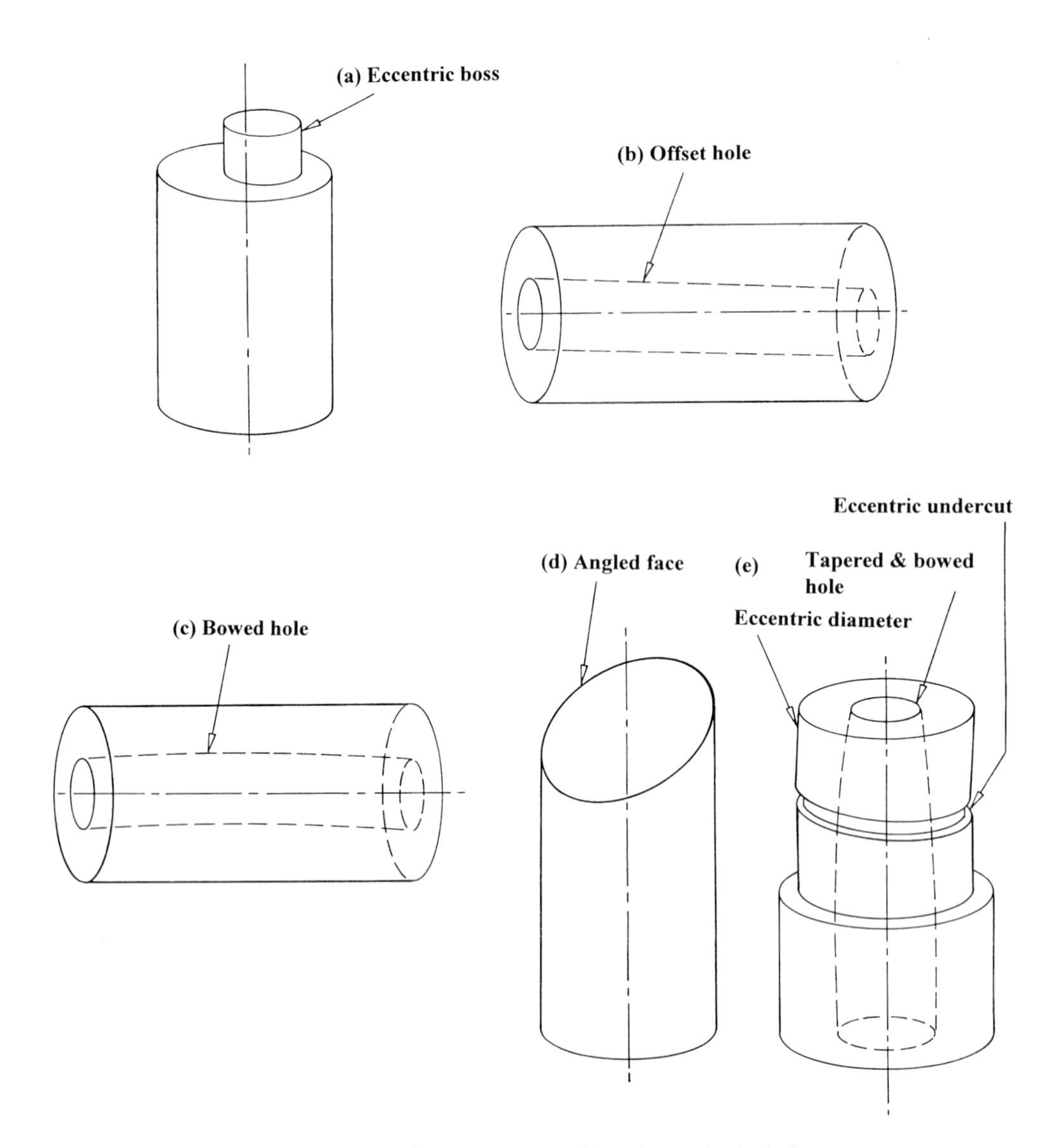

Figure 105. Typical component errors of form that can be checked.

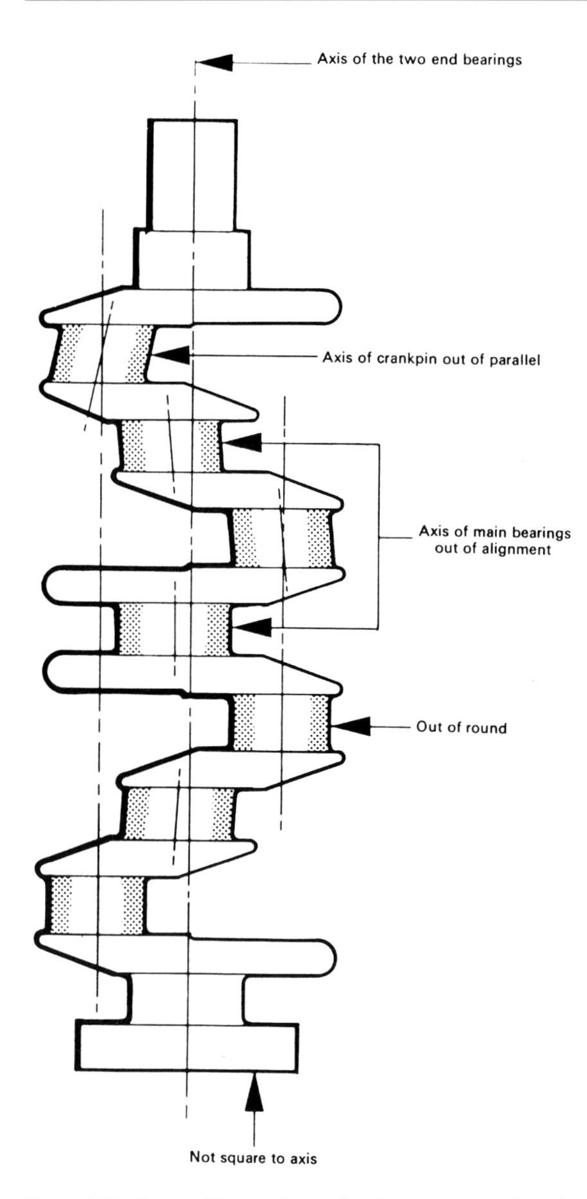

Figure 106. The possible roundness-related errors in a typical crankshaft's geometry. (Courtesy of Taylor Hobson.)

Roundness measurement: how is it achieved?

In general terms, a part can be said to be *round* in a specific cross-section if there exists within the sectional plane a point – namely its centre – from which all other points on the periphery are equidistant. If these geometric conditions are met, then the cross-section will be a perfect circle (see Figure 107 "A"). If the cross-sectional geometry does not equate to a perfect circle, then two conditions that might occur are:

- an elliptical shape – having a major and minor axis of different lengths (Figure 107 "B");
- a trochoidal or three-lobed shape – exhibiting a difference in distance points on the periphery to its centre (Figure 107 "C").

From the lobed shape in Figure 107 "C", if "R1" is the maximum distance of the periphery from the centre and "R2" the minimum distance, then the roundness will be "R1 − R2". When the profile is symmetrical this value is simple to establish; however, specifying the departures from roundness for an irregular profile like the example shown in Figure 108 is only possible if the "centre" of the part can be found, from where the measurements are determined. The question is: "where is the centre?" If an arbitrary point "A" is established where we estimate this point to be somewhere near the true centre, the maximum departures from roundness might be "R1 − R2", but if we were to choose a different centre "B", then the maximum roundness could be "R3 − R4". So, determining the centre from where the variation of the profile is an important factor of roundness assessment and more will be said on this topic later.

To measure roundness, it is often the practice to include a *rotational* factor to this measurement; conversely, diametral measurement is normally undertaken in a *static* manner, perhaps utilising either a micrometer or coordinate measuring machine (CMM). Although a component's roundness and diameter are two distinct parameters, its roundness, or more particularly its departure from roundness, does

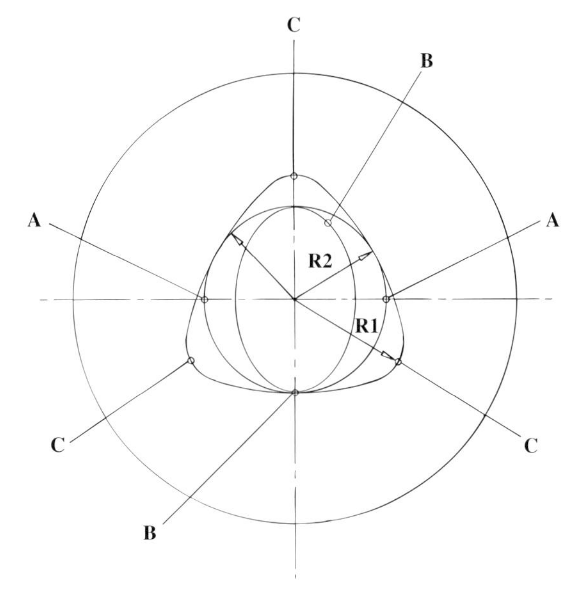

Figure 107. Typical roundness geometric shapes: A, "true" roundness; B, elliptical shape; C, trochoidal (i.e., lobed) shape. NB they all have the same diameter when measured at specific points.

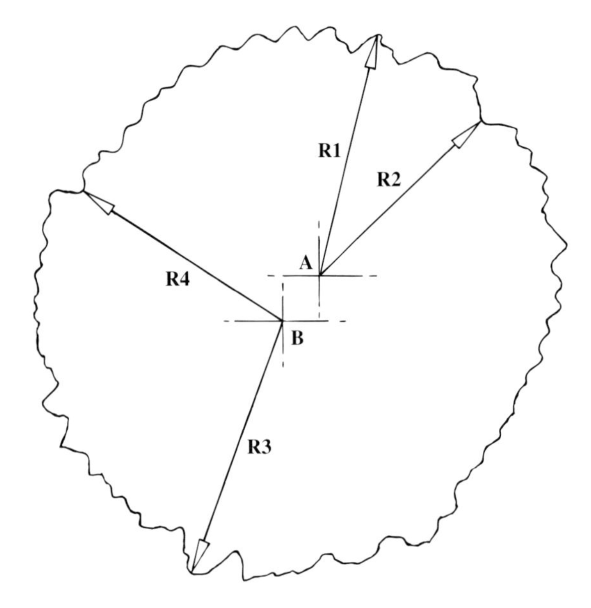

Figure 108. Determination of the centre of a roundness profile is an important factor in predicting its roundness measurement.

have a practical effect on the measurement of its diameter – possibly making these diametral measurements very misleading.

Differences in measured and effective size

The representation of the three-lobed shape illustrated in Figure 109(a, left), has a measured size of exactly 25 mm; therefore it would be expected to fit into an appropriately toleranced hole of ϕ25 mm. Here, the term *measured size* refers to the diameter as measured between a pair of parallel faces by either a micrometer or limit gauge (gap gauge). In reality, the hole must be increased to ϕ28.9 mm before a mating fit can be established (shown in Figure 109a, middle). Similarly, a three-lobed hole (Figure 109a, right) that has a measured size of ϕ25 mm will not allow a truly round ϕ25 mm shaft to fit. In fact, the largest shaft that can be located in this lobed hole has a considerably smaller size, namely of ϕ21.1 mm. Hence, the *effective size* of these two mating components will be ϕ28.9 mm for the shaft and ϕ21.1 mm for the hole. This lobed hole and shaft geometric relationship is still further reinforced in Figure 109(b), where similar differences are exhibited by parts having odd-numbered lobes, so the shape – *roundness* – of a component can affect its size when measured in the conventional way.

Causes of departures from roundness

The determination of a component's roundness necessitates the measurement of irregularities. Such irregularities are normally referred to as undulations, peaks and valleys, and lobes – this latter term describes regularly spaced undulations (which are usually few in number). It is important to realise that undulations and lobes do actually exist on nominally round components and are not simply theoretical concepts, but are attributable to all machining and process-related operations.

The reasons for possible departures from ideal roundness of mechanical parts might be due to a number of process-related factors, such as some poorly maintained bearings in a machine tool, or by deflections of the workpiece resulting from the tool experiencing interrupted cuts or differential material thicknesses. Roundness departures may also occur due to shafts being either ground, or turned which are inadequately supported, causing deflections, or centres that are not correctly aligned.

A particularly difficult problem to avoid in many machining processes is that of lobing, particularly when the component is manufactured from either round bar-stock or ring-type parts that are securely held in either a three- or five-jaw self-centring chuck (for turning and grinding, respectively). The part's periphery will be compressed at the points of contact for individual jaws, giving rise to compressive stresses in the material. After a perfectly circular diameter has been bored, upon its release from the chuck a lobing effect will be seen, this lobing being related to the number of chuck jaws. Centreless grinding operations are yet another source of lobing error, particularly where the original bar stock has a poor and irregular shape, which is exacerbated by the machine's work-rest being incorrectly positioned beneath the centreless ground part. Tool marks on turned parts are not readily described as roundness features, whereas a worn tool, or one which is set incorrectly, could increase the tendency for surface chatter marks on the component, showing up in subsequent roundness measurements. These production variables and their associated process, tool and machine-related factors will be discussed in more depth in this and the following chapter.

Components that are either drawn or extruded take their final shape from the drawing/extrusion dies. Roundness checks on such parts will not only reveal changes of shape of the die as it wears, but can also highlight imperfections that are the result of scoring along the surface, as drawing/extrusion occurs.

(a) Geometric relationships of lobed-shaped, holes and shafts.

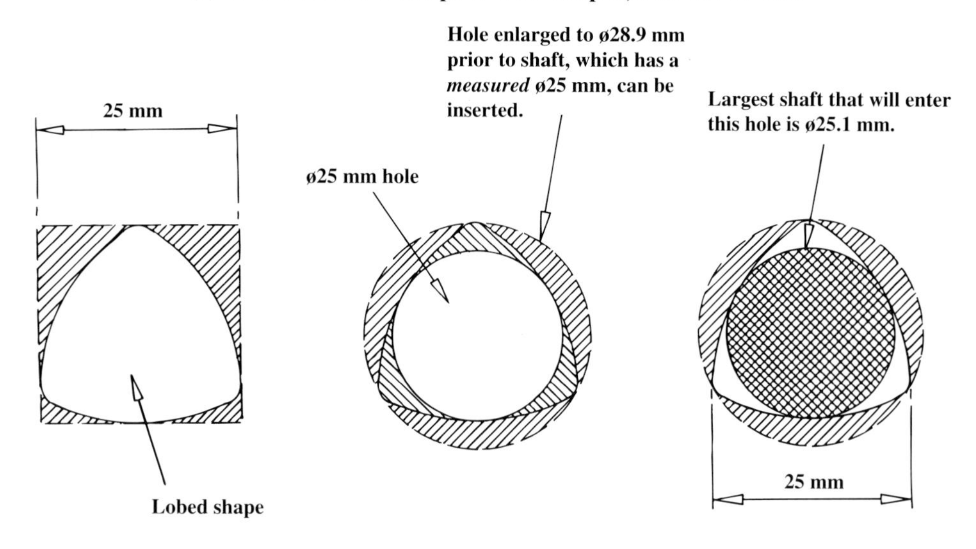

(b) As lobing increases, holes become larger and diameters smaller.

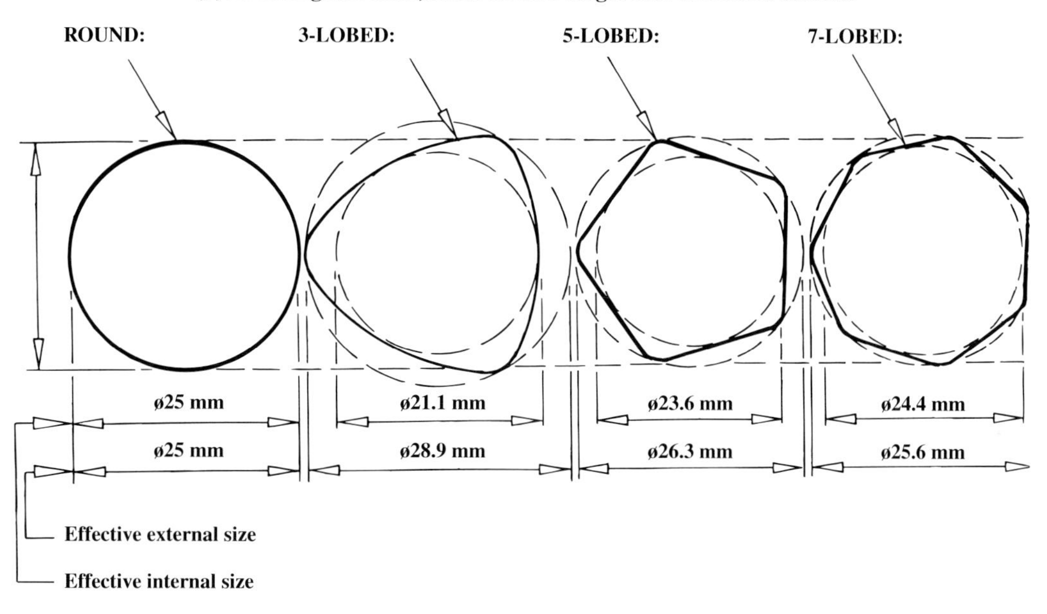

Figure 109. Differences between measured and effective sizes. (Courtesy of Taylor Hobson.)

4.1.1 Roundness measurement: basic approach

In the previous section it was mentioned that departure from roundness cannot always be fully detected, or indeed measured by dial guage *static* techniques (BS3730: Part 3). If access to sophisticated roundness instrumentation is not available, then some degree of elemental roundness measurement by the *rotation* technique can be made. The simplest method that can be adopted is to place the part in a vee-block (BS3731: 1987) and carefully rotate it with a dial gauge or dial test indicator in contact with the surface (Figure 103a, b). A truly round part will show no movement on the dial gauge pointer; however, if the component is not round then irregularities as they contact the vee-block's sides will cause the part to move up or down. These component irregularities (Figure 103a and b) will displace the dial gauge's plunger as the part is slowly rotated; "correction factors" should be applied. The amount by which the gauge pointer is displaced depends on several factors:

- the height of the irregularities – peaks and valleys;
- the angular spacing of the irregularities;
- the vee-block included angle.

The pointer movement will be greatest when the peaks or valleys simultaneously contact the plunger and faces of the vee (Figure 103a). The movement will be smallest when a valley is under the plunger and peaks are contacting the vee-block (Figure 103b) and vice versa. If the included face angle of the jaws utilised to inspect the part differ, then the plunger motion relative to component rotation can cause two distinct conditions to occur:

- in-phase – with a wide vee-block face angle (Figure 103c);
- out-of-phase – with a narrow vee-block face angle (Figure 103d).

NB: See BS3730 Part 3 (Figure 103 "Summit and rider method").

Vee-block roundness measurement is essentially a three-point technique, with several variations in the manner in which it can be applied, such as when inspecting a large external diameter or a similar internal bore.

A simple artefact can be manufactured that incorporates the dial gauge and datum feet, its body shape designed to suit the particular metrological application. Bench centres for longer parts, with centre-drilled ends, are a useful inspection apparatus for assessing either roundness or concentricity of a stepped-diameter shaft, in conjunction with a dial gauge. This technique is further refined when bench centres are employed, as they allow all of the above roundness parameters to be assessed, together with tapered diametrical features, although care must be taken to prevent component gravitational sag for long slender parts. The dial gauge previously mentioned can be replaced by an electronic gauging head and its associated meter unit, the main advantage being that both range switching and data-logging can be undertaken. These latter refinements to radial measurement are necessary when many parts are required to be inspected and for statistical process control/quality control applications. This aspect of multi-gauging can be quite sophisticated, as shown in Figure 110 for a typical modular component feature set-up of the myriad of industrial applications, where a multiple receiver gauge station can simultaneously inspect critical elements of stepped diameters and geometric features. Such electronic indicators equipped with a digital display provide both numeric and analogue visualisation of measurement. A three-colour analogue display (e.g. green, yellow and red), provides clear visual definition of the measurement status, even from a distance, where for example:

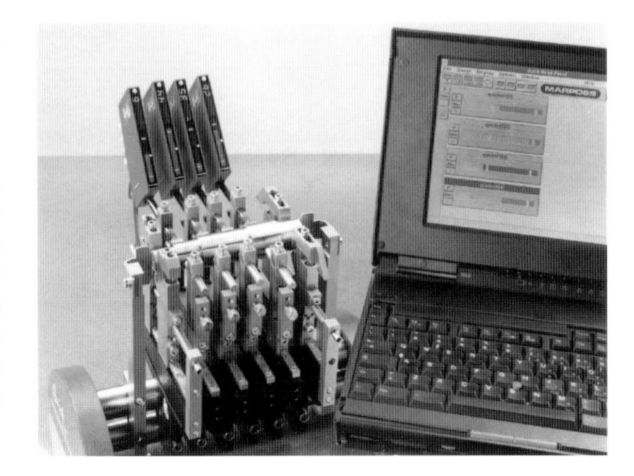

Figure 110. Multidimensional modular receiver gauge and concentricity inspection with data capture and manipulation. [Courtesy of Marposs Ltd.]

- *green* illumination indicates the part is within the tolerance limit;
- *yellow* illumination indicates the part feature is at the tolerance limit;
- *red* illumination indicates measurement is outside the pre-set limits.

Multiple diameter assessment with such modular-constructed receiver gauge elements allows the part's functional performance to be assessed. By arranging the transducer indicators in suitable banks (see Figure 110) a host of the component's geometric features can be simultaneously assessed.

4.2 Roundness: measuring instruments

The design of a roundness instrument (see Figure 111 for a schematic diagram of a rotating worktable type) incorporates a number of extremely accurate parts that are configured to assess specific elements of a part's roundness. The roundness-related features are:

- *rotation* – rotary motion providing the reference axis;
- *rotational axis* – this must be independent of the part being measured;
- *measuring device* – some form of stylus, and a transducer (pick-up), which converts mechanical stylus motion into a proportional electrical

signal, that is then subsequently electronically amplified;

- *indicator* – these roundness instruments generally employ a meter in association with the electrical pick-up for part set-up. The final displayed profile would normally be displayed in either a polar or linear form on the display screen. The polar plot from an integral printer in a dedicated roundness computer produces a permanent hard copy for future reference;

  NB One of the main reasons for not using a meter as an indication of roundness is that the fluctuations of the meter's needle will not only indicate the component error magnitude, but also the eccentricity and ovality caused by an incorrect set-up. This is also true of a polar plot, unless a reference circle is used.
- *reference datum* – this is required for referenced and repeatable roundness measurements;
- *vertical straightness column* – this is not essential but is a desirable feature on a roundness instrument, as it allows measurements to be taken at various heights on a component, to establish the part's roundness characteristics for specific geometric features, such as concentricity, eccentricity and runout or simply roundness at a number of heights;
- *additional feature* – coarse and fine adjustments for part positioning (centring and levelling), which have the advantage of reducing the set-up time (using automatic set-up procedures on some sophisticated instruments).

4.2.1 Types of instrument

Roundness-measuring instruments tend to be of two basic configurations as illustrated in Figure 112. For components of small size and weight, the most suitable configuration is the version based on a rotating table, or *turntable* (Figure 112a). The work-

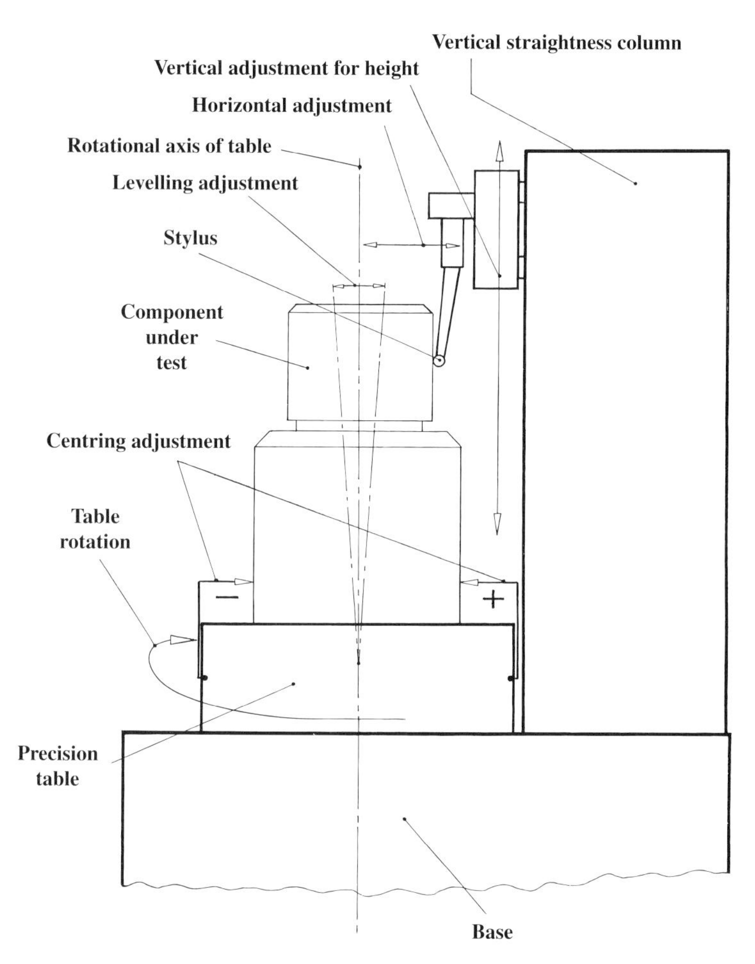

Figure 111. Schematic diagram of the main features of just one configuration for a roundness-testing machine.

piece rotates while the pick-up remains in a set position. For large and notably heavier parts, the *rotating pick-up* type, which rotates the pick-up (Figure 112b), is preferred. Each type of instrument has advantages and is more suitable to certain types of measurement, the choice of instrument being largely dependent on the anticipated measurements to be undertaken and the shape and size of the parts.

The first equipment available commercially, "Talyrond 1" was of the rotating pick-up type. The rotating pick-up design became the model from which all other manufacturers' roundness-measuring instruments were developed. The success of the roundness instrument's design was such that over the years only minor modifications were necessary to enhance the equipment's capabilities. Most of these modifications were necessary to change the electronics from valve amplification to transistor amplification, resulting in the "Talyrond 2" design. Later, integrated circuit electronics replaced the transistors in the design of the "Talyrond 4". The current instrument is illustrated in Figure 113; this instrument has the ability to accommodate an extremely large workpiece size and weight, featuring automatic component centring and levelling - an important feature with parts that are very large and less moveable.

Comparison of instrument types

This section will discuss the reasons why two types of roundness instruments (Figure 112) were developed, explaining design differences and, more importantly, the situations why one type is more suitable than the other for certain kinds of roundness measurement:

- *Rotating pick-up type* (Figure 113).

Operational characteristics of this roundness instrument's configuration include the following:

1 the precision spindle need only carry the comparatively light and constant load of the pick-up - its high accuracy is attainable without excessive cost;
2 as the worktable is not part of the measuring system it can be of substantial construction, so the weight of the part is not a limitation of measuring capacity. Moreover, large parts - like the four-cylinder engine block shown being inspected in Figure 113 - can be of asymmetrical shape, with the bore or surface being measured offset from the centre of gravity of the part. This offset load is not a limitation with this instrument's design; also long shafts, crankshafts and many more large parts can be simply accommodated on the table;
3 in order to speed up the measurement task and minimise uncertainty of measurement, computer-aided centring and levelling is available as an option on the more basic versions, or a standard feature with the sophisticated versions. This latter feature is true for both instrumenntal types.

- *Turntable type* (Figure 114).

Operational characteristics of this instrument's roundness configuration include the following:

1 due to the pick-up not being directly associated with the spindle, this type of the roundness machine is more easily adapted to other measurements in addition to roundness, notably concentricity and alignment. Repositioning the pick-up on a component - transferring it from external diametral measurement to an internal surface - has no effect on the reference axis. Further, pick-up positioning is simple; enabling it to reach into slots or to the undersides of shoulders becomes a straightforward operation, without having to resort to using long, or cranked stylus arms;
2 by incorporating a straight vertical movement of the pick-up, any straightness measurements can be made without modification to the spindle, or its mounting;
3 the turntable weight, together with the part being measured, has to be supported by its spindle bearings, which places a restriction on the weight of the part requiring measurement; moreover, it also limits the amount of offset that can be permitted;
4 computer-aided centring and levelling are standard on the more sophisticated instruments. If the centre of gravity of the workpiece falls into the triangular demarcation marks engraved into the top of the table's surface (see Figure 114), the manual and somewhat time-constraining centring and levelling activity becomes an automatic adjustment function.

4.2.2 Spindle and bearings

In any roundness-measuring instrument, the instrument's spindle is the most vitally important component in its assembly. Here, the instantaneous error of rotation is the key feature of the spindle; since it is from the axis of rotation that all measurements are taken, it must allow only minute deviation

(a) Turntable type of instrument

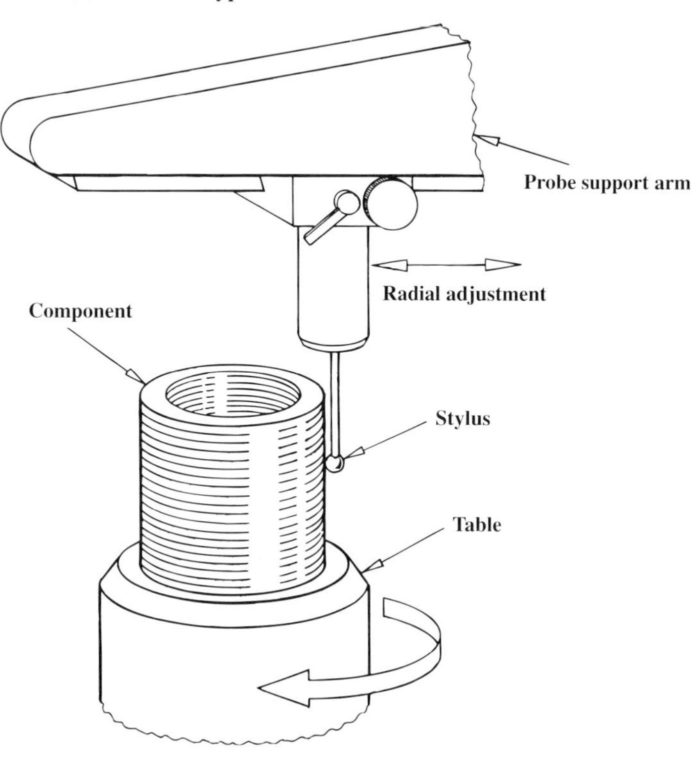

(b) Rotating pick-up type of instrument

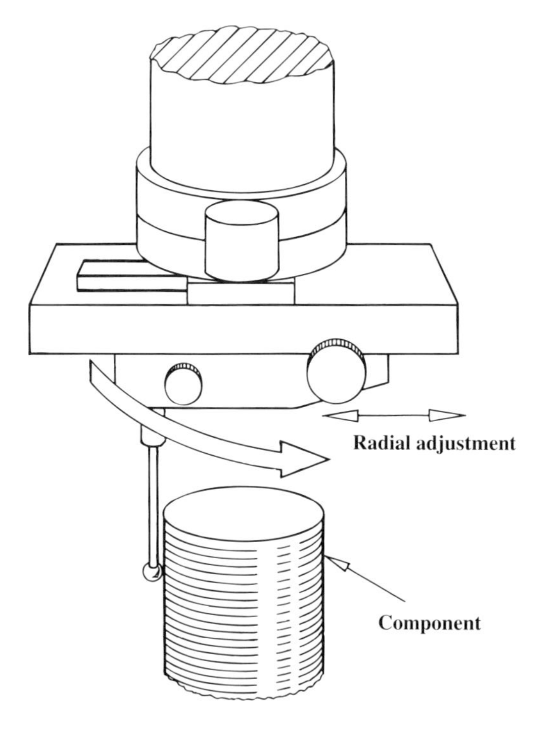

Figure 112. Types of roundness instrument. (Courtesy of Taylor Hobson.)

Figure 113. Large-capacity roundness instrument "Talyrond 4", fully automated for roundness geometry measurements. (Courtesy of Taylor Hobson.)

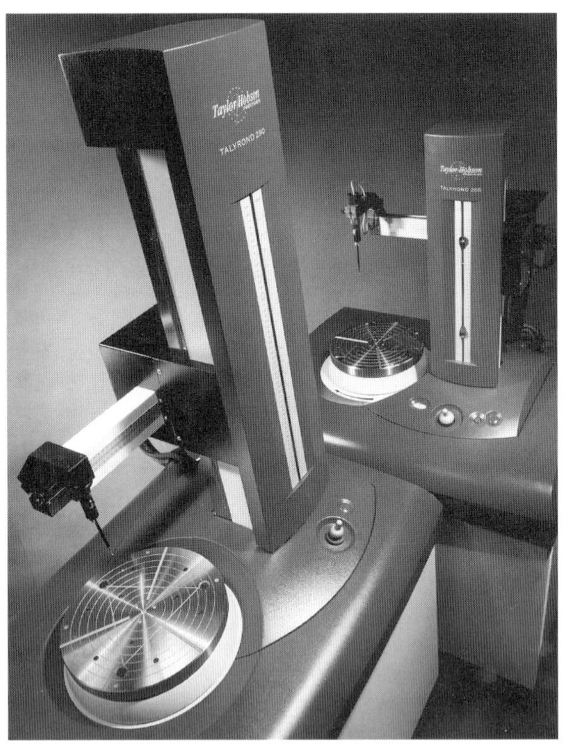

Figure 114. Highly accurate and sophisticated roundness instruments, based on modular design principles. (Courtesy of Taylor Hobson.)

from its fixed centre as the spindle rotates. To achieve this minimal spindle error, not only must the spindle's bearings be of the highest quality, but also they should approach almost perfect roundness.

Several types of bearing design can be employed in the construction of precision roundness spindles; these include:

- *dry bearings* (Figure 115) – these are normally dry, but can be of the type that are lightly lubricated, using a steel ball rotating in a plastic cup for the thrust bearing and pads of the same material for controlling radial movement. This type of bearing requires minimal maintenance, having constant accuracy under varying conditions of speed and load, accepting limited radial loading. Its main limitation is that the load-carrying capacity is restricted;
- *ball bearings* (not illustrated) – these are of a type in which the balls support the spindle both circumferentially and longitudinally. This axial and radial support enables the spindle to be utilised in either a vertical or longitudinal orientation, without loss of accuracy. Moreover, it allows the spindle's kinematics to be quite sophisticated, as it can be traversed axially for straightness measurement and, as the same bearings are employed for either rotary and linear motions, the axis of rotation is also the axis of linear movement;
- *hydrodynamic bearing – oil* (Figure 116) – from the original roundness instrument concept, this type of bearing has predominantly been used for the rotating pick-up instruments. Metal-to-metal contact occurs between the two spherical bearing members when the spindle is stationary, but as the spindle rotation reaches or exceeds 6 rev/min a thin and even film of oil is maintained between the surfaces – due to the hydrodynamic "wedging effect" – resulting from the action of the rotating surfaces. This film of oil is of constant thickness at a steady speed. At either slow speed conditions or heavy loads the oil film may cease to exist. These latter operational conditions means that such hydrodynamic bearings are mainly used where the load is relatively small and rotational speed is constant. One factor that needs

to be considered is that the spindle has to be oiled every day (when in use);

- *hydrostatic bearing – air* (not illustrated) – this is more commonly known as an "air bearing". This type of bearing has been used with considerable success notably on coordinate measuring machines (CMMs) and certain machine tools and instruments, where accuracy of movement must be combined with load-carrying capability, making these bearings ideal for specific types of turntable-type instruments. With these types of hydrostatic bearings the air supply must be clean and dry;
- *hydrostatic bearing – oil* (Figure 117) – in this version of hydrostatic bearing the oil is forced under pressure through a gap or clearance between the spindle housing and its mating spindle. Thus a film of oil is present whether the spindle is stationary or rotating. An oil hydrostatic bearing has much in common with its air-driven counterpart, but due to oil being considerably more viscous in nature to that of air it has several important advantages. These benefits include the following:
  - clearances between the spindle and housing can be greater;

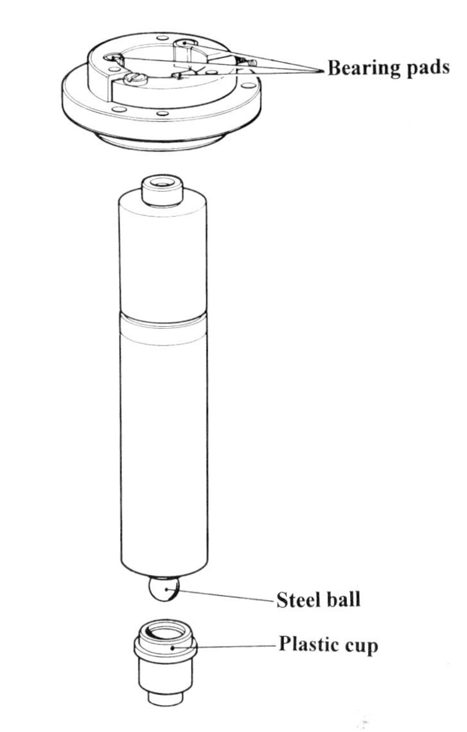

Figure 115. Spindle employing a dry bearing. (Courtesy of Taylor Hobson.)

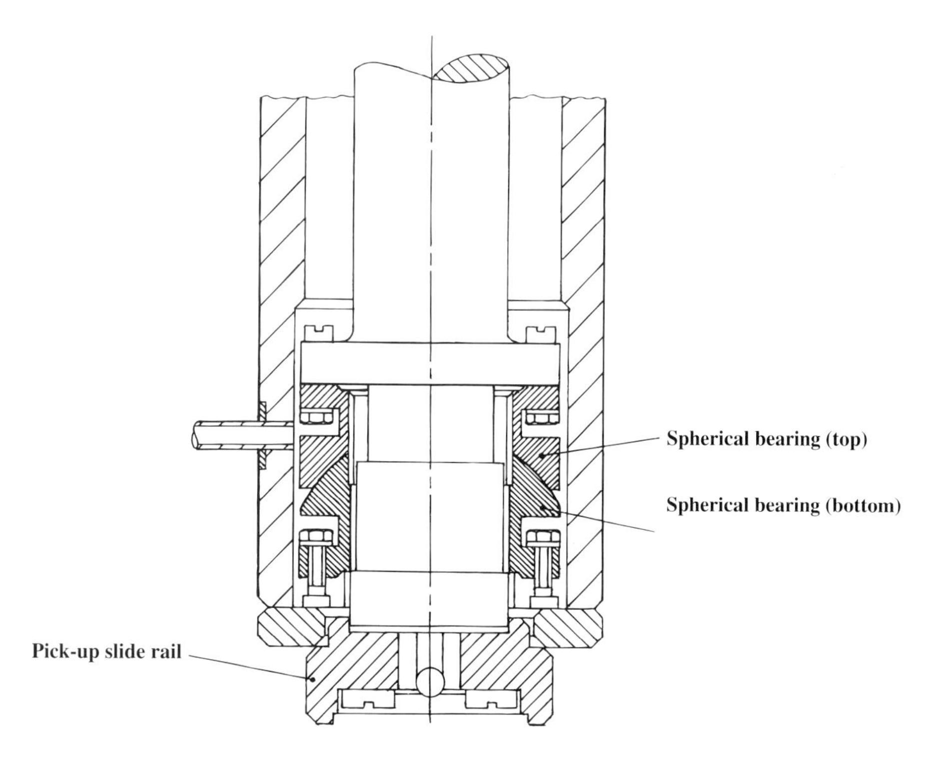

Figure 116. Hydrodynamic bearing for rotating pick-up roundness-testing machine. (Courtesy of Taylor Hobson.)

- the bearing stiffness is increased (meaning that sideways force applied to the bearing has comparatively little effect on the axis of rotation).

In Figure 118 four types of bearings are illustrated that can be fitted to roundness instrument spindles. Table 12 indicates the "order of merit" (ranking) performance of these spindles and Table 13 highlights their respective accuracies and load-carrying capacities.

From these tables the evidence suggests that optimum bearing performance occurs utilising the oil hydrostatic bearing, with respect to:

1 bearing accuracy;
2 vibration damping capacity – instrument vibration is a major factor limiting the level of magnification which can be used;
3 power consumption – not important from the viewpoint of running cost, but crucial because low consumption equates to both less vibration and less heat generation.

Furthermore, this oil hydrostatic bearing exhibits the penultimate performance with respect to low bearing friction and high stiffness. As a result of these characteristics, such oil hydrostatic bearings are the preferred selection for highly sophisticated precision instruments having heavy load-carrying capacities.

4.3 Methods of measurement

As the stylus traces the periphery of the component being inspected, the stylus displacement corresponds to movements *relative to the reference axis of spindle rotation*, hence this motion will also relate to undulations of the trace on the displayed profile as shown in Figure 119. This displayed profile is the same for both the rotating pick-up and turntable type of instruments. On roundness equipment it is necessary to position the part on the instrument so that its centre nominally coincides with the rotational axis of the device. This *centring* operation will vary from one instrument to another, depending on the centring adjustments provided and whether an automatic facility is available. If a perfectly circular component is to be measured, there is little difficulty in achieving successful centring, particularly with the aid of the mechanical device (self-centring iris) shown in Figure 120. Centring is achieved when there are no cyclic deflections of the stylus, as either the part or pick-up rotates. However, if the component

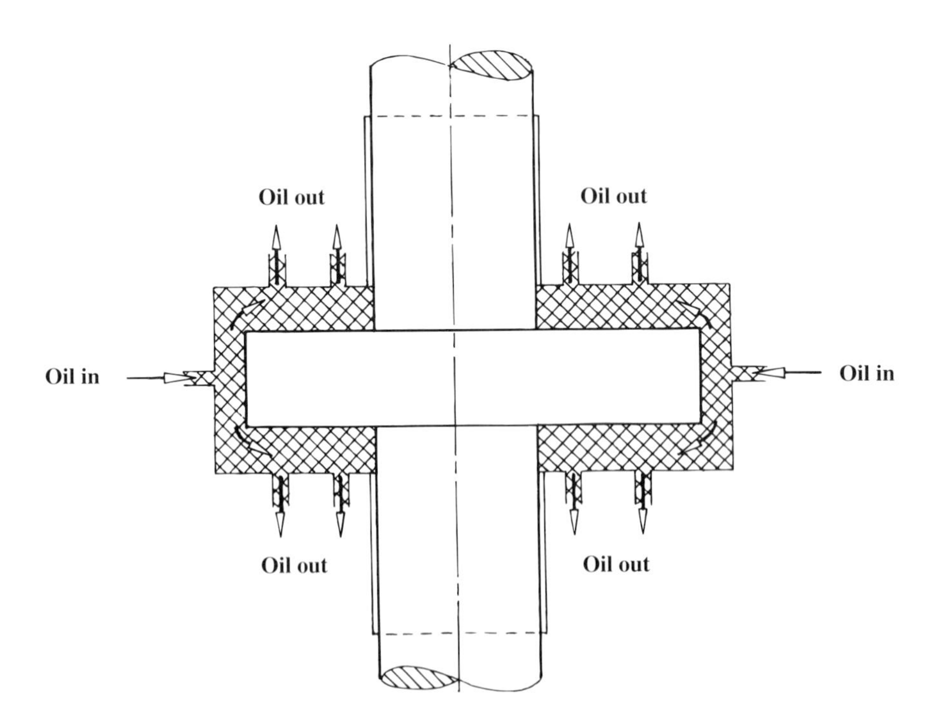

Figure 117. Schematic illustration of the principle of hydrostatic bearings used for turntable-type roundness machines. (Courtesy of Taylor Hobson.)

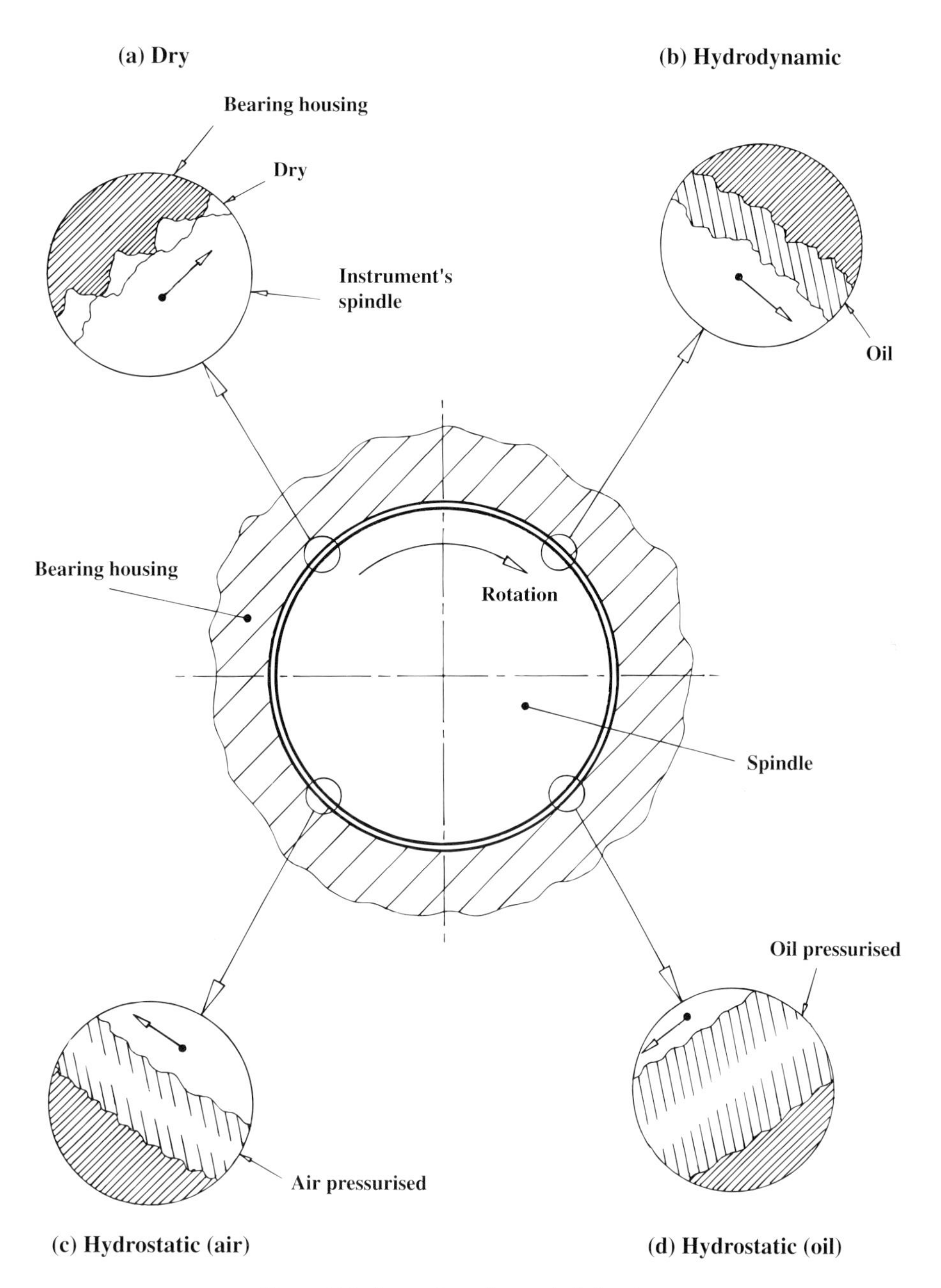

Figure 118. Types of roundness spindle bearings.

Table 12. Comparison of the bearing characteristics (Courtesy of Taylor Hobson)

Ranking	Accuracy	Vibration damping	Power consumption	Friction	Stiffness	Load bearing capacity
Best	C	D	D	C	D	D
↑	D	B	B	D	C	C
↓	B	C	C	B	B	B
Least good	A	A	A	A	A	A

Table 13. Accuracy and load-carrying capacity of various roundness spindle bearings (Courtesy of Taylor Hobson)

Bearing	Maximum load (kg)	Accuracy (μm)
Air	0.025	12
Hydrodynamic	0.100	n/a
Dry	0.030	20
Air	0.040	50
Hydrostatic	0.050	225

has an irregular shape optimum centring is attained when the range of the stylus deflection cannot be further reduced.

Once the roundness profile has been produced at a specific cross-section of the part, then this will allow the cross-sectional plane to be displayed and assessed. What these various polar plot displays show, and how you can obtain information from them regarding its roundness value, will be described in the following section.

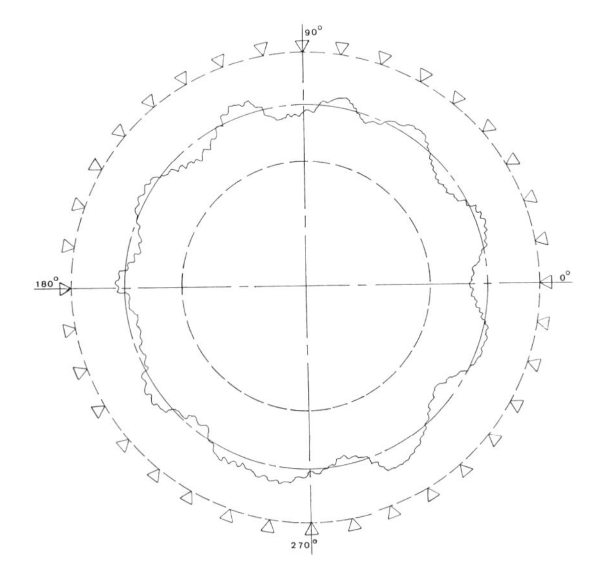

Figure 119. A typical displayed profile indicating the "least squares" fit. (Courtesy of Taylor Hobson.)

4.3.1 Assessment of part geometry

Often a single roundness trace can be rather limiting in its scope of information, as it is only partially informative and gives no information about the overall functioning of the part. However, if several roundness and/or straightness measurements are made they can produce a considerably broader understanding of the component's subsequent functional performance. In other words, the *whole geometry* of the part must be considered, assuming this can be assessed on a single instrument and, preferably, at one setting. The cross-hatched part shown in Figure 121 is used by way of an illustrative example of just some of the measurements that can be obtained on roundness instruments. In the initial set-up, the first activity is to centre and level the component and to align the part to the reference axis of rotation. Of late, with most roundness software, it is possible to remove any eccentricity and ovality due to initial centring and levelling by the operator. From this single set-up, a series of roundness, straightness/flatness traces can be obtained, allowing the following analyses to be calculated:

- Activity A (traces 1 and 2) – these are roundness profiles that can be analysed in cylindricity mode with a datum set between the two polar trace centres. This datum will allow a similar exercise to be carried out between traces 3 and 4, with the cylindricity of these two displayed in addition to the coaxiality of the two cylinders namely, traces 1 and 2 together with 3 and 4.

Figure 120. "Self-centring iris" accessory for speedy set-up and efficient processing of batch production requirements. (Courtesy of Taylor Hobson.)

- Activity B (traces 1, 2, 3 and 4) – can be utilised to establish a new datum running through the whole component.
- Activity C (traces 7, 8 and 9) – having completed Activity B, a flatness profile of trace 9 will not only show the peak-to-valley of the circular trace on the plane, but can also be displayed with its squareness relationship to the defined axis. Flatness traces 7 and 8, can once again be compared for squareness to the established cylinder axis, or for parallelism of their flatness planes.
- Activity D (traces 5 and 6) – the vertical traces (5 and 6) can be displayed in relation to a "least squares" line or "minimum zone straightness". By assessing one of the traces as the datum and comparing it with the trace obtained at 180°, it is possible to display and print out the parallelism of the bore.
- Activity E (trace 10) – using the datum previously derived from traces 1, 2, 3 and 4 any additional roundness profiles can be assessed for eccentricity, concentricity and runout (trace 10).

One of the main features of the modern analysis packages is the flexibility to store data in raw form, which can then be manipulated for additional roundness analysis to give a more precise picture of component accuracy.

In the automotive industry, one of the internal combustion engine's most highly stressed components is its crankshaft. This crankshaft is subjected to continual bending, torsion and shearing loads, in combination with variable rotational speeds, temperatures and oil viscosities. For smooth running and lifetime reliability, it is essential that no additional stresses, strains or vibrations are imposed by the geometry of the bearings. Normally, a main bearing is situated between each crankpin and, upon inspection, a vital check is to ensure that these crankpins are in line, round and straight along their entire length (Figure 106). In a similar fashion, the crankpins must be round and straight, with their axes parallel to that of the main bearings. All these vital measurements can be assessed using a suitable roundness-measuring instrument (see Figure 122).

Horizontal surfaces

In most instruments the pick-up can be positioned to detect vertical displacements of a horizontal surface, such as an end face (Figure 121, trace 9), or a shoulder from a plane that is square to the axis. The whole geometry can include measurements similar to those shown in Activities D and E illustrated in Figure 121. By now it should be appreciated that the trace is taken only around a circular track and, therefore, does not indicate true *flatness* of the surface.

A component feature that has previously been centred which has a spherical geometrical form would give a trace that superficially appears to be flat. However, it is normally only when a series of traces taken at different radii are assessed in relationship to each other that the true value of flatness can be appreciated. This assessment can only be undertaken if the instrument has a known relationship between the horizontal and vertical datum. When assessing the flatness of horizontal surfaces, the probe must not be moved to the surface between traces.

Straightness

Straightness can be measured using a pick-up traversing vertically with the worktable stationary (Figure 121, trace 5). Measurements can be made relative to the line of traverse, which is normally parallel to the roundness axis of rotation; this allows for both the roundness and straightness measurements to have the same datum and can therefore be related to each other without difficulty. Such combinations of component roundness and straightness employing the same initial set-up are another feature of the whole geometry measurement.

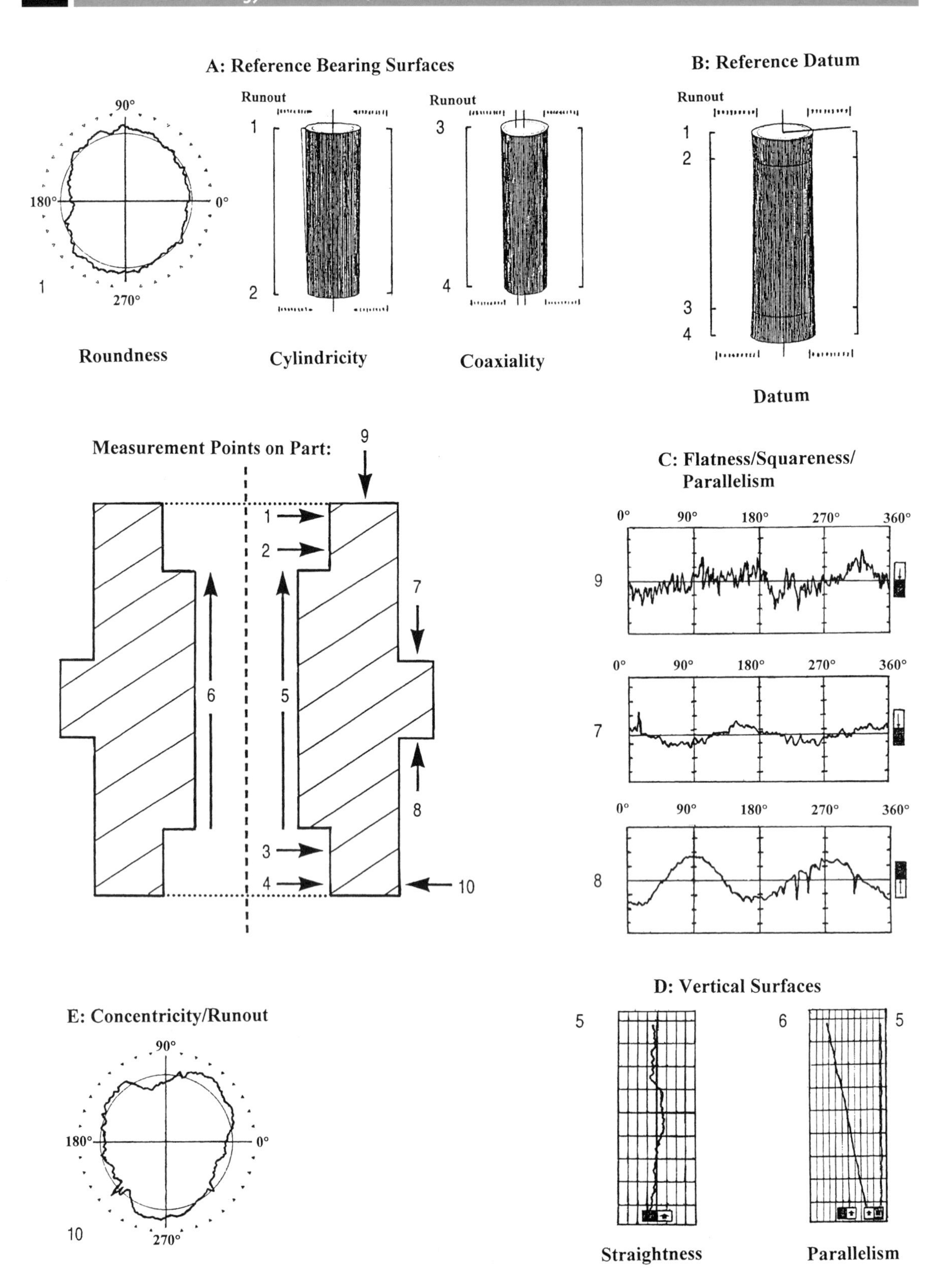

Figure 121. Typical roundness, straightness and functional part measurements that can be undertaken. (Courtesy of Taylor Hobson.)

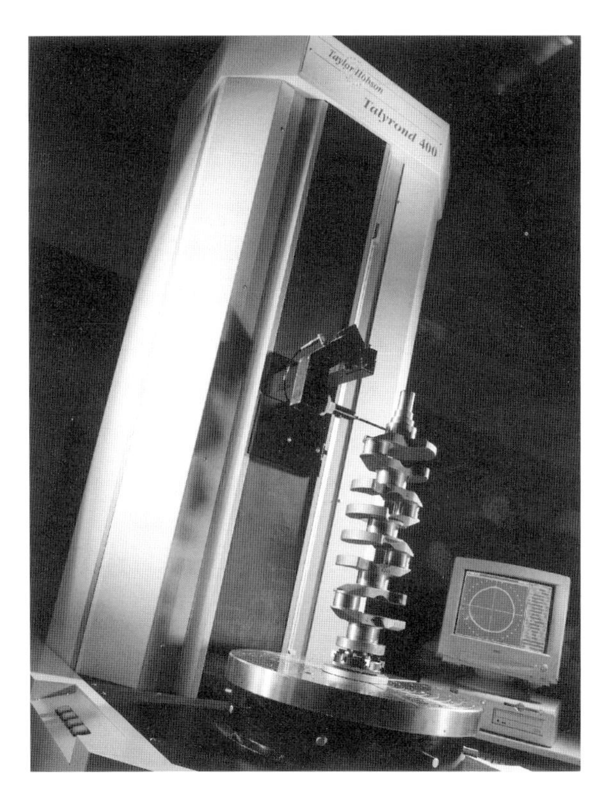

Figure 122. "Talyrond 400" instrument for automated cylindrical form measurement – bearing and alignment assessment of a crankshaft. (Courtesy of Taylor Hobson.)

Interrupted surfaces

Occasionally, some components have a peripheral surface that is broken by slots, splines, keyways or cross-drilled holes whose sides – being interrupted – may catch on the stylus and dislodge or jolt the stylus arm, thereby making it impossible to obtain accurate measurements. These interruptions in the periphery can be overcome by using an attachment to the pick-up, in which a stop is used to limit the depth to which the stylus is allowed to drop into the recess. If this mechanical stylus stop is used, the resulting trace will appear similar to the example illustrated in Figure 123. It can be seen in this case that although the trace does not show any recess features, it is a true representation of the outer profile of the toothed pulley. If computer software is used the unwanted data can then be removed.

Styli for roundness instruments

The only contact between the measuring system and the part is via the stylus. Thus stylus shape is important and influences the information that the pick-up obtains from the surface. By way of illustration, many parts are produced on machine tools and have representative tool feed marks around the circumference, which are not part of the surface geometry (shape) but contribute to the surface texture – more will be said on this topic in Chapter 5. It follows that a sharp stylus (Figure 124) would sink into every one of the machine cavities (cusps), or scratches present and the resulting irregularities displayed could mask the more widely spaced undulations and lobes which contribute to out-of-roundness. As a consequence of the problem associated with using small styli, it is standard practice to employ a relatively large stylus (Figure 124) known as a "Hatchet stylus". This stylus acts as a filter and bridges the gap in the closely spaced surface texture marks (cusps), ensuring that the pick-up is insensitive to these localised surface irregularities. The stylus ball tip diameter selected is dependent on the inspected part's diameter and surface. Typically, various styli are available with diameters ranging from 1 to 4 mm. Such styli are manufactured from either hardened alloy steel or sapphire tips, mounted on an aluminium or carbon fibre shaft. Where a particular standard stylus is unable to interrogate the part successfully, because it may be too short in length, or of the wrong stylus geometry, then special-purpose styli can be produced by the roundness instrument manufacturer.

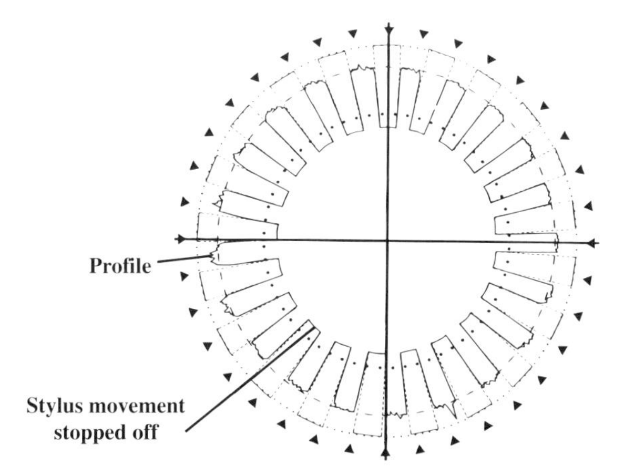

Figure 123. Interrupted surface roundness trace of a toothed pulley. (Courtesy of Taylor Hobson.)

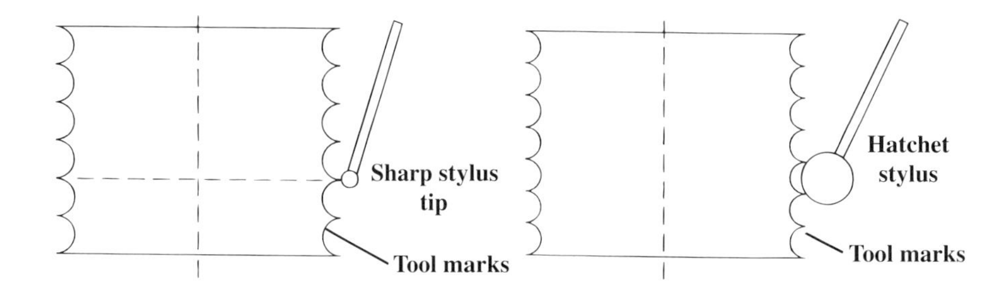

Figure 124. Comparison of various sizes of ball-tipped and hatchet-shaped styli. (Courtesy of Taylor Hobson.)

4.4 Display and interpretation

The majority of roundness instruments can present a displayed profile of the part profile, which in its most elementary form is a graphical trace of polar plot, but can also be displayed on the screen. It is important to recognise what is being displayed and how measurements can be obtained from such displayed profiles. The production of a roundness traced profile from the instrument provides an accurate and assured method of display of profile features, with an expedient means for interpretation of these measured results. Additionally, the displayed profile can be kept as a permanent record of the measured profile. Alternatively, some instruments have the facility to provide a linear form of the roundness displayed profile, which can be constructive in interpreting the waveform's regularity. The polar plot (Figure 119) will minimise conjecture in its interpretation because it is a visual record. It is not simply an enlarged representation of the surface and requires the information depicted to be interpreted by the operator, before it has any meaning.

Analysis of the polar plot is quite straightforward once the underlying graphical principles have been comprehended. The question could be asked: "In what manner is the polar trace's display representative of the real roundness profile?" Prior to a discussion on this point, it is important to fully understand just what the polar plot illustrates and, just as vital, what it does not show.

Four general assumptions can be made relating to such polar plots, which need to be addressed before the correct interpretation can be made:

1 *the displayed profile is not simply a magnification of the cross-section of the component;*
2 *measurements of peak and valley heights are correct, but their shape is visually distorted;*
3 *there is a direct angular relationship between the displayed profile and the component;*
4 *the profile is shown relative to a reference circle.*

A brief discussion on these points now follows.

The trace is not simply a magnification of the cross-section of the component.

Due to the variations in roundness being minute, typically 2 μm on the periphery for a part having a diameter of 200 mm, a magnification of 2000 would be required to ensure that the displayed profile variations are visually large enough and can be measured. For example using a magnification of 2000, an undulation of 2 μm on the component surface would impart a recorded gauge deflection of 4 mm. On a similar scale, if the whole cross-section of 200 mm diameter were to be proportionally enlarged then the diameter of the displayed profile would be 400 m. To press the point still further, this represents a polar plot paper size of 125,664 m^2 – mostly blank – would be necessary on which to record the plot (Figure 125)! Moreover, the circumference of the profile would be 1.25 km, despite the fact that interest on the polar plot is only for an undulation of 4 mm. This highlights the fact that direct enlargement of the whole profile is obviously impracticable and unnecessary, as the only requirement here is to measure the variations on the surface, not its diameter. Hence the obvious strategy is to retain the magnified integral surface variations, but reject the enlarged radius, ensuring that whatever diameter is selected is of a convenient size to display the roundness information given that data is radially suppressed.

Measurements of peak and valley heights are correct, but their shape is visually distorted.

For many scientific, commercial and statistical applications, the *shape* of any trace is often a direct visual interpretation of what is portrayed; for example, the increasing slope of a graph might be used to indicate

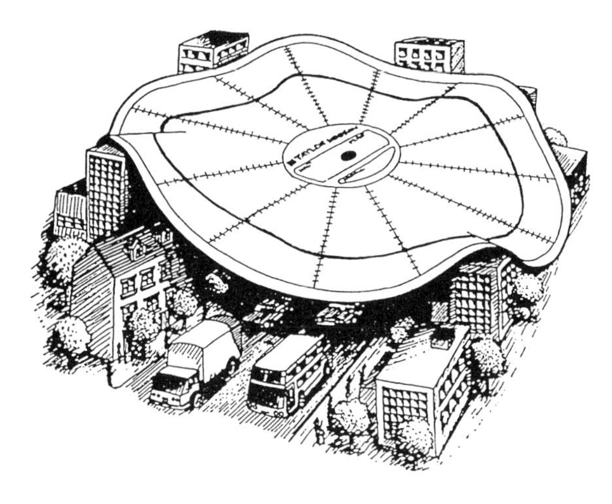

Figure 125. Exaggerated diagram illustrating why one cannot produce a "polar plot" of the simple enlargement of the cross-section of the component. [Courtesy of Taylor Hobson.]

factory output on a week-by-week basis. Conversely, in the case of roundness assessment, it is often difficult at first to comprehend that the visual image of the displayed profile does not give a clue to the shape of the undulations on the part. The reason for any confusion in displayed profile interpretation was just mentioned, this being the result of the radial peak and valley heights being retained at their correct magnification. While the radial enlargement is great, the accompanying circumferential magnification is often less than ×1, this being a reduction resulting from the trace diameter, which must of necessity be smaller than that of the vast majority of parts that are measured. The effect on the polar plot of this disparate magnification can be visually shown by the group of displayed profiles of a four-lobed part, which in these examples has a 3 μm variation in radius and is illustrated in Figure 126. Using a low magnification (Figure 126a) the displayed profile will visually differ only slightly from a perfect circle. At higher magnifications, the radial distance between the outermost and innermost portions of the trace will be proportionally increased (Figure 126b), until at an optimum magnification the displayed profile appears exaggerated, as shown in Figure 126(c). These visual effects emphasise the fact that the displayed profile's shape is not simply a visually enlarged view of the cross-section: the displayed profile does not contain concave portions. This visually based fact can be demonstrated by adjusting the magnification still further to produce a square displayed profile (Figure 126d). In all four displayed profiles the measured maximum departure from roundness (negating the influence of different scales) is identical, namely 3 μm.

Peak shape can also be modified at certain magnifications. By way of a descriptive example of this magnification behaviour, if slight undulations occur at low magnification these undulating features could be recorded as a needle-sharp peaks at different magnifications.

The question that this discussion raises is: "Does this mean that the trace is of no value?" Certainly not, as the first impression of the trace is not always valid and judgement is necessary in the manner of trace interpretation. This lack of visual correspondence between the physical part and its associated trace is probably the reason why a few less knowledgeable engineers avoid using a roundness instrument for assessment, but instead favour the far cruder and considerably less informative dial gauge and vee-block method (Figure 103).

There is a direct angular relationship between the trace and the component.

In many cases, if the roundness instrument has a built-in recorder this allows the traced profile to have a direct angular relationship to the component, because the recorder rotates at the same speed as the spindle. The PC-based roundness instruments have analytical capabilities, with the display showing the data at the 0° angular position of the spindle on the screen at the (conventional) 3 o'clock position. This rotational and angular datuming facility enables the user to relate any roundness angular feature abnormalities to a known spindle position.

The displayed profile is shown relative to a reference circle.

Current roundness instruments can display the measured profile with one of the four reference circles superimposed on it – more will be said on this topic in Section 4.5. This facility allows the operator to obtain, say, a peak-to-valley value based on a pre-selected reference circle centre, by the simple expedient of assessing the largest peak and the deepest valley from the reference. Most software can output the peak-to-valley value, together with eccentricity and the runout of the profile – the latter being related to a selectable datum such as the spindle axis.

4.5 Roundness measurement from the display

Previously it was explained that the roundness of a component was expressed as the difference between

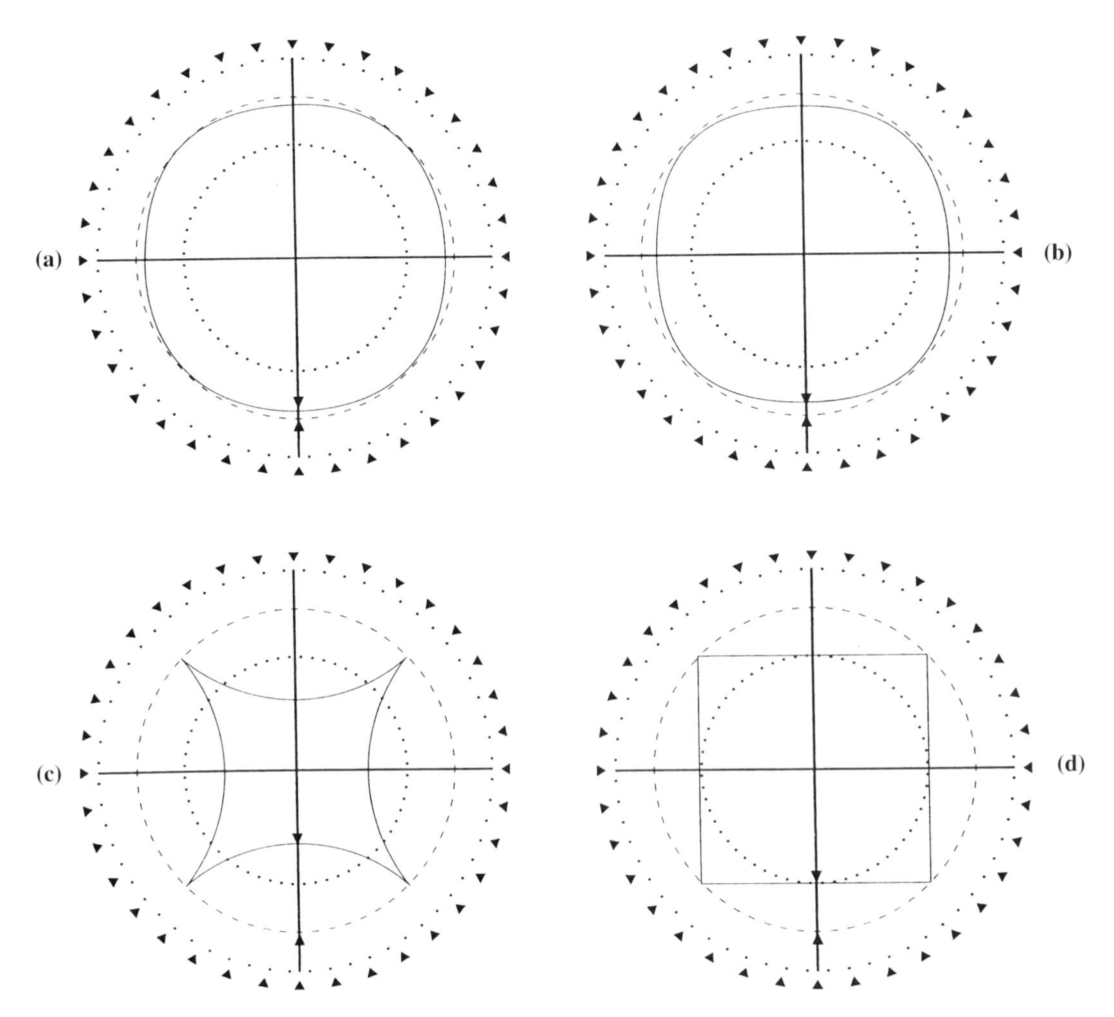

Figure 126. Effects of magnification on the visual shape of the displayed profile. (Courtesy of Taylor Hobson.) The scale indicates that each division represents (a) 5 μm, (b) 2 μm, (c) 1 μm and (d) 0.5 μm on the trace.

the greatest and least distance of the profile from an origin. To enable a measurement of this distance, or more correctly to assist in identifying the distance between two features, it is normally necessary to utilise a reference circle, or a pair of circles, which can be drawn, recorded or otherwise superimposed onto the graph. Positioning the reference circle(s) with respect to the displayed profile is not an arbitrary activity, as its chosen to fulfil specific conditions. Only in this manner can an unambiguous and repeatable value for the departure from roundness of any particular roundness profile be obtained.

The roundness value has historically been calculated by situating a transparent template of radial rings over a graphed profile and with the aid of the radial rings, centralising the profile visually. This manual technique allows the highest peak and the deepest valley to be found and, using the pre-set magnification utilised in this roundness assessment, the departures from roundness can be determined. The difference between the two features on the displayed profile can be numerically assessed. This operator-dependent procedure was somewhat subjective in interpretation and therefore accuracy and repeatability were open to conjecture. The actual positioning of the reference circle cannot readily be determined by manual methods, but this problem can be overcome by software that can display and print out the reference circle(s) on the profile. These software facilities will not only display the peak-to-valley information, but can also indicate its computed centre and any other relevant data to a selected datum.

4.5.1 Roundness reference circles

Standardisation

The measurement of roundness, like the majority of other types of measurement, is subject to national and to international standards. These standards specify the manner in which measurements should be made and how the results are to be expressed – for example, which reference circles are preferred. Conformity to these standards will ensure that any measurements undertaken on different instruments are compatible. This conformity requirement is essential when components are manufactured in one country, or by a subcontractor, that have to be matched to parts made elsewhere.

Manufacturers of roundness-measuring instruments try to ensure that their products closely conform to the requirements of a standard. Thus, when a designer comprehends that a certain roundness tolerance is specified this component can be inspected, measured and either accepted or rejected, according to whether it conforms to tolerances specified unambiguously not only within the company but in fact anywhere in the world.

The standards in Britain, Germany, Japan, America and elsewhere have accepted four reference circles, fulfilling different roundness requirements:

1 *Least squares circle (LSC)* – this circle can be regarded as the most commonly used of the references (Figure 127). Referring to Figure 127, a circle is fitted to the data such that the sum of the squares of the departure of the traced, or modified profile from that circle is at a minimum. Another way of considering LSC is that the area bounded by the profile on one side of the circle is equal to the area bounded on the other side. Still yet another way of taking into consideration the mathematical derivation of LSC is: "The sum of the squares of a sufficient number of equally spaced radial ordinates, measured from the circle to the profile, has minimum value". The departures from roundness is depicted in the lower diagram in Figure 127, illustrated as the radial distance of the maximum peak (P) from the circle, plus the distance of the maximum valley (V) from this circle(Z_q);
2 *Minimum circumscribed circle (MCC)* – this is a circle of minimum radius which will enclose the profile data (Figure 128). The out-of-roundness is then given as the maximum departure (valley) of the profile from this circle and is occasionally referred to as the "ring gauge reference circle". Its centre and radius can be found manually, by trial and error using either a compass, template or, for more exacting precision, instrumentally. Hence, the departures from roundness is the distance of the lowest valley (V) from the circle, once again being equal to P + V, as now the value of P in this case is zero (Z_c);
3 *Minimum zone circles (MZC)* – these are two concentric circles enclosing the traced or measured profile such that their radial departure is a minimum (Figure 129). The roundness value is then given as the radial separation Z_z;
4 *Maximum inscribed circle (MIC)* – this is the circle of maximum radius which will be enclosed by the profile data (Figure 130). The departures from roundness is then given as the departure (peak) of the profile from the circle and is occasionally

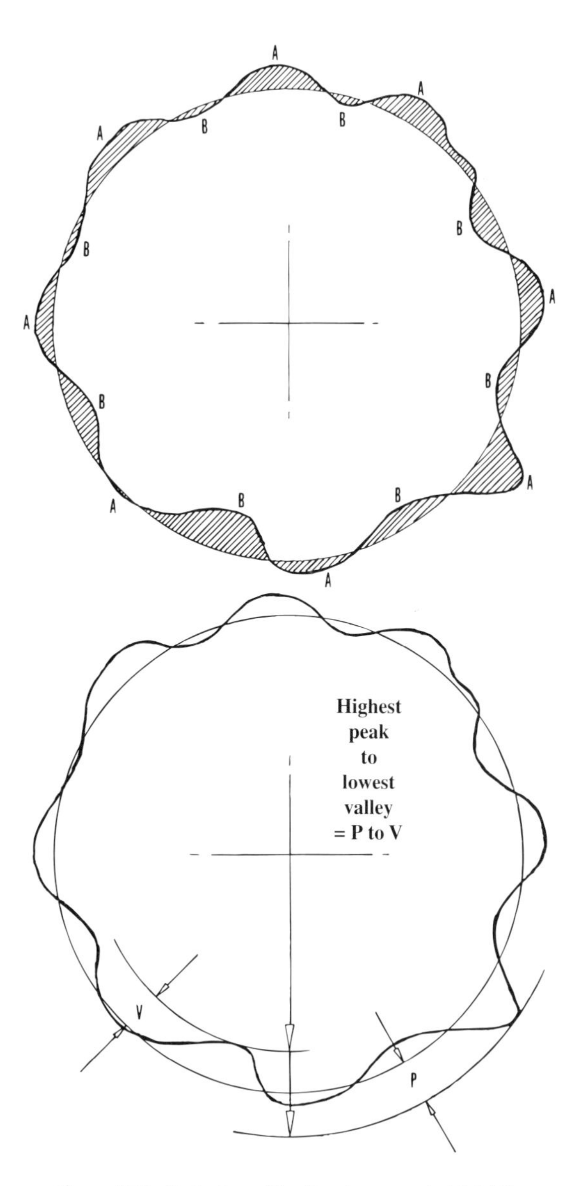

Figure 127. Derivation of the "least squares circle" (LSC).

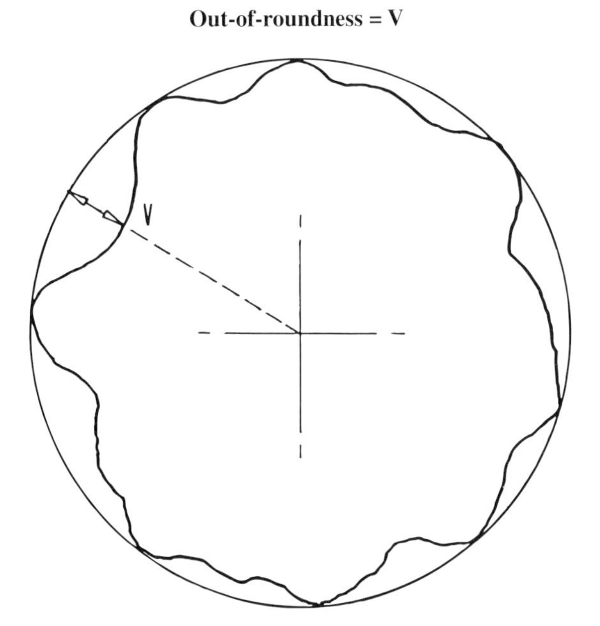

Figure 128. Minimum circumscribed circle (MCC).

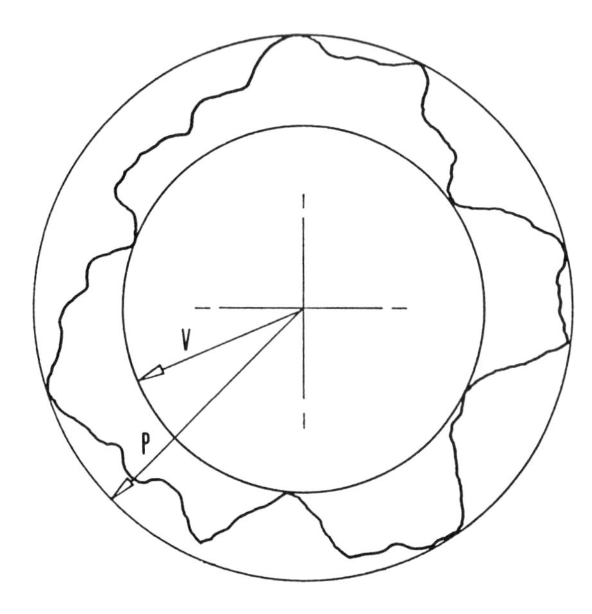

Figure 129. Minimum zone reference circles (MZC).

referred to as the "plug gauge reference circle". Both manual and instrumental analysis techniques can be utilised to establish the MIC. Therefore, the departures from roundness is specified as the height of the largest peak (P) above the circle, a value that can still be regarded as P + V, because now V equates to zero (as no valley exists inside the circle) –(Z_i).

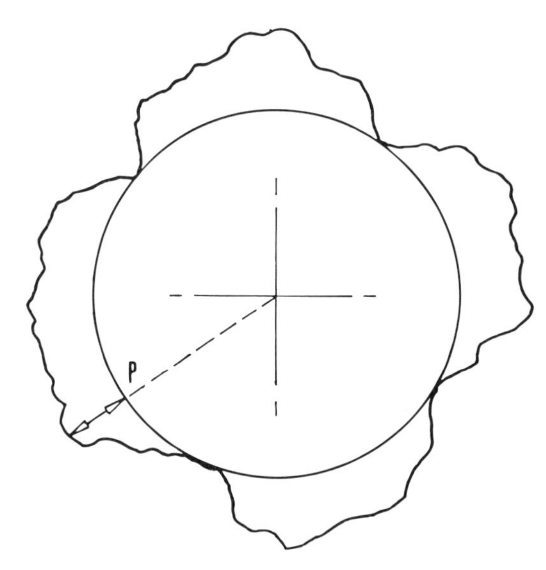

Figure 130. Maximum inscribed circle (MIC).

As mentioned above, both the circumscribed (Figure 128) and inscribed (Figure 130) circles are sometimes known as the "ring" and "plug" gauge circles, respectively, because they simulate these limit gauges in checking a shaft or bore. It has been observed that these terms can be criticised, because when checking roundness features with limit gauging equipment this is principally a three-dimensional check, whereas a roundness profile can only be considered as a two-dimensional cross-section.

The choice of referenc circle will be influenced by the manufacturing process of the part. For example, the LSC is less influenced by scratches than the other techniques and is important in concentricity assessment, etc.

4.5.2 Numerical value of roundness

The maximum peak-to-valley height (Z) gives the component's numerical value of roundness. If this value is obtained from a displayed profile it will range from the highest peak to the deepest valley. For example, if the least squares circle method of roundness assessment is considered (Figure 127), then it is simple to obtain the distance from the circle to the extreme points (largest peak and deepest valley) on the displayed profile. However, when the minimum zone reference circles are employed (Figure 129), then the departures from roundness can be numerically specified as the radial width of the zone measured from the chart between these two circles, divided by the magnification.

It should be noted that several factors need to be addressed when considering any form of roundness interpretation, including the following:

- roundness value will convey no information as to the general shape of the profile or to the number of irregularities that occur around the component's periphery;
- minimum zone is not equal to the maximum peak-to-valley height (compare the displayed profiles in Figure 131a), although in many cases the two will be virtually the same;
- inclusion of an exceptionally large peak or valley (possibly due to a scratch or particle of dirt on the surface) can cause a large alteration in the departures from roundness value (Figure 131b). Today, software can automatically remove these asperities, if required.

Straightness can be measured by the numerical value by the departure from a linear trace, for either the least squares or minimum zone circles (see Figure 121, D5). Reference lines are superimposed onto this linear trace and value calculations are undertaken in

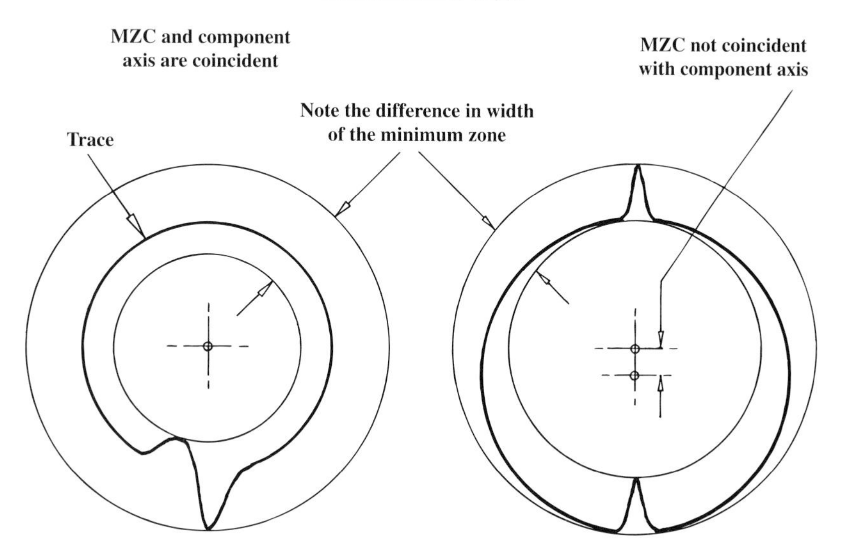
(a) Examples showing how the relative positions of peaks and valleys can influence the "minimum zone width"
MZC and component axis are coincident
MZC not coincident with component axis
Note the difference in width of the minimum zone
Trace

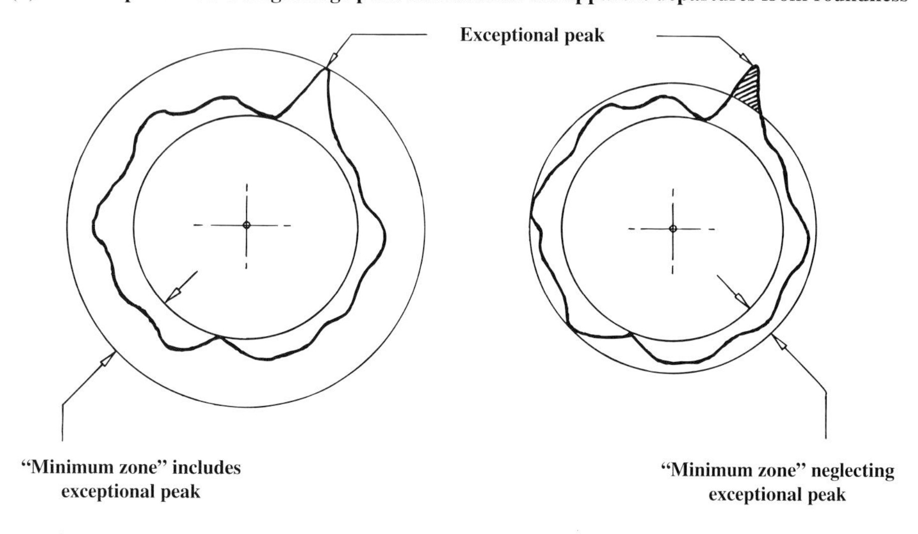
(b) An example of how a single large peak can increase the apparent departures from roundness
Exceptional peak
"Minimum zone" includes exceptional peak
"Minimum zone" neglecting exceptional peak

Figure 131. Cases where a single dominant peak, or its relative position, can increase the "minimum zone" width, hence the apparent departures from roundness. (Courtesy of Taylor Hobson.)

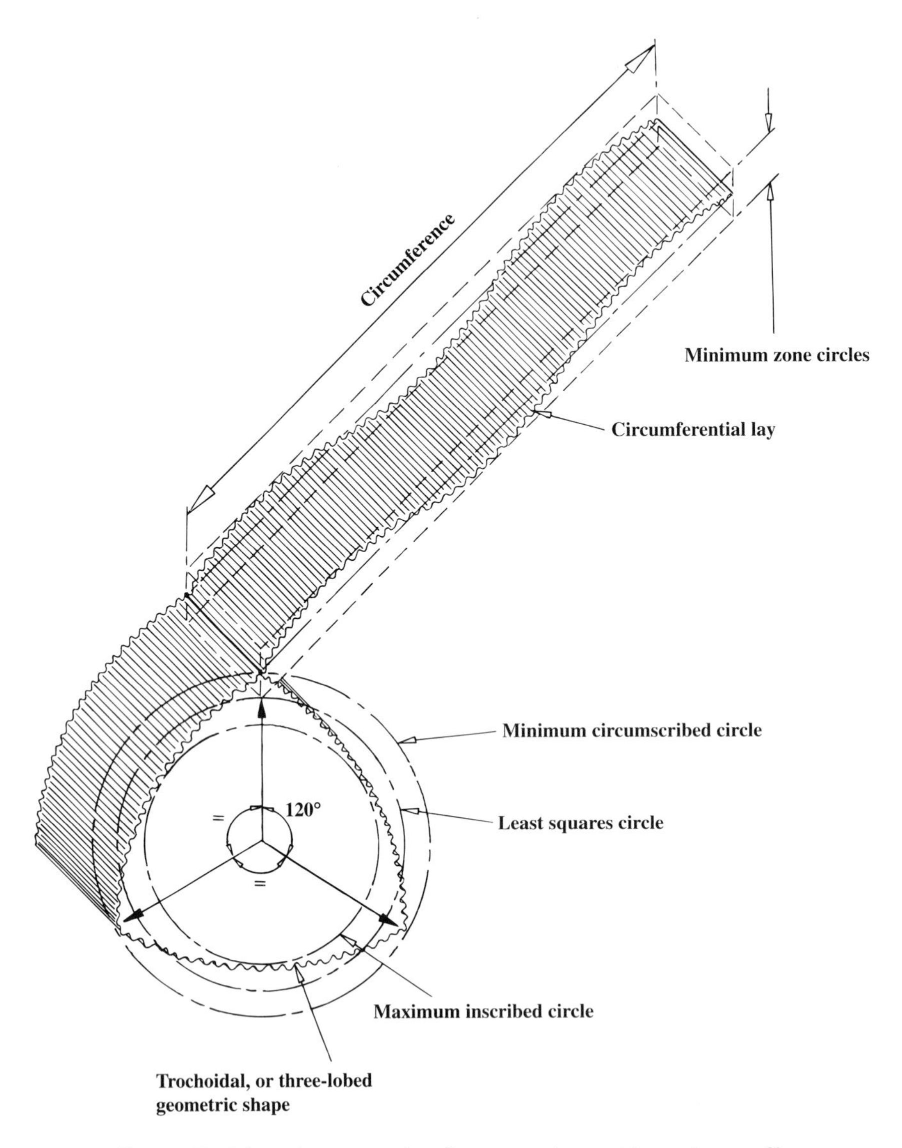

Figure 132. Schematic representation of an unwound geometric roundness profile.

the same manner as a polar plot roundness calculation. The linear plot of a roundness trace can visually highlight any harmonic departures from roundness by the trace's sinusoidal deviation from a straight line, as indicated in Figure 132, where a three-lobed (trochoidal) component geometry is depicted on a traced profile and its associated linear plot.

4.5.3 Filtering and harmonics

Filtering

If a component has a rough surface texture the geometric form of this surface may be concealed by its surface roughness, which would otherwise give misleading information concerning the relative roundness of the part's profile. Mathematical filtering can be applied to the roundness profile, or modified profile, to separate out the roughness from the general form. As an alternative action, the departures from roundness can be suppressed, enabling examination of the periphery for surface irregularities at higher roundness magnifications (see Figure 133). Internationally accepted filters are:

- 1–15 undulations per revolution (upr);
- 1–50 upr;
- 1–150 upr;
- 1–500 upr.

Displayed profile A: Typical example of a four-lobed part with a scale of 5 μm.

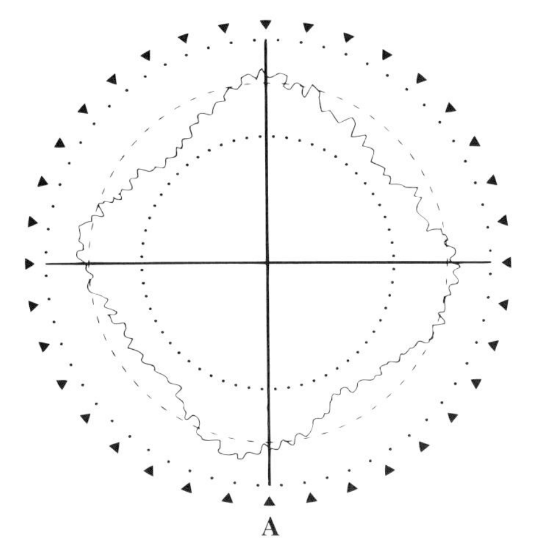

Displayed profile B: The same profile with the lobes filtered out to allow the display to be shown at 1 μm, enabling the smaller irregularities to be more clearly seen.

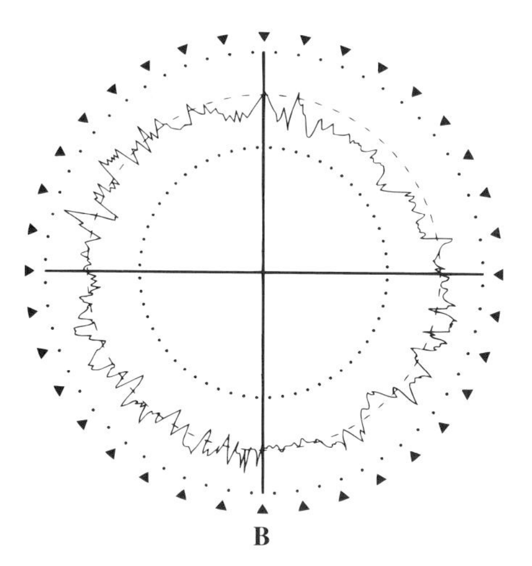

Figure 133. Effect of filters in roundness assessment. (Courtesy of Taylor Hobson.)

Such filters will eliminate any undulations above the pre-selected number. Furthermore, it is also possible to select a bandpass filter which removes the form, typically 15–500 upr, which will then display the surface irregularities with the basic form removed.

Harmonic analysis

Harmonic analysis of a component is an important aspect when attempting to control either the manufacture of the part or its subsequent in-service performance. The component's harmonics are a product of various factors, such as component geometry, part material and its method of manufacture. In general terms, the harmonics of a profile can be grouped together in the manner indicated in Table 14.

On a circular profile the *harmonic* can be thought of as a uniform waveform (sine wave) that is superimposed onto the part's surface (see Figure 134). Unless a "true circle" occurs during roundness measurement, it can be stated that any roundness profile will consist of a series of sine waves that are combined to form the overall roundness shape and its resulting trace. This harmonic feature can be seen in the series of minimum zone circle (MZC) polar traces on an identical part illustrated in Figure 135, where different filters have been utilised to filter out the relative harmonics (a–d) at a series of undulation cut-off values.

The calculation of harmonics in a representative polar trace for the harmonic behaviour exhibited in Figure 134 is comparatively simple to achieve, in essence by counting the number of harmonics, with their respective amplitudes being relatively straightforward to measure. When a more complex harmonic profile occurs, consisting of either one or two harmonics (though typically a component often has between 10 and 12), then it is essential to use software to perform the harmonic analysis. These discrete harmonic features of the part's traced profile are normally calculated by using what are termed *Fast Fourier Transforms* (FFT). The application of FFT algorithms, in effect, breaks down the profile into its constituent waveforms and calculates both the amplitude and phase angle of each harmonic. The results of the FFT study can be visually displayed as either a histogram or a tabulated display on the screen (Figure 136). Figure 136(a) is an example of an FFT harmonic output of the roundness assessment for a component having eight equi-spaced splines on the periphery. The harmonic histogram shows that the relieved areas of the eight-figured spline have a significant effect on the remainder of the component's harmonics; this is evidenced by the large eighth harmonic and that the subsequent harmonics are all multiples of eight. This minor harmonic behaviour is the result of the part's relieved sections inducing "bounce" on the grinding wheel and hence workpiece as the periphery is cylindrically ground. The first harmonic is the result of the setting-up

Table 14. The harmonic behaviour related to either the component manufacturing process or its measurement

Harmonic	Cause
1st (1 upr)	*Function of measurement* – only caused by the setting-up error on the instrument being utilised to measure the *departures from roundness.* The amplitude of this harmonic is equal to the eccentricity of the part, relative to the spindle axis of the roundness instrument
2nd (2 upr)	*Function of measurement or manufacture* – this aspect of harmonics is generally termed *ovality* and can be caused either by a setting-up error of the roundness instrument, or the part being machined out-of-square to its axis of rotation
3rd–7th	*Function of manufacture* – these harmonics are normally introduced by the work-holding technique during manufacture. By way of illustration, if a three-jaw chuck were used to hold the part and excessive clamping force was employed, then upon machining and removal of the clamping forces a three-lobed part would have been produced
15th–upwards	*Function of material and manufacture* – this aspect of harmonic behaviour is usually introduced to the part by either machine instability (self-exciting vibration – *chatter*) or by the reaction of the materials used in the component, cutting tool and lubricant – if used

NB: Higher harmonics could be the result of instrument noise/vibration

upr: undulations per revolution.

(Courtesy of Taylor Hobson).

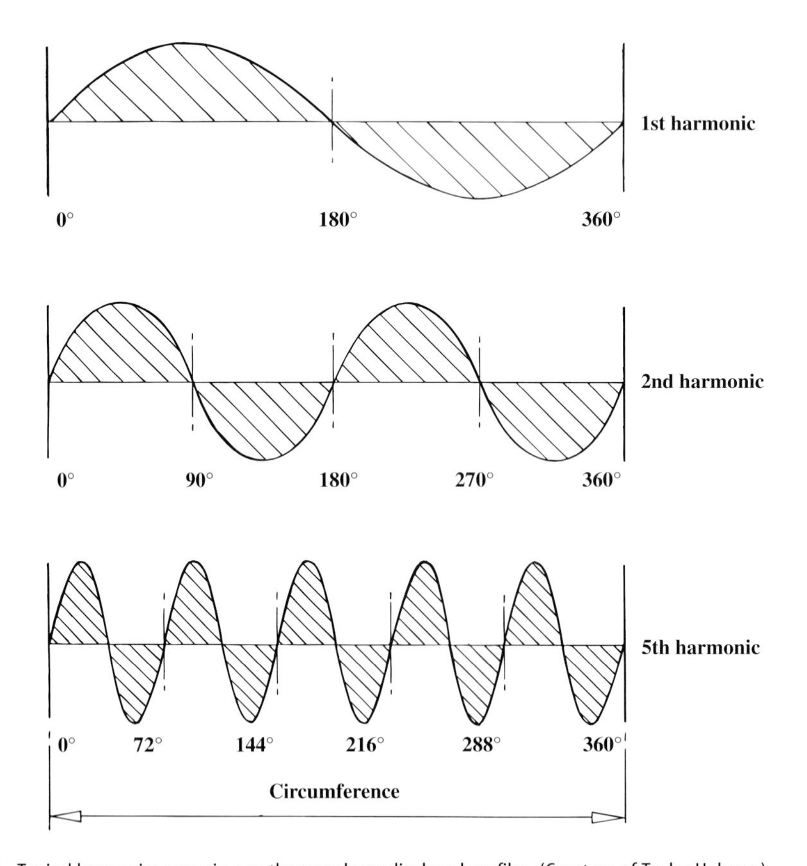

Figure 134. Typical harmonics occurring on the roundness displayed profiles. (Courtesy of Taylor Hobson.)

error (eccentricity of the part to that of the spindle axis), whereas the second harmonic may also be due to setting-up error, in this case probably due to the part not being level, that is, the axis not parallel with spindle axis. Conversely, the small group of harmonics in the region of the 64th upr level is probably due to the influence of the component's geometry on the stiffness of the grinding machine. The harmonic histogram given in this example illustrates the diagnostic capabilities of the software, for interrogation of the departures from roundness condition of the surface profile.

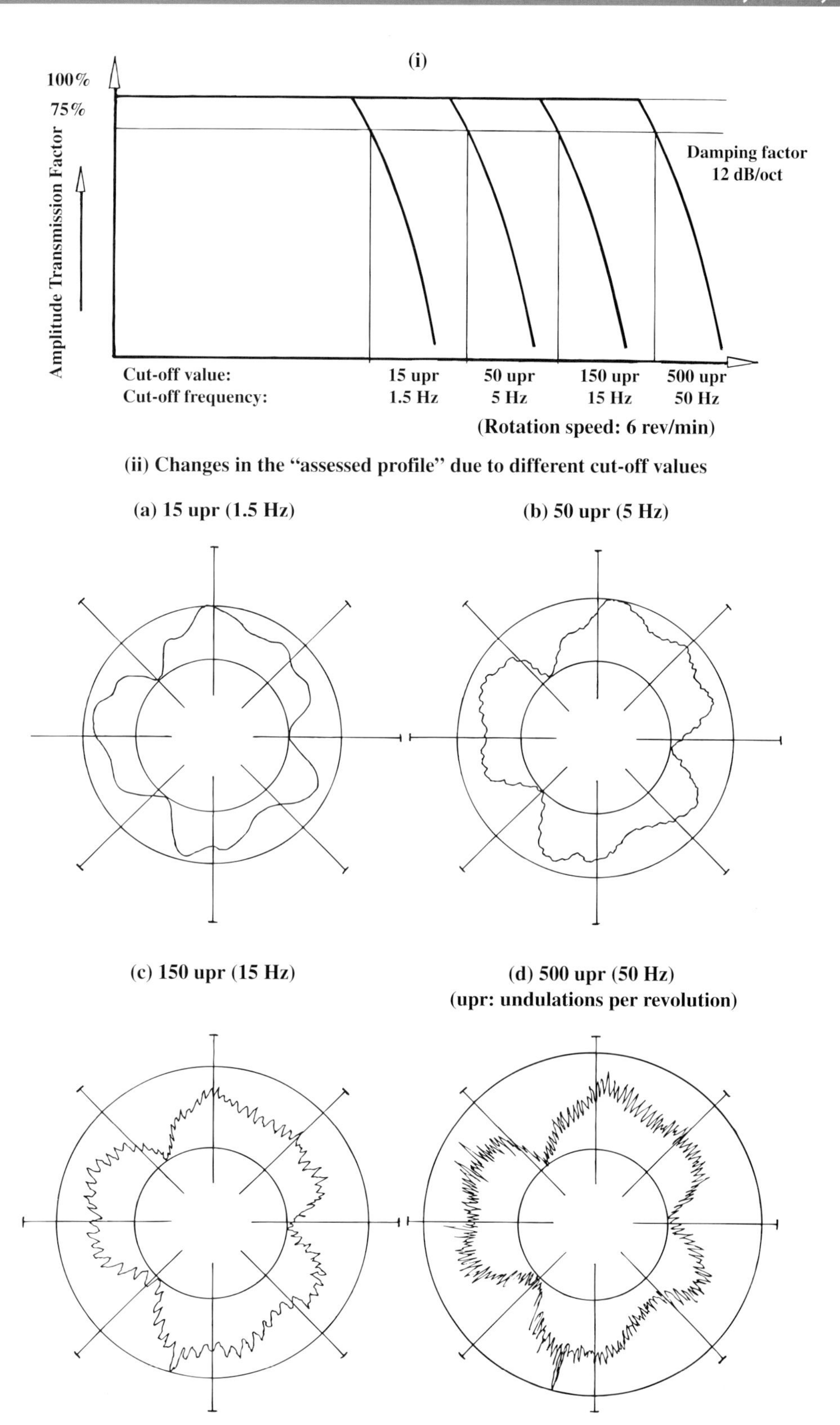

Figure 135. Effect of filtering on an identical profile assessed using the "minimum zone centre" (MZC) at a range of cut-off values.

(a) Typical harmonic histogram (FFT) for a part's roundness profile

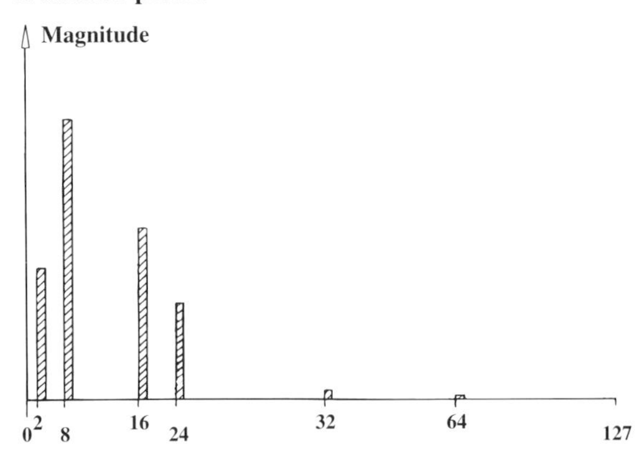

(b) An actual FFT screen display of the harmonics of a part

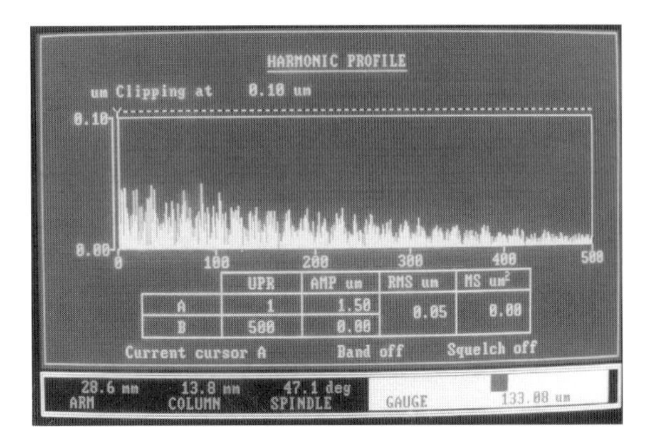

Figure 136. FFT histograms of the harmonics of a part's profile. (Courtesy of Taylor Hobson.)

Root Mean Square (RMS)

The RMS value is exceedingly useful as an analytical tool, as it exhibits to the user the "power" of the harmonics between any two associated harmonic values. This relative RMS value is helpful when attempting to control the part's machining process and the prospective in-service performance of the component. It has demonstrably been shown that if a designer decides on a specific component operating speed, then for its stated band of harmonics the RMS level will indicate whether or not the part is likely to vibrate in service. Therefore, by condition monitoring, say, certain bands in the component's in-cycle rotational speed range, it is possible to stop and reset potential production processing, or in-service performance before any subsequent parts are scrapped.

Mean Square (MS)

Compared to the previously discussed RMS technique, the MS value tends to be more sensitive to *slight variations in the component geometry* and, under these established conditions, it is used in preference to RMS, where a greater degree of control is required. It should be obvious to the reader that

$$MS = (RMS)^2$$

Harmonic analysis techniques are not only employed in the field of surface metrology (surface texture and roundness) but also in a wide variety of inspection applications. In all of these harmonic techniques the fundamental principle is identical, but the methods utilised will vary considerably. In all cases, the techniques used attempt to determine the frequencies or harmonics at which the component will vibrate. Yet another important aspect in the determination of harmonic behaviour is the effect a component will play inducing sympathetic vibration or resonance in adjacent components in an assembly by virtue of its discrete harmonics.

Reasons for filters

The general concepts relating to filtering were previously mentioned at the beginning of this section. However, a basic understanding of profile harmonics is necessary, in order to comprehend the reasons that filters are required. The operation of a filter for whatever purpose, be it for filtering oil, water or even coffee, is simply a means of separating the unwanted from the wanted. Roundness filters are no different. Roundness filters are sensitive to the number of undulations per revolution on a component; with the application of filtering it will progressively eliminate the contribution of specific harmonics to a profile.

In Figure 135(i), the graph depicts a typical set of roundness filter characteristics. As an example of their filtering behaviour, if the 500 upr curve is considered, this indicates that undulations occurring 500 times per revolution can be reduced by 25% in amplitude. The accepted manner of denoting the standard undulation range of filters is 1-15, 1-50, 1-150 and 1-500 upr (as previously described). These filtering ranges are generally termed *low-pass filters*, because they allow lower numbered harmonics (upr) than the maximum number specified in the filter to be *retained* in the profile. By utilising a 500 upr filter, as one moves to the right along this curve to higher numbers of undulations (not shown, but typically including 1-1500–150 Hz; 1-5000–500 Hz), these signals get progressively reduced. Conversely, if one considers values in the opposite direction toward the lower number of undulations their amplitudes are progressively increased, until 100% of the amplitude is allowed through the filter. It should be noted that a relationship exists between undulations per revolution and Hz (rotational speed dependency).

Yet another type of filter is known as a *bandpass filter*; as its name suggests, it has a band of undulations which it allows through. The standard value of such a filter is 15-500 upr. Basically, at 15 undulations per revolution the signals are reduced by 25%, with lower numbers being successively reduced. As the number of undulations is increased, the filter allows more of the amplitude through until at around 150 upr the amplitudes begin to reduce again, until one reaches the 25% reduction point at 500 upr.

An example of the application of filters to a component's roundness profile is given in Figure 137. Here (Figure 137a), a profile is depicted with the two distinct harmonic features, namely with the eighth and 60th harmonics displayed. In Figure 137(b) the 1-15 upr filter has been applied, which has removed the 60th harmonic from the polar trace. Figure 137(c), on the other hand, illustrates the trace with only the 60th harmonic present, because a 15-500 upr filter has been utilised, which removes the eighth harmonic.

Filters have many applications and as a result can be used for a variety of reasons, but fundamentally they are present to "decode/unscramble" the complex harmonic components represented in a roundness profile. The application of filtering enables a detailed examination of individual effects to be assessed; typically the filtering operation can determine machining defects, or a part's in-service function such as predicting the effects of wear.

Gaussian filters

Gaussian filters and, indeed, double Gaussian filters are now available for many roundness and surface texture/form instruments; however, they differ from some current and previous filter types such as the 2CR, RC ISO second order and RC phase-corrected filters in two distinct ways:

- Gaussian filtering does not simulate a specific electronic filter, but is purely a mathematical function which is applied to the profile data;
- the results obtained from a Gaussian filter have no phase shift and therefore offer more realistic results.

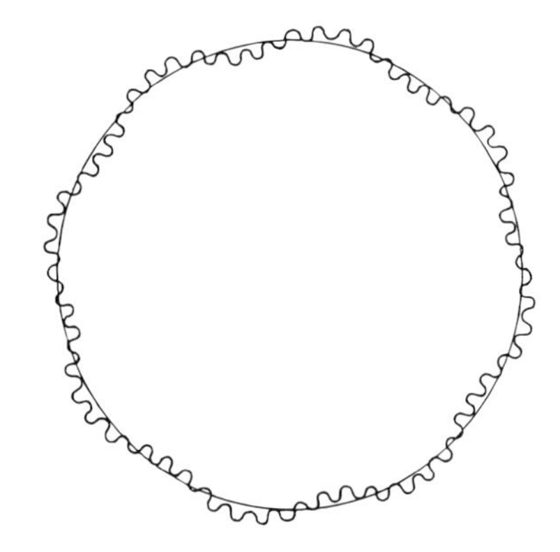

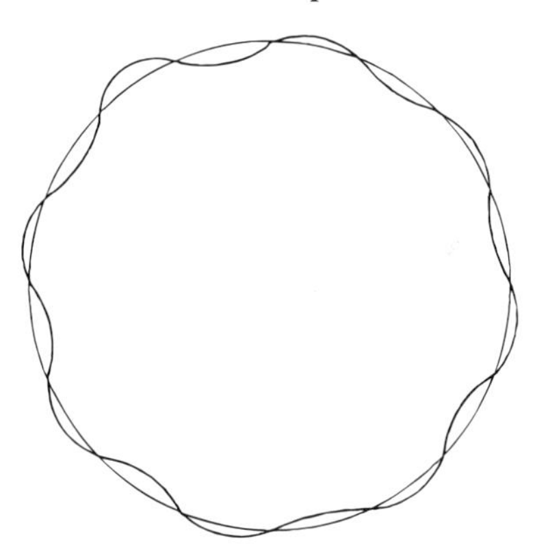

(c) Profile with a filter
of 15–500 upr

Figure 137. Filtering a component's profile for differing harmonic effects. (Courtesy of Taylor Hobson.)

(a) Establishing a "weighting factor"

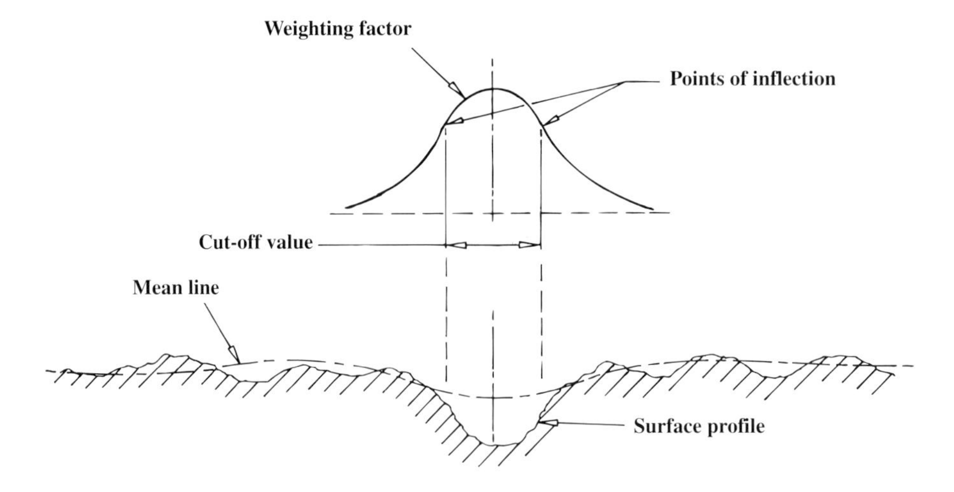

(b) "driving analogy" for different filtering types

Conventional filtering - "driver" unable to detect the contour of the road ahead until the car senses the upward or downward turn of the road and possibly overshoots at the top of a steep rise.

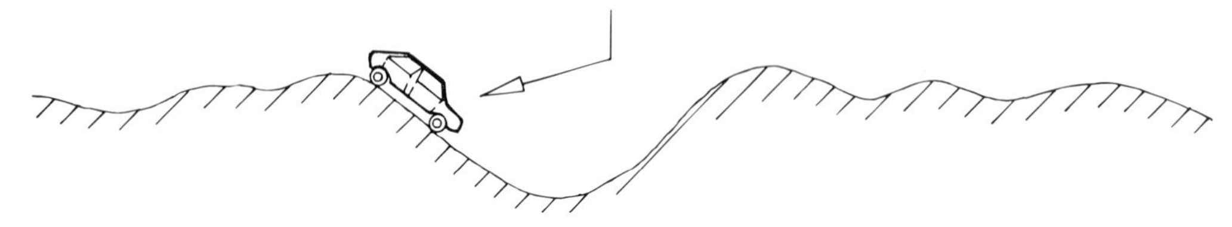

Gaussian type - "driver" has the ability to look, as in daylight and get a better reaction to the profile of the road ahead. In this case, the "driver" can see the general contour of his route.

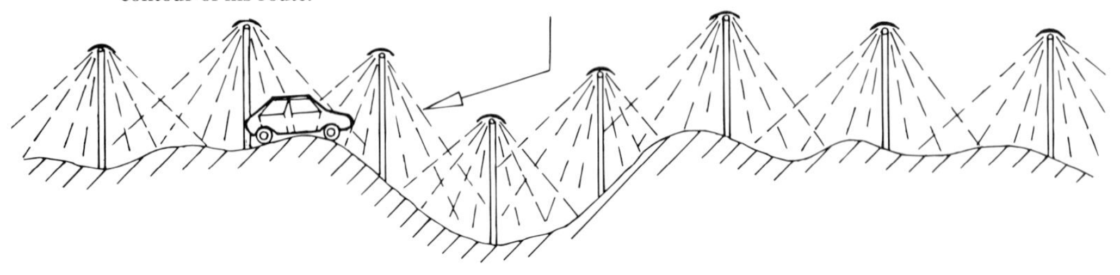

Figure 138. Weighting and filtering of roundness and surface texture instruments. (Courtesy of Taylor Hobson.)

A Gaussian filter has a *weighting factor* which has a Gaussian shape (Figure 138a); this gives the ability to take account of data before *and* after the effective stylus position when calculating the mean line. In *weighting*, one set of numbers are assigned as multipliers to quantities to be averaged, indicating the relative importance of each quantity's contribution to the average. The cut-off can be established by the width of the Gaussian curve and is defined as *the width between the points of inflection* (Figure 138a), namely the two discrete points on the curve where the direction changes. In statistical terminology, such curve inflection points mark the linear distance from the arithmetic mean to each of these points, denoting its standard deviation. In Figure 138(b) a simple but effective pictorial representation is given illustrating the effect of employing a Gaussian filter compared with conventional types.

Slope and windows

The term *slope* refers to the *rate of change* of radius with respect to angle ($dr/d\theta$). In some roundness instrumental software, the *average and maximum slopes* are given with respect to a *window*. The main reason for the measurement of slope would normally be to control component performance. For roundness assessment, the slope values have similar functions to those used in surface texture. The influence of slope in roundness analysis can be appreciated if one considers a large slope value; this will indicate a greater frictional force when utilised in, say, a bearing application. Normally, but not true for every case, if the visual finish is degraded this denotes poor wear and hence reduced in-service characteristics for the part. Displaying the average and the maximum values of slope can also allow the determination of whether the manufacturing process was or is consistent. If an increase in the relative difference between the maximum and average slopes occur, then it may indicate that the production process has changed for one reason or another.

Window assessment

The window variable operator is usually set to 5° making it, in effect, a filter. The window is the aperture from which the slope values are calculated, the calculation method being by the number of data points in this window, which are based on the following relationship:

Number of data points = $n/360 \times 3600$

where n = window selected (degrees) and 3600 = number of data points in the profile (one revolution of the part).

For example, if a window of 5° were selected, then the number of data points used for each assessment will be

$5/360 \times 3600 = 50$ data points

Pressing this point further, if a window of 15° were chosen, then the number of data points used for each calculation would represent

$15/360 \times 3600 = 150$ points

As has been mentioned, each profile typically consists of 3600 data points and, the *slope* or *rate of change* at each point is calculated. The slope at any given data point is derived by taking three adjacent data points on either side of the respective data point – a total of seven data points (which overcomes aliasing) – and then fitting a line through all seven points. From this newly created line (Figure 139a), the slope can be calculated in terms of a rate of change per degree, in μm/degree. Once the slope at each data point in the window has been calculated, then an average value is derived for the window's slope, which is known as *window slope*. In this instance, the sign (positive or negative) of the slope is taken into account, as follows:

Window slope = slope at each data point in window/ number of data points in window

Since the sign of the slope at each data point is considered in this calculation, it is possible for the slopes at each data point to cancel each other out. By way of illustration, if for example (Figure 139b) a profile window of 5° was selected this would represent a window slope of zero μm/degree, whereas a window of 2.5° would give a much larger value, hence introducing a filtering effect. As the 5° window is indexed around the component's profile, the window slope will change and it will not remain at zero μm/degree; similarly the 2.5° window would eventually yield a zero μm/degree. However, the 5° window would give a lower *average slope* value when compared to the 2.5° window.

Once the preliminary window slope has been calculated, then this window is moved around the profile by 1 data point, allowing a new window slope to be automatically calculated (Figure 139c). This procedure is repeated – 1 data point each time – until a window slope has been calculated for all 3600 data points. The average slope represents the mean value for all of the individual window slopes, that is, 3600 window slopes are averaged. The *maximum slope* can be considered to be the largest of the window slopes.

The two diagrams given in Figure 139(d) illustrate the "filtering effect" of the window. For a roundness profile which consists of 6 upr a window of 60° would not be sensitive enough; it would therefore yield a result of zero μm/degree. However, if a window of 30° were selected, this would be most sensitive since it would always be in at least one position where it was at the true maximum slope on the component. This is a "general rule" that should be used in *slope analysis*, or more specifically: "The most *sensitive* window is half the wavelength of the undulations under consideration, and the most *insensitive* window is equal to the wavelength of the undulations under consideration". If, for example, just a small difference occurs between the maximum slope and the average slope, it may suggest that the size of the window needs to be reviewed to determine whether it is sufficiently sensitive.

Therefore, by selecting an appropriate window as a filter and monitoring both the maximum and

(a) Calculating slope

Slope at this data point

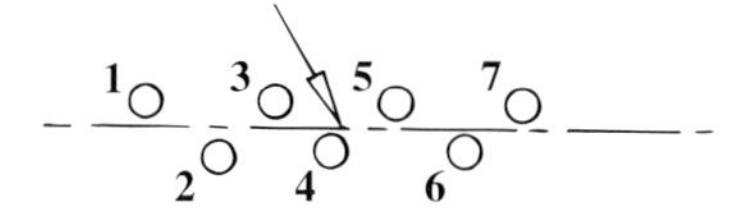

(b) The effect of window size, related to angle

(c) Determination of window slope

Window is rotated through 1 data point.

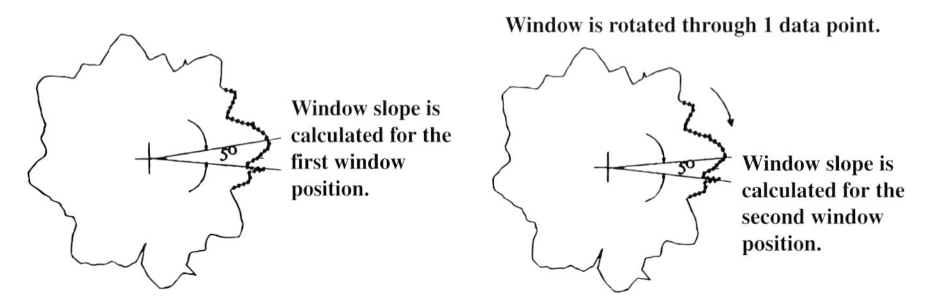

(d) The effect of window

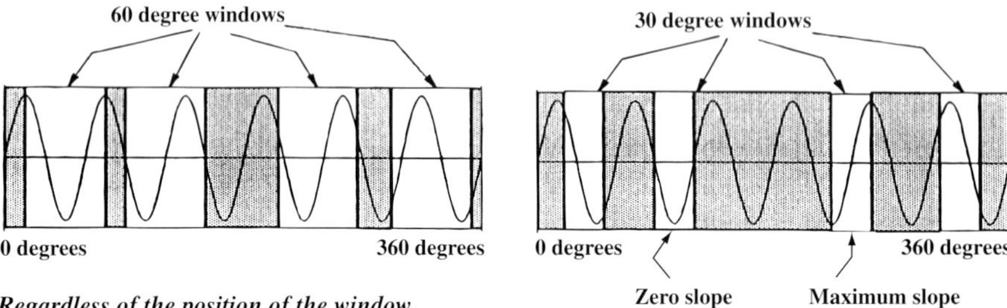

Regardless of the position of the window, the "window slope" is always zero.

On this profile a 30° window will go from a zero slope to a maximum slope.

Figure 139. Effect of "window slope" and the "window filtering effect" on slope and harmonic analysis. (Courtesy of Taylor Hobson.)

average slope values, a high degree of control of component performance can be achieved. For example, if a bearing has high slope values this will tend to be "noisy" and, as such, will be more likely to suffer from premature failure in service. The main purpose in monitoring the maximum slope parameter is so that it acts as an indicator as to whether the *window* is set at a suitable angle. If the *average* and *maximum* values are similar, then it is likely that the *window* is at an angle that is not sufficiently sensitive to the component features.

4.6 Geometric roundness parameters (ISO 1101: 1983/ Ext 1: 1983)

Eccentricity

The parameter *eccentricity* (Figure 140) is a roundness term used to describe *the position of the centre of a profile relative to some datum point.* Eccentricity is a *vector quantity*, in that it has both *magnitude* and *direction*. Moreover, the magnitude of the eccentricity can be simply expressed as the distance between the profile centre and the datum point, whereas the direction is expressed as an angle from this datum point.

Concentricity

The *concentricity* parameter (Figure 141) is similar to that described for the term *eccentricity* (above), but only has *magnitude* and no direction. Concentricity can be described as *the diameter of the circle described by the profile centre when rotated about the datum point.* It may be appreciated that the value of concentricity is *twice* the magnitude of eccentricity.

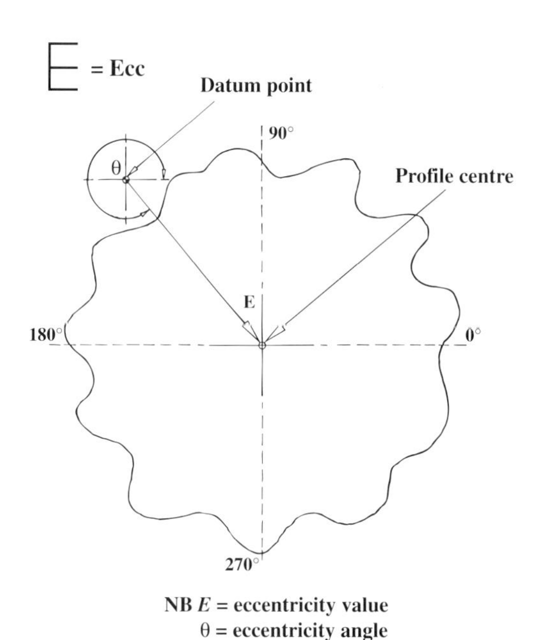

Figure 140. Roundness parameter: eccentricity.

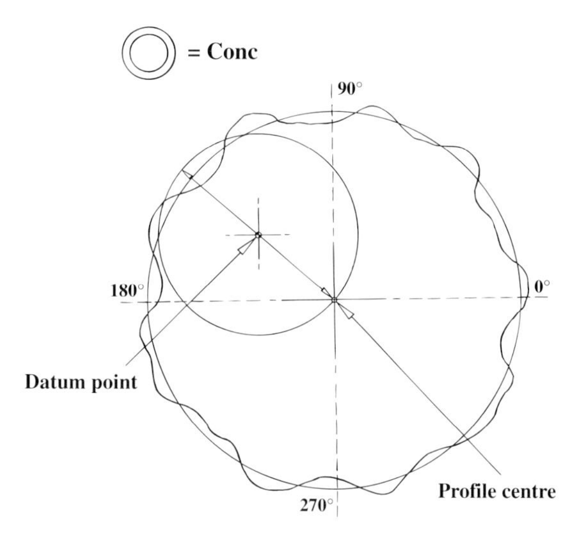

Figure 141. Roundness parameter: concentricity.

Runout

The term *runout* (Figure 142) is occasionally referred to as *total indicated reading* (TIR). This parameter attempts to predict the behaviour of a profile. The definition of runout is: *the radial difference between two concentric circles centred on the datum point and drawn such that one coincides with the nearest and the other coincides with the farthest point on the profile.* As a practical parameter runout is useful, as it combines the effect of both form error and concentricity to indicate the predicted performance when rotated about a datum.

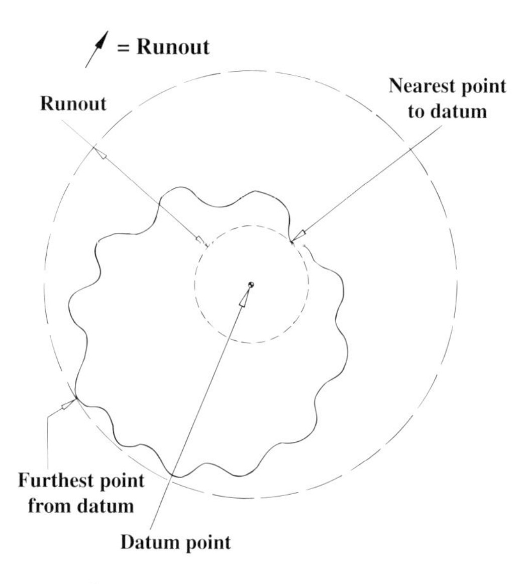

Figure 142. Roundness parameter: runout.

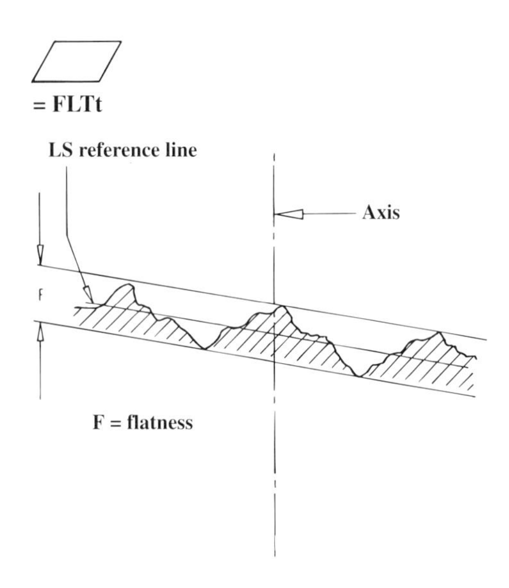

Figure 143. Associated parameter: flatness.

Flatness

In the case of *flatness* (Figure 143) a reference plane is fitted such that its flatness is calculated as *the peak-to-valley departure from that plane.* In roundness reference terminology, either "least squares circle" (LSC) or "minimum zone circles" (MZC) can be utilised for flatness determination.

Squareness

Once an axis has been confirmed, its *squareness* parameter (Figure 144) can be described as: *the minimum axial separation of two parallel planes normal to the reference axis and which totally enclose the reference plane.* The term *squareness* is a useful indicator of the fitment behaviour of two adjacent parts in a precision assembly.

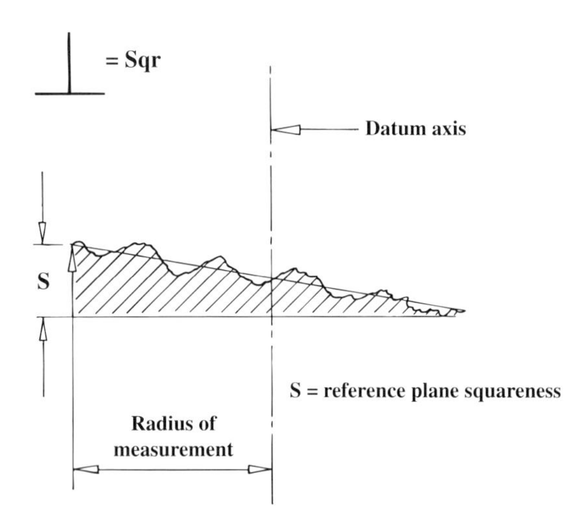

Figure 144. Associated parameter: squareness.

Coaxiality

The *coaxiality* parameter (Figure 145) refers to *the relationship of two cylinder axes, one of which is used as a datum.* This term is useful when attempting to control the relative motional behaviour of axial and roundness alignments between two adjacent diameters.

Cylindricity

The *cylindricity* parameter (Figure 146) refers to *two, or more roundness planes used to produce a cylinder where the radial differences are at a minimum.* This *cylindricity* term will be discussed in more detail in the following section.

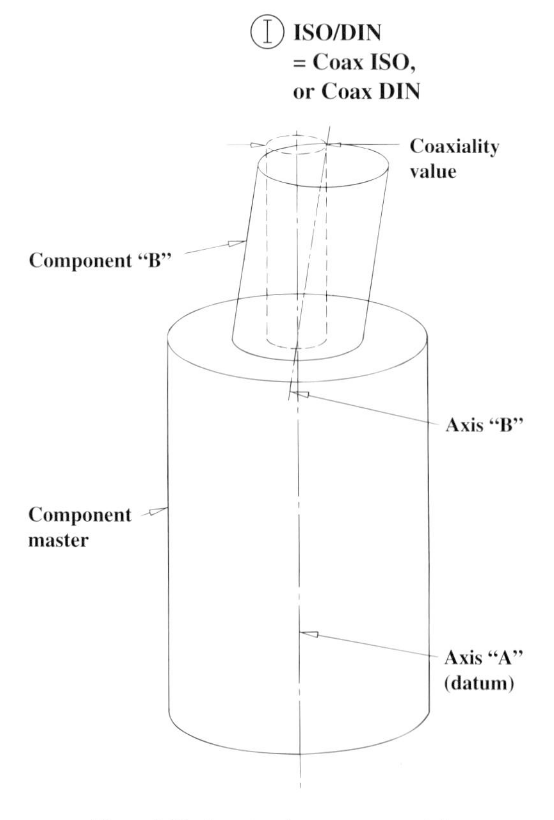

Figure 145. Associated parameter: coaxiality.

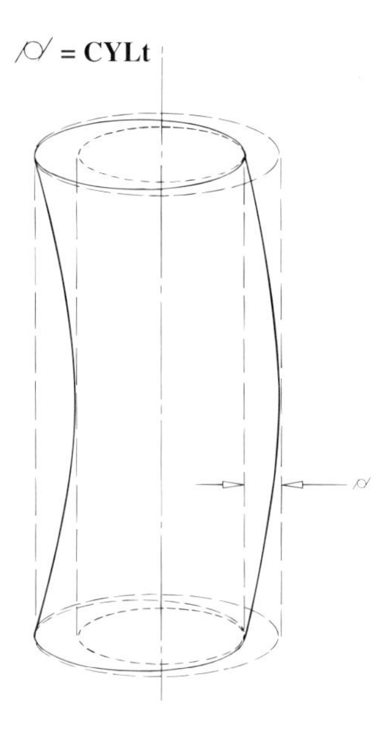

Figure 146. Associated parameter: cylindricity.

4.6.1 Cylindricity

For many mechanical design and assembly applications cylindrical features are crucial as they significantly contribute to the performance of a range of products, typically revolving equipment, transmission and part assemblies, precision metrology equipment, precision gauges and injection moulding dies, enabling them to achieve their intended functions. During the production process several factors can cause a cylindrical feature to depart from its intended geometric shape or profile. A definition of the measure of cylindricity was given in Section 4.6, but a simpler working description may be in order here to appreciate its importance in precision engineering applications. If a perfectly flat plate is inclined at a shallow angle and a parallel cylindrical component is rolled down this plate, if it is truly round then as it rolls there should be no discernible radial/longitudinal motion apparent. However, if the same plate is inclined and now a less-than-cylindrical component is tested that exhibits one of the departures from a cylindrical shape shown in Figure 147, then as the component rolls down this incline it will either wobble or deviate in some radial or longitudinal manner – this condition can be said to be qualitative proof of a lack of cylindricity.

The schematic diagrams in Figure 147 show just some of the combinations of geometric departures from cylindricity, which are invariably caused by their inherent production processing – more will be said on this topic in Chapter 5. As can be seen (Figure 147a), the combined geometric shape is the amalgamation of a lack of straightness, poor roundness and taper. Furthermore, a so-called cylindrical product may also exhibit either curved, waisted or barrelled geometric shape, with variations in longitudinal cross-section superimposed onto the part shape (Figure 147b). For example, a component may have the correct nominal size – diameter – but its form could be incorrect, and therefore its function will be impaired. Classic cases of this problem are encountered in automotive fuel injection systems, where the mating cylindrical components of valve pumps and injectors must not allow a bypass of fuel under high pressure when not required, namely during either deceleration or braking. These fuel injector systems are subject to ever-increasing and stringent tolerances as greater fuel economy becomes the norm.

Cylindricity measurement

In order to measure cylindricity, reference must be made to the basic definitions previously described; namely we require an instrument that will measure radial form or roundness, axial form or vertical straightness, dimensional uniformity or parallelism. Two possible metrological sources are available to perform these measurements:

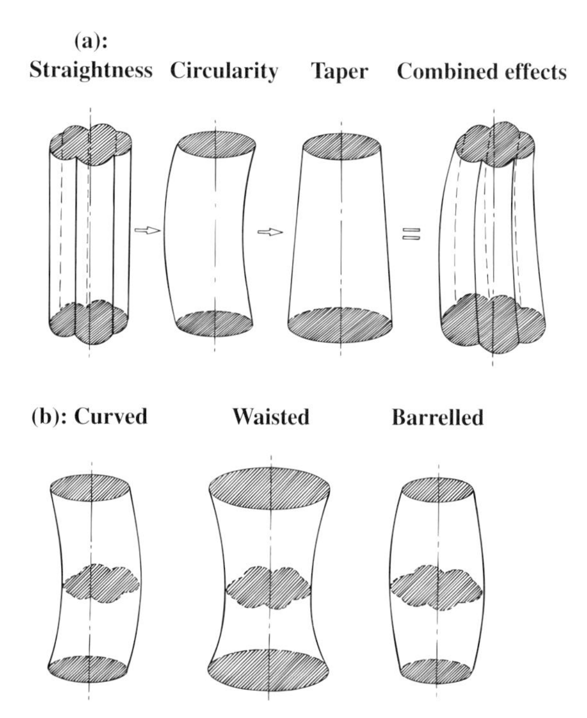

Figure 147. Assessment of the geometry in cylindricity.

1 coordinate-measuring machine (CMM);
2 form/roundness-measuring instrument.

In the former case (CMM), actual volumetric assessment of parameters such as cylindricity requires a considerable number of measurement points to be made, which of necessity takes more time and provide limited accuracy. The exception to these CMMs is when a very expensive CMM is utilised having sophisticated continuous scanning software with rotational datums coupled to analogue probes. Under these operational circumstances, the latter case of using form/roundness-measuring instruments for cylindricity assessment offers an ideal alternative measuring technique.

It is convenient when measuring cylindricity to represent the cylinder by a series of horizontal slices of the radial profiles, where the axial separation is known (Figure 148). Conventional form and roundness-measuring instruments are capable of performing this type of representation, provided certain criteria are observed. The contribution to uncertainty for both radial and axial form has been well documented, but what is sometimes forgotten is that the parallelism of the instrument's spindle with respect to the column axis needs to be addressed. Specifically, such potential spindle-to-column alignment error can influence measurement – more will be said on alignment checking in Chapter 6. These radial slices and their axial displacement factors link the two-dimensional roundness and straightness measurements to the three-dimensional or total form analysis that is required for cylindricity measurement. It is therefore essential that any measuring instrument used to determine cylindricity has the ability to accurately measure radial form and axial form, together with dimensional uniformity.

As mentioned previously, geometric form tolerances and measurements are described in a series of standards, notably ISO 1101: 1983. The inspection procedure and associated algorithm to determine the precise measurement requirement are not provided in these standards. However, based on the

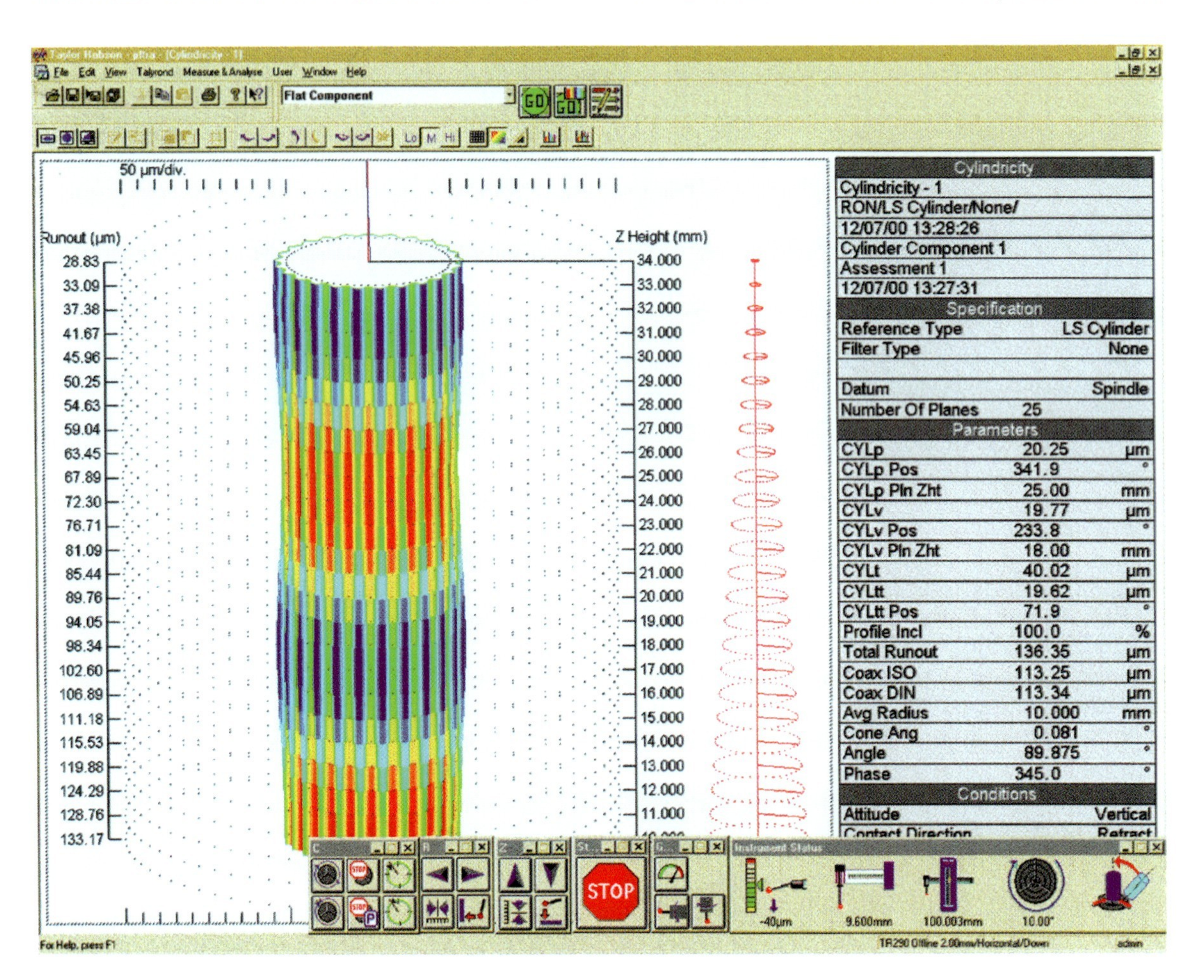

Figure 148. Wire cage cylindricity display, utilising Windows™-based software. (Courtesy of Taylor Hobson.)

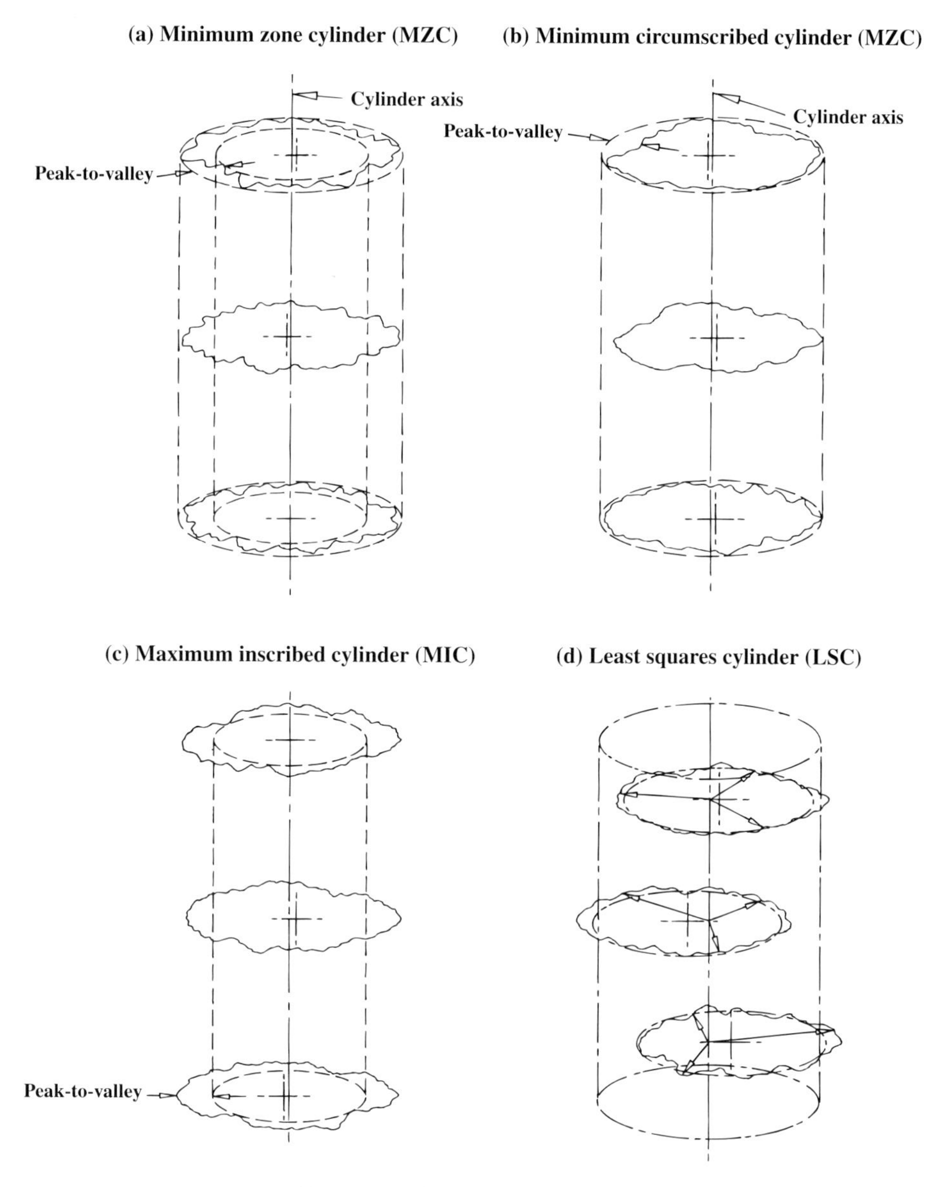

Figure 149. The four reference cylinders.

standards definitions, the cylindricity of a feature is normally assessed in two stages: firstly, a reference cylinder is obtained utilising data obtained from measurement of the profile; and, secondly, these measured points are then compared with the reference cylinder, allowing the derivation of cylindricity. There are four types of reference cylinders to which data can be fitted:

- minimum zone cylinder (MZC; Figure 149a);
- minimum circumscribed cylinder (MCC; Figure 149b);
- maximum inscribed cylinder (MIC; Figure 149c);
- least squares cylinder (LSC; Figure 149d).

A brief review will now be given for each cylindricity method of assessment.

Minimum zone cylinder (MZC).

The MZC technique requires information on the total radial separation of two concentric cylinders (having the same axis) which totally enclose the data generated during measurement and are positioned such that their minimum radial separation is known.

Due to this MZC assessment method it places equal emphasis on the internal and external profile imperfections to determine the reference axis. The main industrial application for MZC would be where running (clearance) fits are required.

Minimum circumscribed cylinder (MCC).

The MCC technique relates to the minimum radius that fully encloses the profile data.

This MCC method can be utilised on precision parts where the surface of the outside diameter is important, such as for attribute sampling inspection techniques, typified by inspecting internal bores by limit gauging applications (plug gauges).

Maximum inscribed cylinder (MIC).

This differs only marginally from the MCC, in that the first terms are reversed in the following equation, such that this MIC technique is one of where the maximum radius that fully fits into the profile data occurs.

Hence, in the case of the reference cylinder for MCC, it is the cylinder that encompasses all the data points and has the smallest radius; conversely, the MIC reference cylinder does not enclose any data points, but simply has the largest radius. The MIC method might be used for female mating part assembly techniques, or for inspection by limit gauging using ring gauges.

It should be noted that in all cases for the above reference cylinders they are extremely sensitive to asperities, such as that caused by dirt or scratches on the surface. These arbitrarily induced asperity effects can promote axial drift and the solution to this problem will be dealt with later.

Least squares cylinder (LSC; Figure 149d).

Unlike the previous reference cylinders, which are in essence a combination of mathematical algorithms and to a lesser extent iterative trial-and-error techniques, the LSC can be more specifically defined.

This LSC method is probably the most popular approach for defining cylindricity because of the ease of computation and the uniqueness of the solution. From a practical viewpoint, the LSC is constructed from two or more roundness planes at different heights and a "best fit" line is fitted through the centres of the various planes. This line is termed a *least squares axis*. The least squares cylinder is then calculated by taking the average radial departure of all the measured data from the least squares axis. It should be observed that the planes are not shifted to the least squares axis but remain eccentric. Furthermore, unlike the three previously described "cylinder fits", the axis of the LSC is very stable and much less influenced by the effects of asperities.

4.6.2 Cylindricity measurement techniques

A question frequently raised concerning the measurement of cylindricity measurement, is: "How many planes are required to define a cylinder?" In theory, the minimum number of planes is two, but with such a small number this can lead to problems. For example, if a frustum (truncated cone – one of the geometric shapes in Figure 147a) was to be measured, two planes will detect any angled taper. However, if the shape appeared in the geometric form of those given in Figure 147(b), then this number of planes would be insufficient to detect, say, the waisted or barrelled shapes. Basically, the number of planes utilised must be adequate to detect the variation in the vertical (axial) form. Once again, to define an axis a minimum of two planes is required, although to detect variations in axial form, or dimensional uniformity, three to five planes would be more realistic. The trade-off here must be between adequately representing a component and the time taken to compute the "best fit" reference cylinder, although with the latest instruments having significantly greater computing power than previously was the case this statement has less relevance. It should be emphasised, however, that the time factor in this calculation significantly increases with the number of planes and/or data points per plane selected, especially when calculating the MZC, MIC and MCC reference cylinders.

In any form of cylindricity profile assessment, it is important to establish the relative radius of all the planes, because this can lead to constraints with some stylus instruments. For example, if the only means of measuring radial departure of the system is by the pick-up, then the stylus must not decouple (lose contact) with the component, or the radial reference will be lost. This is particularly relevant if the pick-up had to be retracted to avoid collision with a shouldered feature on the part. This problem is somewhat mitigated against by the latest roundness instrumentation, due to the fact that invariably they have an optical scale attached to the radial arm, meaning that it can be moved, yet still preserve the relative radius between the measuring planes. Moreover, in this arrangement, the accuracy will now depend on the resolution of the radial arm grating and not on the resolving accuracy of the pick-up.

4.6.3 Cylindrical measurement problems

If a true cylinder should be tilted (Figure 145, smaller-diameter boss) when measured, the result will appear as an ellipse; therefore it is essential that any cylinder has been levelled prior to measurement. However, it may be the case that a second cylinder has to be measured relative to the first (Figure 145, for coaxiality); in this case re-levelling is not the answer, as the primary datum will be lost. In this situation it is possible for the computer to correct for the tilt by calculating this tilt, then orienting the axis and noting the radius of the second cylinder. At this stage the computer will compensate for any error by removing the cylinder tilt ovality for each radial plane, prior to performing the cylinder fit. The measuring strategy of removing the second harmonic term or applying a 2 upr filter is not adequate, as any true ovality in the component will also be removed.

For results to have any meaning in roundness measurement and particularly in the case of cylindricity, various factors have to be considered; therefore before any actual measurement of cylindricity occurs on the instrument, three important rules must be addressed to obtain an accurate result:

1 the spindle should continuously rotate during measurement, which ensures that any slight shifts during the start and stop of the spindle are eliminated. This operating condition, although not so critical in roundness and concentricity measurements, increases in importance with the measurement of roundness at different planes, particularly when cylindricity tolerances approaching 0.5 μm are specified;
2 the maximum accuracy of the pick-up is maintained by ensuring that it is in contact with the workpiece at all times. If a radial arm transducer is fitted the resolution of measurement will be reduced by at least one order of magnitude worse than the pick-up;
3 the component normally requires centring and levelling with respect to the spindle axis. How well the part should be centred and levelled is related to the roundness-related value that has been previously specified on the drawing for the workpiece. As a general rule of thumb, *the centring and levelling should be within 10 times the peak-to-valley values to be measured.* Any residual error that occurs can be automatically taken into account by the software. This automatic compensation ensures that cylinder errors are kept within the gauge range; moreover, they allow the appropriate resolution to be selected.

Good metrology rules of practice are extremely important to cylindricity measurement. The major reason for strict adherence to these metrological rules is that cylindricity measurement is dependent on two axes, namely the axis of rotation of the spindle and a linear axis. Therefore, the relationship between these axes must be known and stable. The dependence on two axes for measurement of cylindricity establishes a system requirement for the straightness unit to be set parallel to the axis of the rotating spindle. Any misalignment of the vertical straightness datum relative to that of the axis of the spindle will result in a perfectly straight workpiece appearing tapered. The "coning error" of the spindle will also have a detrimental effect on the measured result, with a perfect cylinder giving the appearance of a cone. Furthermore, if the form error of the instrument's column – its straightness – is significant, then this will have an effect on the part's measured result and influence its peak-to-valley value.

The MCC, MZC and MIC axes are very sensitive to asperities and whenever possible they should be avoided for datums, because if a large uncharacteristic asperity occurs in one of the measuring planes this will result in a datum shift, which will unduly influence the overall cylindrical assessment. Conversely, the LSC axis is insensitive to asperities and is recommended as a datum whenever possible. The reason for this insensitivity asperity is that asperities have little or very marginal effect to affect a datum in an LSC. It follows that asperities such as debris (dirt), or scratches will still affect the peak-to-valley results regardless of which reference cylinder is utilised. It is therefore important to clean both the workpiece and stylus before measurement.

A factor that needs consideration particularly with reference to the MIC and MCC (cylinders) is their inherent instability on tapered components (Figure 150). For example, if an MCC were fitted to a tapered component then it must touch in at least three positions, notably at two points on the largest diameter and one on the smallest. Under these circumstances, the cylinder axis will be "skewed" from that of the component's axis. To compound this skewness problem, there would not be a unique position, thus making its value somewhat limited in assessment. The one case where this skewed axis movement would have significance is when it is required to establish the amount of "play/clearance" that could be obtained with a tapered shaft in a bore, or vice versa, if employing the MIC.

In all these measurement techniques that are used to determine a component's cylindricity, "sampling techniques" are utilised. Such "sampling" is essential, since it is clearly impractical to measure every single point on the surface of a cylinder and, even if this were technically feasible, the calculation itself would be prohibitive.

The effect of the selection of a number of planes to be used on a cylinder has already been discussed and, when widely spaced, results in their inability to detect geometric cylindrically based shapes such as the curved, waisted or barrelled figures (Figure 147b). A potential measurement problem can occur in the number of data points used in each plane, because if insufficient numbers are used then this can mean that certain profile detail will be lost, such as occasional dominant peaks and valleys. This loss of surface detail resulting from inadequate numbers of data points means that both inaccuracies and instability may result.

There are numerous ways in which the data can be displayed by sophisticated software, notably the multiplane screen image depicted in Figure 151. This figure illustrates roundness on a component at differing diametral points, indicating runout (μm) at these pre-selected locations. Conversely, Figure 152 shows a three-dimensional "bed-type" plot of an LSC component.

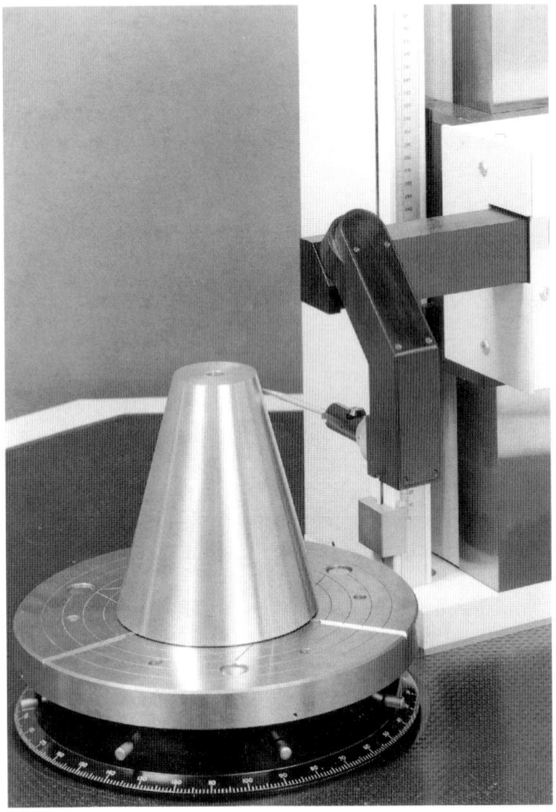

Figure 150. Assessment of a tapered component. The axes of the column and taper should be parallel, thus avoiding any potential measurement error. (Courtesy of Taylor Hobson.)

4.7 Non-contact spherical and roundness assessment

Non-contact measurement of roundness is valid and has been possible for some time. A brief discussion of the range of instrumentation that has been developed is worthwhile, albeit if discussed in depth, this would significantly increase the scope of the current work and be somewhat counter-productive. Despite these self-imposed objections, it is worth reviewing two techniques: one interferometry-based instrumental method for determining sphericity of journal balls and another employing an optical sensor in conjunction with a displacement probe for applications in both roundness and sphericity assessment.

4.7.1 Sphericity interferometer

The sphericity interferometer illustrated in Figure 153 utilises in this version a monochromatic light source along its optical path; however, as an alternative, it would be possible to construct a similar instrument with laser-based illumination. As the instrument's name implies, the optical system utilises the effect of light interference to compare the curvature of the reflecting spherical surface under test, typically a precision ball, with a curved reference surface. The interferometric principle is as follows: when light rays are reflected between two surfaces that are not parallel, the difference in path lengths at particular portions of the spherical part's surface causes phase changes in the light reflected back to the observer. As a result of these light path differences, some rays are cancelled out while others augment each other. This interference behaviour causes patterns of alternate dark and light lines to appear across the image (Figure 153, bottom). The shape and spacing of these interference fringes, as they are usually known, will depend on the regularity and spacing of the two surfaces.

In this particular optical configuration of the sphericity interferometer, the light is reflected between the semi-reflecting surface of the precision ball and a hemispherical reference optic, the resulting interference pattern being directly related to variations in curvature on the surface of the ball.

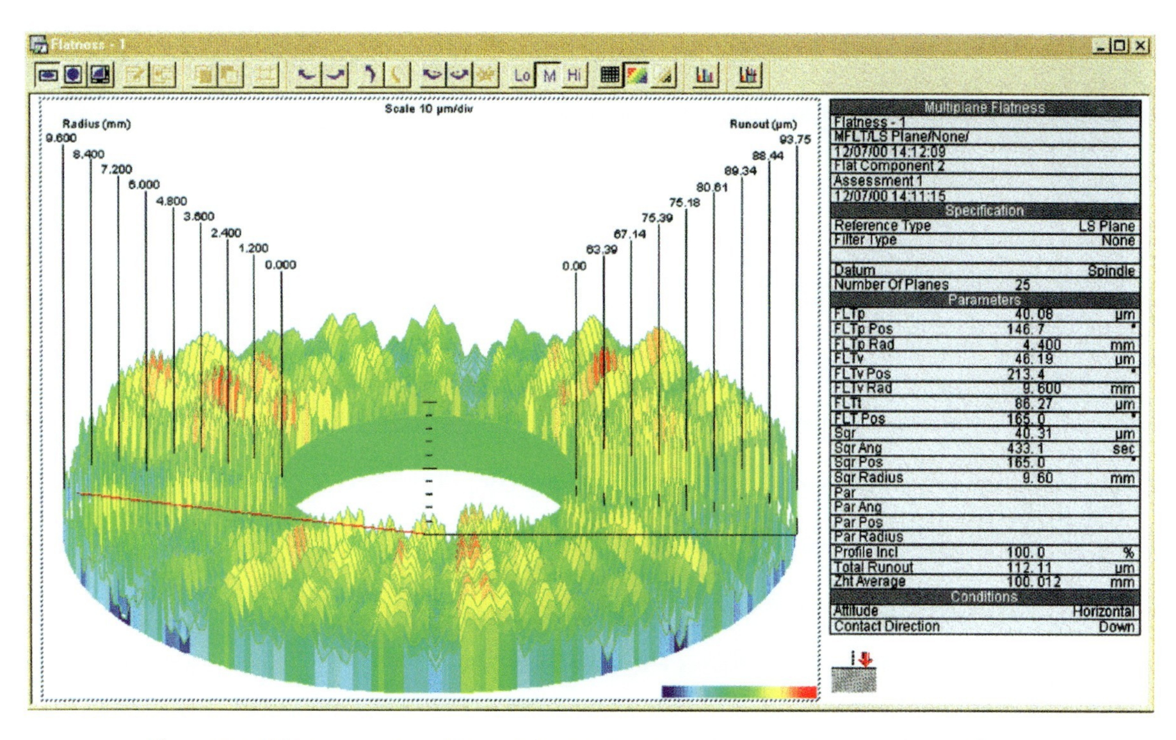

Figure 151. Multiplane roundness/flatness display, Windows™-based software. (Courtesy of Taylor Hobson.)

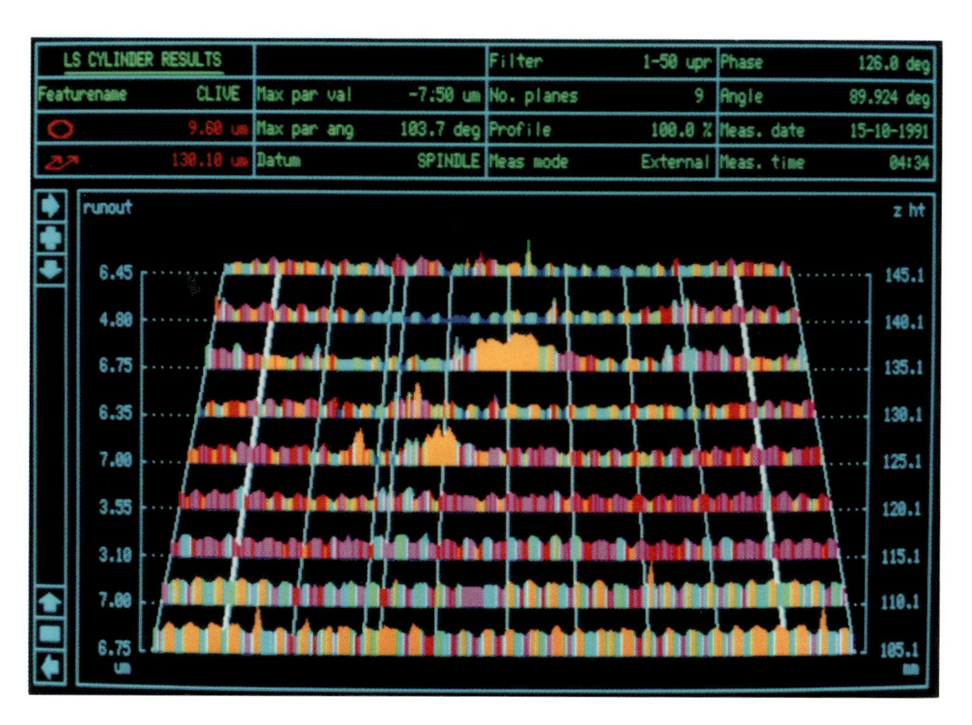

Figure 152. A three-dimensional "bed-type" plot of least squares cylinder results. (Courtesy of Taylor Hobson.)

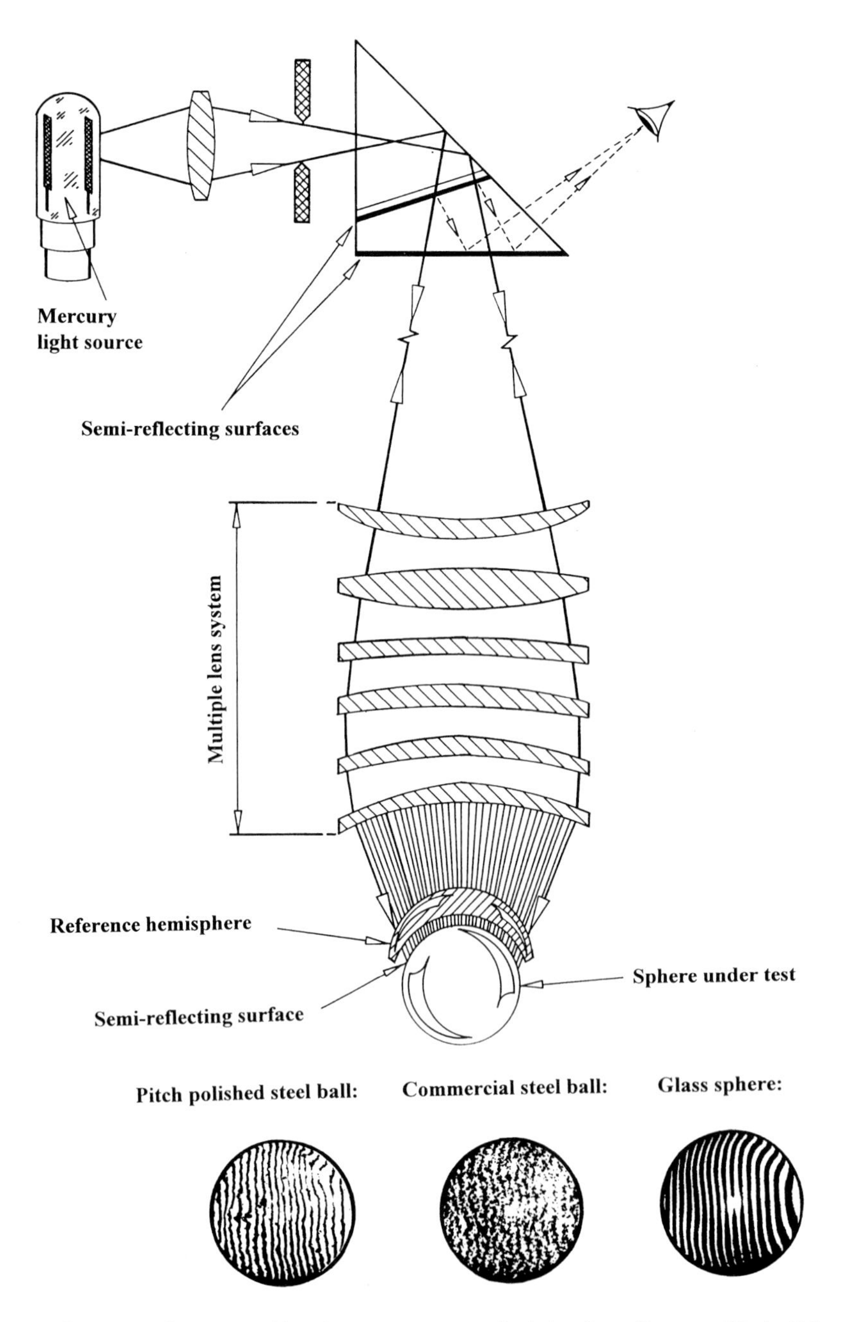

Figure 153. Sphericity interferometer and interference patterns on spherical surfaces. (Courtesy of Taylor Hobson.)

This type of optical interferometer can be considered very much a single-purpose instrument of limited application, but it does illustrate one method of non-contact roundness measurement. It is particularly relevant for use on precision highly polished soft surfaces, which could be unintentionally scratched by the action of a stylus-based roundness instrument. Moreover, it is limited in the diameter range of spheres the technique can measure.

4.7.2 Spherical and roundness assessment by error separation

Components produced to nanometric accuracy are becoming increasingly important of late, with the demand for on-machine measurement to improve manufactured parts. The separation of potential workpiece roundness errors from those imparted by the spindle error from the machine tool are important for on-machine inspection. For some years, non-contact measurement utilising *error separation techniques* employing multiple probes to perform on-machine measurement has been undergoing development. As has been previously mentioned, any form of single-plane roundness measurement involves the product of several essential parameters; these are:

- *workpiece roundness error* – intrinsic errors generated during the production processing technique which are superimposed onto the part;
- *kinematic motions* – of the *X*- and *Y*-directional components.

In 1972, Donaldson introduced a "Ball reversal technique" for error separation, which is analogous to that of "Straightedge reversal". This "Ball reversal technique" (see Figure 242, Appendix C) occurs, where the radial motion – given by R(θ) and the ball roundness – given by B(θ) are in the prescribed relationship as shown. Therefore, the angular information is assured to be derived from the spindle position, the indicator output and the two positions $l_n(\theta)$ would produce the following relationships:

For the two positions:

$$l_1(\theta) = R(\theta) + B(\theta)$$
$$l_2(\theta) = -R(\theta) + B(\theta)$$

and

$$R(\theta) = \frac{l_1(\theta) - l_2(\theta)}{2}$$

$$B(\theta) = \frac{l_1(\theta) + l_2(\theta)}{2}$$

It is worth noting that this technique gives the errors of the test ball exactlly, together with the radial motion of the spindle along the indicatorr (sensor) test direction. If the orthogonal component of the spindle radial motion is needed. It can be found from a single (added) test with the indicator and ball rotated 90°, using the known ball roundness error. Furthermore, with a known roundness error, the ball can then be employed in any of the axis of rotation test configurations.

Most of the sensors used for non-contact roundness assessment have employed a *three-point method*, normally using three displacement probes that are strategically positioned to measure the workpiece profile. The problem with this triple-sensor technique is that high-frequency components of roundness error cannot be accurately measured due to harmonic suppression. Even when more than three displacement probes are used, this suppression problem cannot be entirely eliminated. Relatively recently, however, it has been proposed that several multi-probe methods will eliminate much of the harmonic suppression, more specifically, by adopting an *orthogonal mixed method* (Figure 154a); this virtually overcomes any high-frequency suppression. In the latter arrangement, only single displacement and angle probes are used; they are arranged at an angular orientation of 90°, which has been found to yield good operational characteristics. This orthogonal mixed method system configuration is shown in a working arrangement for sphericity/roundness assessment in Figure 154(b). The measurement system consists of assessment of workpiece for roundness by use of angle and displacement probes, an air-bearing spindle motor-drive, optical rotary encoder, together with an analogue-to-digital converter and PC. In the present configuration the displacement and angle probes are angular-oriented at 90° with respect to each other. The roundness specimen (workpiece), is mounted on the air spindle and the round/spherical profile can be sampled by the probes as it rotates. In this current configuration, the probe's output signals are input to a PC by a 12-bit A/D converter, as a trigger signal. To avoid errors that might otherwise be introduced as a result of delay due to the sampling time, output signals are simultaneously sampled. The workpiece can be adjusted in both the *X*- and *Y*-directions (via adjustment screws), enabling any eccentric error to be removed, allowing the workpiece to be inspected within the measurement proximity range of the sensors. Each probe is mounted on individual *XYZ* microstages allowing the relative positions of each sensor to be set at the desired orientation to the workpiece. Results from this orthogonal mixed method instrumentation are given in Figure 154(c, d), indicating for part (c) the measured roundness errors of two distinct measurements and the repeatability error between them. From the test trials (c-ii) the roundness error was around 60 nm, with a repeatability error of 5 nm. The largest individual repeatability error (c-iii) occurred at the third-order harmonic, namely 0.6 nm, this may have been the result of initial set-up positioning error. Conversely, the relatively large measured spindle errors in this configuration (d) for the repetition trials gave a spindle error of 800 nm, with a difference of 140 nm. It should be mentioned that a major contributing factor to this spindle error was vibration components. If a comparison is made between the plotted results in Figure 154 for this instrumental configuration (small differences: c-ii and c-iii; against large differences: d-ii and d-iii), then the roundness error (c) is separated from the large spindle error (d) with high repeatability, thus confirming the effectiveness of the orthogonal mixed method of non-contact roundness assessment.

(a) Roundness measurement by the orthogonal mixed method

Probe 1: Displacement probe
Probe 2: Angle probe

$e_Y(\theta)$ $e_X(\theta)$ Y Probe 1 θ P A Probe 2 B O X R_r Workpiece

(b) Experimental system for roundness

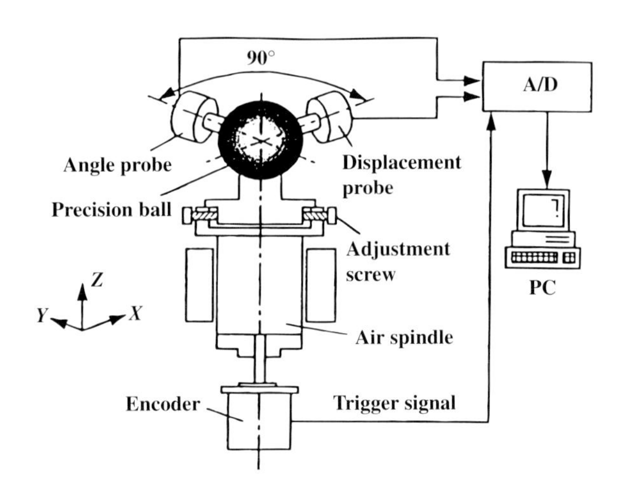

(c-i) polar plot of roundness error $r(\theta)$ and repeatability error $\Delta r(\theta)$

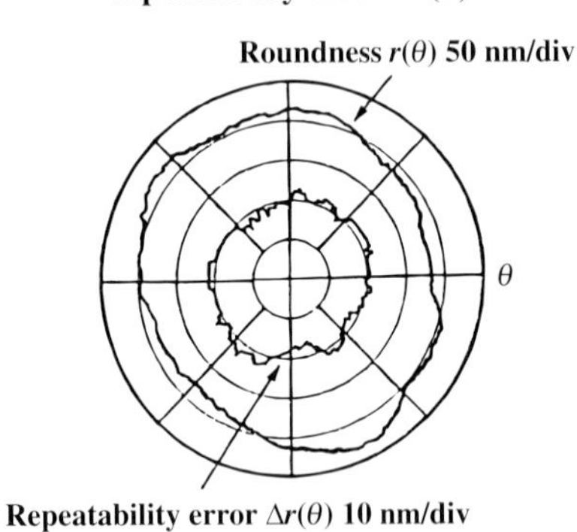

(d-i) Polar plot of spindle error $c_x(\theta)$ and difference $\Delta c_x(\theta)$

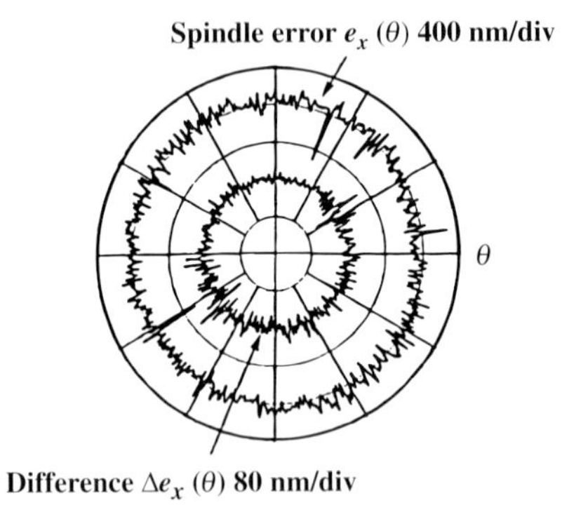

(c-ii) Rectilinear plot of roundness error $r(\theta)$ and repeatability error $\Delta r(\theta)$

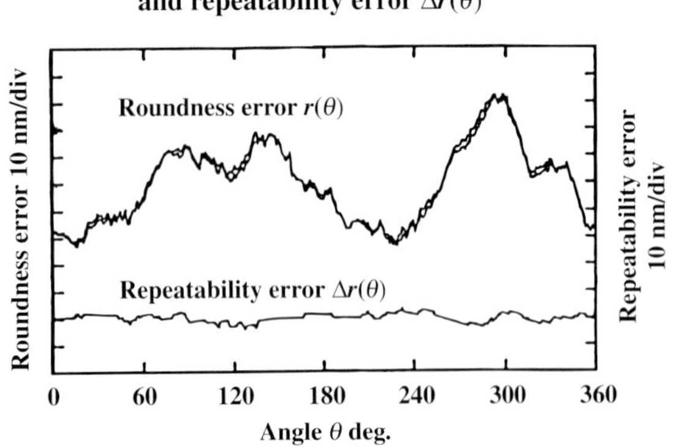

(d-ii) Polar plot of spindle error $c_x(\theta)$ and difference $\Delta c_x(\theta)$

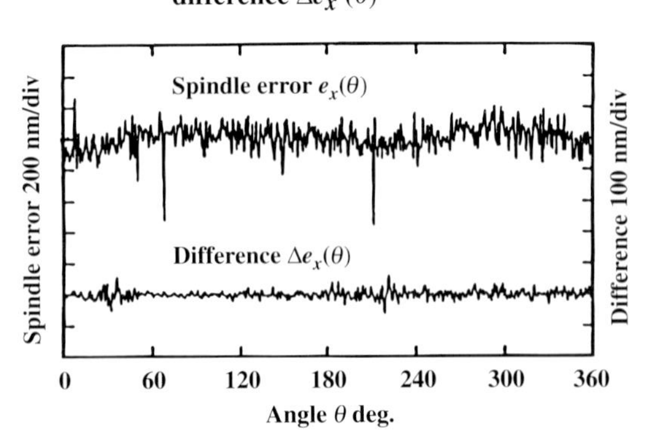

Figure 154. Measured roundness and spindle error in error separation by the *orthogonal mixed method*. [After Gao *et al.*, (1997).]

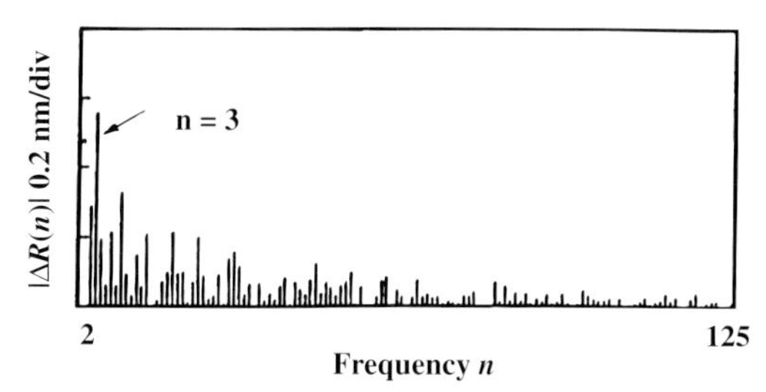

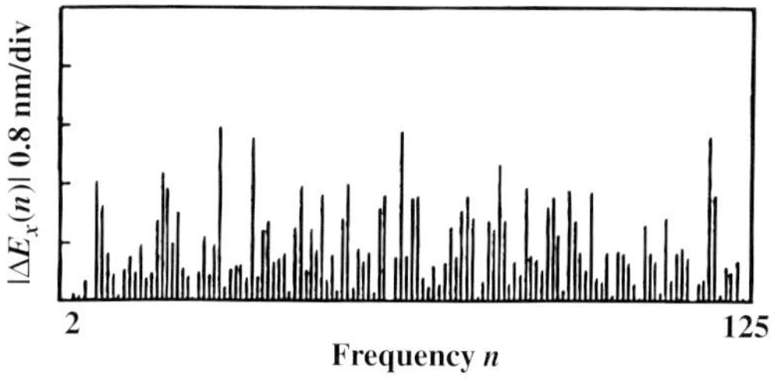

Figure 154. (*continued*)

This chapter has briefly reviewed some of the wide range of concepts, methods and techniques for roundness assessment, although for reasons of space it has not been possible to discuss much of the other valid instrumentation equipment that is currently available, but simply to indicate the potentialities of roundness measurement to the reader. In the following chapter the *surface integrity* aspect of modifications to both surfaces and roundness will be pursued. Attempts will be made to try and explain how, why, where and when the surface condition can be influenced by either its production processing conditions, or through some form of material phase changes, together with any plastic deformation via mechanical working that is present.

References

Journals and conference papers

Carpinetti, L.C.R. and Chetwynd, D.G. A new strategy for inspection roundness feature. *Precision Engineering* 16(4), 1994, 283–289.

Castle, P. How to measure roundness. *American Machinist* December 1993, 41–43.

Chetwynd, D.G. and Siddall, G.J. Improving the accuracy of roundness measurement. *Journal of Physics E: Scientific Instruments* 9, 1976, 537–544.

Chetwynd, D.G. and Phillipson, P.H. An investigation of reference criteria used in roundness measurement. *Journal of Physics E: Scientific Instruments* 13, 1980, 538.

Chien, A.Y. Appropriate harmonic models for roundness profiles with equivalent mean square energy value. *Wear* 77, 1982, 247–252.

Chou, S.-Y. and Sun, C-W. Assessing cylindricity for oblique cylindrical features. *International Journal of Machine Tools and Manufacture* 40, 2000, 327–341.

Chou, S.-Y. and Woo, T.C. On characterizing circularity. *Journal of Design and Manufacture* 4, 1994, 253-263.

Dawson, D.J. Cylindricity and its measurement. *International Journal of Machine Tools and Manufacture* 32(1/2), 1992, 247–253.

Donaldson, R.R. A simple method for separating spindle error form test ball roundness. Annals of the CIRP, Vol 21, 1972, 125–126.

Estler, W.T., Evans, C.J. and Shao, L.Z. Uncertainty estimation for multiposition form error metrology. *Precision Engineering* 21(2/3), 1997, 72–82.

Evans, C.J., Hocken, R. and Estler, W.T *Self-calibration: Reversal, redundancy, error, separation, and 'Absolute Testing'*. Annals of the CIRP, 45, 1996, 617–634.

Fan, Y., Zhang, S. and Xu, W. Kinematic and mathematical research on three-point method for in-process measurement and its applications in engineering. *Proceedings of the 7th International Precision Engineering Seminar*, Kobe, 1993, 318–328.

Gao, W., Kiyono and S. Nomura, T. A new multiprobe method of roundness measurements. *Precision Engineering* 19(1), 1996, 37–45.

Gao, W., Kiyono, S. and Sugawara, T. High-accuracy roundness measurement by a new error separation method. *Precision Engineering* 21(2/3), 1997, 123–133.

Giardini, W. and Ha, J. Measurement, characterization and volume determination of approximately spherical objects. *Measurement Science and Technology* 5, 1994, 1048–1052.

Gleason, E. and Schwenke, H. Spindless instrument for the roundness measurement of precision balls. *Proceedings of the ASPE 11th Annual Meeting*, 1996, 167–171.

Henke, R.P. et al. Methods for evaluating of systematic geometric deviations in machined parts and their relationships to process variables. *Precision Engineering* 23, 1999, 273–292.

Huang, J. An exact solution for the roundness evaluation problems. *Precision Engineering* 23, 1999, 2–8.

Huang, S.T., Fan, K.C. and John, H. A new minimum zone method for evaluating straightness errors. *Precision Engineering* 15, 1993, 158–165.

Kakino, Y. and Kitazawa, J. In situ measurement of cylindricity. *Annals of the CIRP* 27(1), 1978, 371–375.

Kanada, T. Computation of sphericity from minimum circumscribing and maximum inscribing centers by means of simulation data and downhill simplex method. *International Journal of the Japanese Society for Precision Engineering* 30(3), 253–258.

Kato, H., Nakano, Y. and Nomura, Y. Development of in-situ measuring system of circularity in precision cylindrical grinding. International Journal of the Japanese Society for Precision Engineering 24, 1990, 130–135.

Kiyono, S. and Gao, W. Profile measurement of machined surface with a new differential method. *Precision Engineering* 16, 1994, 212–218.

Krystek, M. Formfiletrung Durch Splines. *Proceedings of the IXth International Oberflächenkolloquium*, Chemnitz, Germany, 1996.

Krystek, M. Bias error in phase correct 2rc filtering in roundness measurement. *Measurement* 18(2), 1996, 123–127.

Krystek, M. Discrete L-spline filtering in roundness measurements. *Measurement* 18(2), 1996, 129–138.

Lai, H.Y. and Chen, I.H. Minimum zone evaluation of circles and cylinders. *Precision Engineering* 36(4), 1996, 435–441.

Lai, H-Y., Jywe, W-Y., Chen, C-K. and Liu, C-H. Precision modelling of form errors for cylindricity evaluation using genetic algorithms. *International Journal of Machine Tools and Manufacture* 24, 2000, 310–319.

Linxiang, C. The measurement accuracy of the multistep method in the error separation technique. *Journal of Physics E: Scientific Instruments* 22, 1989, 903–906.

Linxiang, C., Hong, W., Xiongua, L. and Qinghong, S. Full-harmonic error separation technique. *Measurement Science and Technology* 3, 1992, 1129–1132.

Moore, D. Design considerations in multiprobe roundness measurement. *Journal of Physics E: Scientific Instruments* 9, 1989, 339–343.

Murthy, T.S.R. and Abdin, S.Z. Minimum zone evaluation of surfaces. *Proceedings of MTDR Conference* 20, 1980, 123–136.

Novaski, O. and Barczak, A.L.C. Utilization of Voronoi diagrams for circularity algorithms. *Precision Engineering* 22, 1997, 188–195.

Ozono, S. On a new method of roundness measurement based on the three points method. *International Conference on Production Engineering*, Tokyo, 1974, 457–462.

Radhakrishnan, S., Ventura, J.A. and Ramaswamy, S.E. The minimax cylinder estimation problem. *Journal of Manufacturing Systems* 17(2), 1998, 97–106.

Reason, R.E. The report on reference lines for roughness and roundness. *Annals of the CIRP* 2, 1962, 96.

Reeve, C.P. *The calibration of a roundness standard.* National Engineering Laboratory/National Bureau of Standards Report. NBSIR79-1758, June 1979.

Shunmugam, M.S. New approach for evaluating form errors of engineering surfaces. *Computer-Aided Design* 19(7), 1986, 233–238.

Spragg, R.C. Eccentricity: a technique for measurement of roundness and concentricity. *The Engineer* September 1968, 440.

Sonozaki, S. and Fujiwara, H. Simultaneous measurement of cylindrical parts profile and rotating accuracy using multi-three-point-method. International Journal of the Japanese Society for Precision Engineering 23, 1989, 286–291.

Suen, D.-S. and Chang, C.N. Interval polynomial regression by use of a neural network for minimum zone problems. *Measurement Science and Technology* 9, 1998, 913–921.

Tong, S. Two-step method without harmonics suppression in error separation. *Measurement Science and Technology* 7, 1996, 1563–1568.

Tsukada, T. et al. An evaluation of form errors of cylindrical machined parts by a spiral tracing method. *Proceedings of the 18th MTDR Conference* 1977, 529–535.

Tsukada, T. and Kanada, T. Minimum zone evaluation of cylindricity deviation by some optimization techniques. *Bulletin of the Japanese Society for Precision Engineering* 19, 1985, 18–23.

Tsukada, T., Kanada, T. and Liu, S. Method for the evaluation of form errors of conic tapered parts. *Precision Engineering* 10(1), 1988, 8–12.

Udupa, G. and Ngoi, B.K.A. Form error characterisation by an optical profiler. *International Journal of Manufacturing Technology* 17, 2001, 114–124.

Wan, Q., Yang, S., Sun, Y. and Che, R. A new method on the evaluation of roundness errors. *Proceedings of ISMTII'96*, Japan, September–October 1996, 174–176.

Whitehouse, D.J. A best fit reference line for use in partial arcs. *Journal of Physics E: Scientific Instruments* 6, 1973, 921–924.

Whitehouse, D.J. Some theoretical aspects of error separation techniques in surface metrology. *Journal of Physics E: Scientific Instruments* 9, 1976, 531–536.

Whitehouse, D.J. Radial deviation gauge. *Precision Engineering* 9, 1987, 201.

Zhang, G.X. and Wang, R.K. Four-point method of roundness and spindle error measurements. *Annals of the CIRP* 42(1), 1993, 593–596.

Zhang, H., Yun, H. and Li, J. An on-line measuring method of workpiece diameter based on the principle of three-sensor error separation. *Proceedings of IEEE*, 3, 1990, 1308–1312.

Zhang, Q., Fan, K.C. and Zhu, L. Evaluation method for spatial straightness errors based on minimum zone condition. *Precision Engineering* 23, 1999, 264–272.

Books, booklets and guides

Dagnall, H. *Let's Talk Roundness.* Taylor Hobson Pneumo, November 1996.

Galyer, J.F.W. and Shotbolt, C.R. Measurement of surface texture and roundness. Ch. 9 in *Metrology for Engineers.* Cassell, 1990.

ISO R1101. *Technical Drawings: Geometrical Tolerancing.* International Organization for Standardization, Geneva, 1983.

Nakazawa, H. *Principles of Precision Engineering.* Oxford Science, 1994.

Reason, R.E. *Report on the Measurement of Roundness.* Rank Taylor Hobson, 1966.

Spragg, R.C. *Methods for the Assessment of Departures from Roundness.* BS 3730: 1964.

Thomas, G.G. Roundness. Ch. 9 in *Engineering Metrology.* Butterworths, 1974.

Whitehouse, D.J. *Handbook of Surface Metrology.* Institute of Physics, Bristol, 1994.

5 Machined surface integrity

"Ars longa, vita brevis."

Translation
"The life is so short, the craft so long to learn."
(*Aphorisms*, I, i; Hippocrates, 460–357 BC)

Outline on surface integrity

Up to now, the discussion in the text has been principally concerned with the surface topography, or roundness of components. This topographical information is valid but disguises the fact that the subsurface material layers may have been fundamentally altered during the production process by either forming or generating the part. The term *surface integrity* has been coined to describe the *altered material zone* (AMZ), for the localised subsurface layers that differ from those of the bulk material. These AMZs are in reality a sequence of zones that correspond to a range of subsurface features and alterations, which can take the form of modifications promoted by metallurgical residual stresses, plastic deformations, chemical changes and modifications in hardness. They may not all be present at every instant of time, but any one of them can become dominant, depending on whether the production process and operating environment vary in their relative relationships and magnitudes.

The whole concept of surface integrity and its various generating mechanisms coupled to the production process is known as a *unit event*. If this surface integrity unit event is considered, it may promote a better understanding of the mechanisms behind the alterations and at the same time explain both how and why the surface occurred. Moreover, it enables the designer to develop surfaces that can be engineered for specific in-service requirements. The unit event generation mechanisms can be thought of as three discrete types of effect: mechanical, thermal and chemical. Such mechanisms always occur within the subsurface to some degree, their dominance altering depending upon various factors such as the component material, processing conditions and time. However, it is often unclear what the relative distinction might be between these subsurface generating mechanisms. Due to the fact that a final component's surface integrity can be quite diverse and the result of a range of interrelated factors, it is necessary to reclassify these surface integrity affects. Therefore, it is often more appropriate to subdivide the production processes into five classes - chemical, mechanical, mechano-thermal, thermo-mechanical and thermal - the order in which they are listed reflecting their respective power density per unit area. An increase in power density from the chemical end of the series will result in an augmented level of thermal energy entering the surface, affecting the potential for greater thermal damage, which in turn would lead to poorer surface integrity of the part. It should be emphasised that the "chemical mechanism" is dominant across all classes of production processes to some degree and that surfaces react with their immediate environment via oxidation, adsorption and so on. Due to these complex interrelated reactions it could be said that the ultimate generating mechanism would be chemical in nature. Thus, the unit event plays some role in influencing both the component's surface and subsurface in a variety of ways, dependent on its generating mechanism. A complex relationship of zones could also be present; typically the complete range would consist of the *chemical affected layer* (CAL) and *mechanically affected layer* (MAL), while a *heat affected layer* (HAL) occurs together with a *stress affected layer* (SAL). Thus considering each in turn:

- CAL - the result of chemical surface changes from the production process, or by post-process exposure to local environment;
- MAL - will result from factors such as material bulk transportation (deposits, laps, folds and plastic deformation);
- HAL - concerned with factors like phase transformations, thermal cracking and retempering;
- SAL - results from residual stresses by a combination of mechanical and these events.

The following section discusses many of these surface-related features in more depth.

5.1 Introduction

The need to satisfy the demands of sophisticated component performance, reliability and longevity for a range of industrial, commercial and military applications where critical parts are subjected to severe conditions of stress, temperature and in hostile environments has become increasingly important in recent years. This problem is exacerbated by reductions in component section size, in response to the designer's need to reduce weight yet still retain mechanical strength, which means that the part's in-service performance is strongly influenced by its surface condition. Coupled to these component performance requirements, there have been continued increases in the development and use of heat- and corrosion-resistant alloys. One of the principal design factors today is that of *dynamic loading* and accordingly a component's design capabilities are often limited by its inherent fatigue characteristics, particularly for structural material applications. If a review is undertaken of stressed parts and structural members that have failed in situ, their service histories indicate that fatigue failures - a major factor - almost without exception nucleated from a site on, or in close proximity to, the surface of the component. Moreover, it has been recognised that the component's surface is a primary factor in the determination of susceptibility to attack and premature failure due to low stress corrosion resistance.

Component hardness has been severely tested of late, and to expand on this theme by inappropriate and somewhat less-than-informed treatment from some inexperienced designers the following example highlights the problem. Many medium carbon steels which had in the past been utilised at low levels of both strength and hardness are of late being subjected to stringent heat treatment processing to obtain very high strength values, coupled, in certain applications, to greater hardness for use as structural parts. Such inappropriate component heat treatment processing at these extremes goes well beyond the demarcation boundaries initially envisioned for this material, which can severely limit the part's performance in-service. From the discussion above, it can be appreciated that highly stressed components are influenced by at least two important material properties that are significantly surface-oriented. Therefore, whenever these component qualities are imperative, significant attention should be made to the characteristics of the surface in its pre-selected working environment.

Demands by the designer for many of today's components have meant that they are often subjected to *machining* to maintain specific surface texture and/or roundness requirements, together with holding dimensional tolerances. The term *surface integrity* was coined in the mid-to-late 1960s out of the need to express a component's potential in-service performance from a more stringent and exhaustive viewpoint. Surface integrity has been coined by two of the principal "champions" of the subject (although many people working in this field of research consider themselves to be originators of this important concept). The principal researchers were namely, Michael Field and John F. Kahles of Metcut Research Associates Inc. in the United States, who defined surface integrity as *the inherent, or enhanced condition of a surface produced in a machining, or other surface generation operation.* This important topic of surface integrity involves both the study and control of two material-related factors. These requirements can be succinctly stated as:

1 surface roughness, or topography;
2 the metallurgy of the surface region.

First to be considered is the component's surface texture, which principally governs the surface roughness (a topic considered in Chapters 1 and 2), this being in essence a measure of its surface topography. Secondly, a more complex inter-relationship is present consisting of plastic deformations, metallurgical phase changes and surface stresses, which constitutes the nature of the region termed the *surface layer*; in the current chapter this is principally the result of machining production processes. The nature of this altered surface region resulting from machining has been shown to exhibit a strong influence on the component's subsequent mechanical properties. Many highly stressed component failures, or potential failure modes, are associated with the part's inherent machining-to-service demands; this is more apparent in some materials than others under particular machining process conditions. Below are listed just some of the production processes, materials and circumstances in which surface integrity problems may arise in surface regions:

- abusively drilled or deep-drilled/bored holes with poor coolant supply, which may introduce metallurgical changes in the material, such as untempered martensite ("white layers") most notably in steels and some of their alloys;
- machining-induced residual stresses promoting component distortion, fatigue and the possibility of stress corrosion;
- thin component distortion, as a direct result of surface-induced machining stresses;
- grinding cracks resulting from either a "loaded/dull" grinding wheel (from various causes), promoting abusive grinding conditions, this being a particular problem on cast nickel-based alloys;
- grinding burns occurring on high-strength parts which can result in lower fatigue life for highly stressed components, typically hydraulic and stressed members on aircraft landing gear;
- the influence of cutting fluids while machining, specifically with water-based coolants on certain types of materials having stress corrosion properties, but notably in the case of titanium;
- workpiece processing by electrical discharge machining (EDM), or electrochemical milling (ECM) can result in lowering of the material's fatigue strength.

Surface alterations

Any surface that has had some form of value-added activity carried out either to shape the component from raw stock or to manipulate it from forged billets or castings, typified by the following production techniques –

1 *conventional machining* – drilling, milling, turning, boring and reaming, and so on;
2 *non-conventional machining* – EDM, ECM, laser and abrasive water jet, and the like;
3 *production processing* – hot or cold working;

– will to a greater or lesser degree have some type of surface alteration present. The former two, and most notably that resulting from "conventional machining"-induced surface integrity alterations, result from the cutter's action at a zone which includes localised surface material removal. The "conventionally machined" component's surface layers and topography will often include some of the following features:

- the results of hot or cold plastic deformation;
- the by-products of a tool's built-up edge (BUE), including laps, tears and crevice-like defects;
- recrystallisation of the surface;
- localised phase transformations;
- surface micro- and macro-cracking;
- the distribution of surface layer residual stresses.

"Non-conventionally-machined" surface layers, on the other hand, may include some of the above, together with the following:

- chemical absorption of elements (typified by hydrogen, or halogens) resulting in embrittlement of the surface;
- a remelted metal deposit/spattered coating on the surface, deposited during EDM, laser machining or electron beam fabrication.

More particularly, in the case of machining operations, the principal causes for these surface alterations are localised high temperatures or gradients at or near the cutting zone, together with accompanying localised plastic deformation as the tool point "ploughs" or "burnishes" along the recently machined surface. As a result of these highly localised temperatures in the tool's cutting vicinity, chemical reactions and possibly subsequent absorption into the machined surface may occur.

Materials are rarely homogeneous, higher-strength alloys often being sensitive to property variations in the main resulting from the presence of minute quantities of either an element or trace elements. In recent years complex heat treatments have been developed to optimise an alloy, but this also highlights the fact that such alloys are prone to thermal gradient sensitivity.

The amalgamation of material properties and process energies will influence component structural integrity. Often design data has been conceived from information collated on the operational performance of specially ground and polished material specimens having a pre-selected heat treatment/thermal history. From a practical viewpoint, any designer should consider how surface effects and property changes have been affected by the part's previous processing, prior to its shipment. This

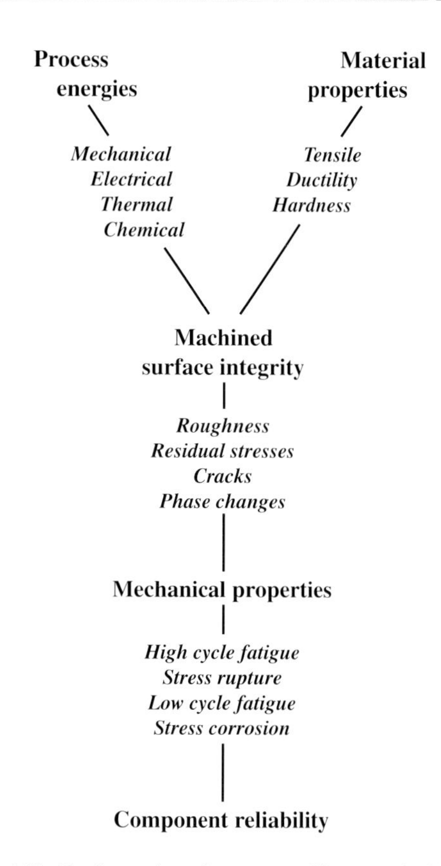

Figure 155. The interaction of processes with materials. [Source: Bellows *et al.*, 1975.]

component processing sequence is as critical as the base metallurgical characteristics, due to its inherent overall performance impacting on the surface integrity. Previously, this problem of in-service performance was overcome by considerably greater leeway on component design factors. The current stringent need to conserve material, from energy and cost-saving aspects, has led to material utilisation closer to the original design-testing criteria, thereby confirming the fact that a greater consideration of surface integrity requirements is demanded.

In order to understand the material and process interactions and potential failure modes, Figure 155 indicates their relationship to subsequent component reliability, while Tables 15(a and b) highlight just some of the materials having potential surface alterations introduced by machining operations.

5.2 The machined surface

When considering a machined surface, it can be the result of either an *abusive* or *gentle* machining

Table 15(a). Summary of the potential surface alterations occurring as a result of metal removal production processes

Material	Material removal methods			
	Conventional		Non-conventional	
	Milling, drilling, or turning	Grinding	EDM	ECM and ChM
Steels				
Non-hardenable	Roughness Plastic deform. Laps and tears	Roughness Plastic deform.	Roughness Microcracks Recast metal	Roughness Selective etch Intergranular attack
Hardenable (alloy)	Roughness Plastic deform. Laps and tears Microcracks Untemp. mart. Overtemp. mart.	Roughness Plastic deform. Microcracks Untemp. mart. Overtemp. mart.	Roughness Microcracks Recast metal Untemp. mart. Overtemp. mart.	Roughness Selective etch Intergran. attack
Tool steel (D2)	Roughness Plastic deform. Laps and tears Microcracks Untemp. mart. Overtemp. mart.	Roughness Plastic deform. Microcracks Untemp. mart. Overtemp. mart.	Roughness Microcracks Recast. metal Untemp. mart. Overtemp. mart.	Roughness Selective etch Intergran. attack
Stainless steels				
Martensitic (416 grade)	Roughness Plastic deform. Laps and tears Microcracks Untemp. mart. Overtemp. mart.	Roughness Plastic deform. Microcracks Untemp. mart. Overtemp. mart.	Roughness Microcracks Recast metal Untemp. mart. Overtemp. mart.	Roughness Selective etch Intergran. attack
Austenitic (316 grade)	Roughness Plastic deform. Laps and tears	Roughness Plastic deform. Recast metal	Roughness Microcracks Intergran. attack	Roughness Selective etch
Precipitation hardening	Roughness Plastic deform.	Roughness Plastic deform.	Roughness Microcracks	Roughness Selective etch
Laps and tears Over-ageing	Over-ageing	Recast metal Over-ageing	Intergran. attack	
Maraging (250 grade)	Roughness Plastic deform. Laps and tears	Roughness Plastic deform. Resolutioning	Roughness Recast metal Resolutioning	Roughness Selective etch Intergran. attack

After Bellows et al. (1975).

regime, these being directly related to the cutting process and associated feeds and speeds. However, this is not the complete picture for surface integrity, as many other interactions influence the surface during either its forming or generating process. Machining, being a complex relationship of interrelated factors, affects the outcome of the production process, which is depicted schematically in Figure 156. Here, for one of the less complex machining operations, namely turning, the surface integrity – from only a simplistic viewpoint – is shown grouped as follows:

- *surface condition* – surface texture & roundness;
- *micro-structural changes* – microcracks, etc;

Table 15(b). Summary of the potential surface alterations occurring as a result of metal removal production processes

Material	Material removal methods			
	Conventional		Non-conventional	
	Milling, drilling, or turning	Grinding	EDM	ECM and ChM
Nickel and cobalt-based alloy				
Inconel (718) Rene (41)	Roughness Plastic deform. Laps and tears Microcracks	Roughness Plastic deform. Microcracks	Roughness Microcracks Recast metal	Roughness Intergran. attack Selective etch
Titanium alloy				
(Grade 5: Ti-6A1-4V)	Roughness Plastic deform. Laps and tears	Roughness Plastic deform. Microcracks	Roughness Microcracks Recast metal	Roughness Selective etch Intergran. attack
Refractory alloy				
(Moly TZM)	Roughness Laps and tears Microcracks	Roughness Microcracks	Roughness Microcracks	Roughness Selective etch Intergran. attack
Tungsten				
(Pressed and sintered)	Roughness Laps and tears Microcracks	Roughness Microcracks	Roughness Microcracks	Roughness Selective etch Microcracks Intergran. attack

Abbreviations: (Machining) EDM: electrical discharge machining; ECM: electrochemical milling; ChM: chemical milling; (Metallurgy) Plastic deform.: plastic deformation: Untemp. mart.: untempered martensite; Overtemp. mart.: overtempered martensite.

After Bellows et al. (1975).

- *surface displacement* – bulk material transportation and residual stresses;
- *surface/subsurface micro-hardness* – plastic deformation and residual stresses.

Machined surfaces are even more complex than seems at first glance; their performance can be influenced by external layers (chemical transformations and plastic deformations), and/or internal zones (metallurgical transformations and residual stresses). For example, the anisotropic (periodic) turned surface depicted in Figure 157 is influenced by the tool tip geometry and the regularity of the cusps (peaks and valleys) is dominated by the pre-selected feedrate. A range of other micro-topography features may also be present, superimposed onto the machined surface, such as tool wear, vibrational influences and, to a lesser extent, machine tool-induced errors. In the circumferential direction the *lay* is both periodic and regular, albeit this round surface is generated by turning which will have harmonic departures from roundness characteristics present.

The exposed sterile surface (Figure 157) occurs as a result of highly localised temperatures and transient, but cleanly cut metal resulting from this machining, which will instantaneously oxidise and adsorb contaminants. The outermost adsorbate layer is often termed the *Beilby layer*; it is approximately 1 μm thick and consists of many complex factors, notably the presence of hydrocarbons and water vapour, that originated in the coolant or atmospheric environment, respectively. Underneath this metallic surface there is normally a plastically strained region that may have been metallographically and metallurgically altered. The depth of this layer will vary significantly, but is typically in the region of 10 μm, according to the plastic deformation induced by the passage of the sharp tool over the surface and, is exacerbated by the metallurgical composition of the metallic substrate. This plastic

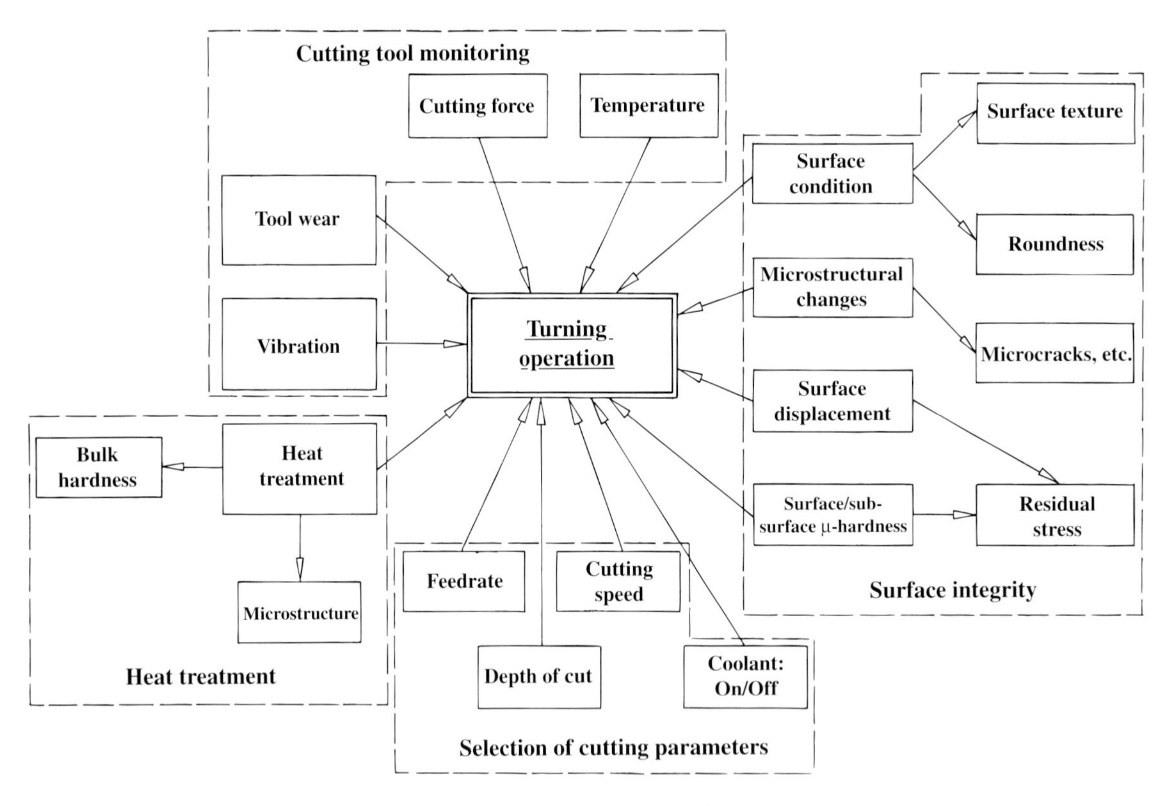

Figure 156. Major factors that influence the output from a turning operation.

deformation depth, as a result of residual stresses, can in certain circumstances penetrate to fractions of a millimetre. As an approximation, the depth of hardness penetration is about 50% of that for the residual stress penetration depth, whereas the depth of observational plastic deformation is around 50% greater than this penetration. This theme of residual stresses resulting from machining production processes will be the subject of the following section.

5.2.1 Residual stresses in machined surfaces

The residual stresses in a component are a function of the previous material process route, in combination with its machining history. The fact that residual stress levels are present may either enhance or, more likely, impair the functional behaviour of a machined workpiece. Internal stresses in a component are generally unstable and over a period of reasonable time can produce alterations in either dimensional size or geometry. For example, in the case of certain alloy gauge block materials, over a period of greater than 25 years their length has been reported to have changed by approximately 0.8 μm with little in the way of wear through usage, which is a serious dimensional error that cannot be ignored.

The physical condition of critical components requires to be known, including an understanding of surface layer residual stresses, prior to in-service applications. As has been previously alluded to, the process of machining parts will generate functionally relevant surfaces that have some significance for the development of the physical state of the surface, influencing the somewhat unpredictable distribution of its residual stresses. In many industrial applications the properties of the component's surface dominate the functional behaviour of the part in service. The effects of residual stresses on mechanical and electrical components are summarised in Figure 158(a). This family tree of factors that affect usage is not complete, as there are notable omissions such as optical, acoustical and thermal results of residual stresses, which are outside the current narrative. Each of the residual stress effects indicated in Figure 158(a) will now be briefly reviewed.

Residual stress deformations

Any residual stresses acting within a body (component) occur without external forces or moments. Internal forces form a system that is presently in a state of equilibrium and if sections of this body are removed – by machining – the equilibrium status is

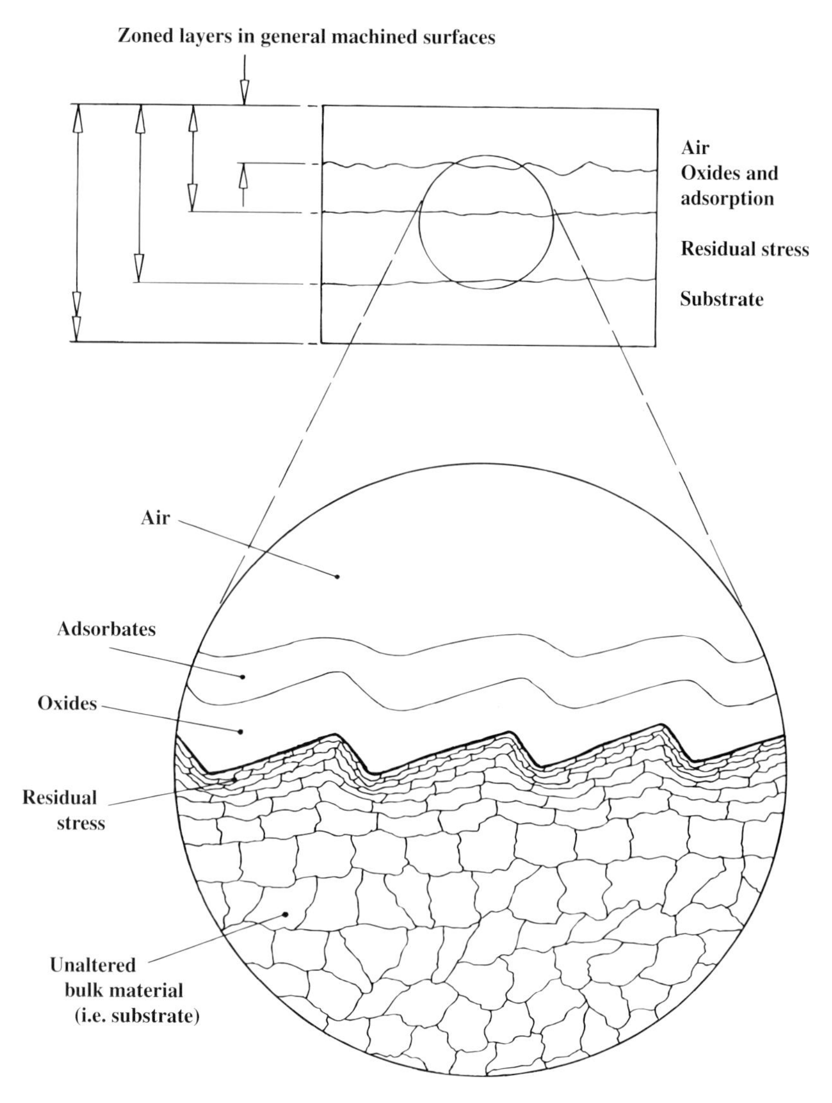

Figure 157. Cross-section of an anisotropic (i.e., periodic) surface illustrating the surface contaminants (oxides and adsorbates) together with subsurface plastic deformation (the residual stress zone) and unaffected substrate.

usually disturbed, resulting in potential deformation. This distortion resulting from the machining condition is well known to industrial engineers when, for example, machining one side of a thin component. If either a forging or casting has not been heat treated for stress relief and requires asymmetrical machining, it will deform somewhat after unclamping from its work-holding device in the machine tool. In order to alleviate this problem an experienced machinist will release the clamping forces after roughing cuts so that stressed surfaces are equalised, prior to taking a finishing pass. Component deformation is approximately proportional to the removed cross-section of material. Any further finishing is usually concerned with removing only a thin layer of material, minimising any detrimental effects of residual stresses resulting from the previous production processing route.

The release of internal residual stresses must not be confused with the input of such stresses by machining, as indicated in Figure 158(b). The machining process generates residual stresses by plastic deformation (Figure 159) or metallurgical transformations. As mentioned, residual stress depths can reach several hundredths of millimetres, which on large parts will cause little in the way of component distortion when compared to the same effect on thinner parts. For example, if a single-point asymetrical "light" machining operation such as planing is undertaken on a 1000 mm long bar of 20 mm thickness, the deflection will be around 1 mm. While machining, or immediately afterwards,

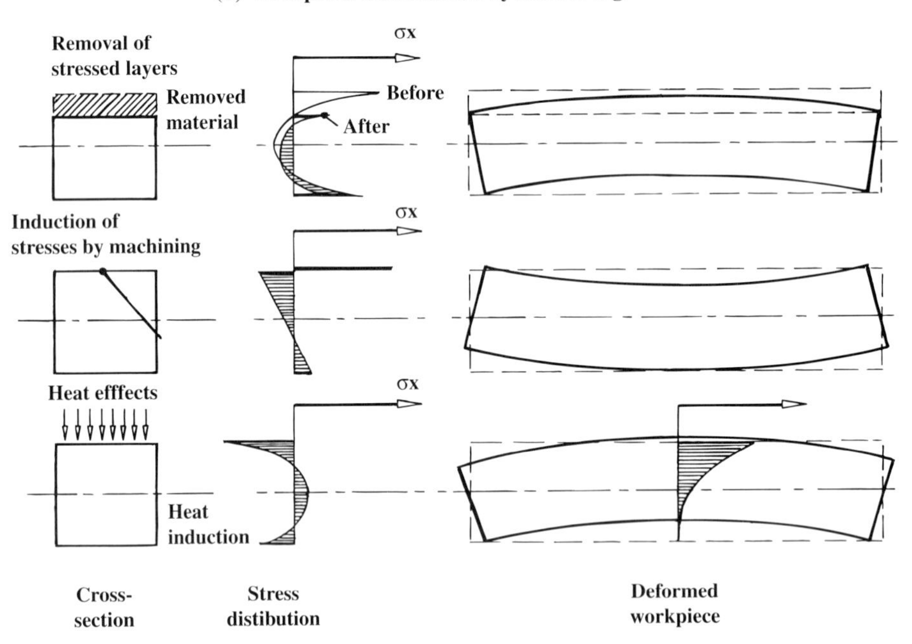

Figure 158. Effects of residual stress and deformation of a workpiece by machining. [After Brinksmeier et al., 1982.]

thermal influences can promote further deformations. This workpiece machining can impart asymmetrical heating to the component. As a result of the machining-induced temperature profile, stresses occurring from thermal expansion are generated which cause additional deformation; this condition can be somewhat alleviated by subjecting the part to subsequent temperature compensation (Figure 158b). Moreover, by vibrating the part at its natural frequency – after rough machining (in situ), this allows a finish cut to to be taken and then unclamped avoiding distortion.

Influence on static strength

From a macroscopic viewpoint, residual stresses act in a similar way to that of a pre-stressed state for the material. Any material that can be deformed will have a characteristic yield point; therefore if residual stresses are present they will influence the level of the part's yield strength. In the case of a buckling load the critical stress is always lowered in the presence of residual stresses.

In many industrial applications, increasing the component's strength by pre-stressing is extensively employed. For example, extrusion dies are often fabricated from two cylinders that have the ability to pre-stress the inner part, enabling the extrusion material to be held at a less critical state during the operation. This technique has been used to good effect in high-pressure physics and significantly extends the pre-stressing process.

Influence on dynamic strength

The influence of residual stresses on fatigue strength has been cited and proven over many years. By way of illustration, in a surface grinding operation an AISI 4340 steel component has its cutting speed varied from 60 m/min to 180 m/min. The variables in this example were for various wheel hardnesses and coolant applications, together with either coarse or fine dressing; the resultant maximum residual stresses will be modified ranging from 80 to 1000 N/mm^2, indicating how processing conditions will significantly affect induced stresses. Furthermore, if the roughness of the machined surface increases, this decreases fatigue strength, as expected, although this textural influence for longitudinal or traverse grinding is considerably smaller than for that of the surface roughness state. Fatigue fractures are normally initiated at the weakest point; thus it is not the average stress value that is important but the local extreme. This localised stress concentration value is particularly important for those machining processes that generate a scattering of stress distributions.

In the example of case-hardened gears (16% Mn, 5% Cr), the effect of residual stresses resulting from abusive grinding operations causing burning (without surface cracking) will lead to a 25% decrease in the durable flank pressure in the dynamic conditions associated with gearing applications. Under such abusive grinding operations, the surface residual stresses of the tooth flank in either the axial or radial direction tend to be highly tensile in nature.

Chemical resistance

If a corrosive environment occurs and certain metals are subjected to stresses on exposure to this atmosphere over a period of time, then *stress corrosion* may be observed. Conditions conducive to stress corrosion arise because of specific material sensitivity, in association with accompanying surface tensile stresses in the presence of a corrosive medium. Typically, if an austenitic chromium-nickel steel surface, after tension, is submerged in a solution of sodium chloride and sulphuric acid, cracks will appear due to stress corrosion. Several theories have been expounded as to what is the dominant process during stress corrosion. For example, the *electrochemical hypothesis* assumes that potential differences occur between the precipitations at the grain boundaries, or within the grains and the adjacent matrix, causing the precipitates to be dissolved anodically. Tensile stresses tend to open up cracks developed in this manner, promoting rupturing of the surface film at the root of the crack, causing the "pure metal" to be attacked by the corrosive environment. In a chemical plant, one of the main failure causes is stress corrosion; in a typical year around 60% of documented failures occur by corrosion, with greater than 20% of such detected corrosion failures being attributable to stress corrosion.

Magnetisation

For a ferromagnetic body, the magnetic properties depend on its physical state. In many industrial applications these dependencies are utilised for testing and measuring purposes. For instance, micro-cracks act as a disturbance in the material's magnetic flux and can therefore influence both the flux distribution and intensity, which can then be employed to detect surface faults or those internally. These magnetisation effects may be used to investigate changes in the crystallographic structure of a ferromagnetic material. Furthermore, the coercive force is extremely sensitive to hardening effects; as a result it can be used for in-process measurements during machining operations. A component's magnetic properties can be directly influenced by residual stresses together with any localised disturbances of the crystallographic structure. The surface texture of, say, a soft magnetic nickel alloy situated in an application for the magnetic head of a machine determines the information density of the tape by the focusability of the head. This relationship is dependent on the width of the undisturbed zone, including the air gap between the poles, with the pole surfaces being machined by a lapping process. If any local dislocations are present – tilting of the lattice structure – this will result in distinctive contrasts inside the grains. The soft magnetic properties of the material will potentially result in degraded head focusability as a result of these micro-mechanical lattice influences.

Finally, residual stresses in the machined surface layers can generate deformations that affect both static and dynamic strength, together with the magnetic and chemical properties of the surface. In order to quantify a machined surface it is necessary to determine the residual stress tensor. This tensor can be thought of as being caused by the action of residual strains and from further strains that result from the component being subjected to compatibility conditions. In machining processes residual stresses tend to be generated by mechanical, thermal and transformation factors; thus characteristic stress distributions are developed that amalgamate in practice across a variable and wide range. For example, shot peening introduces an isotropic compressive stress tensor into the surface, as does rolling, but with distinct principal directions (anisotropic surface behaviour) these are typical of mechanical influencing processes, whereas EDM affects the part in a thermal manner, introducing equal tensile elements to the surface integrity.

5.2.2 Tribological cutting effect on surface

When machining a component's surface the resulting machined topography will be significantly influenced by the wide range of fixed and variable cutting parameters. The ability to modify and improve this cutting process is based on the judgement and experience of the operator, or on the inherent capabilities of the computer-aided manufacturing (CAM) package. The scope to modify the cutting process is considerable, with the outcome improving the surface integrity, often at the expense of production output, part quality or tool life. Tool companies expend considerable time and effort in developing tooling geometries to efficiently cut material and evacuate chips from the cutting zone. Many cutting inserts today have highly complex surface profiles, often having a particular edge preparation – T-hone, parabolic, etc. – either to improve tool wear characteristics or to facilitate the retention of a multi-coating for cutting inserts. Honing of the tool's rake face allows a sharp cutting edge to result, which has the ability to cleave through the workpiece material in an efficient manner. The outcome of machining with a highly positive and honed insert geometry is that the component's surface exhibits little in the way of subsurface plastic deformation, on materials that are not prone to work-hardening. When machining with a sharp cutting tool, where the corner intersection between the rake and clearance faces meet, the tool's edge will become progressively more rounded because of the tribological action. This "rounding effect" will mean that at a certain height above the rounded portion of the tool's chip/tool interface it will cut material and, below this contact point, it will plough the surface. Ploughing is highly inefficient and will plastically compress the surface, acting in a "burnishing-like" manner, as depicted in Figure 159. Burnishing is an abusive regime of machining, although it is often intentionally used to produce subsurface work-hardening to improve the wearing ability of the surface, yet at the same time a burnishing tool will improve the finish, by flattening local asperities and cusps, from a previous machining pass. Sometimes a material that can work-harden will have a machined surface that gives the appearance of a high-quality surface texture, which in reality disguises the fact that considerable work-hardening has taken place due to the burnishing effect of tool flank wear. A surface of this unstable type can create reliability problems at a later stage in the component's life, due to the instability of the subsurface promoting potential for fatigue.

Machinability testing

In Figure 160 are shown the results from an *accelerated machining test* performed on Fe-Cu-C powder

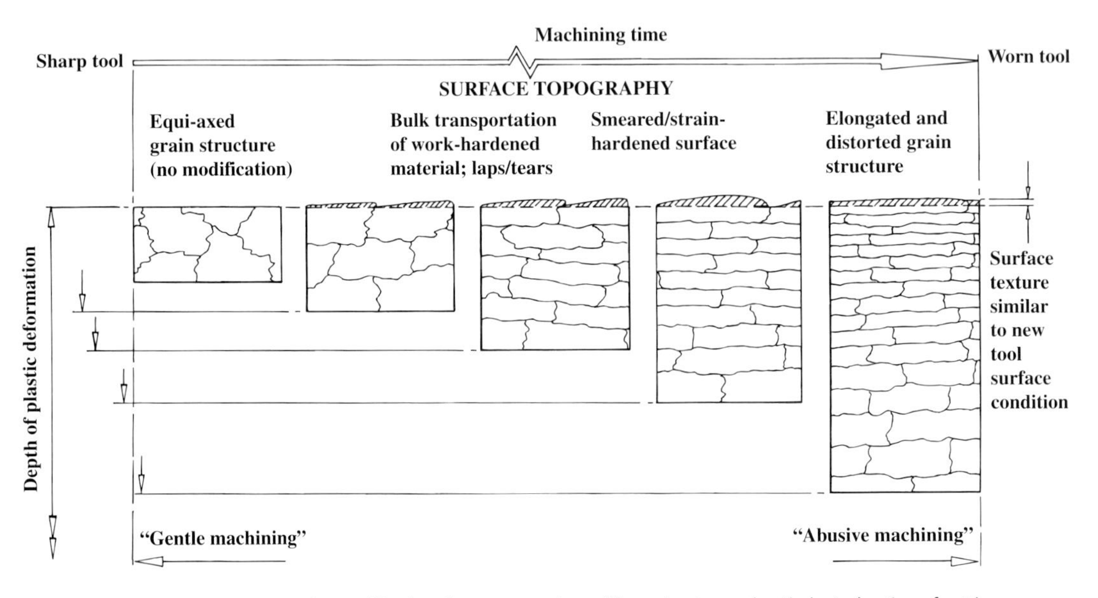

Figure 159. Distortion to the machined surface topography and integrity due to the tibological action of cutting.

(a) Schematic illustration of "accelerated machining test": 10 facing passes then surface texture and insert wear checked

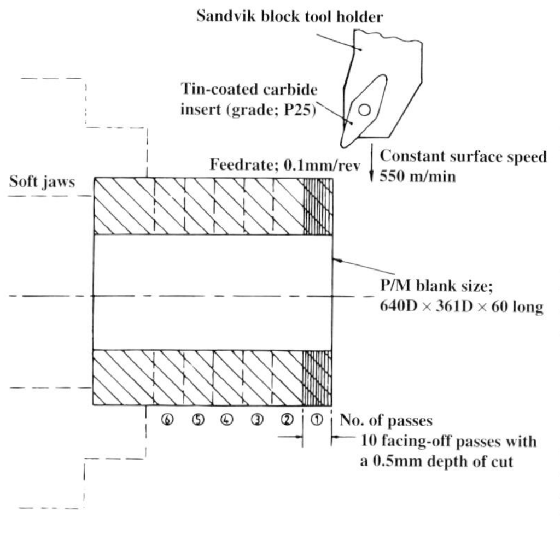

(b) Tabulated results from turning test procedure

	Specimen			
	1	2	3	4
0.10 mm/rev feedrate				
Ra, μm	0.762	0.780	0.910	0.890
Rsk, μm	0.7	0.6	0.8	0.4
Rku, μm	2.5	2.8	3.1	2.2
Rtm, μm	3.6	4.1	4.4	3.7
Sm, μm	75.3	69.5	97.5	90.3
DELQ	7.5°	7.4°	8.4°	5.2°
LSC, μm	28	12	20	10
0.25 mm/rev feedrate				
Ra, μm	5.5	5.9	5.5	6.2
Rsk, μm	0.5	0.6	0.7	0.5
Rku, μm	2.0	2.3	2.3	2.1
Rtm, μm	22	26	23	25
Sm, μm	247	232	247	248
DELQ	13.6°	16.1°	14.3°	15.6°
LSC, μm	46	18	26	20
0.40 mm/rev feedrate				
Ra, μm	16.3	16.8	16.1	15.7
Rsk, μm	0.2	0	0.1	0.1
Rku, μm	1.8	2.0	2.0	1.9
Rtm, μm	64	71	71	60
Sm, μm	401	398	402	397
DELQ	22.5°	23.1°	23.2°	22.4°
LSC, μm	60	20	48	22

(c) Graphical plots of machined surface texture and flank wear

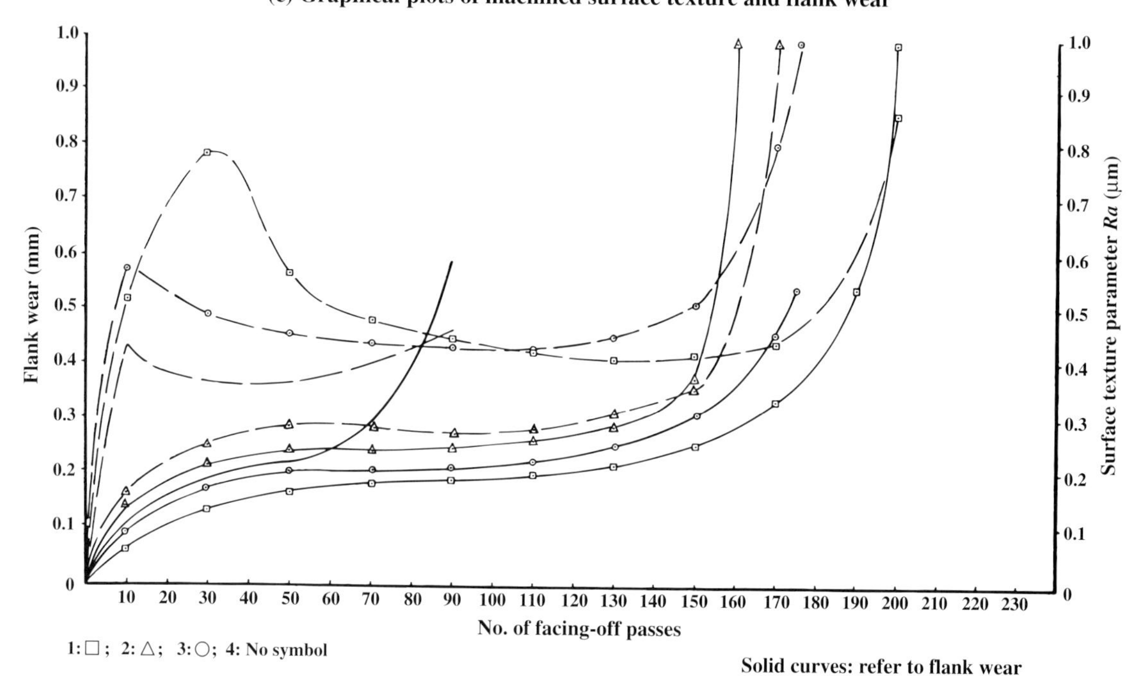

Solid curves: refer to flank wear
Broken curves: refer to surface texture (*Ra*)

Figure 160. An "accelerated machining test" for assessment of tool wear and surface texture.

metallurgy (PM) compacts that were subjected to a face-turning operation at constant surface speed on a CNC turning centre (Figure 160a). The results of four different powder mixtures are illustrated in Figure 160(b), with graphical plots in Figure 160(c). This particular accelerated machining test has the objective of assessing the relative machinability of components based on a relatively short time scale (termed "ranking"), this being an amalgamation of the previous *degraded tool* and *rapid facing* tests, devised by the author. The test quickly establishes the relative machinability of differing specimens, with a slightly degraded tool and marginally faster rotational speed for the facing operation, this being undertaken with a constant surface speed feed function engaged on the turning centre.

The tool life and wear curves are shown for the four specimens in Figure 160(c), with the tool life in particular showing the anticipated three stages of wear (*Taylor curves*) associated with steady-state cutting operations. These reported wear stages are the initial edge breakdown (tool point rounding), steady-state wear as the flank progressively degenerates while in-cut, and finally catastrophic tool edge breakdown as the edge completely fails. The elemental additives to these powder compacts represent the major variable in the cutting process, this interplay of elements and their relative proportions within the sample significantly altering the shape of the wear curves (some specimens are more abrasive than others).

Superimposed onto Figure 160(b) are the circumferential surface finish results for each testpiece. It is curious to note from these graphical plots that originally a high *Ra* value was established during the early stages of the operation during the initial tool edge breakdown; however, this surface *Ra* value dramatically improves during the successive facing passes, regardless of metallurgical composition. This improvement is thought to occur because the flank wear that is present on the cutting insert produces a burnishing effect on the machined surface topography. Burnishing produces plastic deformation and in essence compresses the surface asperities and localised surface layers, giving rise to an improvement in the recorded surface texture. However, a surface condition of this type is misleading, as both the surface topography and subsurface integrity are locally extremely work hardened. This work hardening continues to progressively degenerate until the insert edge completely fails; furthermore at the latter stages of the flank's life the surface texture is exacerbated. A secondary effect of this burnishing action by the insert's flank is the transport of workpiece material and its subsequent deposition in a highly work-hardened state at new sites on the previously machined surface. This *smearing* of the machined PM surface is highly undesirable and should be the cause of workpiece rejection, despite the fact that all the metrological indications state otherwise. The reason for part rejection in this case would be because of surface pore closure, unique to PM compacts, which could make these parts useless for porous bearing applications.

5.2.3 Micro-hardness testing

Utilising micro-hardness testing techniques for surface integrity assessments produces an indirect comparison of the potential residual stress in the mechanically affected layer (MAL). Such destructive techniques require samples to be sectioned – the sectioning technique and the reasons for undertaking it will be discussed shortly – although the method should be used with caution, as any hardness plots obtained cannot show whether forces are compressive or tensile in nature. A major benefit from utilising micro-hardness testing is that it allows metallographically prepared specimen features to be optically investigated and assessed with the test sample in situ on the instrument. A range of micro-hardness tests are currently available, the indentation being of two distinct types:

1 the geometric shape of the impression left after a hard indentor has been pressed into the surface feature to be measured – typified by Vickers and Knoop indentors (Figure 161b);
2 where a indentor penetrates into a specimen and relative depths to which the indentor penetrates are the measure of the specimen's hardness – typified by the Rockwell indentor.

Vickers hardness

The faces of the square-based diamond pyramid for a Vickers indentor are inclined at an angle of 136°; this angle approximates the most desirable ratio of indentation size to that of a ball diameter in Brinell hardness testing. Because of the shape of the indentor it is often known as the *diamond-pyramid hardness test*. It is defined as *the load divided by the surface area of the indentation*. The Vickers hardness number (VPN) may be determined from the following equation:

$$\text{VPN} = \frac{2P \sin(\theta/2)}{L^2}$$

$$= 1.854\left(\frac{P}{L^2}\right)$$

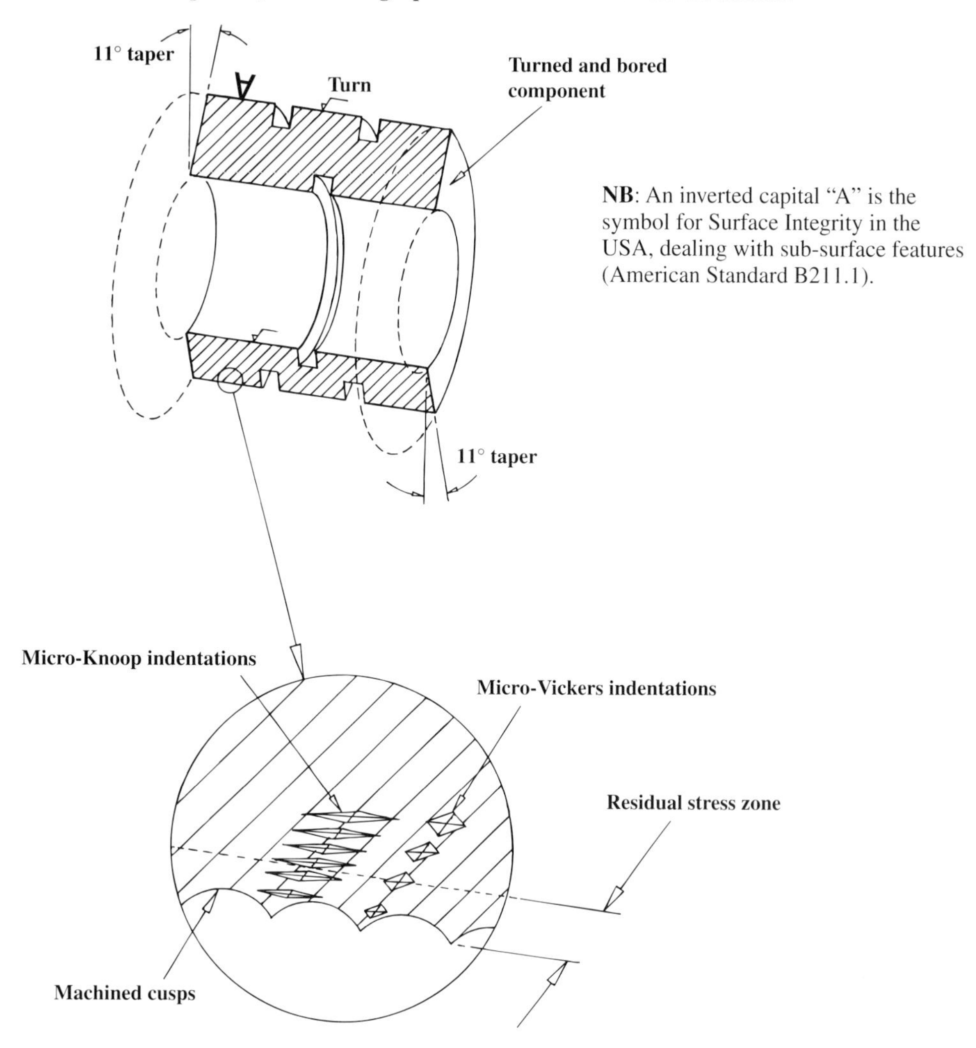
(a) Tapered section of component, for metallographical and micro-hardness assessment
11° taper
Turn
Turned and bored component
NB: An inverted capital "A" is the symbol for Surface Integrity in the USA, dealing with sub-surface features (American Standard B211.1).
11° taper
Micro-Knoop indentations
Micro-Vickers indentations
Residual stress zone
Machined cusps
(b) Detail of "micro-hardness footprinting" across the taper section

Figure 161. Tapered sectioning of a component, for metallographical inspection and micro-hardness "footprinting".

where P = applied load (kg), L = average length of diagonals (mm) and θ = angle between opposite faces of diamond (136°).

The diamond indentation in the surface would ideally be a square shape with diagonals from corner to corner; however, a "pin-cushion" indentation (concave sides to the square) would occur due to sinking into the material's surface; conversely a "barrelled" indentation (convex sides to the square) occurs by ridging. The test has found wide acceptance and can be utilised from relatively soft surfaces up to the hardest surfaces yet produced.

Knoop hardness

The Knoop indentor has more complex facets to its diamond indentor, having angles of 130° (short diagonal) and 172.5° (long diagonal) respectively. This arrangement of the indentor geometry leaves a significantly narrower and longer impression (approximate diagonal ratio of 7:1) in the surface than that of the Vickers indentation. The Knoop hardness indentation test was developed by the National Bureau of Standards (USA), the Knoop hardness number (KHN) being defined as *the applied load divided by the unrecovered projected area of the indentation*. The following equation relates to the surface indentation:

$$\text{KHN} = \frac{P}{A_p} = \frac{P}{L^2C}$$

where P = applied load (kg), A_p = unrecovered projected area of indentation (mm^2), L = length of long diagonal (mm) and C = a constant for each indentor supplied by manufacturer.

The unique shape of the Knoop indentor enables it to position adjacent indentations much closer together than, for example, the Vickers indentor, allowing measurement of a steep hardness gradient (Figure 161b). Moreover, the Knoop's long diagonal length of indentation at a given depth is only 15% of that for an equivalent Vickers impression of identical length. These width and depth advantages can be used to good effect when assessing surfaces with a thin layer, by utilising the *footprinting technique*. This footprinting entails taking closely packed indentations across a specimen's transverse section of subsurface features (Figure 161b) to determine the hardness profile, or taking hardness readings into a normal surface, but increasing the pre-load, until the thin layer is pierced, thereby gaining an indication of the depth of the hardened layer.

Rockwell hardness

The Rockwell Hardness test was invented in 1919 by Stanley P. Rockwell and is a widely used testing procedure, particularly in North America. The test employs a variety of scales and either hardened steel balls (for softer materials) or a diamond cone (120°) indentor with a slightly rounded point termed a *Brale indentor* – the latter being used for micro-hardness testing. The operation differs from the previously discussed methods in that hardness is correlated with the depth of penetration by exerting a pre-load that reduces the amount of surface preparation, while at the same time minimising the tendency for the indentor to ridge or sink. After pre-load has been applied, then the major load is engaged and the depth of indentation is automatically recorded on a suitable dial gauge scale; the scale is reversed, so that a high hardness corresponds to small indentor depth, relating to a high hardness number. The whole process can be automated for movement of the indentor to obtain a pre-selected surface hardness profile across the specimen being tested.

Due to the fact that the Rockwell hardness value is dependent on the load and indentor, it is important to specify the combination that is employed. This is achieved by prefixing the hardness number with a letter from the alphabet, indicating the particular combination of load and indentor for the hardness scale utilised, as illustrated in Table 16.

Table 16. Rockwell hardness test indentor scales, geometries, loads and prefixes

Rockwell test	Indentor	Shape of indentor	Load (kg)	Prefix
A			60	HR_A
C	Diamond	120° cone	150	HR_C
D			100	HR_D
B			100	HR_B
F	Hardened steel sphere	1/16″ ϕ ball	60	HR_F
G			150	HR_G
E	Hardened steel sphere	1/8″ ϕ ball	100	HR_E

Micro-hardness applications

In Figure 162 is illustrated how important information resulting from work hardening related to localised plastic deformation and thus, residual stresses, due to the machining process and tool geometry can be established. Here, the drilling of "blind holes" by a jobber drill with a conventional drill point geometry (118°) and split-point drill with web-thinning and modified point angle (135°) has been assessed, employing consistent cutting data. The indirect measurement of the residual stresses in the work-hardened region is confirmed by obtaining hardness plots through a "taper section" (more on this topic in the following subsection) at three pre-defined hole depths ("slice levels"). The results of the drilling process show increases in hardness in the proximity of the hole's surface, their values steadily decreasing the further such readings are taken into the substrate, indicating the limit of the localised work-hardening effect. More specifically, as the jobber or split-point drill penetrated into these compacts, the local hardnesses were modified as follows:

- jobber drill with a penetration at low feed (Figure 162a) – the mean micro-hardness increased from 193 H_V @ top, by 16% and 33% at a depth of 10mm and 20 mm respectively;
- jobber drill with a penetration at high feed (Figure 162c) – the mean micro-hardness increased from 248 H_V @ top, by 10% and 15% at a depth of 10 mm and 20 mm respectively;
- split-point drill with a penetration at low feed (Figure 162b) – the mean micro-hardness

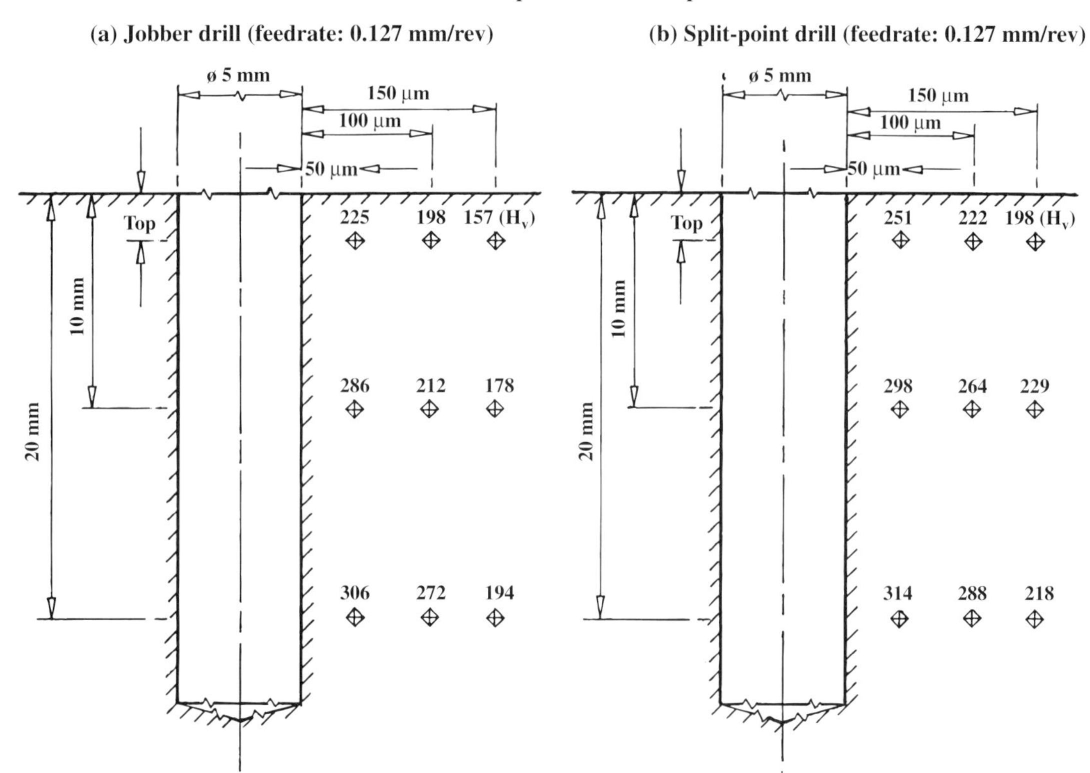

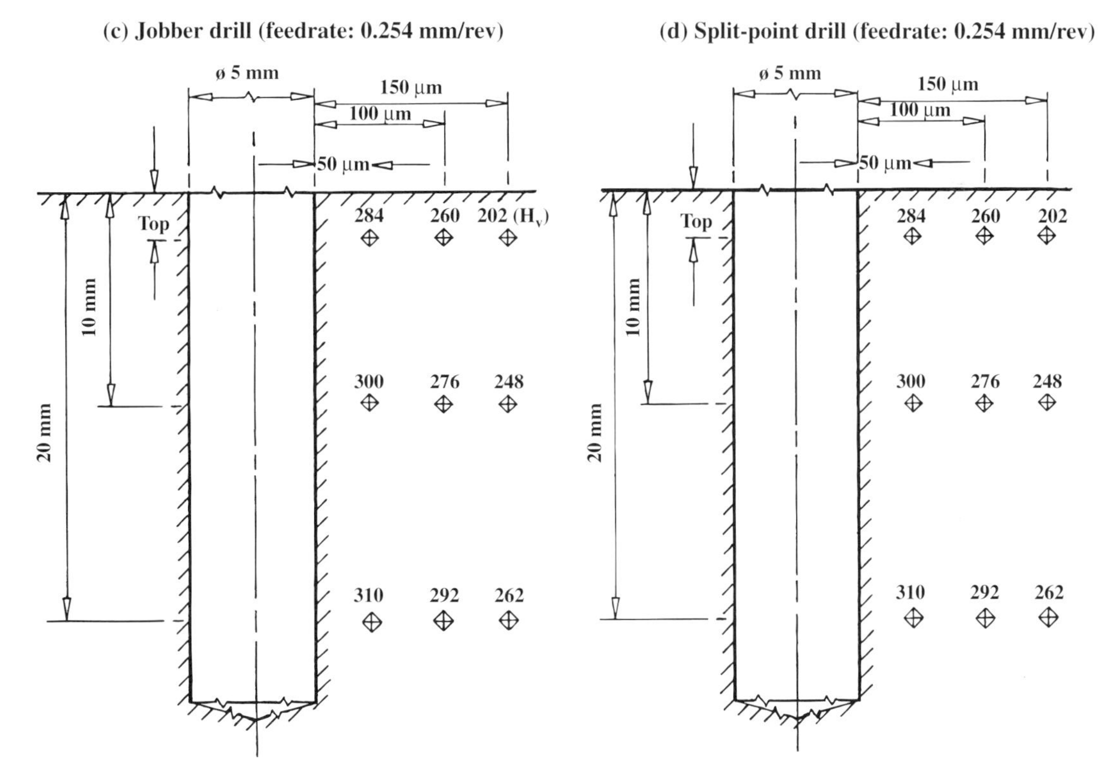

Figure 162. Influence on localised residual hardness, near the drilled hole edge, with variations in drill geometries and component hardnesses.

increased from 234 H_V @ top, by 13% and 8% at a depth of 10 mm and 20 mm respectively;

- split-point drill with a penetration at high feed (Figure 162d) – the mean micro-hardness increased from 264 H_V top, by 11% and 16% at a depth of 10 mm and 20 mm respectively.

These hardness data plots illustrate that for a given feedrate and, in nearly all cases, the jobber drill did not promote significant plastic deformation in the hole's surface region when compared with to a split-point drill. This hardness increase globally ranged from 6–7% at higher feeds to 21% at the lower feedrate. The mean difference in micro-hardness, at its extremes, ranged from 47% from the top surface of the low-compaction specimen to the bottom of the high-compaction variety, with this difference increasing to 56% in the case of the high-feedrate/compaction specimens at the deepest hole. The overall increases in plastic deformation values from the jobber to the split-point drills are thought to be due to three factors:

1. the greater bulk hardness of the high-compaction PM varieties increasing the overall hardnesses of the compacts after drilling;
2. as the point angle increased from 118° (jobber drill), to 135° (split-point drill), this induced greater plastic deformation – hardness – in the hole's surface region;
3. as penetration of the respective drill types occurred, the abusive regime of drilling also increased because of higher drilling temperatures, toward the hole bottom, and this was exacerbated by poor ejection of the work-hardened chips and compounded by the drill's margin influence on the hole's wall – "burnishing effects".

In all of the above drilling operations, the PM compacts were drilled without coolant, which if utilised would have reduced the abusive hole generation procedure considerably, towards a "gentle regime". However, this discussion illustrates the information that can be gleaned from undertaking hardness data studies and the effect that abusive machining plays in just one aspect of a machinability/surface integrity study.

Taper sectioning

In order to improve the metallographical assessment of a sectioned surface and its subsurface features, while achieving greater discrimination in hardness profile analysis across the viewed area – after appropriate polishing and etching – *tapered sectioning* can be employed. Interrogation of subsurface features such as phase transformation, plastic flow zones, localised cracking, and bulk transportation of material (redeposited material, re-entrant angles and under-cutting) can be undertaken that would otherwise have been missed if only profilometry assessment had been used. As its name implies, "tapered sectioning" overcomes the limitation of perpendicular sectioning by modifying the magnification of the subsurface features without undue distortion to the transversely cut surface topography. The procedure is illustrated in Figure 161(a), where previous production processes, in this case turning and boring, have been produced and the subsurface detail would otherwise have been lost (as these modifications may occur in a relatively small and localised surface zone), if a conventional perpendicular section had been taken. Taper sectioning can overcome the potential distortion problem, by selecting a section angle that increases the vertical magnification without unduly influencing the metallographical and topological features of interest. Figure 161(a) illustrates the transverse component cutting procedure, showing that an 11° section has been made. This 11° sectional cut improves assessment discrimination by increasing the vertical section magnification by around five times. Thus, the taper section angle (TSA) will be 79°, the vertical magnification being obtained from the following expression:

$$TSM = \text{secant}\ (TSA)$$

where TSM = taper section magnification and TSA = taper section angle.

Such vertical magnification by taper sectioning, increases the ability to obtain valid hardness information on, for example, a machined surface, by taking staggered hardness readings across the cut, polished and etched surface as illustrated in Figure 161(b). This subsurface feature of interest that has been plastically deformed/mechanically altered is often quite small in width, somewhat less than 0.1 mm wide, requiring a small micro-hardness indentor such as the Knoop, as this indentor geometry and associated readings can be more closely packed than the equivalent micro-Vickers impressions. It is normally advisable to obtain several micro-hardness readings in the subsurface area of interest, as an otherwise spurious result can heavily influence the overall hardness profile result; therefore it is advisable to take several localised hardness readings and then average them to obtain mean values to minimise such influences.

Caution and some degree of care must be taken when attempting to section a metallurgically altered subsurface workpiece feature prior to assessment, as when sectioning the component heat can be accidentally induced into the transversely cut surface by the previous process of cutting and polishing the surface for etching and then testing. This undesirable heat induction due to sample preparation could "swamp" the potential localised subsurface hardness values and give significantly misleading and unrepresentative hardness profile results.

5.2.4 Surface cracks and "white layers"

Surface cracks

Any cracks that are present at the free surface which extend into the substrate are potential sites for premature component failure for highly stressed components. It has been shown in the UK railway industry in recent years that despite track being precision machined and occasionally inspected by non-destructive testing (NDT) techniques, instances have occurred when the rails (employed on high-speed corners) have delaminated, thus causing a passenger train to lose contact and crash, resulting in significant loss of life. The method of machining can contribute to the susceptibility of surfaces to fail. In the case of milling operations, it has been recognised that conventional or up-cut milling (Figure 163a) can introduced surface tensile residual stresses into the surface layers of a component. If this component is then subjected to both an arduous and potentially fatigue-inducing environment, then these tensile layers can open up and may result in premature failure. Conversely, a machined component that has been climb-milled or up-cut milled (Figure 163b) will induce surface compressive residual stresses. This type of stress concentration has invariably been shown to remain closed and avoid crack growth, under identical circumstances to those previously mentioned. It has been recommended for CNC milling and machining centre applications that climb-milling on a component not only generates a compressive stress into the surface but also has the added benefit of normally drawing less spindle power.

Both craters and pits in a machined surface do not pose too great a problem in terms of influencing fatigue life of parts, because their depth-to-width ratios tend to be shallow, although from the cosmetic appearance viewpoint they are unacceptable. Cracks are normally classified as either being micro- or macro-cracks; these cracks have depth-to-width ratios greater than four; typically they can promote:

- a reduction in mechanical strength, fatigue and creep;
- an increase in the susceptibility to stress-corrosion;
- increased probability of surface material breakout and generation of debris;
- higher wear rates resulting from "three-body generation";
- surface delamination and fatigue.

Cracks may be considered as either separations or narrow ruptures that interrupt the surface continuity and usually include sharp edges, severe directional changes or both. Macro-cracks can normally be visually seen unaided, whereas micro-cracks require microscopic examination. Often complex metallurgical interactions are compounded with an abusive regime in the production processing that contributed to the unacceptable surface condition. A crack's origin can be the result of some extremely complex phenomena; typically they can be promoted by an inter-granular attack that might be exacerbated by surface dissolution, via chemical processes. Whenever preferential intergranular attack takes place, this additionally introduces a grain boundary network of micro-cracks that can extend beneath the surface, following the underlying grain boundaries. Micro-crack sites cannot be ignored, as they can affect the component's functional performance because they act as a potential source for macroscopic crack failure. Once a crack has been generated it cannot be successfully re-sealed, owing to subsequent contamination and chemical reactions. Just because a manufactured surface visually appears to be perfectly smooth and of high integrity, it may not be without fault and should be treated as a potential failure site.

"White layers"

Certain ferrous-based work-hardening materials, if they have been subjected to an abusive regime, say, of machining, from a metallurgical viewpoint can exhibit an undesirable and unwanted trait termed "white layers". This metallurgically unstable subsurface, refers to localised hard surface layers that, when etched and viewed through a microscope, appear as white, featureless and bland metallographical areas, hence the term white layer (Figure 164). Previously, they were known as "white phase", "white etching" or "hard etching" and, depending upon the variety of white layering production, several other classifications have been identified. The classification depends upon whether a mechanical, chemical or thermal event has transpired, which in turn directly relates to factors such as strain, strain rate, heating and cooling rates, plus environmental conditions.

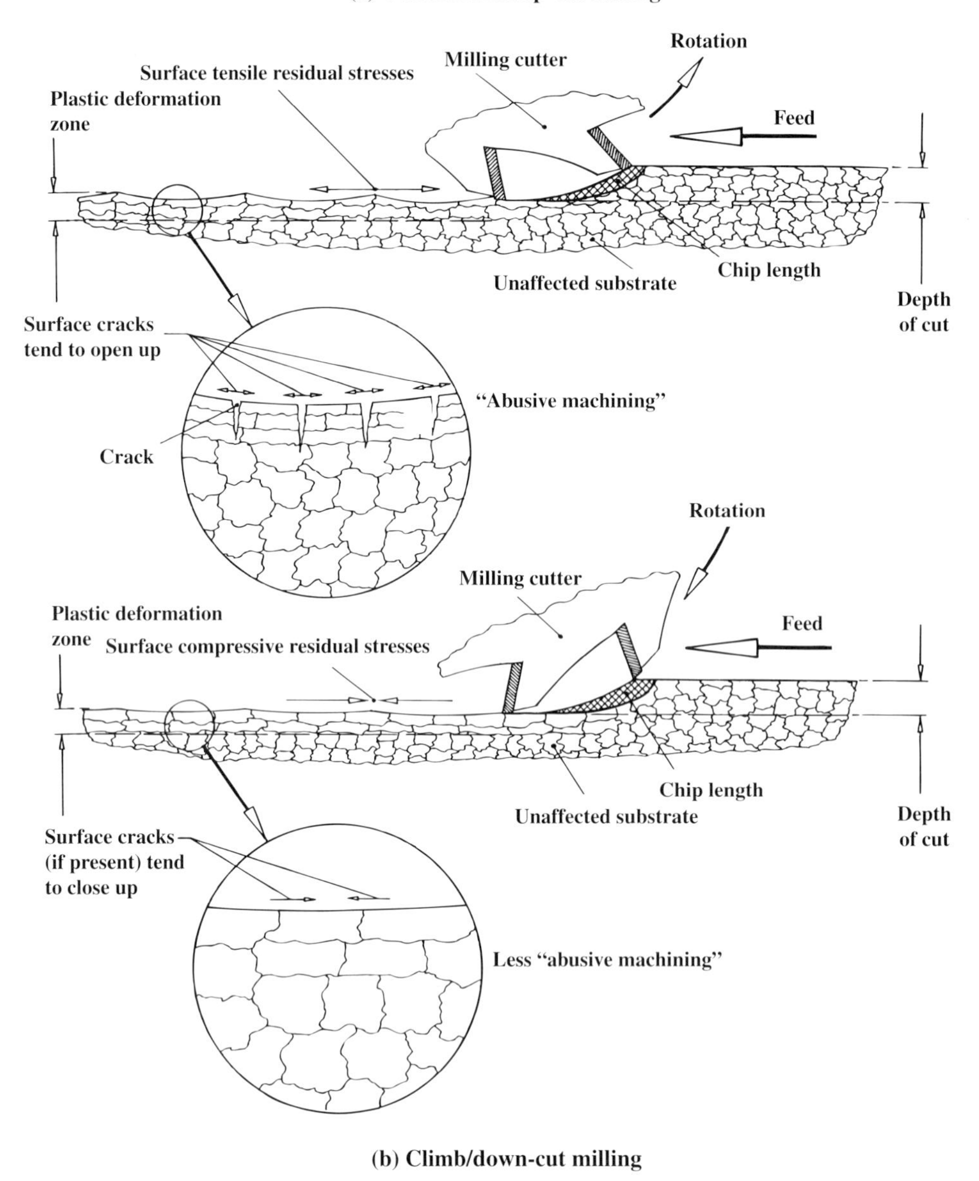

Figure 163. Fatigue characteristics of the surface region being influenced by the mode of machining.

In Figure 164(b) a white layer exists beneath a recast and redeposited layer, resulting from abusive drilling with a dull drill. Due to the fact that the recast layer has a similar metallurgy to that of the white layer, the delineation of their respective zones is not clearly defined. Underneath the white layer is a complex metallurgical zone consisting of some white layering, untempered martensite (UTM), together with over-tempered martensite (OTM), while beneath these layers the bulk material of the substrate remains unaffected. White layers tend to be very hard and in this example (Figure 164b) the recorded hardness was 62 HR_C, with a significantly softer layer beneath this zone. In this situation, the white layer-generating unit was probably a combination of both thermal and mechanical events.

The thickness of a white layer in the drilling example is strongly influenced by both the potential plastic deformation that can be created in this vicinity and to a lesser degree by the thermal influence of the passage of the dull drill as it penetrates through the workpiece. The problem is compounded by lack of

(a) Surface produced by a new drill, indicating no surface alterations

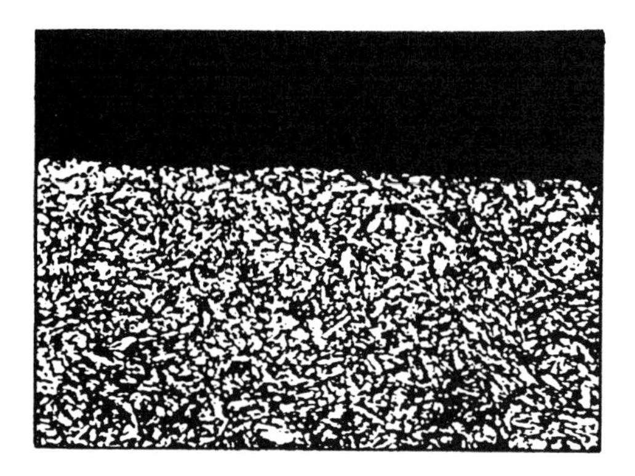

(b) Surface produced by a dull drill, indicating "white layering" together with surface cracking. NB untempered martensitic phase of the workpiece 62 HRc

(c) Blind hole drilled under an "abusive" machining regime, indicating the "white layer" at hole's bottom

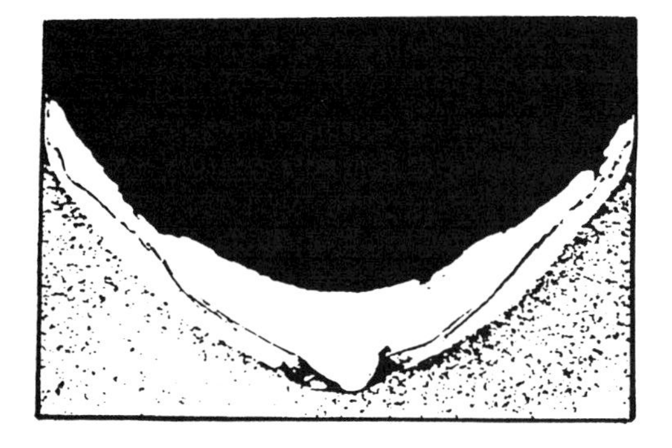

Figure 164. Surface integrity resulting from an "abusive" drilling machining regime. [Source: Field and Kahles, 1971.]

coolant during the hole's generation; this creates very high levels of friction at the drill margins because of inefficiency in the dull drill's cutting lips. Virtually all tooling, even the most sharp, have a finite tip radius of approximately 8 μm, which results in increased wear that can transform the surface metallurgy by thermo-mechanical generation. The point has already been made that many machining processes impart residual stresses into the machined surface layer; this can significantly increase the tendency toward component distortion. For example, when a face-milling operation occurred on 4340 steel that was quenched and tempered to a bulk hardness of 52 HR_C, a tensile residual stress was the less-favoured machined surface condition, as illustrated in Figure 165. Even with cutting inserts on a sharp tool, a certain degree of tensile residual stress was apparent in the immediate surface region, directly under the surface (zone width of around 50 μm); this stress concentration changed to one of compression. After successive milling passes, the flank wear increased in proportion to the Taylor wear curves:

Taylor's general cutting tool wear relationship:
$C = VT^n$

where V = cutting speed (m/min), T = tool life (minutes) and C and n are constants, being tool- and material-derived.

As the cutter became steadily more worn and the flank wear increased, as did the subsurface compressive layer, so this would lead to even greater potential for distortion.

When the forces involved in the production process exceed the flow stress, plastic deformation occurs and the structure is deformed. In the case of ductile materials, the plastic flow can create a range of degenerative surface topography characteristics, including burrs, laps, a residue from the tool's built-up edge (BUE) and other unwanted debris deposits. If this deformation becomes severe because of excessive plastic flow, any grains adjacent to the surface may become fragmented to such an extent that little or no structure can be metallographically resolved, hence white layering will result. Normally, a white layer region extends to quite a small depth beneath the surface, typically varying between 10 and 100 μm, depending upon the severity of the abusive regime of surface generation. Although the graph illustrated in Figure 165 has residual stress along its vertical axis, it is possible to superimpose a microhardness axis here; this is because the shape of the hardness profile plot closely follows that of residual stress. Care must be taken when employing this analogous treatment, as residual stress and hardness are two distinct quantitative values. As has been mentioned, the hardness profile closely follows that of the residual stress graph; however, instead of a tensile stress at the surface resulting from the by-product of a machining process, the subsurface layer here could equally be one of compression.

If a sharp or new tool is employed with little if any flank wear land present (Figure 165) in the

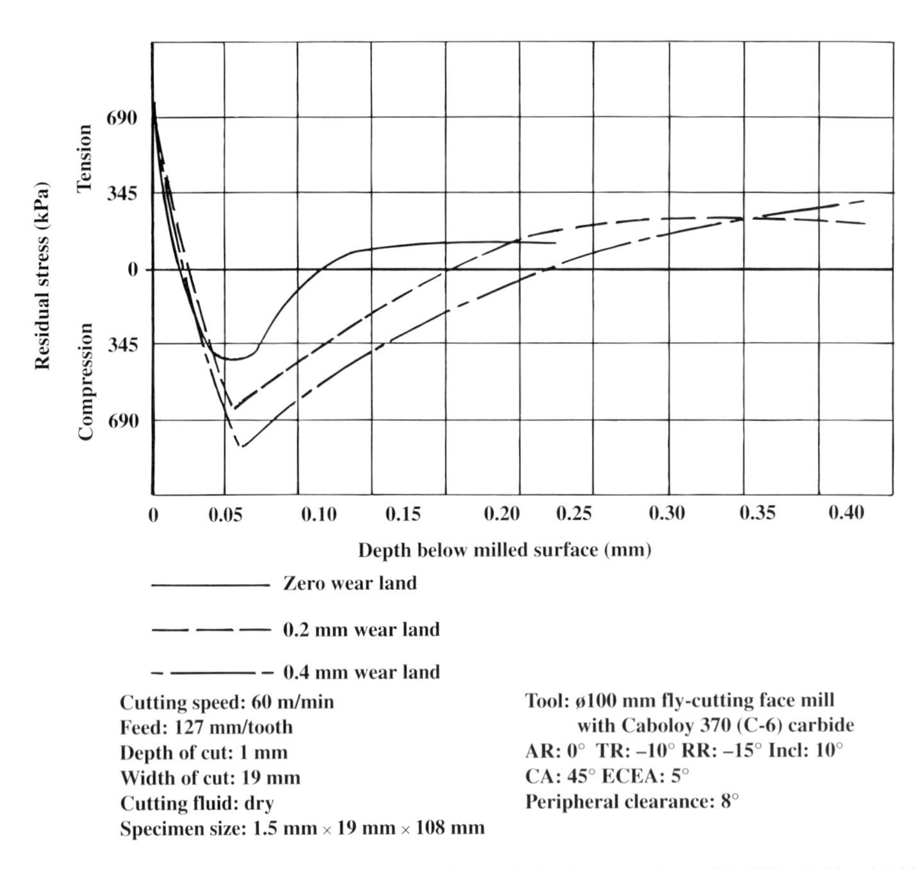

Figure 165. Residual stress in a milled surface of 4340 steel quenched and tempered to 52 HR_C. [After Field and Kahles 1971.]

machining of a potential work-hardening material, then some plastic deformation will be present in the grains immediately adjacent to the surface. A certain amount of over-tempering may be associated with this machining operation that extends a little way into the material's substrate. In this example, as the tool's flank wear increases from 0.2 to 0.4 mm during successive in-cut operations, then significant over-tempering occurs, with the effect extending considerably further into the substrate, until it diminishes. Such significant plastic deformation of the substrate combined with the unstable metallurgical condition may possibly lead to premature failure of a highly stressed component member in service, which might have a catastrophic influence on an assembly, with disastrous consequences well beyond its original and intended purpose.

Electrical discharge machining (EDM) can initiate a "thermal event" such as white layering and its associated martensitic transformations, which exist as recast and redeposited layers. Due to the fact that the recast layer has a similar metallurgical formation to that of a white layer, the delineation between them is not readily apparent. Beneath this white layer the existence of a metallurgically modified heat-affected zone (HAZ) normally occurs, this being a product of either (UTM) or (OTM). As the effect of either of these HAZ's conditions diminishes further into the workpiece, then the original substrate condition will predominate. The white layer effect tends to be significantly harder than the substrate, with typical values for the substrate being 50 HR_C, while the UTM can quite easily reach hardnesses of >62 HR_C, with a local tensile residual stress present. This purely thermal event that occurs during the EDM process results generally from a non-contact process, occurring as a result of the extremely rapid heating and cooling of the EDM continuous cyclical operation. The white layer is strongly dependent upon the energy density, resulting from the interplay of two process parameters, namely the current magnitude and associated pulse duration. As both of these parameters increase, so does the depth of these metallurgically altered states.

The influence of the unwanted by-products of the EDM process can, on some electrical discharge machines, be minimised by switching on a *surface integrity (SI) generator* (Figure 166). The advantages of utilising this SI generator are significant and include the following:

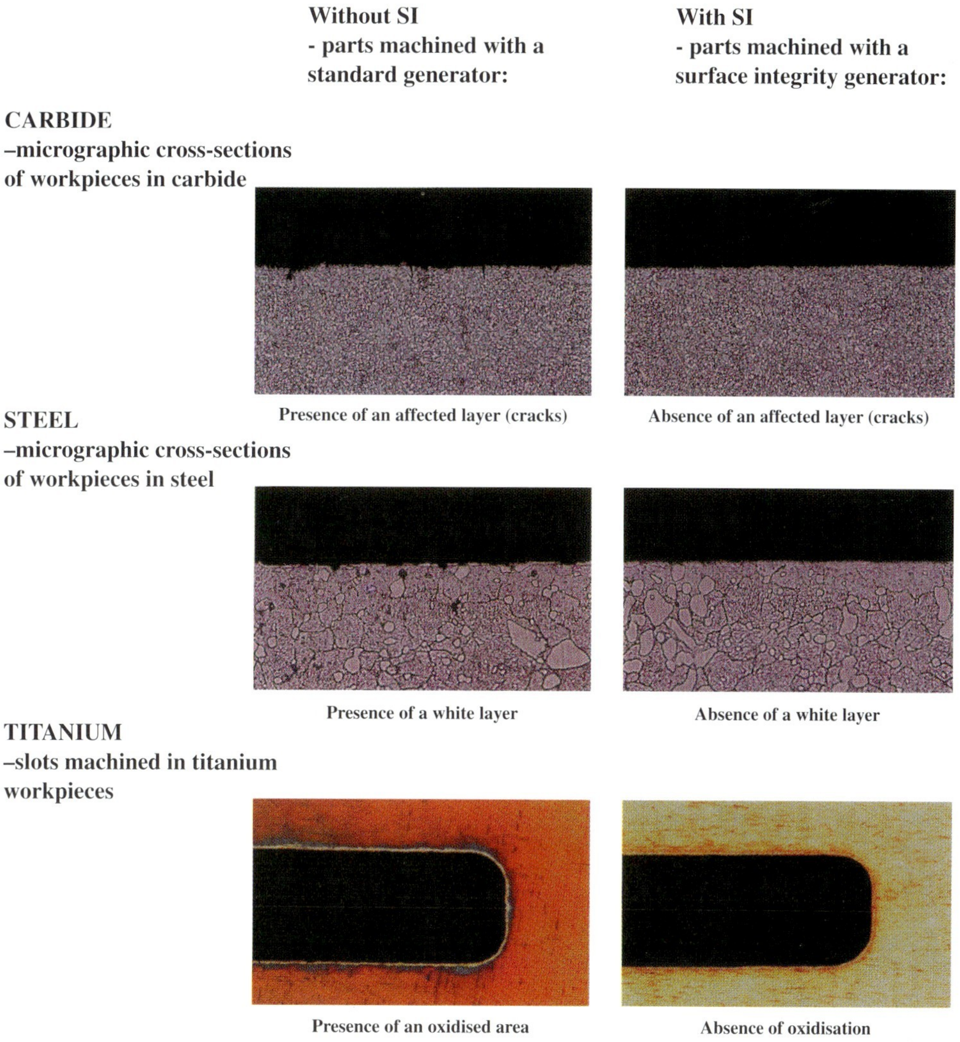

Without SI
- parts machined with a standard generator:

With SI
- parts machined with a surface integrity generator:

CARBIDE
–micrographic cross-sections of workpieces in carbide

Presence of an affected layer (cracks)

Absence of an affected layer (cracks)

STEEL
–micrographic cross-sections of workpieces in steel

Presence of a white layer

Absence of a white layer

TITANIUM
–slots machined in titanium workpieces

Presence of an oxidised area

Absence of oxidisation

Figure 166. Surface integrity of electrical discharge machining (EDM) with/without an SI generator. [Courtesy of Charmilles Technology.]

- surface integrity of the material improves through a reduction of unwanted and undesirable metallurgical effects;
- using the SI generator, the electrical discharge machine tool allows fine surface finishes $\geqslant Ra$ 0.2 μm up to component heights of 100 mm; conversely an Ra of <0.1 μm can be achieved at around 5 mm component heights);
- output from the whole production process can be increased, while obtaining the equivalent level of surface roughness and integrity;
- as a result of utilising the SI generator, the component's parallelism is improved for high workpieces.

If the electrical discharge machine does not have enhanced features such as the SI generator discussed above, then the consequences are for a poor surface integrity workpiece to result. These problems may include *shorter tooling life*, through impaired mechanical characteristics of the metal; *reduced resistance to wear*, promoting surface fragility on, for example, dies and punches; and *bad adherence of surface covering*, on materials such as titanium nitrate (TiN) and titanium carbide (TiC). Other metallurgically related surface integrity EDM problems may include the following:

- *Thermal damage* – during the process a large proportion of energy contained in the spark is transmitted to the material being machined, causing localised heating. The temperature difference inside the material generates constraints, resulting in surface micro-cracks;

- *Corrosion effects* – once micro-cracking has occurred, small quantities of water stagnate in these cracks and are not deionised. Hence, the properties of these minute reservoirs of water are modified and the liquid corrodes the material by dissolution;
- *Electrolysis* – during the EDM process, an electric current passes through the water, which in turn increases the quantity of OH^- and H_3O^+ ions, exacerbating the surface integrity still further.

The use in EDM of enhanced control features, typified by these SI generators, significantly reduces the likelihood of white layering and other unwanted process-induced surface defects.

The process of grinding, whether of the *surface*, *cylindrical* or *centreless* variety, can introduce a range of surface topographical effects and sub-surface integrity modifications, depending upon whether the operation was of a gentle or abusive nature. In Figure 167 the abrasion performance of a conventional grinding wheel consists of a number of variable factors; these include its *abrasive*, *grit size*, *grade*, *structure* and *bond* type. Today, grinding wheels can be manufactured from ultra-hard abrasives, but for brevity and to simply illustrate the problems that occur in grinding operations only the conventional grinding wheels will be addressed here, to highlight their influence in the overall grinding process. Prior to this discussion of the relative merits of wheel selection and their anticipated effect on the ground surface topography/integrity, it is worth iterating that several factors influence grinding wheel selection; however, they are outside

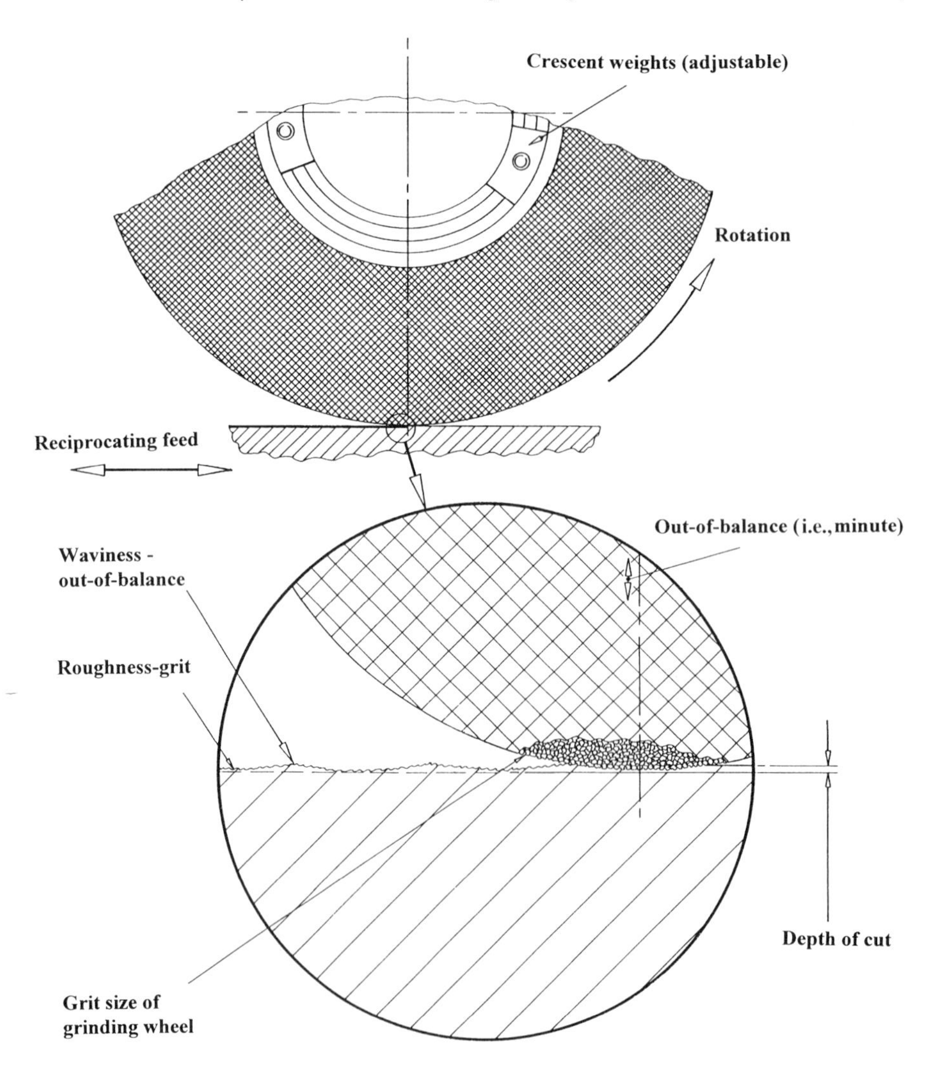

Figure 167. Influence on the ground surface promoted by both the wheel's grit size and out-of-balance effects, when surface grinding.

the scope of the current topic, but nevertheless are worthy of a mention. These grinding factors are the workpiece material, condition of the machine tool, wheel speed, work speed and angle of contact between work and wheel.

A conventional wheel's abrasive make-up tends to be of two distinct types – either aluminium oxide or silicon carbide – the former abrasive wheels, being softer than those of silicon carbide, are employed for harder workpiece materials, as new grains are broken and sharp cutting edges are continuously exposed. The grinding wheel's grit size influences the ground surface as illustrated in Figure 167, as does the wheel out-of-balance. The wheel out-of-balance in surface grinding causes it to "bounce" across the ground surface, introducing medium-frequency waviness, the relative surface roughness being a function of the grit size that is superimposed onto this waviness. If a fine or small grit is selected, then the anticipated ground surface will also be smoother than if a coarser grit had been selected. Both an isotropic and anisotropic surface can be produced by the action of grinding. Wheel grade selection is important as this can affect the overall strength of the wheel, enabling it to be used for either low or high production output. The structure of the grinding wheel is an indication of the proportion of bond to abrasive, an open-structure or porous wheel having around 30% bond, whereas a closed wheel may only have 10% porosity. Bonds can vary considerably, rubber-bonded wheels being the strongest, whereas shellac bonds are used for fine surface-finishing operations. Generally, a good guide as to the type of wheel and grinding conditions to recommend in obtaining efficient workpiece output and surface topographies would be as follows:

- *"harder work, softer wheel"* – relating to workpiece production;
- *"larger angle of contact, softer wheel"* – relating in the main to either internal or external cylindrical and centreless grinding operations;
- *"good rigid machine, softer wheel"* – relates to machine tool rigidity and its associated vibrational tendencies, which may be superimposed onto the workpiece surface;
- *"high wheel speed, softer wheel"* – this refers to the fact that when wheel speed is high the work moves a relatively small distance while the abrasive grains pass, hence its grinding forces are low;
- *"high work speed, harder wheel"* – when the work speed is high an abrasive grain has to remove more workpiece material during each pass, which increases the grinding forces.

The result of all of these variables in the grinding process will be to introduce inconsistent surface integrity effects in combination with a range of residual stresses, depending upon whether the grinding process was gentle or abusive, the latter regime significantly increasing residual stresses in the workpiece surface layers. Under an abusive regime, the dominant feature will be of a thermal nature in combination with tensile stresses in the localised surface layers. Equally, if a gentle grinding regime occurs with copious amounts of coolant present, then reduced friction results and the "unit event" tends toward the mechanical, promoting a compressive residual stress in the subsurface. If the grinding conditions utilise more conventional cutting and grinding data, this introduces a thermal event that has residual stress magnitudes approaching those of the abusive regime, but tend to be of shallower depth below the surface. The peak residual stress in conventional grinding virtually coincides with that from abusive grinding regime and the fatigue endurance limits are almost identical. It has been reported that there is an inverse relationship between the fatigue limit and peak residual stress.

In Figure 168 white layering is shown to be present in the subsurface regions for both surface and cylindrical grinding operations. Even if the selection of wheel, work and associated cutting data are optimised, then if the wheel is not "self-sharpening" and the wheel wear pattern is toward the outer edges, rather than at the centre of the wheel's width, it will "glaze". A glazed wheel will introduce abusive machining and in its worst condition it can become "loaded" with totally inefficient cutting, further exacerbating the workpiece surface integrity. Therefore under either a glazed or loaded wheel local hardness will increase, as will the white layering tendency, which further degrades the workpiece surface texture and integrity.

Altered material layers

In order to gain an overview of the altered material layers (AMLs) that occur in diverse surface and subsurface topographical features ranging from differing metallurgical processes, mechanical applications and uses, see Table 17, which highlights their particular influence on functional performance.

In the majority of cases shown in Table 17 the influence of these subsurface defects tends to be of notable importance, specifically with respect to an abusive regime producing a machined white layer. In some instances the altered material zone (AMZ) can affect component in-service performance in a variety of ways. For example, where the in-service tribological situations produce either redeposited or recast layers in the surface region, it has been known that such defects will influence wear and affect reli-

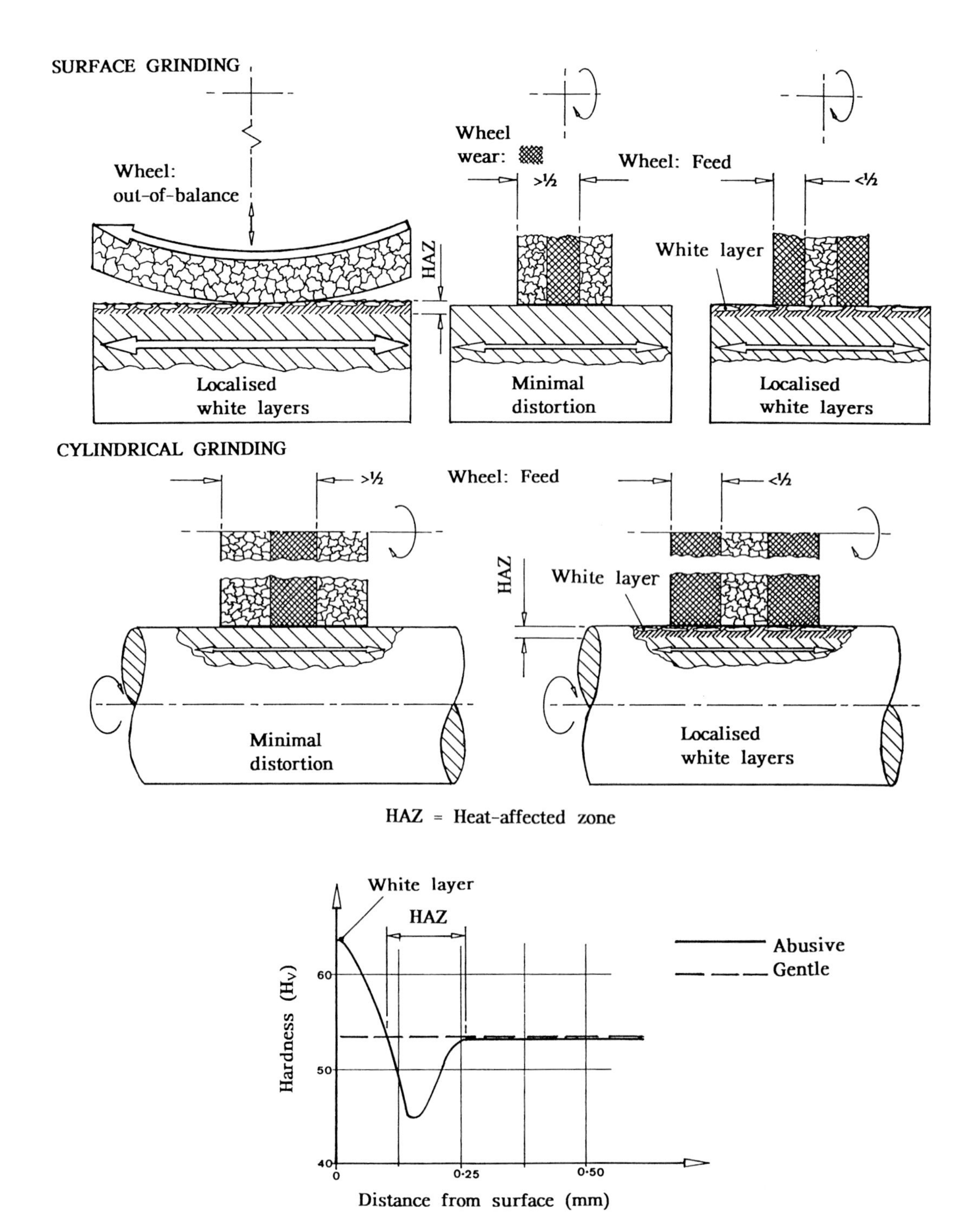

Figure 168. Effect on machined surface integrity during surface/cylindrical grinding of high-strength steels and alloys.

Table 17. The influence of subsurface features on function

Surface integrity: subsurface features												
Function	Metallurgy					Deformation				Deposits		Stress
	UTM or WL	OTM rev	Aust	IGA	WL	Plast defn	Burrs	Cracks	Tears and laps	Tool frags	Redp matl	Res stress
Wear	■		◇	◇	■	■	■	■	■	■	■	■
Strength	■	■	■	■	■	■	■	■	■	◇	◇	■
Chemical attack	□			■	■		■	■	■	◇		■
Fatigue	■	■	◇	■	■	■	■	■	■		◇	■
Magnetism					◇	◇						
Bearings	■			◇	■	■	■	■	■	■	□	
Seals	◇			◇	◇	◇	■	■	■	■	□	
Friction	□						■	■	■	■	◇	
Bonding and adhesion	◇			◇			■	■	■	■	■	■
Forming	■				■	■	■	■	■	■	◇	■

Key: ■: strong influence on function; □: some influence on function; ◇: possible influence on function.

Abbreviations: UTM: untempered martensite; OTM: over-tempered martensite; Aust rev: austenitic reversion; IGA: interangular attack; WL: white layer; Plast defn: plastic deformation; Tool frags: tool fragments; Redp matl: redeposited material; Res stress: residual stress.

After Griffiths et al. (2001).

ability. This often undetected subsurface condition degrades the functional performance; because they are the product of hard, brittle and unstable layers with tensile residual stresses present, combined with an acute change to the bulk substrate, they are likely to *spall* (delaminate and break away). Conversely, if a subsurface feature produces severe plastic deformation, evidence has shown that in the case of the die and tool industry some dies benefit from increased life due to this enhanced abrasion resistance. Table 17 indicates to the design engineer that simply selecting a production process without an intimate knowledge of how it is to be manufactured will affect the subsequent part's performance. Moreover, due regard must be given to the component's potential subsurface state, as this condition will inevitably lead to problems when the part is utilised in service with its potential reliability being impaired.

5.2.5 Machined surface topography

Turning

When a component has been machined this can be considered to be a permanent record of the "unit event" and, as such, can be diagnostically investigated to obtain invaluable information of the production process and the data utilised in its manufacture. The term "roughness", relating to most surfaces and particularly a turned surface, refers to any surface undulations that are regular/periodic (anisotropic) that are deemed typical for that surface condition. Such roughness can be considered to be the constant and predictable nature of the surface and not the random variations and disturbances that may be present overlaying the fundamental surface geometry. Conversely, a machined surface's topography describes the real surface including the texture and any disturbances present; therefore if there are no surface disturbances then the topography equates to its texture or roughness. Further, if no disturbances occur within the macro/micro-textures of this machined surface, then its finish can be considered as the quantified description of the topography defining the basic structure.

In Figure 169, if consideration is given to the "idealised" turned surface topography, then for constant tool nose geometry and an undeformed chip thickness (often wrongly termed the "depth of cut" in turning), as the feed per revolution is increased the surface texture is degraded. The residual cusps that periodically occur on the turned surface after the tool's passage along the part are the product of two phenomena: the so-called "moving step effect", in conjunction with the associated condition of the "emerging diameter". This relationship increases the notable height of the cusps with larger feed per revolution and diminishes in height with reduced feeds. If proportionally larger feeds per revolution are selected, this increases the residual influence of the tool nose contact region on the

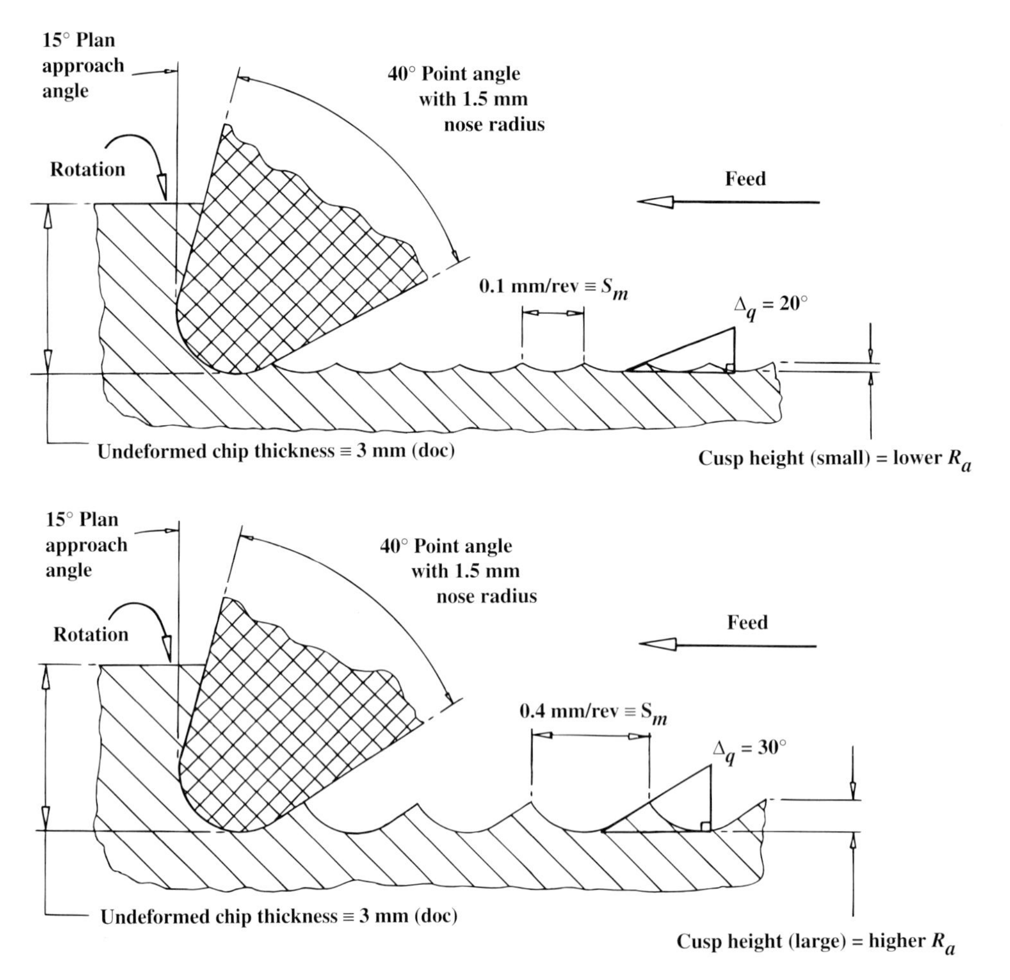

Figure 169. How feedrate influences the machined cusp/surface roughness value S_m and its effect on the waviness parameter Δ_q.

surface and as a result heightens the turned cusps, promoting a larger recorded value in *Ra* and greater angles for Δq values. The opposite effect takes place in the case where feed per revolution is reduced. Namely, as smaller tool nose contact region occurs, this results in a smaller cusp height and accompanying *Ra*; this also gives a shallower Δq due to the partial curvature of the tool nose tending to zero as it approaches tangency with the workpiece axis. The height, profile shape and periodicity of the cusps resulting from the feed and tool geometry significantly influence the measurement and magnitude of the surface topography, and hence resultant parameters. Therefore if the *Ra* value alone was utilised, it cannot adequately describe the nature of the surface topography condition in any meaningful way.

If the cutting data is standardised (rotational speed, feedrate and undeformed chip thickness are constant) and only the tool nose geometry is changed, then the resulting turned surface topography will markedly differ. This topography difference is shown to good effect when turning ferrous PM compacts as depicted in Figure 170, where two extremes of cutting insert nose radii are utilised. In Figure 170(a) the nose radius is 0.8 mm, whereas in Figure 170(b) it has been increased to the equivalent of a 6 mm nose radius, by employing a large button-type of insert geometry. In the case of Figure 170(a), it is visually apparent that the periodicity of the surface is indicated by the regularity of these cusps and, despite the fact that a new tooling insert was employed in this turning operation, the surface shows significant signs of tears, laps and burrs. By way of comparison in Figure 170(b), the surface topography appears significantly smoother in profile, with no noticeable cusps present, although the overall surface topography is marred by similar tears and so on which could be cause for the part's potential rejection. This smoother surface due to the significantly larger cutting insert tool nose radius is not unexpected, as the feedrate was considerably smaller than the nose curvature of the insert; therefore it could blend out and as a result obliterate any potential cusps on the turned surface. This technique of improving the surface by using a large tool nose geometry has been employed by precision turners over the years and can be summarised in terms of the resultant cusp height by the following relationship:

$$H = r - \left(r^2 - \frac{f^2}{4}\right)^{1/2}$$

Therefore

$$Ra = \frac{f^2}{32r}$$

where H = maximum peak-to-valley height, r = tool nose radius and f = feedrate.

In the literature there are traditional formulae for predicting both H and Ra; they are produced by a given feedrate f and tool nose radius r, for rounded geometry turning inserts. With some cutting inserts having almost a sharp point, when in effect the nose radius equates to zero. For finish turning feedrates in conjunction with a rounded or button insert (Figure 170b) the relationship for Ra derived above gives an approximate agreement. Although the value of Ra is an average by definition, this occurs through a combination of both feedrates and nose radii, allowing optimised conditions to be achieved. In reality, the value obtained from the Ra calculation predicts a smoother surface topography than actually occurs in industrial conditions.

Milling

Milling operations represent the main production processes for the manufacture of aerospace and many high-quality precision parts requiring prismatic features to be incorporated into their design. Basically, a prismatic feature on a component consists of a machined face or facet – not necessarily at 90° to an adjacent surface – which normally acts as a location face or datum in a precision assembly. Predominantly, the aerospace industry would tend to manufacture these prismatic features from wrought material by utilising one of two machining strategies:

1 face milling – allowing sizeable surface areas to be generated by large-diameter face mills, running at conventional speeds and feedrates;
2 high-speed milling (HSM) – employing smaller-diameter end-milling cutters rotating at very high peripheral speed and moving at higher feedrates, enabling HSM machining centres to cover similar areas even quicker than their counterparts described in the former case.

In fact, many more industrial processes and applications tend to be utilising the latter approach of HSM. Major production benefits accrue from milling surfaces by this technique, not least of which is the minimal cutter deflection – enabling thin-

(a) Turned surface illustrating partial tool nose geometry, from a 0.8 mm tool nose radiused insert

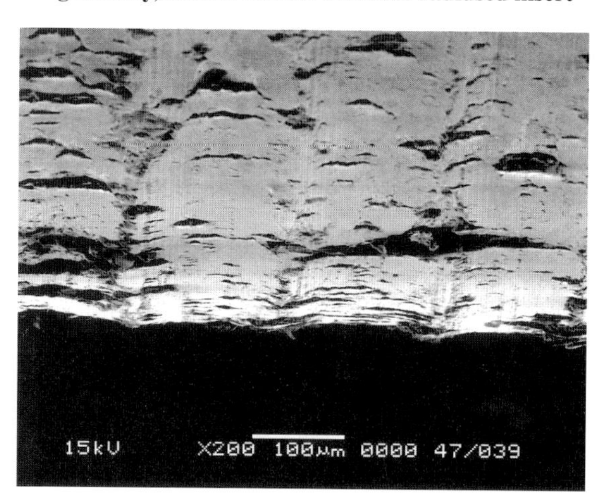

(b) Identical cutting data as in (a), but now utilising a button-style insert

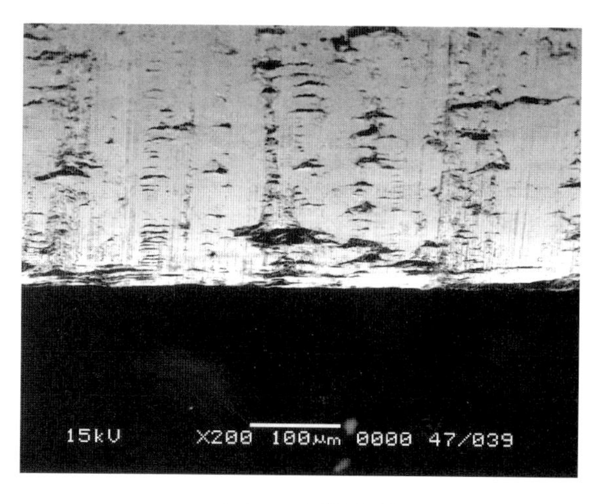

Instrument:JSM-5600
Acc.Voltage: kV 15
Photo mag. ×200
S.E. image
Vacuum: Pa
WD: 46

Figure 170. SEM photomicrographs of typical turned ferrous P/M surface topography with constant feeds and speeds, but varying tool nose geometries. [Courtesy of Jeol (UK) Ltd.]

walled machining to be undertaken – together with negligible subsurface damage inflicted by the HSM strategy and in many cases enhanced milled surface texture.

The end- or face-milling process is an interrupted cutting operation that imparts an isotropic surface topography, as schematically illustrated in Figure 171. If in Figure 171 (top) stock is removed by milling, the resultant milled surface exhibits quite a complex surface topography, due to the "recutting effect" of the cutter's trailing edge as it moves over the surface at a periodic and set speed. This periodic surface topography will not be the case if some form of milling incorporating adaptive control constraint (ACC) is employed. The reason for this variation is that as the cutter progresses over the surface – in, for example, torque-controlled machining (TCM) – its torque is monitored and the stock height may vary. The adaptive control system constrains the machine tool's feed function, thereby protecting the cutter from damage, but more importantly from a surface's viewpoint, the feed varies and hence the milled surface topography will also vary. Returning to the previous scenario using a standardised feedrate, the rotation in combination with the feed for a given cut-off will change the milled surface topography, introducing various cusp height effects along the surface. Here, the periodic nature of the surface topography is regular (*Sm*), but its periodicity changes according to whether the surface is measured at the centre or at some point across the surface; this in turn modifies the relative height of the cusps. Conversely, across the milled surface at arbitrary positions denoted in these examples as "X–X" and "Y–Y" (Figure 171), the topography fluctuates at a predetermined quantifiable interval depending upon where the surface trace was positioned. Milled surfaces with a non-directional or undefined *lay*, as is the case for the recutting effects introduced by either end- or face-milling operations, should not simply have an *Ra* quoted on the engineering drawing. Such design information concerning the milled surface becomes somewhat meaningless and at best only indicates the worst

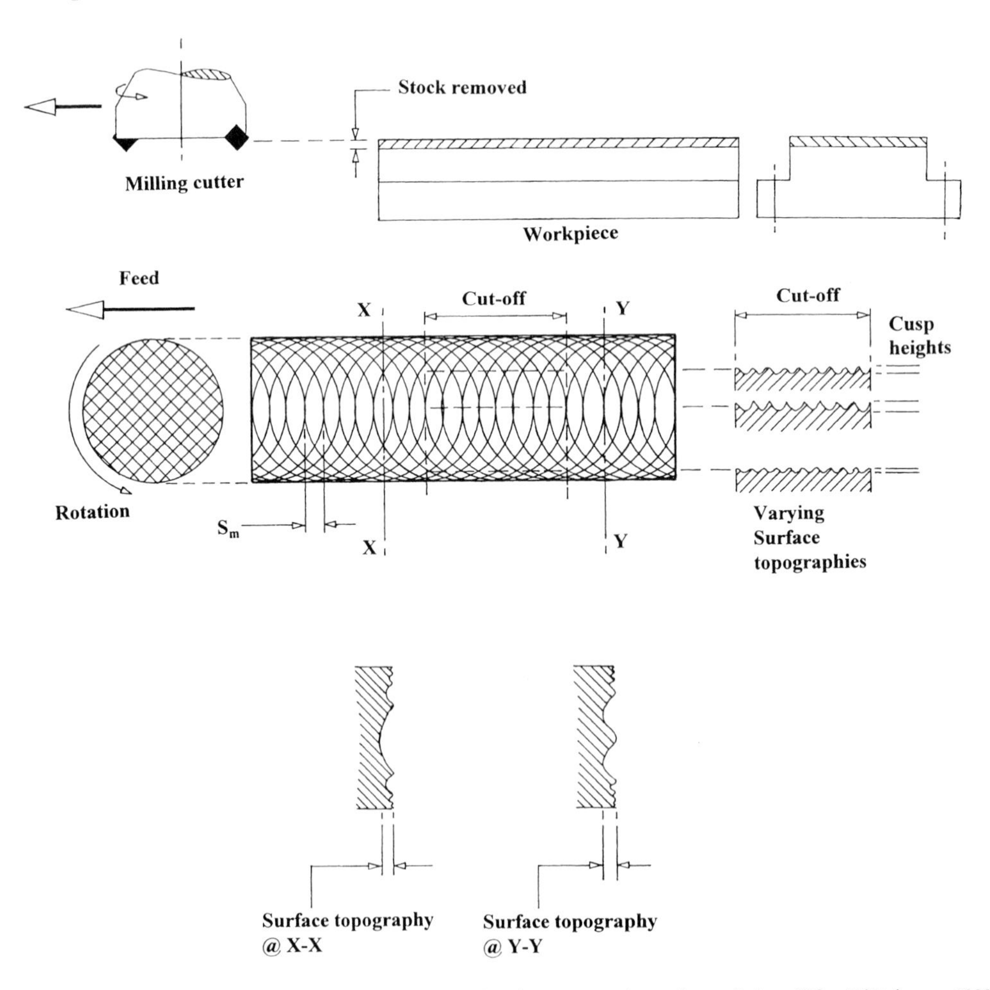

Figure 171. Milling operations impart an isotropic machined surface topography to the workpiece. [After Whitehouse, 1999.]

Category (enlarged profile view)	Examples of kind of variation	Possible reasons for variation
1 Variations in shape 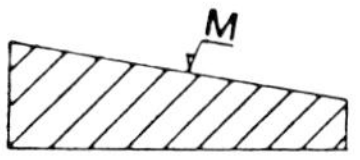 (M)	Unevenness Out of round	Errors in the guideways of the machine tool or of the component Incorrect clamping of the component Distortion due to hardening Wear
2 Waviness 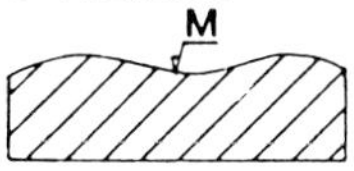 (M)	Waves	Eccentric clamping or errors in shape of a cutter Vibrations of the machine, tool or component
3 Grooving 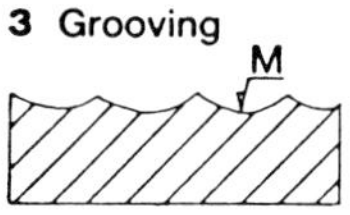 (M)	Grooves	Shape of the cutting edge, feed or supply of the tool
4 Minor imperfections 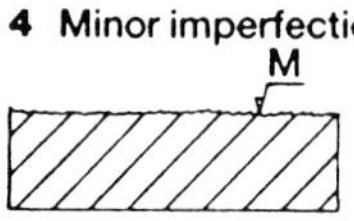 (M)	Scores Flakes Arches	How chips are produced (tear chip, shearing chip, built-up edge), deformation of the component at the sand-blasting, serration (after galvanic treatment)
5 (Cannot be shown on a picture)	Structure	Crystallising processes, alterations of surface by chemical influence (e.g. corrosion)
6 (Cannot be shown on a picture)	Lattice structure of the material	Physical and chemical reactions, in the structure of the material, strains and shearing strains in the crystal lattice
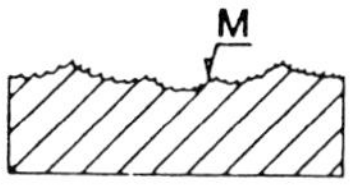 (M)	There is a general overlapping of categories 1 to 4	

Figure 172. Typical surfaces obtained by face or peripheral milling operations.

surface roughness that the designer will tolerate, saying nothing about the surface topography or its functional performance.

In Figure 172 several milled surface topography categories are indicated, with examples of the kind of topographical variation that might occur and the possible reasons for this disparity. Milling operations can be performed in a number of different ways and the resultant milled surface topography is heavily influenced by the numerous varieties in cutter insert geometry, plan approach angles and tool path cutting data selected, as simply shown in the few examples depicted in Figure 173. Not only will the relationship of these interdependent milling factors influence the surface topography but also, as was shown previously in Figure 163, the subsurface integrity is heavily influenced by residual stresses induced by the cutter's passage over the surface being milled at conventional cutting data. This problem of subsurface modification is one reason for the upsurge in the popularity of HSM strategies, as the technique imparts minimal influence on the plastic deformation of this subsurface region after milling. It is often recognised that the recutting effect introduces a certain amount of unpredictability to the functional performance of milled surfaces in critical and highly stressed component environments. Therefore, "wiper inserts" are sometimes incorporated into the face-milling cutter body (Figure 173b), where they are accurately preset to few micrometres below those of the normal cutting inserts (50–100 μm). Firstly, these wipers will minimise cusp effects, thereby enhancing the surface topography, and secondly, they gently remove a small depth from the transient milled surface, minimising the plastically altered subsurface layer.

The term *spindle camber* describes a slight inclination of the milling spindle axis normal to that of the machine's table, or zero inclination (Figure 174). This camber technique is used to avoid the recutting effect on the surface previously mentioned and, additionally, it minimises cutter insert wear, particularly at the cutter insert's periphery. In reality, the spindle camber tends to be very slight, generally only amounting to between 100 and 300 μm over a length of 1000 mm. If this camber is converted to angular measurements, the value ranges between 20 and 60 seconds of arc, this effect being greatly exaggerated in Figure 174 (top).

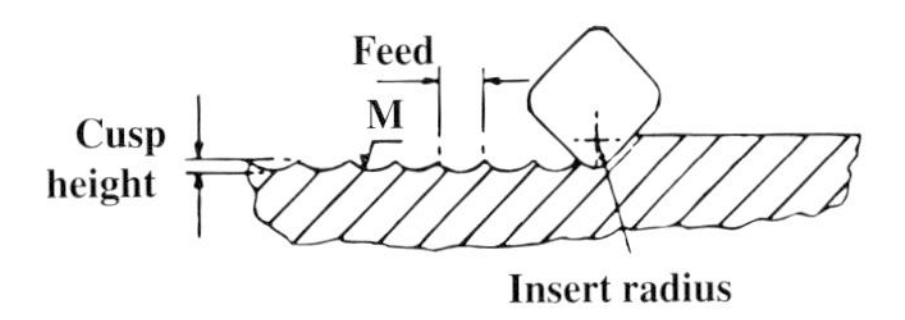

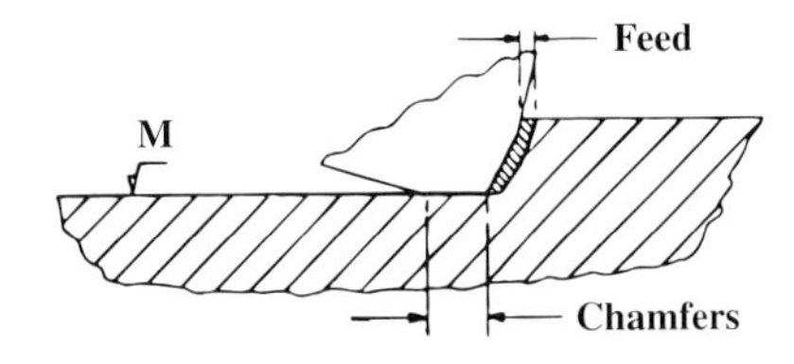

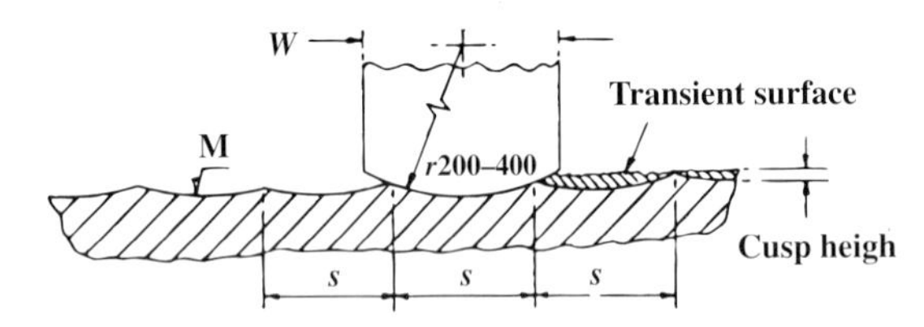

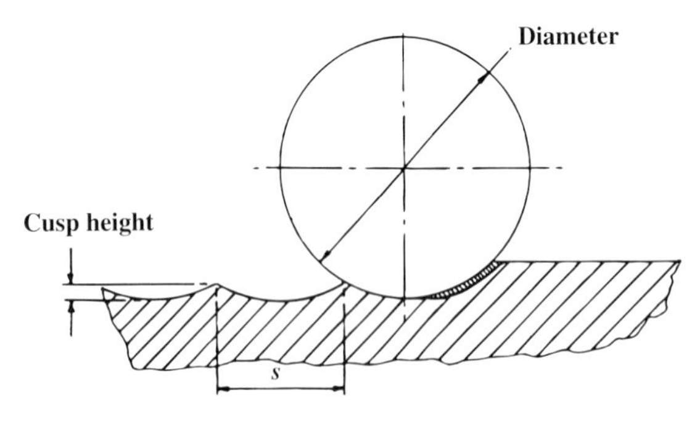

Figure 173. Examples of variations in the surface topography resulting from milled surfaces.

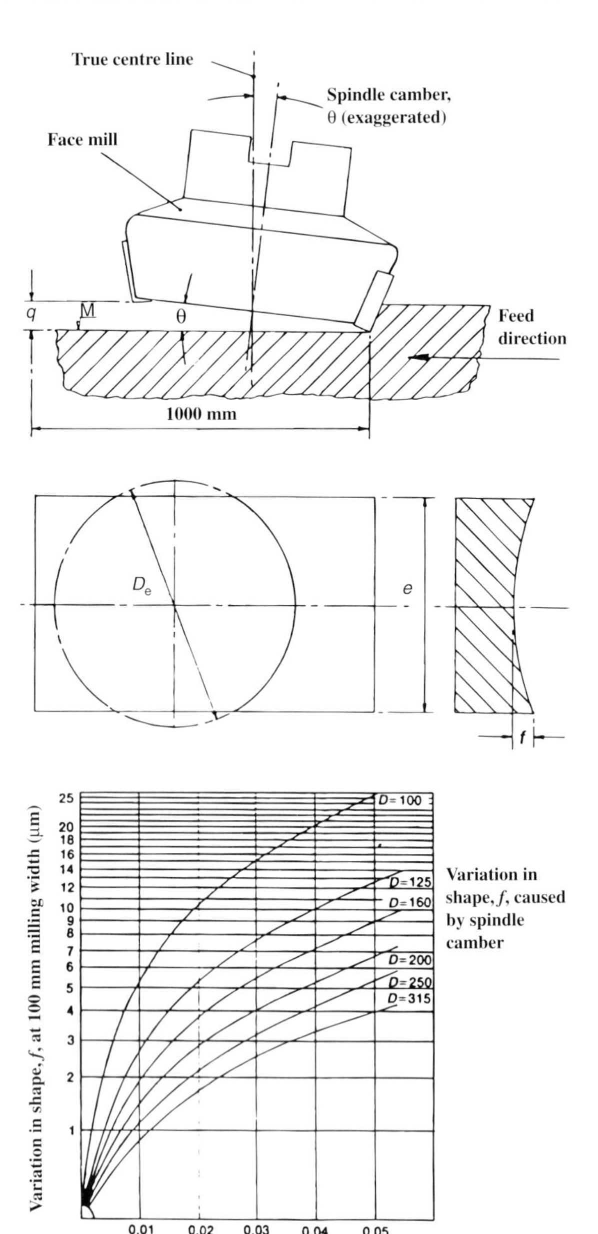

Figure 174. Influence of spindle's relative squareness on concavity of the workpiece surface.

When machining with a relatively large-diameter face-milling cutter, a concave surface is generated as shown in Figure 174 (middle): clearly in this case it is not possible to produce a plane flat surface. The surface concavity generated by the spindle camber depends upon the relationship between the cutter diameter, the width of the workpiece surface and the depth (mm) of this concavity f", which can be calculated using the Kirchner–Schulz formula as follows:

Surface concavity:

$$f = \frac{q}{1000}\left[\frac{D_e}{2} - \left(\frac{D_e^2}{4} - \frac{e^2}{4}\right)^{1/2}\right]$$

where D_e = effective diameter of the cutting circle, e = width of workpiece surface and q = 1000.tan θ, where θ is the spindle camber.

Rather than calculating the surface concavity f", a reasonable estimate can be obtained from the graph in Figure 174 (bottom), showing the variation in surface shape for a variety of spindle cambers and face-milling cutter diameters. These concave surface modifications introduced by the spindle cambering effect are never large deviations from the "true" plane surface. Even under the extreme conditions of, for example, a small cutter diameter (100 mm) and a large q value (50 μm), the deviation only amounts to 25 μm over a workpiece width of 100 mm, this being well within the allowed tolerances for most commercial situations.

Drilling

By far the most popular machining process is that of drilling holes in components and, apart from the more recent designs based on indexable/solid drills (Delta and U-drills), where high penetration rates can be achieved, the twist drill has hardly changed. The twist drill's basic construction was conceived during the latter part of the American Civil War (1864), when Steven Morse perfected its design.

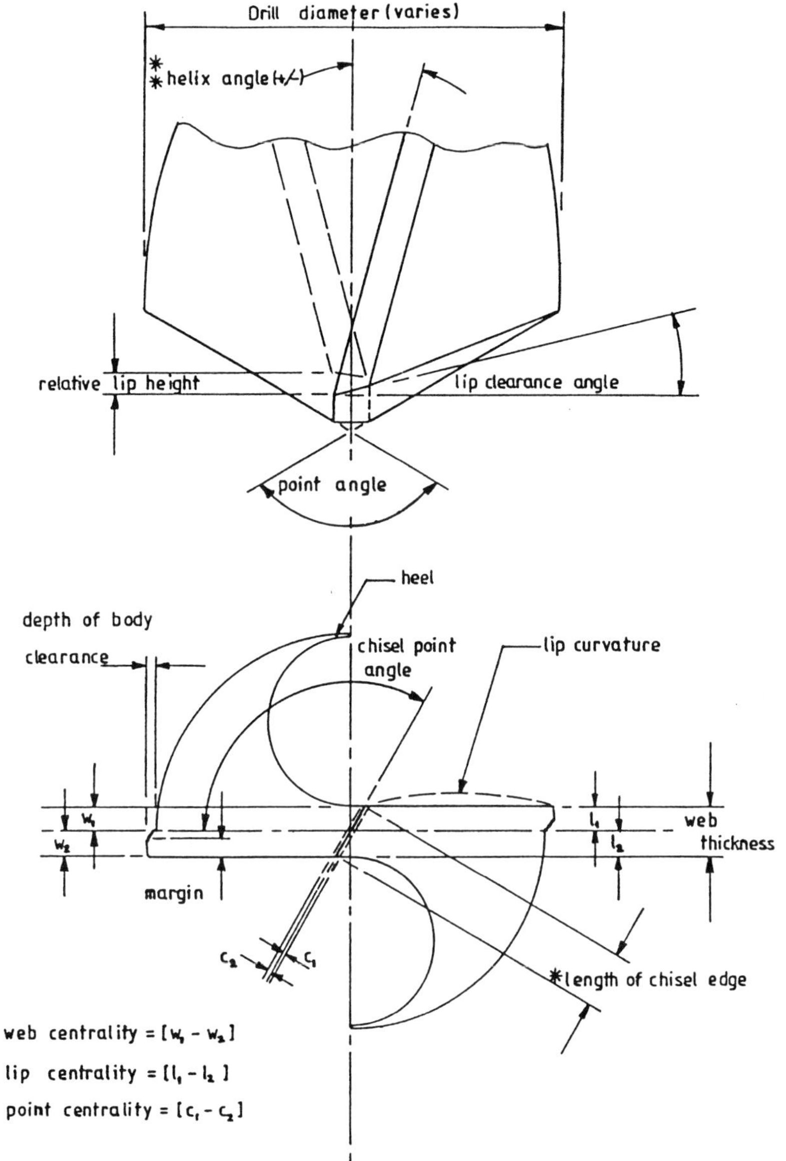

Figure 175. Drilling geometry and chisel point shape: influence on hole generation.

Despite the relatively simple design of the most popular type of drill, this being the twist drill, its material removal mechanism is quite complex and a slight digression to give some understanding of how the hole generation technique occurs will aid in the interpretation of hole surfaces. Near the bottom of the drill's flutes where the radii intersect with the chisel edge (Figure 175), the drill's clearance surfaces form a cutting rake surface that is highly negative in geometry. As the centre of the drill is approached, the tool's action resembles that of a "blunt wedge-shaped indentor" (Figure 176). Under the chisel edge the cutting process is inefficient and a region of severe deformation occurs. These deformed products are extruded, then wiped away into the flutes, whereupon they intermingle with the main cutting edge chips (from the lips). The chisel edge in a conventionally ground twist drill has no "true point", this being a major source for a hole's dimensional inaccuracy. The twist drill geometry of a conventional point is shown in Figure 175, together with associated nomenclature for critical features and tolerance boundaries. From the relatively complex geometry and dimensional characteristics shown in Figure 175, the obtainable accuracy of holes generated while drilling is dependent upon previously grinding the drill geometry to within prescribed limits. Any variations in geometry and dimensions, such as dissimilar lip lengths and angles, or chisel point not centralised, will have a profound effect on both the hole dimensional accuracy and roundness. "Helical wandering" of the drill as it passes through the component will occur with these geometric drill inaccuracies present (Figures 177 and 178). Hole accuracy, particularly the "bell-mouthing effect" (Figure 177), is minimised by previously centre-drilling prior to drilling to size. One of the main causes of a "bell-mouthed hole" is inconsistency in either the lip lengths/angles (Figure 175), or employing a straight (conventional) chisel point, or a combination of both. This "bell-mouthing" at workpiece surface entry by the drill is attributed to the chisel point. This effect is produced by a line contact as the point initially touches the component surface, causing it to "walk" until the drill feed/penetration stabilises at the outer corners (margin) entering the part and guiding the drill (Figures 177 and 178). Such degraded hole effects are exacerbated by longer series drills because of the "rigidity rule", which states that *cutter rigidity decreases by the square of the distance*. Namely, if a drill is twice as long, then it is four times less rigid. It follows that the greater the drill penetration into the workpiece, the progressively larger the deflection and the further from the true axis will be the drill's subsequent path. This drill deflection is compounded when the drill point geometry is not symmetrical and the cutting load and torque on the lips vary (Figure 176). Galloway defined the drilled hole slope angle ϕ and hence the hole's subsequent roundness and circularity in the following manner:

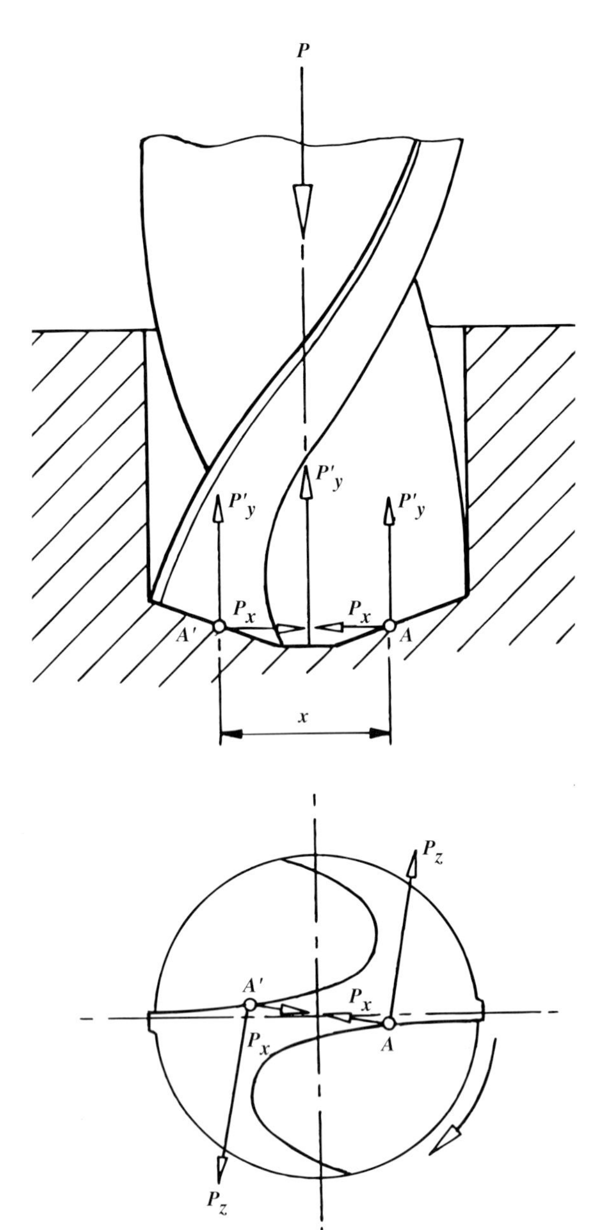

Figure 176. Cutting force generation and the effect that thrust and torque play in hole quality.

Slope in drilled hole:

$$(\phi) = \frac{3}{2} l \frac{R}{T} \left(l - \frac{I}{(k \tan k)\, l} \right)$$

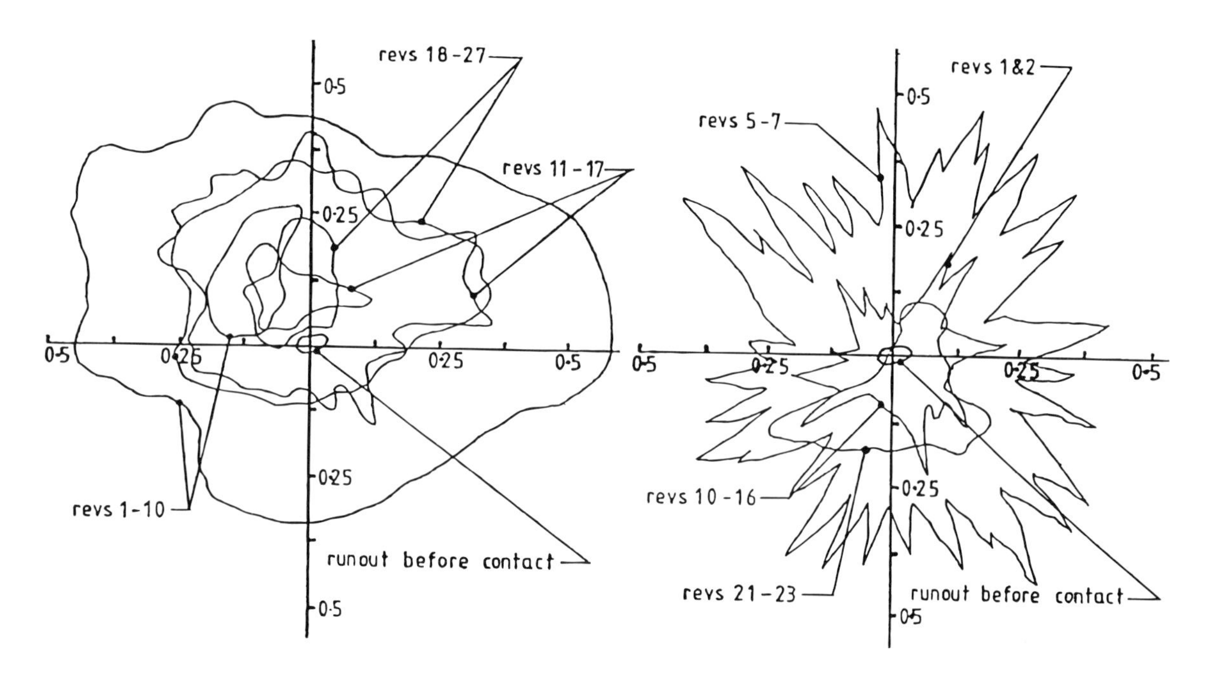

Figure 177. The roundness of a hole is influenced by the initial drill's progression into the part, causing "bell-mouthing" [After Galloway, 1957].

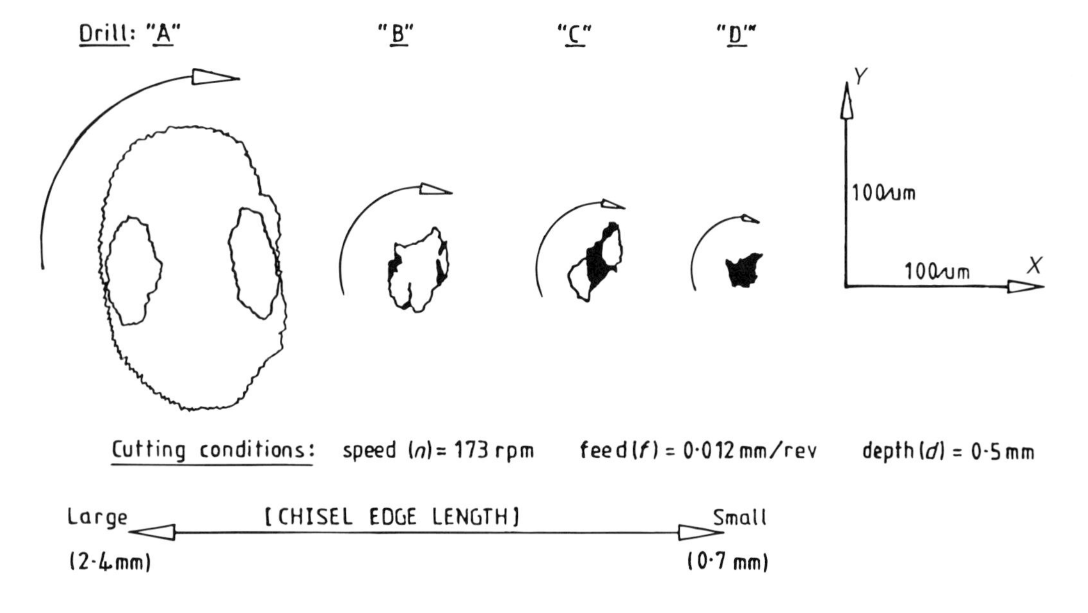

Figure 178. Influence played by an increasingly larger chisel point on the resulting Lissajous figures of drill deflection, as hole depth is generated. [After Onikura *et al.*, 1984.]

where l = length of deflected element, R = ratio of the transverse reaction at the drill point, T = thrust force, I = system's "moment of inertia" and $k = \sqrt{(T/E)I}$.

As this slope suggests, the error is initiated when the chisel edge begins to penetrate into the workpiece, and unless the feed is discontinued or the error is corrected in some way, the magnitude of deflection will increase as drill penetration continues. Typically, the magnitude of deflection can reach up to 60 μm under exaggerated conditions and this could be the source of component rejection.

Due to the drill cutting edges and lip lengths (Figure 175) not being ground correctly, the cutting forces – thrust and torque (Figure 176) – will be both unbalanced and uneven. This imbalance can lead to the drill spiralling down as it penetrates the workpiece, causing the surface profile to exhibit the classical effect of a "saw-toothed profile" on the resultant surface topography. This "saw-tooth effect" on the topography is the result of the partial geometry of the lip corner and its adjacent margin, repeating itself at the periodicity of the feedrate, which could cause problems if the hole is subjected to cyclical stresses in service.

The chisel edge has been shown to play a crucial role in drilling hole quality, not least of which is the result of the inefficient behaviour of the chisel point's extruding operation as the hole is generated. If the chisel point on the twist drill with respect to its diameter is relatively large, then the Lissajous figures, as depicted in Figure 178, will show large fluctuations in the drill's axis of rotation; conversely at a smaller chisel point size, this motional behaviour is significantly smaller. One of the reasons why the chisel point length and its profile play such a major role in producing "bell-mouthing" of the hole at entry is the result of much greater thrust forces, with a proportional increase in the smaller torque values (Figure 176). When the forces on the drill increase as a result of the higher cutting force and torque levels, this tends to cause the twist drill to unwind slightly. This subsidiary drill behaviour introduces the potential for dynamic instability, resulting in an increased tendency toward vibration and additional harmonic out-of-roundness effects, coupled to the likelihood of a degraded drilled surface topography. If a through-drilled hole is required rather than a "blind hole", then there is the potential for the generation of "bell-mouthing" at the exit point of the workpiece, resulting from the so-called "trepanning effect" as the drill breaks through the part's underside.

5.2.6 Machined roundness

Harmonic departures from roundness and surface texture correction

In the previous section it was mentioned that operations such as drilling can impart a range of hole-generating mechanisms that could be the cause for part rejection, such as roundness, cylindricity and surface topography errors. In this section roundness will be the main point of discussion, as it will be seen that machining operations and their accompanying tool geometries can seriously influence the resultant shape of the roundness profile. If, for example, a drill has an asymmetric cutting action (unbalanced), then this condition is likely to impart changes to the generated hole as penetration continues through the part. In Figure 179(a) just such a situation is depicted, where at the top an almost three-lobed roundness profile is shown; this becomes more regular in the middle, but the lobes are partially rotated through an angle, and at the bottom the polar plot has become almost round. This improved roundness at deeper depths of penetration is probably the result of increased support by the drill's margin as it progresses through the workpiece.

In order to improve the drilled hole quality in a number of significant ways, such as in its apparent roundness, surface texture and integrity, together with the correction for any geometric errors present, then reaming may well be employed. Drilling, being generally considered to an abusive regime, will inevitably introduce more undesirable errors during hole generation than the complementary and more gentle reaming operation. The advantage of utilising reamers after the drilling process is shown to good effect in Figure 179(b), where several different reamers are utilised on identical drilled holes, with the roundness being assessed at the same vertical "slice level". On the left-hand illustration, a six-edged carbide-tipped machine reamer of the conventional type was firmly supported in an appropriate adaptor that directly fitted into the spindle nose taper. This positive location and reamer restraint mean that the reamer will introduce into the hole any inaccuracies apparent in the machine's spindle, although even here the accuracy is a significant improvement – five times better in fact – than the previously drilled hole (Figure 179a). If the reamer is allowed to "float" (having lateral/radial play) in the machine's spindle, any spindle errors here will not be superimposed onto the reamed surface and the reamer, with its longer cutting edges than the previous drill, will correct for many of the geometric inaccuracies introduced during the drilling operation. In Figure 179(b, middle) a "float-

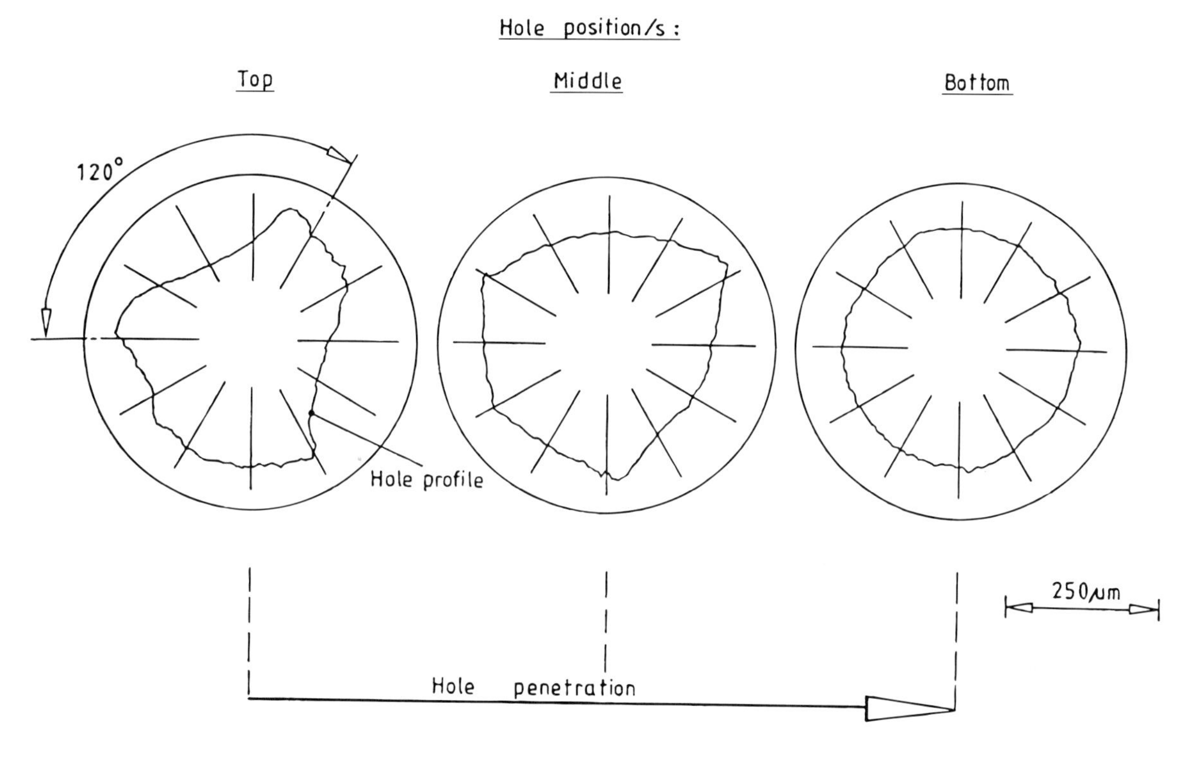

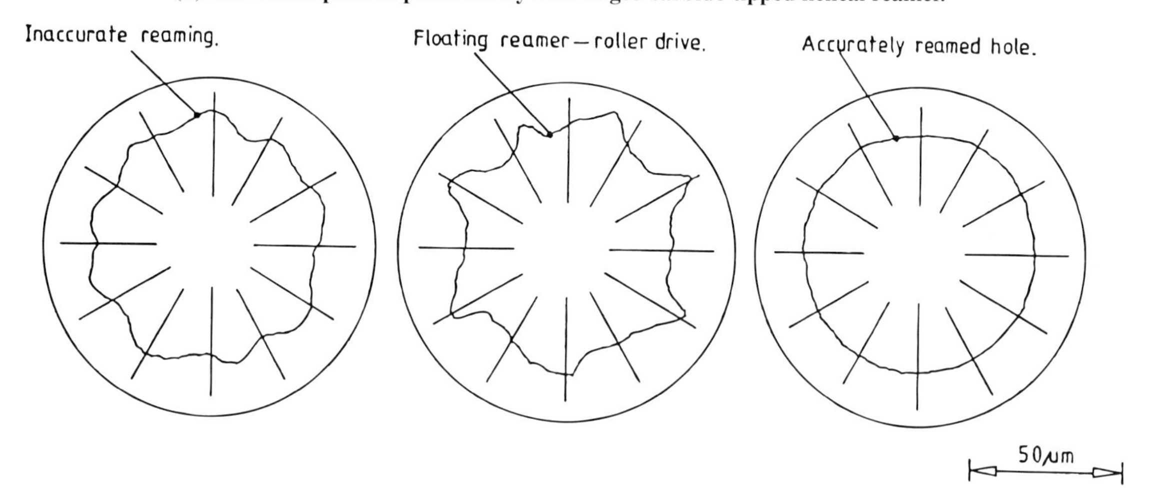

Figure 179. The effect of harmonic departures from roundness change with different drilling hole depths or reamer types.

ing reamer" replaced the conventional type and was driven through the drilled hole at too high a penetration rate, introducing sixth upr harmonic equating to the equispaced six-edged reamer. In Figure 179(b, right), an identical reamer was used to ream a similar hole, but at a slower penetration rate, producing an accurately reamed hole with minimal harmonic disturbance.

In some machining cases, the errors introduced by the drilling process are anticipated to be quite large, as shown by the displayed profiles in Figure 180 (bottom left), illustrating the relative radial movement in the displayed profiles and their associated harmonics. Hence, it is often prudent to introduce a boring operation prior to reaming (Figure 180). After previously drilling a slightly

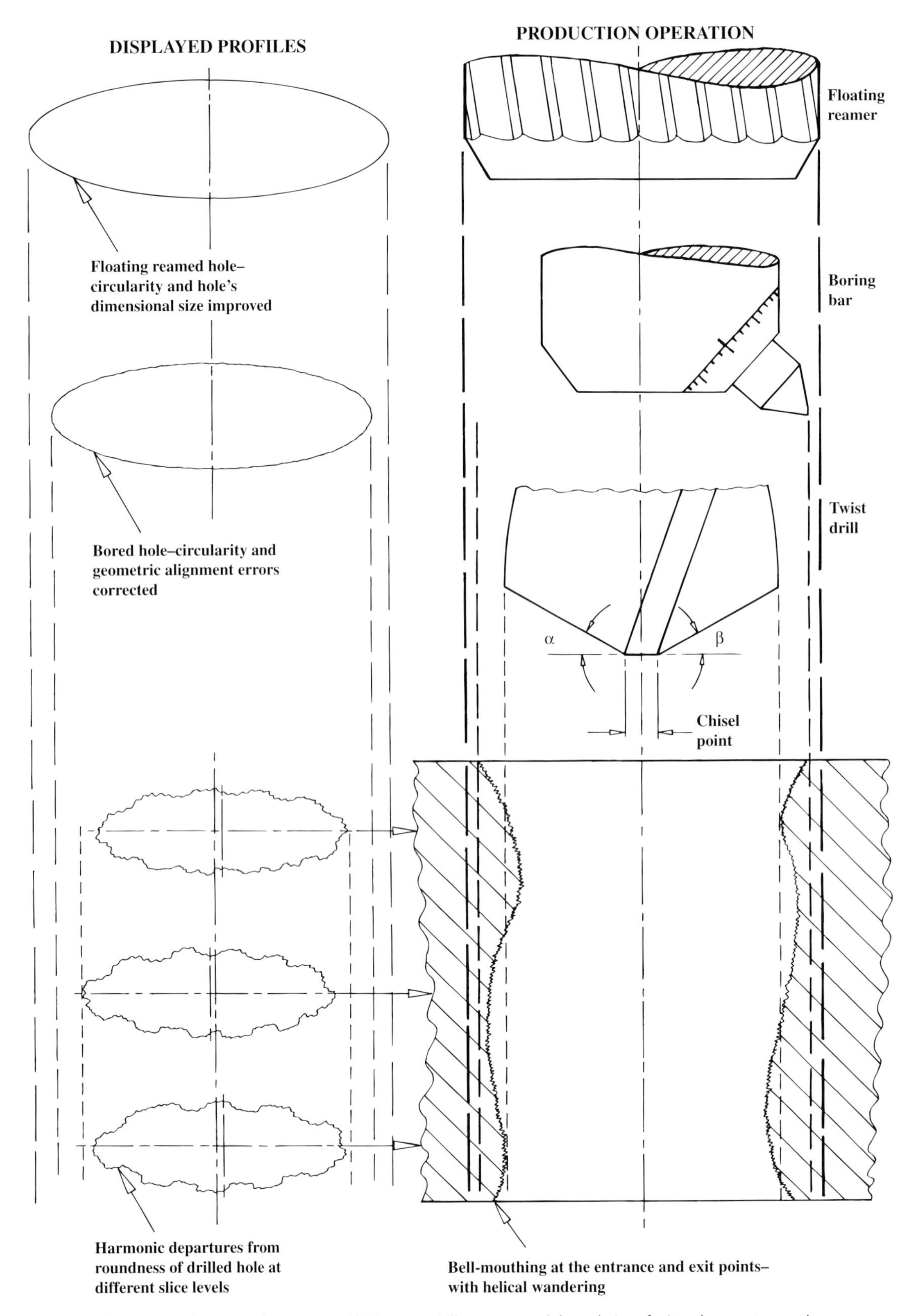

Figure 180. Exaggerated errors caused by incorrect drill geometry and the technique for its subsequent correction.

smaller predetermined diameter hole, a boring operation will then correct any abnormalities (roundness and helical wandering effects) as indicated in Figure 180 (left, towards the top). A boring operation does not follow the hole's contour and, as such, will eliminate these drill-induced errors by machining away. Finally, the reaming operation will impart the finished dimensional size to the hole and simultaneously improve both the circularity and surface texture (see polar plot in Figure 180 (top left).

In order to improve the boring process still further, adjustable twin- or triple-boring heads with indexable inserts can be utilised. These special-purpose multi-boring tools balance out the cutting forces associated with the boring operation and, in so doing, improve surface texture and hole accuracy, together with repeatability and consistency of bored holes. In fact, triple-boring heads in particular can sometimes eliminate the need for the succeeding reaming operation, giving a significant saving in the value-added machining costs.

Reaming technology has radically improved in recent years, reamer design and development consisting of asymmetric replaceable blades having a floating action with accompanying burnishing pads for cutter edge support. Reamers of this level of sophistication are typically employed in the high-volume end of production for automotive engine part manufacture, such as when reaming out the high silicon aluminium alloy camshaft bearing supports. In this production situation the high silicon content in the aluminium alloy produces ideal (small) grain-refining effects. As a result of additions of silicon, it can introduce hard and abrasive inclusions, requiring either a cemented carbide or synthetic diamond reamer blade to be fitted, with guidance through the camshaft's location length provided by support from the bearing pads. Today, "hard reaming" is emerging as an important technique in some industries. Hard reaming allows components made from materials such as austempered ductile iron (ADI) to be machined directly after initial heat treatment, introducing not only an appreciable cost saving but also enhancing the surface texture and roundness levels of the finished part. Moreover, it has been alluded to previously that the machine tool's spindle and to a lesser extent its overall structural configuration can impart significant errors to the manufactured component, due to a variety of machine tool-induced inaccuracies. These can be identified as:

- *spindle imbalance* – introducing dynamic lower-frequency harmonics on the part;
- *cutter forces* – that dynamically affect the machining process, causing a series of high-frequency harmonics to be superimposed on the lower-frequency harmonic resulting from imbalance;
- *thermal growth effects* – changing the both the spindle's growth (axially) and modifications of an elastic nature to the relative axis orthogonalities of the machine tool;
- *working clearances and motor drive configurations* – these are necessary to allow for relative thermal growth and component "running fits" within the spindle assembly, which are compounded by arrangement of the motor drive system.

This latter feature of spindle inaccuracy is present in headstocks on either lathes or turning centres, as illustrated in Figure 181. Direct-driven headstocks for turning centres offer considerable advantages over their belt-driven counterparts. The working clearances and drive considerations found on conventional belt-drive machines suffer from a combination of the effects of spindle motor drive plus associated drive belts on these clearances. This headstock's motor drive arrangement causes an undulating and irregular harmonic rotational motion on the work-holding equipment, which is then translated onto the resultant roundness and surface texture of the workpiece as turning operations occur. The influence of this irregular harmonic rotational belt-driven rotation can be gained from the schematic representation shown in Figure 181, where a "tumbling three-lobed harmonic" shape is reproduced on the workpiece. The irregular but periodic nature of the rotational action of the headstock is reproduced on the workpiece by a series of kinematic combinations of headstock rotation and linear motion supplied by the cutting tool. If a direct-drive headstock configuration is used instead of the conventional belt-drive variety, then there is virtually no harmonic influence on the part and more consistent turned components will result, in terms of their geometrical and linear dimensions.

Direct-drive systems offer other benefits, not least of which are lower maintenance problems, as there is no need to periodically adjust belt tension and the bearings operate under a more consistent and uniform loading, with better thermal growth characteristics, resulting in higher spindle accuracy and improved damping capacity. These positive merits in using a direct-drive spindle promote significant improvements in the overall harmonic roundness throughout the complete length of larger turned parts, with less variation in the medium-to-large frequency surface texture components, resulting from this turning operation.

If one ignores the fact that either a conventional or direct-drive system might be incorporated into the headstock's design, then the problems associated with the former type can be ignored. Instead it is possible to concentrate purely on the rotational

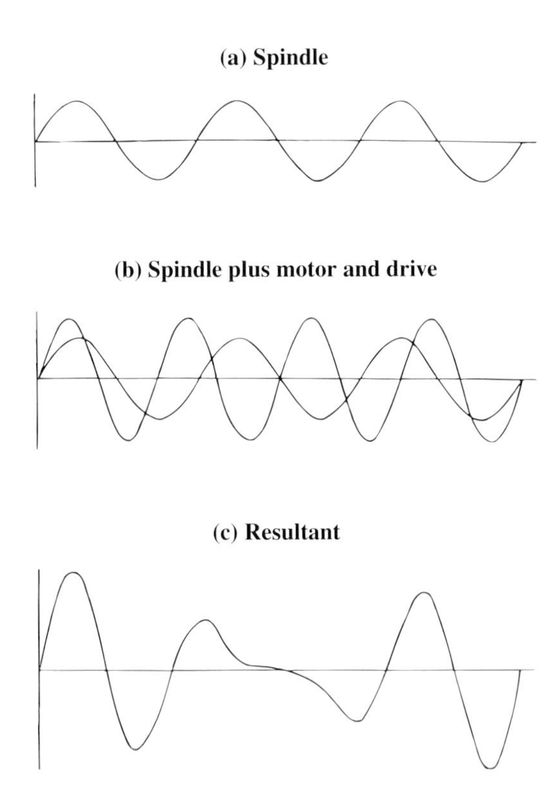

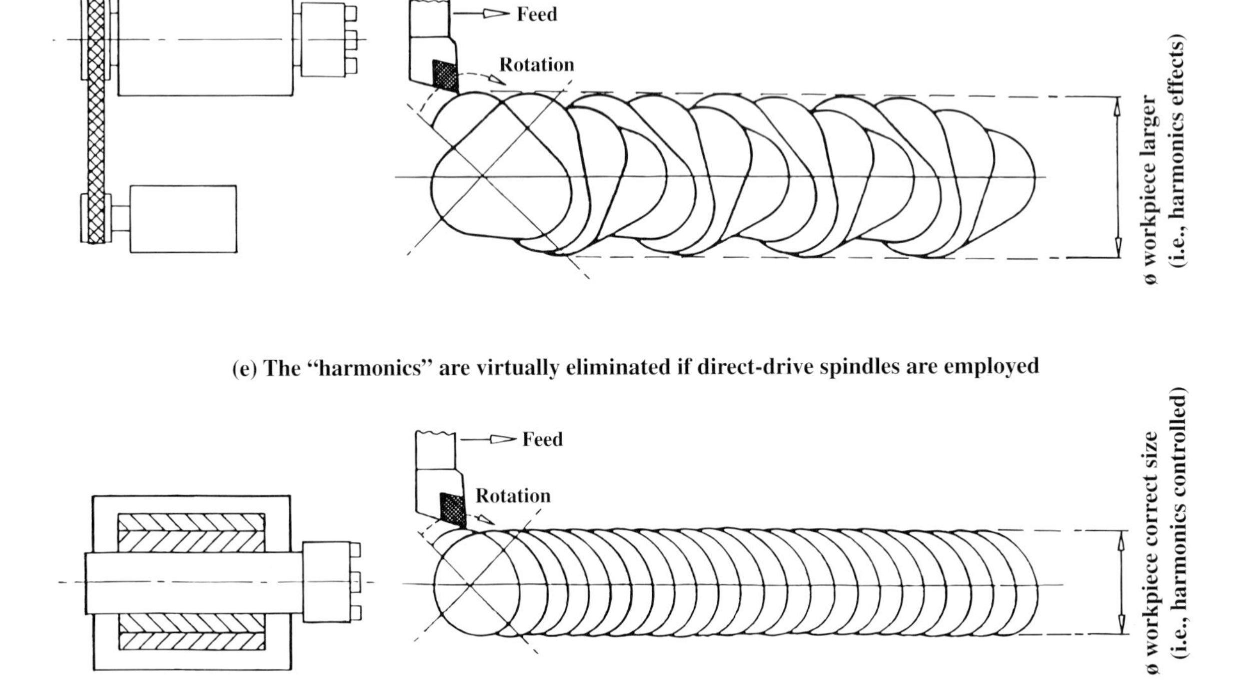

Figure 181. Improvement in roundness quality, via "harmonic suppression" occurs if direct-drive spindles are utilised.

effect of the workpiece, allowing the problem to be somewhat simplified. In Figure 182 the overall machine–workpiece–tool system can be isolated to consider the simple effect of a tool that is either inadequately supported or the unlikely occurrence of too small a cross-section making it under-strength. The main cutting force in turning operations is the tangential force, resulting from several factors, such as:

- *resistance to rotation* – caused by the workpiece material's shear strength;
- *undeformed chip thickness* – resulting from the radial depth of cut selected;
- *orientation of cutter rake angle geometry* – this being a combination of either a positive, neutral or negative rake angle, plus the effect of tool edge preparation (if any) and the shape and size of the tool nose radius;
- *feedrate* – the feedrate, in combination with the depth of cut, will heavily influence the size of the effective chip thickness and play a dominant role in affecting the resultant turned surface texture.

In the top diagram of Figure 182 the tangential force is simplistically shown contacting the cutting insert at the point. This cutting force application causes a large bending moment to occur at the pivot, or fulcrum point, as shown. The resultant dynamic action is represented in an exaggerated form in the lower diagram of Figure 182, where the tool is elastically deflected in a downward manner by this bending moment. As the resistance to deflection increases with the tool's downward direction, this causes increased pressure from the inherent tool body mechanical strength, enabling a certain degree of recovery and hence a partial upward motion of the tool. This cyclical upward and downward tool point motion is repeated at a periodic medium frequency, causing sinusoidal motional effects to be reproduced on the turned surface. High-frequency harmonic behaviour can be superimposed onto the medium-frequency harmonics and can be shown to good effect in the power spectrum analysis of harmonic behaviour during machining. This subject will be discussed later in a relevant section in this chapter.

Boring operations tend to be somewhat less rigid than turning operations because the rigidity decreases by the cube of the distance, as the following equation predicts:

$$f_0 = \frac{\pi\sqrt{3\,EI}}{L^3(M_t + 0.23M_b)}$$

where f_0 = normal force acting on the "free end" of the cantilever, M_b = modulus of elasticity of the bar, $E.I$ = flexural stiffness (I = cross-sectional moment of inertia), M_t = boring bar mass and L = length of cantilever.

Therefore, for large length-to-diameter ratios, the boring bar's rigidity will influence either the harmonics of the hole profile or its surface texture. If the boring tool is too long and unsupported by either bearing/burnishing pads, or via additional end support – as is normally the case for "line-boring" operations – then geometric hole and form deviations will inevitably occur.

If moderate bar overhangs are utilised in the hole generation process, then the operation can correct hole deviations resulting from either the manufacturing process (such as cored holes in castings) or drilled hole deviations (promoted by a drilling operation). In Figure 183 a schematic representation of the helical wandering of the drilled hole is depicted, along with the correction for geometric error resulting from boring hole generation. Notably, as the drill progresses through the workpiece material, minute variations in its geometry cause the drill to helically wander, producing a regular but undesirable eccentricity to the hole, which needs to be corrected by another operation such as boring. This hole correction is necessary because the drill's centre line follows the path as indicated, "visiting" the four quadrant points as it spirally progresses through the part. Hole eccentricity along with harmonic departures from roundness can be excessive if lip lengths and drill point angles are off-centre. Moreover, the combination of drilling faults produces drill out-of-balance forces; these can result in an oversized hole, this effect being exacerbated by high-volume production demands and necessitating significantly greater drill penetration rates. The cross-hatched circular areas represent the excess stock material removed in the passage of the boring bar along the hole's length, as the boring insert corrects for harmonic departures from roundness. In this case, if the boring bar is relatively rigid it will not be unduly affected by the variation in material wall thickness as it bores out the eccentricity caused by the drill's wandering action.

Cutting forces and tool geometry affects on roundness

The cutting forces resulting from a combination of the material's shear strength, undeformed chip thickness, tool geometry and accompanying nose radius in turning operations have a significant affect on the harmonic departures from roundness of the component. In order to show the effect of these variables in the cutting generation process and to simplify the

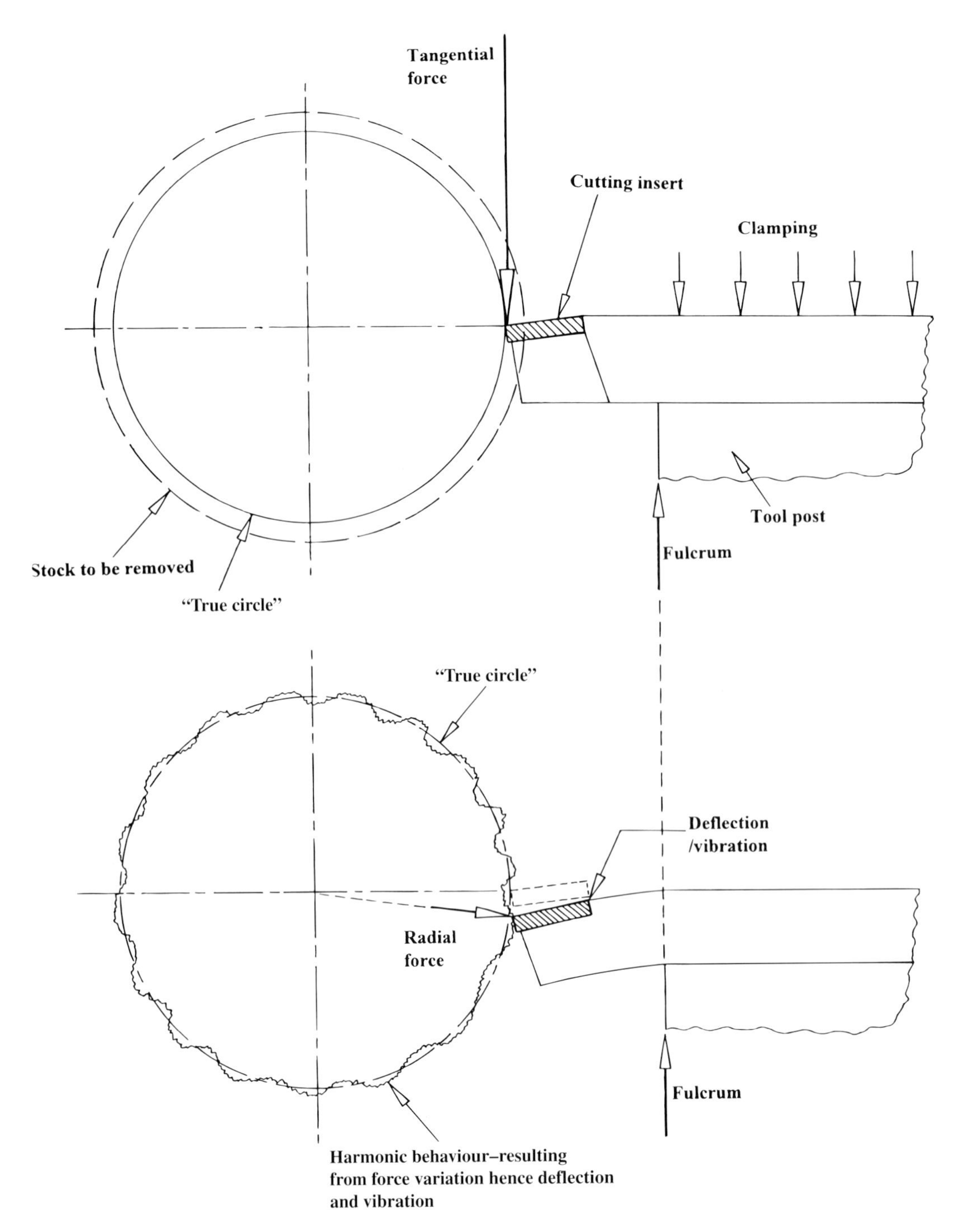

Figure 182. Harmonic departures from roundness resulting from rigidity/damping effects while turning.

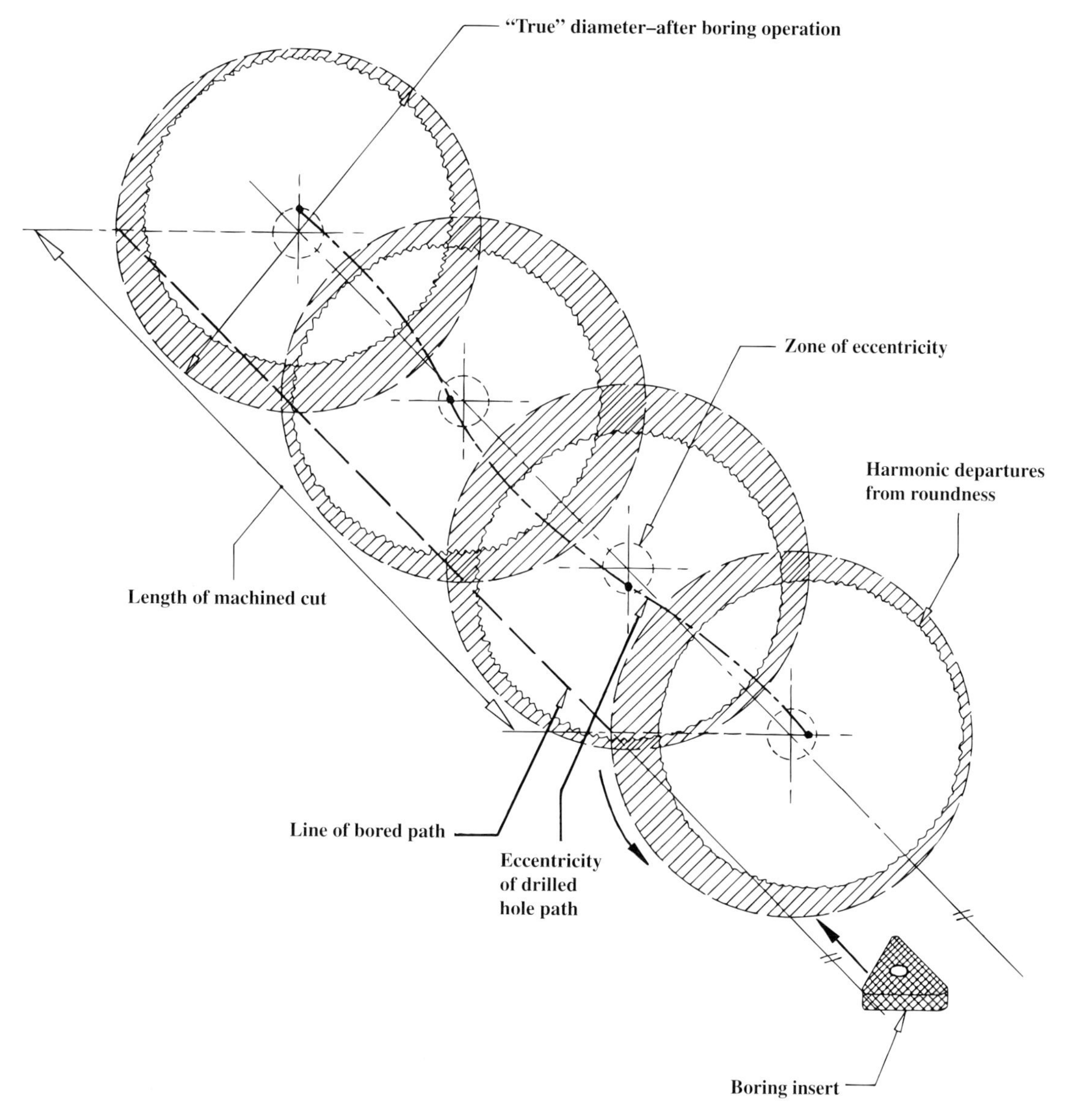

Figure 183. Harmonic and geometric corrections by a boring operation previously caused by the helical drift of the drill's path through the workpiece.

discussion, only external-diameter turning operations will be discussed. In Figure 184 a typical precision turning operation is shown partially completed, with the "light-turning and finishing" tool progressing along the part. The turning situation here shows that a long slender part is held in some work-holding device such as a collet, or a chuck at the headstock end, with additional support supplied by the "dead centre" held in the tailstock. The general trend today is to fit a "rotating centre" in the tailstock, but this can introduce its own eccentric error into the turning process. Rotating centres have become popular because of the high rotational workpiece speeds that can be employed, but this is at the expense of less rigidity within the work-holding arrangement.

As the orthogonally oriented insert (zero plan approach angle) turns along the part, a "moving step" is seen to be present as the "emerging diameter" occurs to the desired dimensional size. However, if a very high quality part is required, then it is necessary to look at the turning operation more critically as some unexpected and unwanted features are present in the final part. As the turning insert has an orthogonal orientation to the axis of rotation of the part, it might be thought that no radial force component occurs, but this is not the case as the tool nose radius can introduce a radial force that can affect the turned surface. The radial force has little effect on the part's harmonics close to the support provided by the tailstock, as shown by the cross-sectional harmonic and surface texture effect indicated in section "C-C" (Figure 184). As the tool progresses along the part the influence of the contribution of the tailstock's support lessens and hence the effect of the radial force component increases, as shown in section "B-B" and to a greater extent in section "A-A". Here, the harmonic departures from roundness is significant and has been recognised for many years by precision turners. Such experienced machinists fit either a "fixed-steady" or the more preferable "moving-steady" close to the tool cutting zone – on the opposite side of the workpiece – to counteract this radial force problem and to minimise component eccentricity/runout. On many sophisticated turning centres "programmable-steadies" can be employed to overcome the problem of turning long length-to-diameter ratios, where there is a significant tendency to introduce negative out-of-roundness effects into the production process. If the turning centre is equipped with twin turrets, then the technique of "balanced turning", using a tool situated in the top and bottom turret with one cutting edge slightly ahead of the other, virtually eliminates the radial force affects.

If the tool is inadvertently set either too low or high then this will cause either "barrelling" along the part, or the "candlestick effect", as it progressively narrows toward its centre, before increasing in size in the same manner as the tool travels along the workpiece's length. Similar effects can occur if the tailstock is not directly aligned to coincide with the

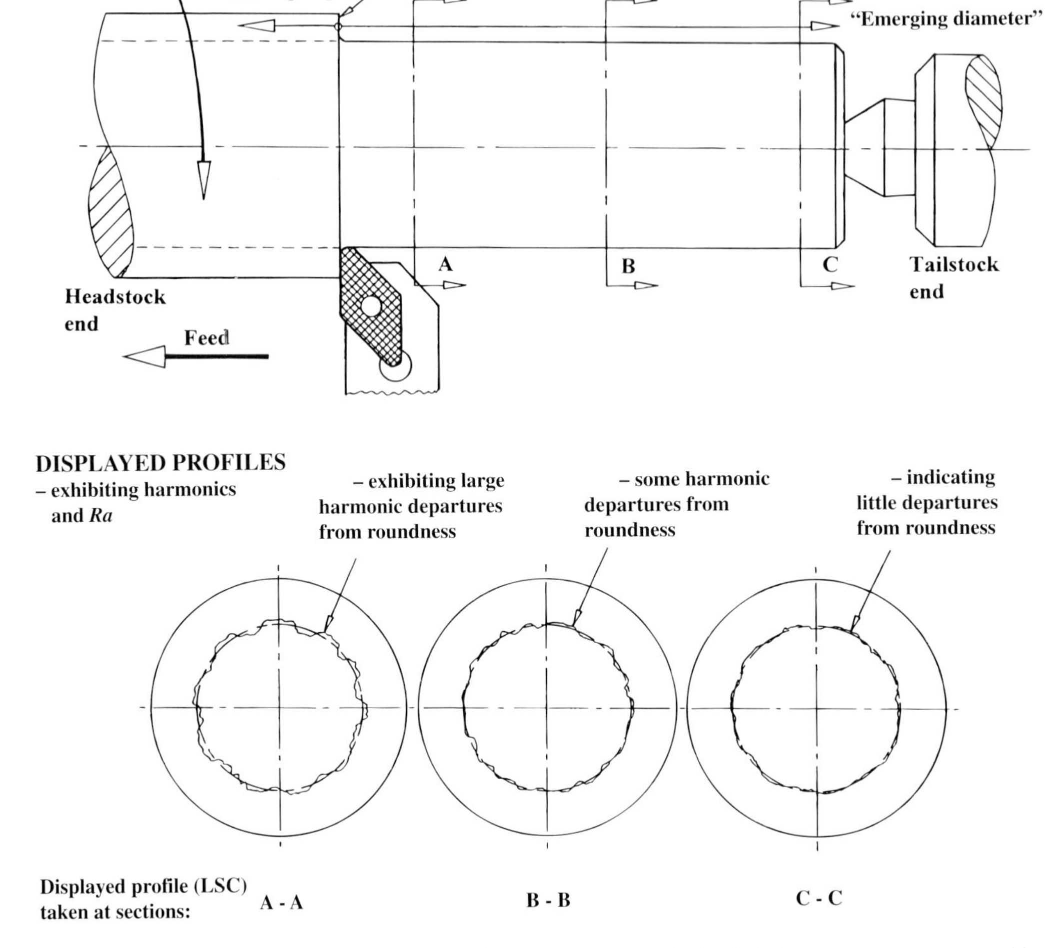

Figure 184. Effect of a combination of several and factors influencing roundness, including lack of support and a non-integral headstock spindle drive system.

headstocks centre. These setting-induced errors compound the problem occurring as a result of the inadequately supported cutting process. The machine tool alignment problem can be simply rectified by using calibration artefact-based or laser-based techniques, but this topic is outside the scope of the present discussion.

Cutting forces can be directly linked to the geometric shape of the cutting insert, this effect being illustrated in Figure 185; in these cutting force representations the tangential or cutting force is not considered in the diagrams. If the overall cutting conditions remain the same – rotational speed, feedrate, undeformed chip thickness, workpiece material and insert rake angle – then the only variable in the process will be the shape of the cutting insert. The forces significantly vary in magnitude due to the plan approach angle. In the case of the orthogonal insert (0°) plan approach angle the axial force dominates, this being associated directly with the feedrate. A very small radial force is present, resulting from the small nose radius on the insert, and its effect on the harmonic trace (A-A) is virtually negligible. If a triangular cutting insert geometry is utilised, in this case with a 15° plan approach angle, then there is a slight reduction in the axial force component and a corresponding increase in its radial counterpart. This minor increase in radial force in combination with a slightly longer cutting edge in contact with the workpiece's transient

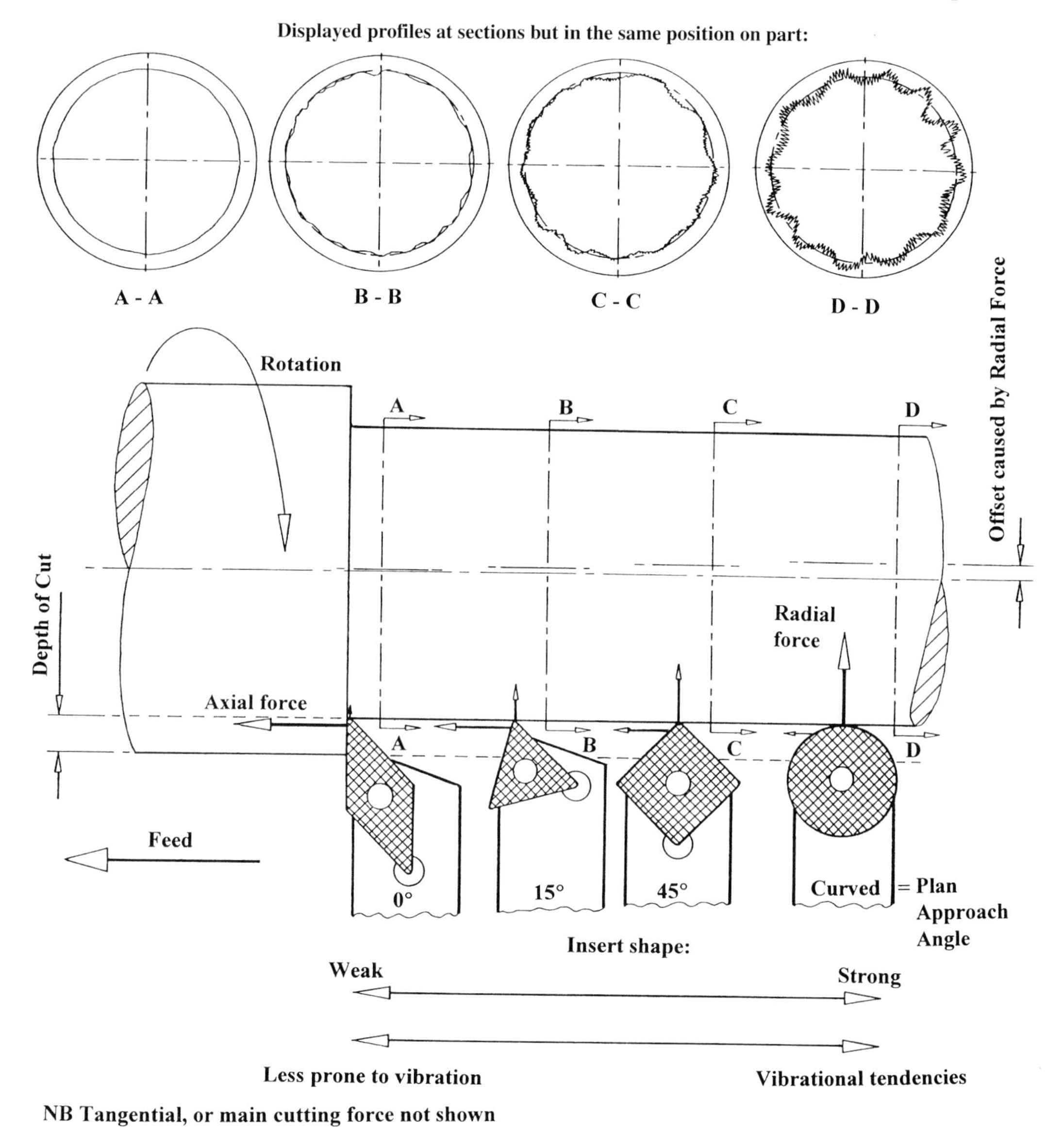

Figure 185. The effect that insert shape and its approach angle has in influencing the cutting forces, hence harmonic departures from roundness.

surface (the workpiece surface to be removed during the next rotation) leads to a marginal increase in harmonics on the polar trace (B-B). As the obliquity of the insert's plan approach angle increases, as shown by the square-shaped cutting insert inclined at an angle of 45°, then the axial and radial force components equalise. The considerable radial force component in this case has a significant effect on the polar plot as shown in C-C, where the harmonics have increased along with a notable vibration tendency that is superimposed onto the primary harmonic. An increase in vibration during cutting is the result of two principle factors: firstly, the length of the transient surface has increased; and secondly, as a result of this first condition, the part rigidity is compromised, leading to this exacerbated workpiece roundness.

If a round insert geometry is selected, this leads to a vast increase in the radial force component, which in turn causes significant harmonic out-of-roundness in the resulting polar plot (D-D). Here, the polar plot shows major sinusoidal tool motion (see Figure 182, lower diagrammatic representation) with a significant increase in vibration present on

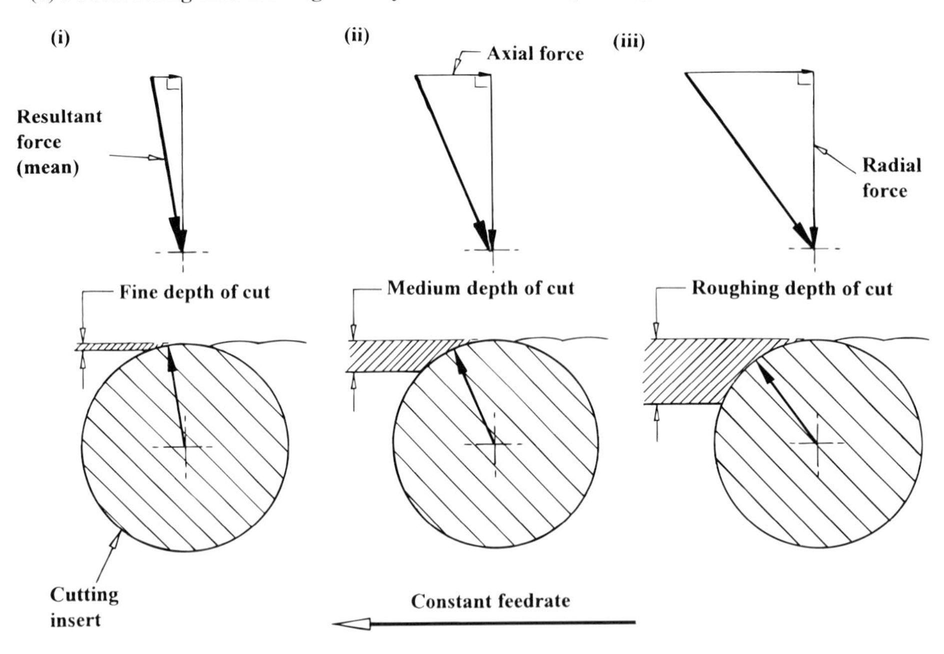

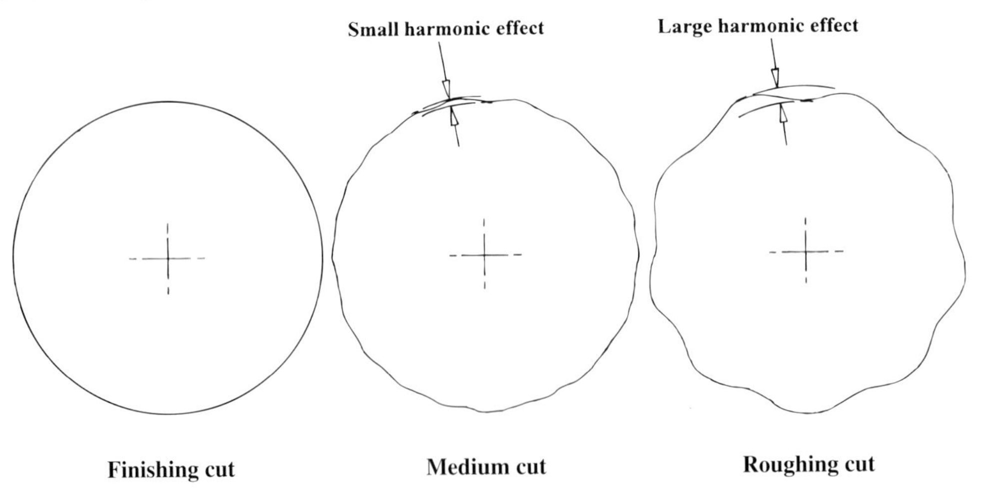

Figure 186. Harmonic departures from roundness resulting from the radial cutting force pressure on the component.

this trace. The reason for this notable increase in harmonics using round inserts is the product of several factors. The transient surface in contact with the tool has markedly increased and the plan approach angle at the tangency point is at maximum; the combination of these two factors leads to a momentous deterioration in workpiece rigidity and, as a result, vibration especially is increased.

The problem associated with the harmonic behaviour of cutting insert geometry is of less importance for large stock removal operations; roughing cuts as a strong insert can be employed, hence the use of either square or round inserts. If vibration is a potential problem, then the light turning and facing insert of the "trapezoid geometry" shown in this case, with a 0° plan approach angle, may be the production solution.

The round geometry insert is worthy of closer investigation and a more detailed sketch of its effect on the harmonic machining behaviour is indicated in Figure 186. Here, identical rotational speed, workpiece material but differing undeformed chip thicknesses (depth of cuts) are utilised to isolate variability in the machining process. In Figure 186(ai) a small undeformed chip thickness is used, creating a significant radial force in combination with a minute axial force component, leading to the polar plot shown in Figure 186(b, for a "finishing cut"). The argument put forward above, concerning the influence of the large radial force component leading to poor harmonics in Figure 185, is not contradicted here, because the cut depth is very small, hence transient surface contact will also be of little influence.

To give an impression of the influence that the cut depth plays in the magnitude of cutting component forces, the following discussion needs consideration. As the undeformed chip thickness increases from the left in Figure 186(ai, aii to aiii), then the arc of contact along the transient surface also proportionally increases. This larger effective cut length creates a significant departure from roundness harmonic behaviour on the polar plot, as depicted in Figure 186(b, for a "roughing cut").

Interpolation and its effect on harmonics

In CNC milling operations circular features on prismatic components such as bosses can be produced in a number of ways, most notably by milling using the circular interpolation function. This CNC function allows accurate circular control of two slideways simultaneously, while the cutter mills around the part, as illustrated in Figures 187 and 188. Cutter rigidity plays an important role in the quality of the final machined feature, being based on the "rigidity square rule", which states: *milling cutter rigidity decreases by the square of the distance from its holder.* In reality what this is attempting to infer is that if a cutter was originally 50 mm long and a cutter of twice this length is fitted instead (100 mm long), then the rigidity of the complete assembly will be four times less rigid. This lack of milling cutter rigidity will cause a number of unwanted effects to appear on the final part feature. Cutter deflection will introduce distortion to a square-milled shoulder and harmonic variation, degrading the departure from roundness of the workpiece, as illustrated in Figure 187. In order to introduce minimal changes to the milled profile cutter lengths need to be kept to a minimum, conducive with correct operational practices.

In order to fully appreciate the significance of the schematic diagram shown in Figure 188, featuring milling by circular interpolation and its relationship to the circular feature produced, a slight digression into basic machine tool-induced errors is necessary. Machine tools that are typified by the popular three-axis vertical machining centre configuration of conventional orthogonally orientated slideways (square to one another) can introduce considerable error into the quality of the final part. It has been established that three orthogonal slideways – *X*- and *Y*-axes in the horizontal plane, together with the *Z*-axis in the vertical plane – can introduce up to 21 kinematic errors into the cutting process. The kinematics are quite complex for any machine tool having the ability to move all its axes simultaneously, but these small though important errors can be simplistically said to occur as a result of:

- *six linear motions* – produced by displacement of the forward and backward motion of the *X*-, *Y*- and *Z*-axes slideway movements, introducing particular non-linearities into slideway positioning;
- *three rotational motions* – yaw, pitch and roll on each axis. *Yaw* is the side-to-side "crabbing motion" along the slideway. *Pitch* occurs through waviness in the slideway, introducing a backward and forward rocking (pitching) action, normal to the slideway, as the moving element traverses along the axis. *Roll* may be introduced by two adjacent ways on the slideway not being coincident (laying in the same respective plane), causing an upward and downward pivoting action with respect to the "line-of-sight" along the axis, as the moving element traverses along its length;
- *three squareness errors* – these errors occur due to the fact that each axis may not be normal (square) to one another.

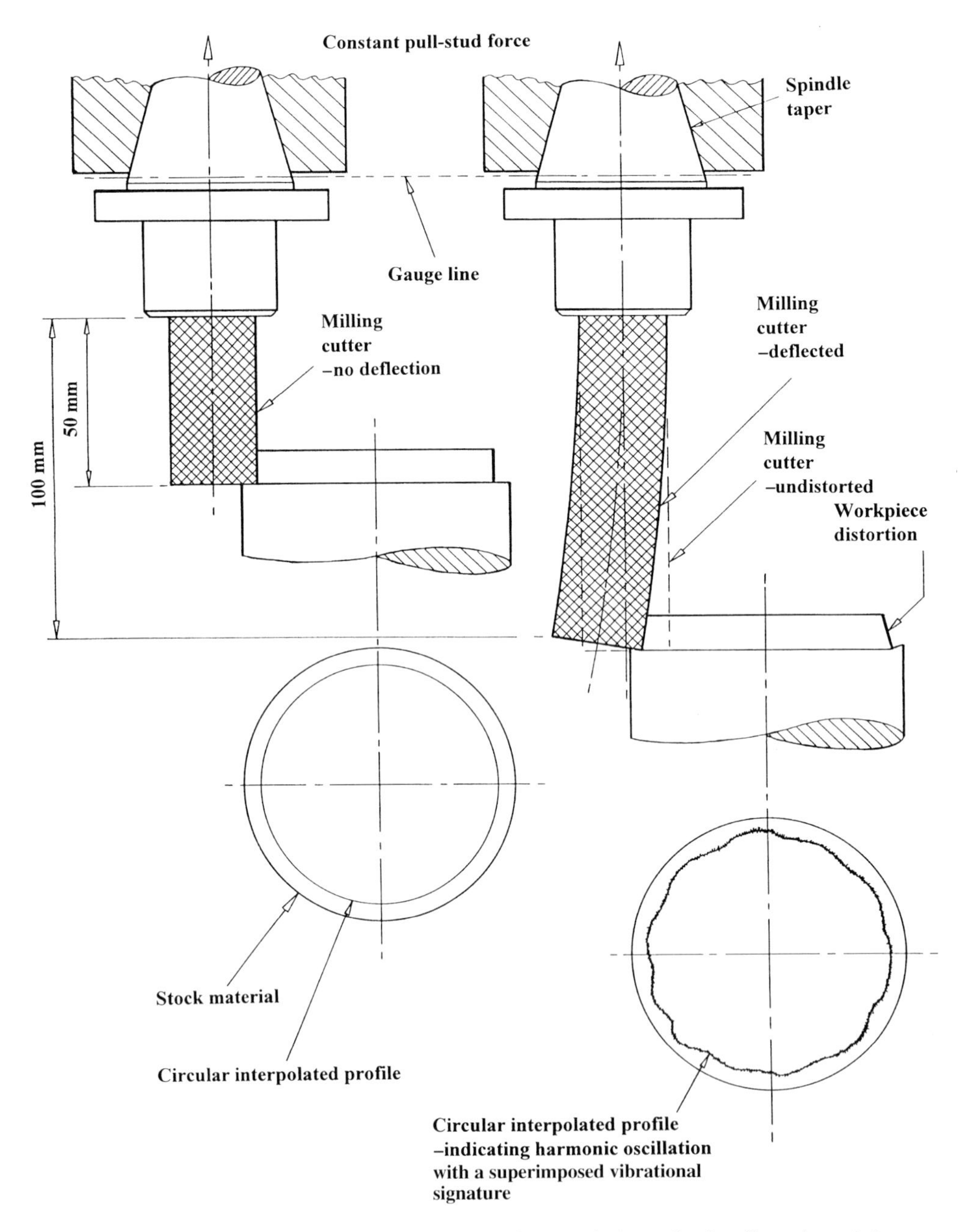

Figure 187. Exaggerated effect of cutter length on the resultant circular interpolated profile on the workpiece.

These kinematic machine errors can be appreciably reduced by the application of calibration through laser-based techniques, or to a lesser extent via ballbar artefact-based methods. These machine tool calibration techniques are outside the scope of the present discussion; the same can be said of the thermally induced errors and how they can also influence the machined part surface and profiling qualities. Moreover, *error-mapping techniques* and in-process control by an associated *dynamic error compensation system* have been shown to extensively reduce the effects of the variety of errors that can be present on the machine tool, but once again this topic is outside the scope of the present discussion.

Considering the circular interpolation of milled profiles as shown in Figure 188, the departures from roundness of the workpiece is a function of the previously discussed kinematic machine-induced and thermally induced errors, together with load-induced errors. The diagrammatic representation

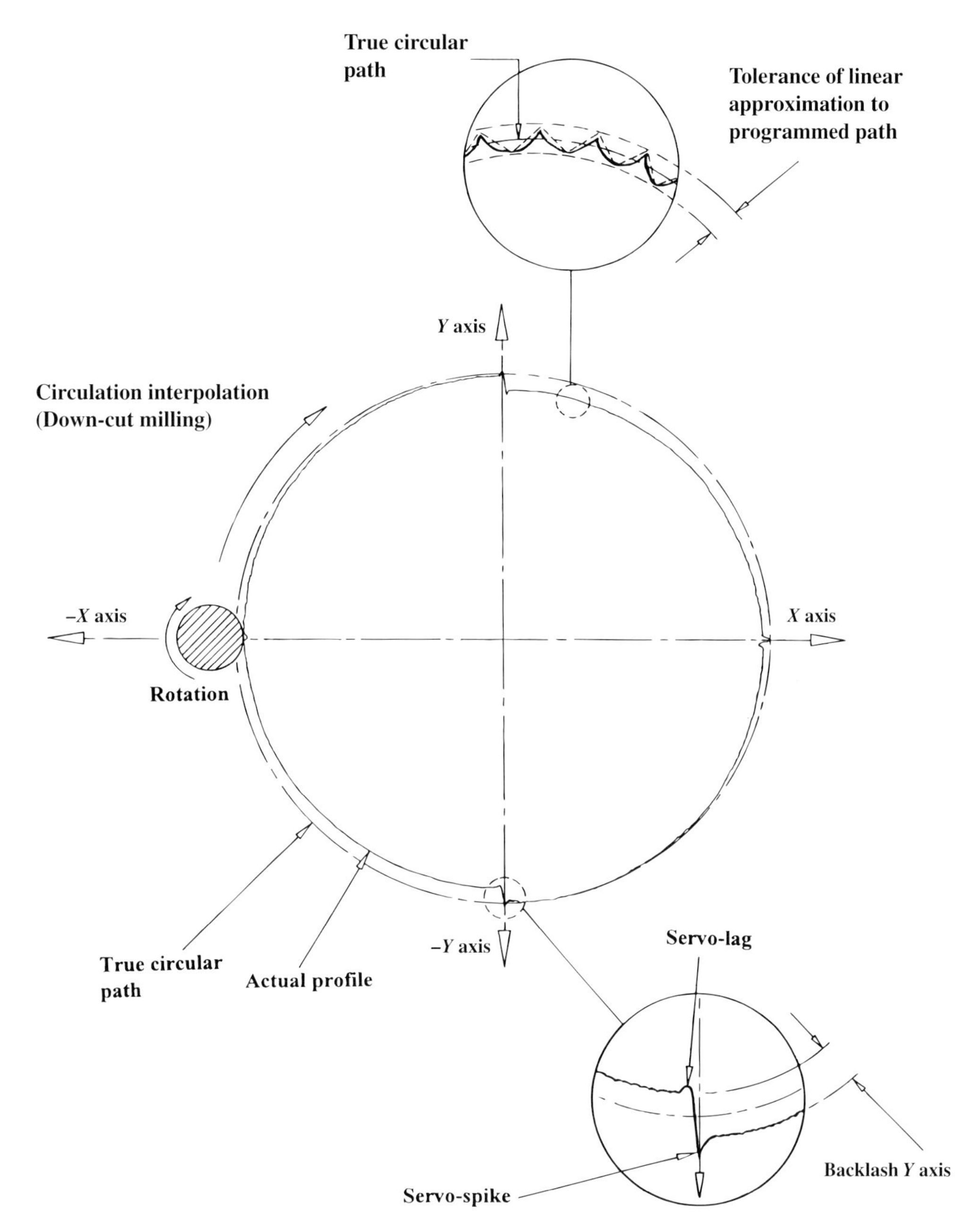

Figure 188. Generated circular interpolation exaggerating the calibration/machine tool errors associated with climb milling.

shown in Figure 188 indicates that several of these errors on the circular profile of the workpiece are present. As the simultaneous motions of the two axes occur (to produce the circular feature), then by increasing the interpolation speed for the cutter the roundness will degrade; this is the result of several interrelated factors, some of the most notable being:

- *servo-spikes* – these occur at the axis transition points at their respective 90° angular intervals. Servo-spikes result from a reversal of one of the axes at this angular position and a motor power surge (spike) occurs with a corresponding localised slack here, as take-up begins (see Figure 188 at the quadrant points on the circular feature);
- *backlash* – this is present here in some form at the axis reversal position, because of the forward and backward motion at the axis transition points, resulting in "play" (backlash) in the ballscrew (attached to and controlling the slideway

motion). This backlash is present despite the fact that a recirculating ballscrew on each axis has been utilised that should have previously been pre-loaded, which was meant to negate such backlash problems;

- *servo-errors* – when both axes are moving simultaneously, their respective linear speeds should have been matched, allowing either a perfect circular feature or partial arc to be reproduced. If any non-synchronised motion occurs (servo-mismatch) between these two axes, then an elliptical profile normally set at a 45° angle would occur. If the contouring interpolation is changed from a clockwise to an anti-clockwise circular feed motion, then the angular elliptical profile shape will "mirror-image" (flip) that of the opposite profile;
- *squareness* – if orthogonality (squareness) is not maintained between the two axes, then this will result in yet another similar milled angular elliptical profile (as discussed above, for servo-mismatch), but the difference here is that the shape orientation does not change (that is, reverse its shape) if either clockwise or anti-clockwise feed motion is applied.

Considerably more factors can affect a milled circular interpolated shape and many of these machine tool-induced errors can be diagnostically interrogated by using dynamic artefacts, such as the ballbar. Such instrumentation can find the sources of error, list their respective magnitudes and apply corrections that can be fed into the CNC controller to nullify the errors, enabling circular contouring milling performance to be significantly enhanced as a result.

5.2.7 Power spectrum analysis of machined surfaces

In many situations of a stochastic (variable) machining nature, the power spectrum is a useful aid in process control monitoring of the cutting capabilities and gives a good interpretation of the anticipated surface topography. Once again the turning operation, because of its relative simplicity of a single-point machining operation, will be used to show how the power spectrum can be employed to identify tribological factors such as tool wear and how this influences the surface topography of the workpiece. In Figure 189 the main feature here is that it is possible to identify and then quantify tool wear by an examination of the power spectrum. To the left-hand side of Figure 189(a) a tool is shown when new, then it progressively degrades with time (in-cut) until it becomes completely worn out. In Figure 189(b, top) the surface profile appears as a series of periodic and regular machined cusps, indicating that an efficient turning operation has occurred, which can be confirmed by the power spectrum in Figure 189(c). This turning operation's power spectrum, as expected, indicates distinct frequencies, the fundamental frequency relating to the feed spacing of the cusps and the harmonics to the non-sinusoidal tool geometry – its shape. After successive passes along consecutive workpieces the tool begins to show signs of localised insert flank and face damage, with the ratio of the height of these harmonics to that of the fundamental frequency increasing. This harmonic increase is a direct result of the imposition of the insert's imprint in the turned surface. The increased harmonic activity here occurs due to the wear scars that correspondingly occur on the insert's cutting edges; these are then faithfully reproduced on the surface topography. Moreover, at this position the baseline of the fundamental frequency can be expected to rise; this increase will probably be because of random effects due to the formation of arbitrary chips and the turned surface exhibiting micro-fracturing. If it is adjudged that these surface random effects of machining are significant, it is probably a more appropriate strategy to consider their examination by utilising the auto-correlation function.

In Figure 190 (right) harmonic information within the power spectrum is depicted, in this case to the left-hand side of the fundamental frequency (feed). Here, periodicities occur which are of considerably greater wavelengths than that of the feed. These harmonics are the result of machine tool-induced problems, such as bearing chatter and slideway errors, together with other unwanted effects. The position of the harmonics related to the bearing problems appear closest to that of the fundamental frequency, whereas the very long wavelengths, for example, can be attributed to form errors in the slideway, or perhaps are the result of structural distortion of the machine tool itself.

A major advantage of using the power spectrum as a diagnostic aid is that it can separate out the process-related problems, as depicted by the power spectrum diagram shown at the bottom left-hand side of Figure 190. Hence, to the left of the fundamental frequency (feed) in this diagram, in region "A", the harmonics of the machine tool problems can be isolated out; and to the right of region "B" those of the feed frequency can be established, with the workpiece material properties occurring displaced even further to the right in region "C". The separation into discrete power spectrum frequencies can be established, although such diagnostic interpretation cannot be readily detected in surface profiles at

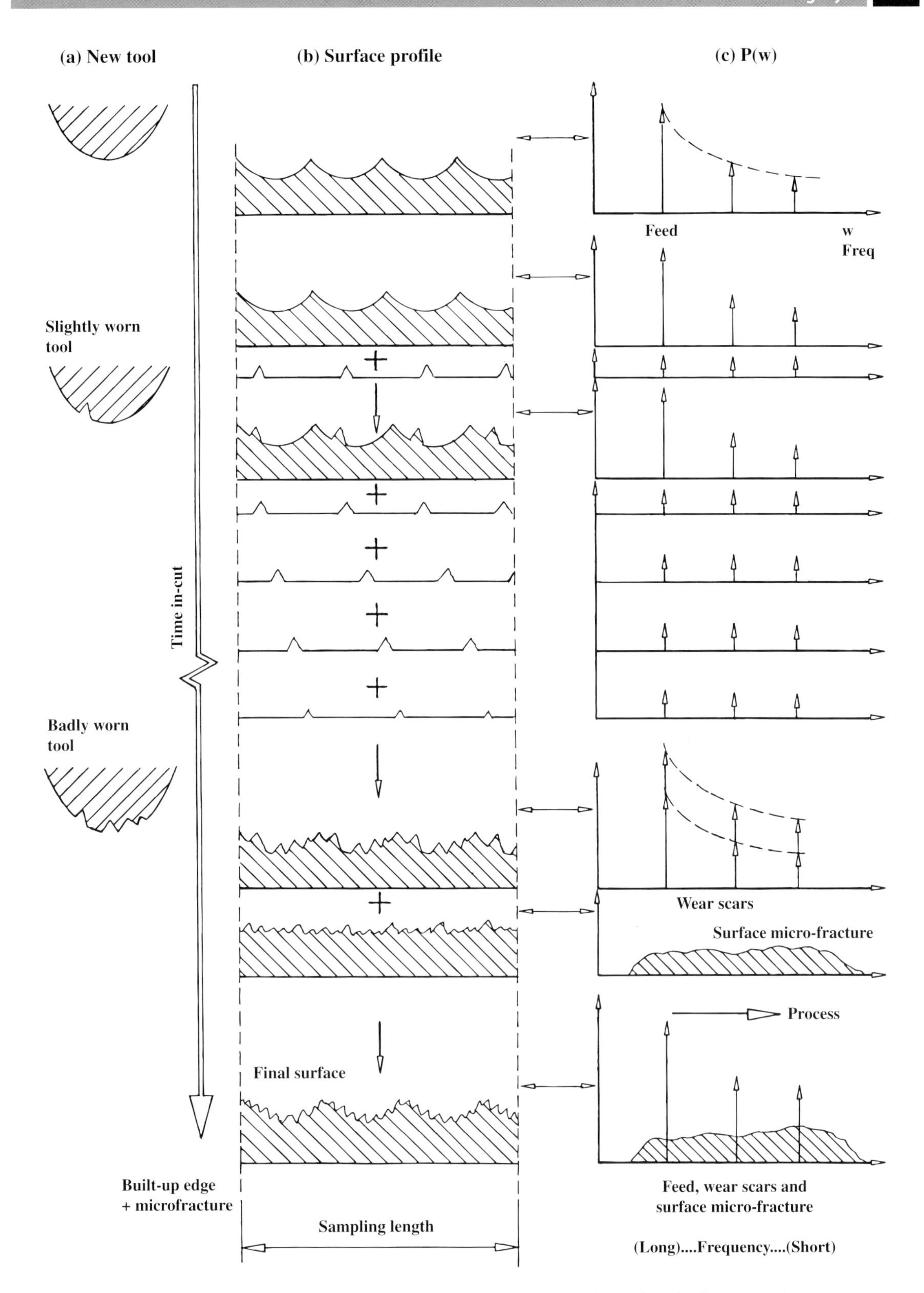

Figure 189. Power spectrum analysis harmonics for a machined surface. [After Whitehouse, 1997.]

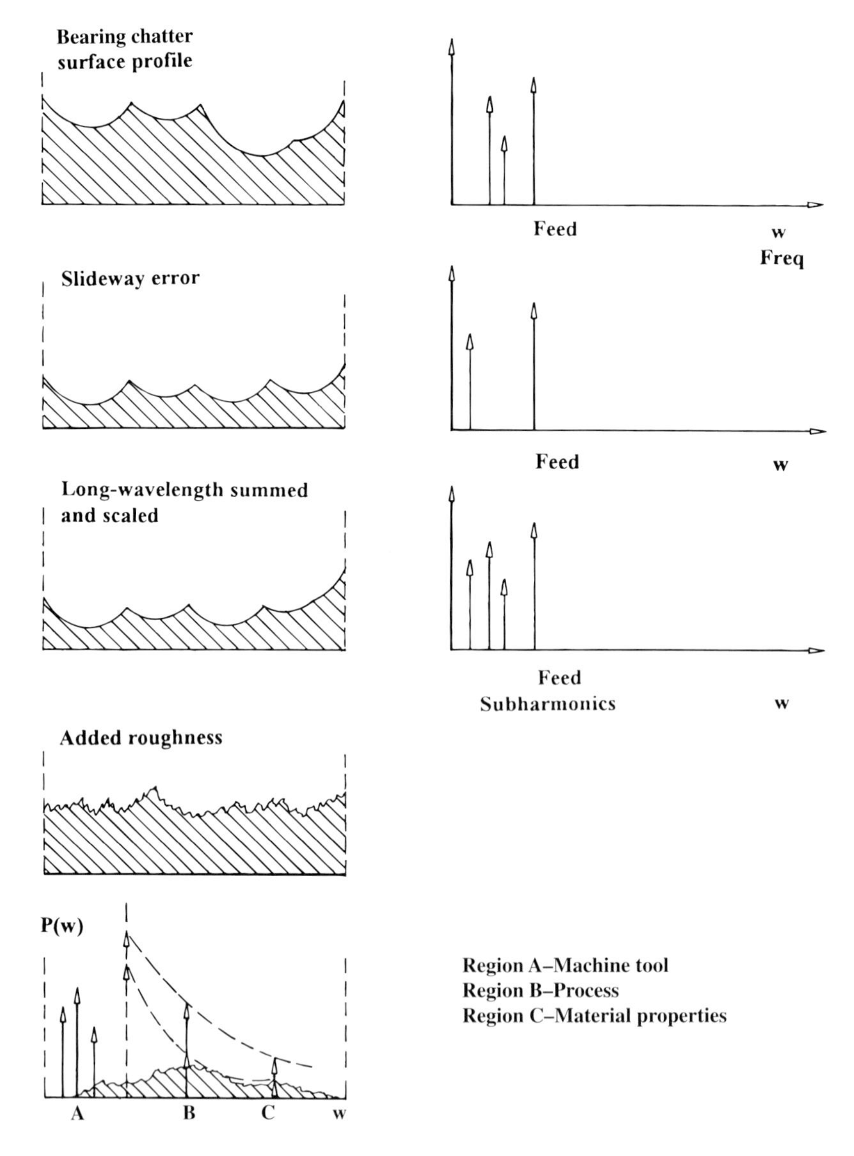

Figure 190. Power spectrum analysis, sub-harmonic and total spectrum. [After Whitehouse, 1997.]

this initial stage in the life of the tool by the senses – either visually or by its sound during machining. This application of the power spectrum to process-related manufacturing problems gives an early warning of the performance of the complete production process, enabling remedial action to be taken to rectify a minor problem before it becomes critical. A note of caution is required here, as there are occasions when even the application and interpretation of the power spectrum may prove inadequate to identify problems in production. This limitation to the use of such powerful diagnostic aids can arise in situations where either time or space varies; furthermore, if workpiece variability through unanticipated defects is present in the production batches this will nullify the technique.

5.2.8 Manufacturing process envelopes

The principal feature of manufacturing process envelopes and, indeed, for many amplitude distribution curves is that they can be approximated by the so-called "beta function" (Figure 191a). Here, the function has two parameters that are independent of one another, enabling them to be utilised as a means of characterisation. The notation a is the allocated weighting for the profile ordinates measured from the lowest valley and above, with the notation b being given to weighting the profile from the highest peak down. Hence, peaks and valleys have accordingly different weights. One of the problems

(a) The beta function: (i) symmetrical and (ii) asymmetrical case. (After Whitehouse, 1994)

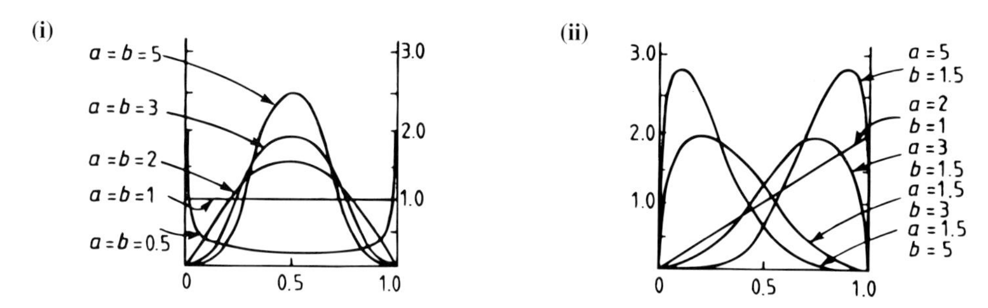

(b) Skewness and kurtosis relationship, illustrating different manufacturing process envelopes

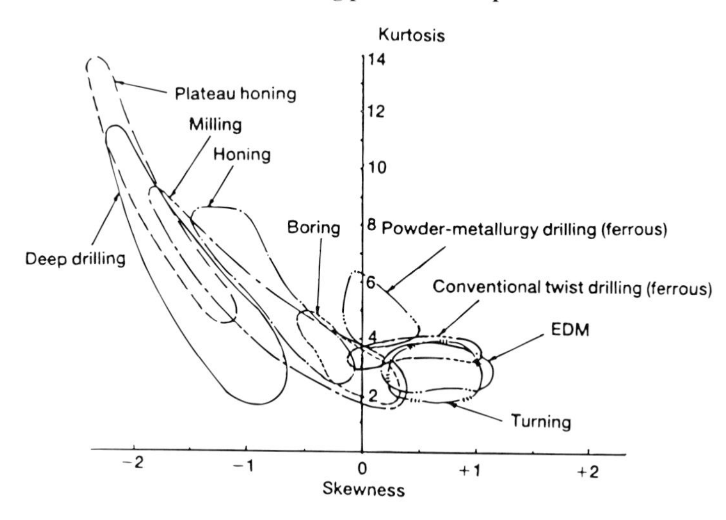

(c) Skewness: kurtosis graph, showing the functional envelopes

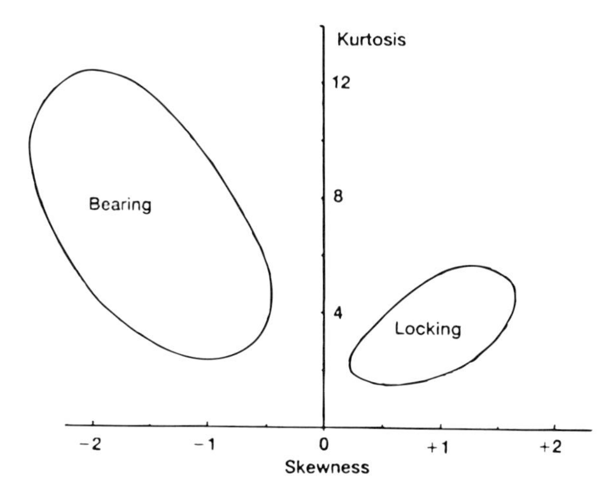

Figure 191. "Beta function" and typical production process envelopes.

that has arisen from using this technique for a topographical profile, which has discredited it for certain applications, is how and in what manner can one determine a and b. The beta function is normally defined within a set range of $0 \rightarrow 1$, being expressed in the following manner:

$$\beta(a, b) = \int_0^1 z^{a-1}(a - z)^{b-1}\,dz$$

If by changing the range of the beta function equation above, from $0 \rightarrow 1$ to $Rp + Rv$, or indeed with Rt, then substituting σ (the standard deviation of the distribution) with Rq, the beta function parameters a and b become

$$a = Rv\left(\frac{Rv\,Rp - Rq^2}{Rt\,Rq^2}\right)$$

$$b = Rp\left(\frac{Rv\,Rp - Rq^2}{Rt\,Rq^2}\right)$$

The reality that any dominant peak or valley within the assessment length is only raised to a unit power infers additional stability over the skewness/kurtosis approach. The trouble with this method is in accurately determining sound results from Rv and Rp, which confirms the problem that obtaining information from peak/valley measurement and then deriving valid information from them can be fraught with difficulties. In Figure 191(a) both symmetrical (i) and asymmetrical (ii) graphs for the beta function are illustrated, based on a class of Pearson distributions. In the symmetrical case the skewness equates to zero; conversely, for an asymmetrical series of results, skewness can be either positively or negatively skewed. Nevertheless, even allowing for these limitations, an example of a range of the manufacturing process envelopes that can be produced using this technique is illustrated in Figure 191(b). Here the production processes can be simplistically classified and grouped into either a "bearing" or "locking" surface topography (Figure 191c). This bearing/locking grouping shows that certain production processes can achieve specific functional surfaces for industrial applications. However, the general picture is not as distinct as the grouping shown in Figure 191(c), because certain processes cannot only run across each classification but may fall between both the bearing and locking envelopes. A typical "intermediate envelope" that normally occurs in just this vicinity of the skewness/kurtosis graph would be machined surfaces manufactured by the PM route, as shown in Figure 192. Here, in both a twist and split-point drilling operation, the porous nature of the PM compacts creates pores that are open to the "free surface". Moreover, with the remnants of the periodic saw-toothed profile (produced by the effect of the partial lip and margin of the drill), this gives a negative skewness via the pores and positive skewness introduced by drilled cusps (saw-tooth effect). In Figure 192 the dispersion or scatter of the results for individual envelopes varies, depending upon the frequency of the feedrate and the type of drill employed. Generally, the finer the feedrate the less the scatter, this also being true for many turning and milling operations. The geometry of the drill's lips, but more specifically the profile of the chisel edge and whether it is of the self-centring type, such as will be the case for a "split-point" drill, creates less drill deflection and hence the results introduce smaller scatter to the envelope. These machining conditions introduce an intermediate-type of manufacturing envelope that is neither a "bearing" nor a "locking" variety.

Ternary manufacturing envelopes (TMEs)

In machining operations the dominant factor that influences both the resultant cutting forces and the surface topography has been shown to be the tool's feedrate. In Figures 193 and 194 the feedrate, in conjunction with the principal factors of surface texture (Ra) and roundness (by least squares circle – LSC), are used to define the limits for these TMEs. Utilising these diverse surface texture, roundness and processing parameters (feedrate) for the major axes on the ternary diagram enable a machined surface to be characterised in a new manner. As might be anticipated, these TMEs differ markedly from the more usual and restricted manufacturing envelopes previously alluded to above; the latter type of skewness and kurtosis axes might otherwise mask crucial information. The "TME approach" gives a pseudo three-dimensional graph that can be exploited to illustrate how the influence of changing a parameter such as feedrate modifies the relationship of the surface texture and roundness values in the final machined result. Furthermore, such a three-dimensional graph can be exploited to illustrate how the influence of changing the feedrate modifies the interrelationship of surface texture and roundness values.

As an example of the usefulness of this TME approach to complex machining analysis, Figures 193 and 194 for dissimilar production processes are presented. The TME graph for a particular range of turning and boring processes, indicated in Figure 193, shows how at low feedrate (0.10 mm/rev) the surface texture is closely confined to a relatively small spread of values – nominally around 0.5–1.5 μm Ra, whereas its associated roundness lies

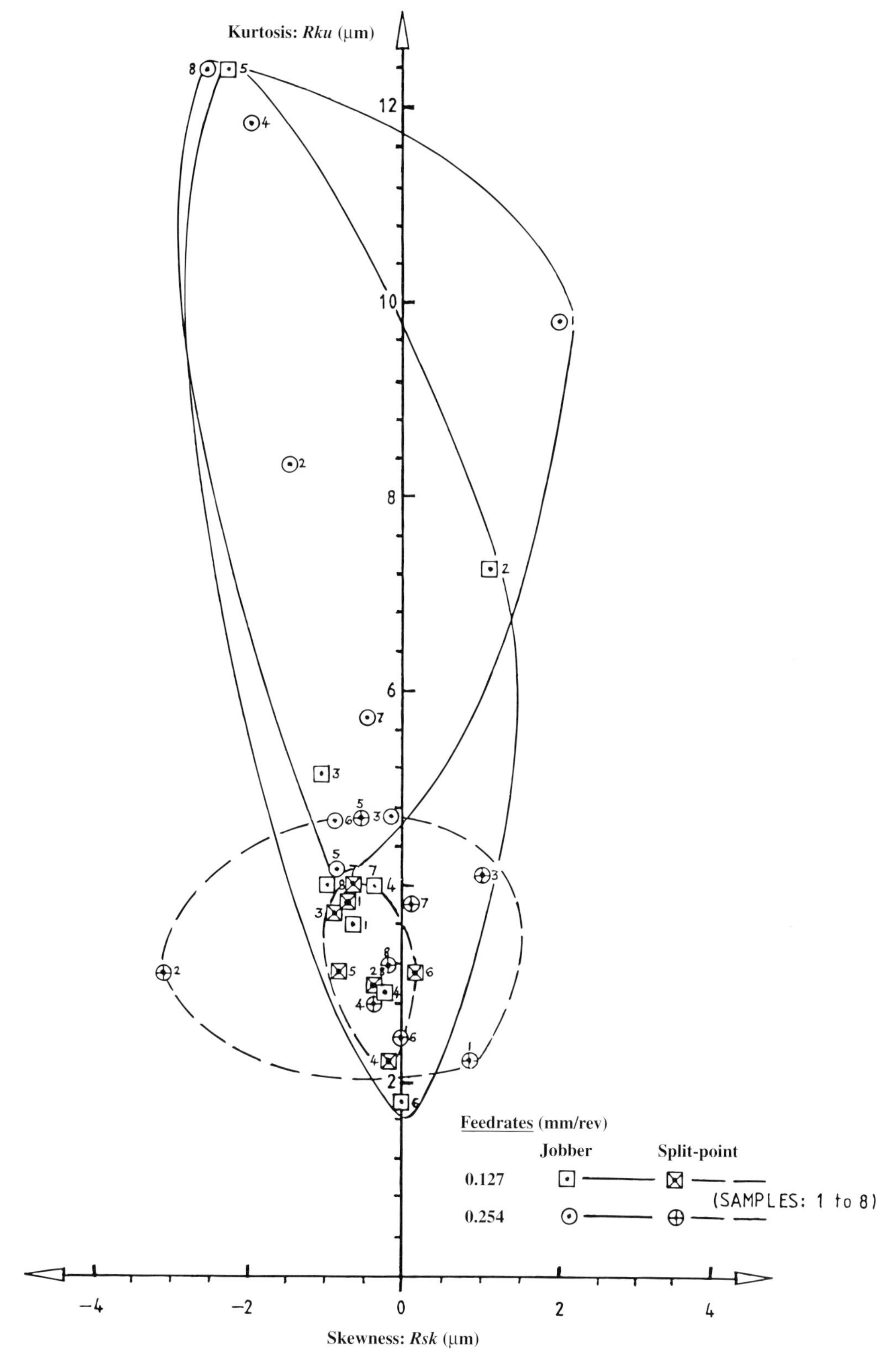

Figure 192. Manufacturing envelopes for jobber and split-point drilling operations: skewness v. kurtosis.

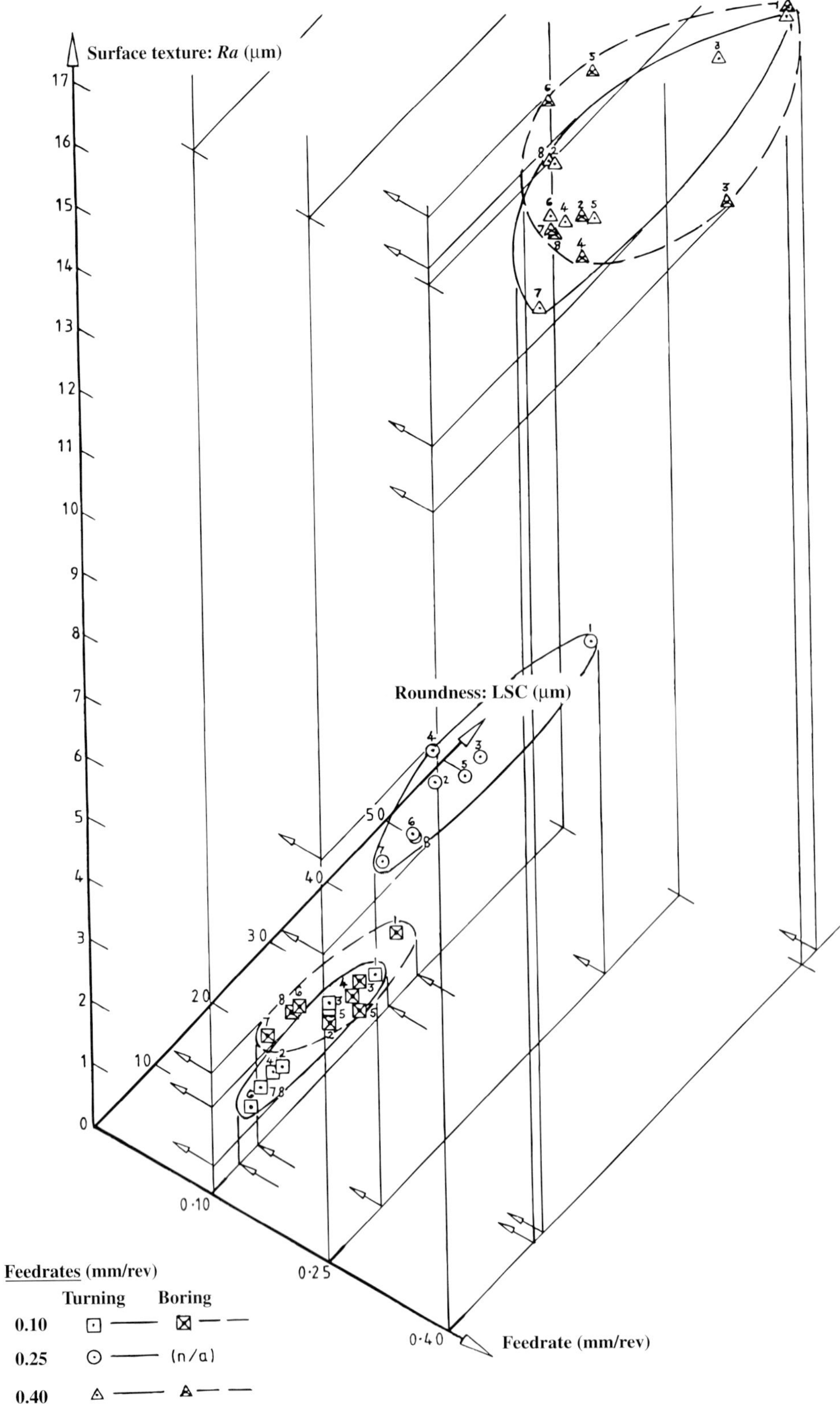

Figure 193. Ternary manufacturing envelopes of the production processes: turning and boring – feedrate, roundness and surface texture.

between approximately 5 and 30 μm LSC. As the feedrate increased in an arithmetic progression to 0.25 mm/rev, the range of the surface texture bandwidth proportionally expanded to 1.5 at approximately 5–6.5 μm *Ra*, with a corresponding increase in roundness from 8 to 48 μm LSC, giving a proportional bandwidth of 1.6. As the feedrate was raised even higher, to 0.40 mm/rev, it was not surprising to note that this also produced increases in both the surface roughness and its proportional bandwidth, with similar values with respect to its roundness.

Similar trends occurred for the boring operation on Figure 193, but here only two feedrates were employed. At low feedrate (0.10 mm/rev) the surface texture range was approximately 0.5–2 μm *Ra*, with roundness being between 8 and 36 μm LSC. Once the boring feedrate had been raised to 0.4 mm/rev, this produced an increase in the surface texture to around 13–18 μm *Ra*, with a corresponding proportional bandwidth of 33.33. Furthermore, the roundness also degraded to 15–63 μm LSC, which resulted in the bandwidth being proportionally increased to 1.7. Discussion of these proportional bandwidth differences can be used to explain why either the outside diameter (OD) turning or internal diameter (ID) boring operational performances vary. In the case of the OD turning operation (a schematic of the turning process is shown in Figure 169), as the feedrate is progressively increased both the surface texture and roundness will degrade. This expected deterioration occurs because as the feedrate increases it results in coarser machined surface topography, and as the cutting forces are proportionally higher this provides a destabilised cutting-edge condition, which subsequently leads to greater departures from roundness, resulting in secondary harmonic roundness error. There was a larger proportional bandwidth expansion for surface texture as a result of the influence of feedrate rather than for departures from roundness. This bandwidth increased in the same fashion to the increases in respective feedrates, namely in an arithmetic progression – lengthening at the rate of 1.5. Such increases in dispersion of resultant machining data can be substantiated by earlier work, which suggests that there is a geometric relationship between the insert nose geometry and selected feedrate on the respective machined surface texture. However, this proportional behaviour did not extend itself to machined roundness values, which are a function of a more complex interactive cutting path action. Such action is promoted by combined rotational and linear motion (as the workpiece rotates and the tool is fed along the part), leading to differential primary, secondary and tertiary (an so on) harmonic departures from roundness. The machined surface topography, indicated through the *Ra* surface texture trace, is only the result of linear motion (being caused by the feeding motion along the part) and it is not surprising to note that a direct relationship exists between selected feedrates and tool nose geometries.

The boring proportional bandwidth can be explained, allowing for its overhang, which must be present to avoid fouling on the workpiece. This overhang creates a reduction in stiffness based on the cube of the bar's overhanging length; when compared to the process of turning this is its OD operational counterpart. A boring tool introduces a "phase-damping" effect on the out-of-roundness and, as such, reduces the overall magnitude of the LSC value obtained. When utilising the 0.40 mm/rev feedrate, boring in this case will generate a proportional bandwidth increase in LSC of 1.7, while an equivalent OD turning operation produces a value of 1.92. The more compliant and less rigid boring bar will have a greater tendency to deflect the cut, leading to slight tool bending termed a "spring cut", introducing a form of "phase-damping" behaviour that tends to smooth and blend out the harmonic factors. Conversely, the opposite is true for the behaviour of the proportional bandwidth in machined surface texture parameters, where an increase was observed in boring with respect to turning at identical feedrates. This difference can be explained in terms of the associated boring bar overhang and its reduction in rigidity. Increased tool deflection would tend to be superimposed onto the bored surface topography, leading to greater dispersion in data points, which then increases the proportional bandwidth if compared to a similar turning operation, with both cutting inserts having identical nose radii.

The drilling TME depicted in Figure 194 shows a limited range of drilling operations using both an uncoated jobber drill and a TiN-coated split-point drill. In this case the TME shows how the generation process of dynamically drilling holes influences the simultaneous procedure of surface roughness and roundness production at different feedrates. The normal technique is to consider each metrological condition in isolation and thereby potentionally missing some vital information. In Figure 194a "comparison of extremes" can be made to establish how differing tool lengths combined with changes in point geometry influence the resulting surface texture and roundness parameters. Most surprisingly, at a 0.127 mm/rev feedrate the split-point drill's mean range of surface texture was worse than that obtained with a jobber drill, ranging over 2.8 μm *Ra*; this was possibly due to the "rounding effect" on the drill's lips and point edges. Equally, the LSC roundness range showed that the jobber drill was far worse than that for the split-point, its variability increasing by 33%, being present over 30 μm LSC. What this variation illustrates is that the jobber

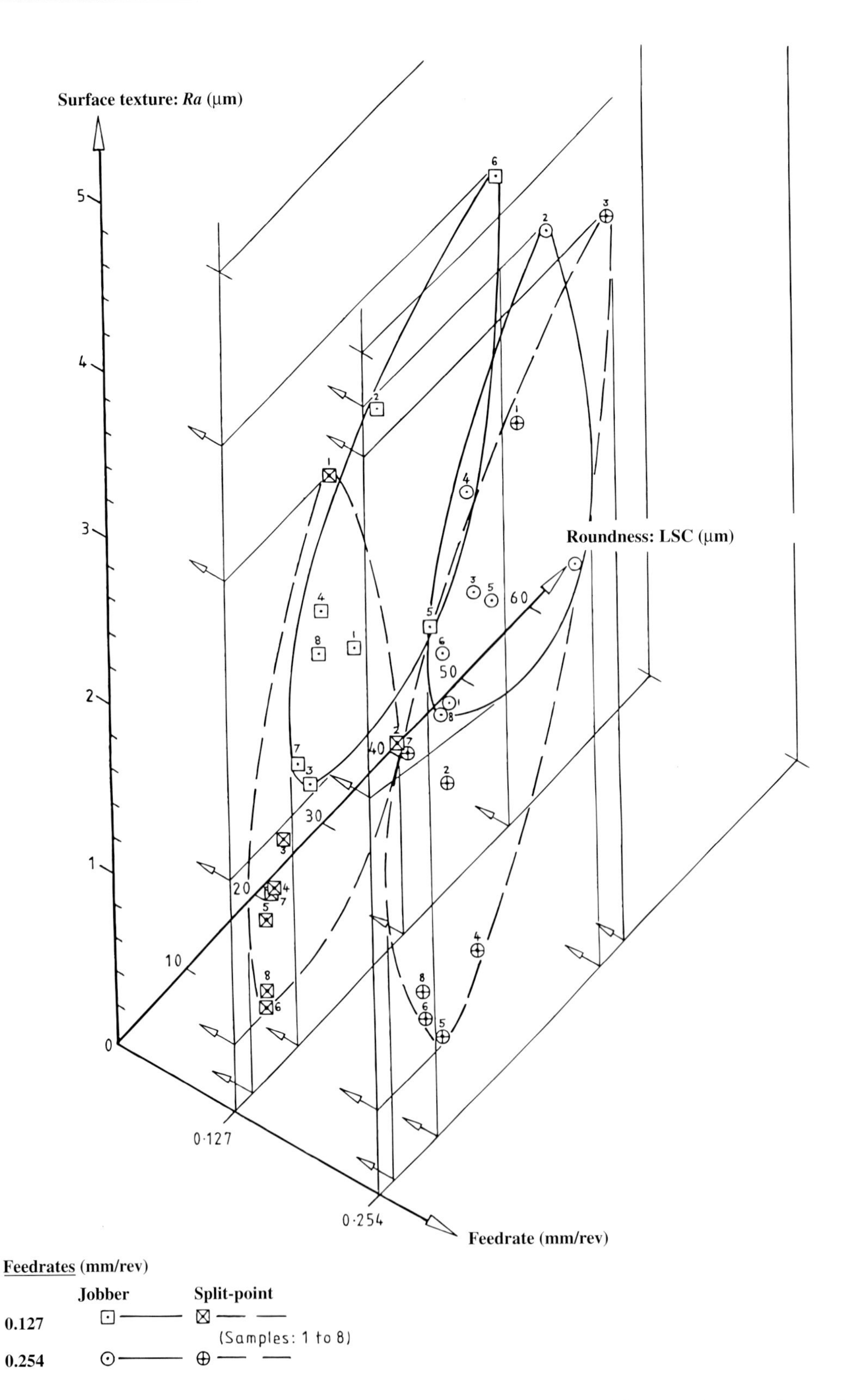

Figure 194. Ternary manufacturing envelopes of production processes: jobber and split-point drilling – feedrate, roundness and surface texture.

drill, at low feed, had a smaller and more consistent spread of results for surface roughness than that of the split-point drill, although its level of recorded roughness was higher. This drilled hole variability suggests that the slightly larger drill margin on the jobber drill burnished the resulting saw-toothed profile topography and, in so doing, reduced the *Ra* range. Hole roundness deteriorated using the jobber drill. This deterioration came as no surprise, as its longer and flatter chisel-edge profile, combined with a greater overall drill length to that of the split-point, inevitably meant that the drill produced a degraded self-centring action. Coupled to the jobber's greater length, reduced drill integral rigidity occurred; this led to inferior drilled hole roundness. Once the feedrate was increased to 0.254 mm/rev, a similar trend occurred for drilled hole roughness. The split-point drill's range was 41% larger and of similar magnitude to that of the jobber – ranging over an *Ra* of 3.9 μm – illustrating that drill burnishing was present in this case at higher feedrates. However, the trend for roundness at this higher feedrate was reversed. The split-point drill showed a larger error in the range of departures from roundness, being 25% worse, coupled to a similar magnitude – ranging over an LSC of 32 μm – indicating that at higher feedrates the self-centring action was a negligible factor.

If a comparison is now made of the surface texture and roundness parameters against feedrate, in the first instance for the jobber drill and then considering the effects produced by the split-point, a significantly different picture emerges for the drilling process. When utilising the jobber drill, and as the feedrate increases from 0.127 to 0.254 mm/rev, there is a slight improvement in the range of surface texture from 2.6 to 2.3 μm *Ra*, for the low-to-high feedrates, respectively. This 11.5% improvement in the spread of *Ra* at higher feedrate is somewhat misleading as the magnitude is greater, which conforms to the published literature. This partial yet slightly confusing improvement in surface texture using the jobber drill at high feedrate also occurred with respect to roundness, as the LSC ranged from 8 to 39 μm at high feedrate. Once again, it would seem that the wider jobber drill's margin influenced the burnishing both linearly with respect to variance in asperities and circumferentially, by minimising harmonic departures from roundness. The drilling behaviour of the split-point was more predictable. In the case of surface roughness, this ranged from 2.8 μm at low feedrate to 3.9 μm at high feed, with similar magnitudes with respect to *Ra* levels. The trend in roundness variations was much the same; at low feedrate the range was 20 μm, while at high feed it degraded to 32 μm, with similar magnitude in LSC level. This degradation is wholly consistent with increasing penetration rates.

The in-depth discussion on the use of the TME technique with their related production processes and associated metrological parameters, together with subsequent results above, was included to illustrate how machined components and specific processes can be combined into pseudo-three-dimensional diagrams. Such graphs may give a new insight into what constitutes a machined surface, indicating that surface texture parameters alone should not be separated out from those of roundness. The combination of these previously disassociated surface texture and roundness parameters may offer a new approach in the manner that the machined surface condition is assessed, allowing an appreciation of the whole surface and not just some associated parameters in isolation.

5.3 Surface engineering

Historical context

The currently accepted definition of the subject of "surface engineering" was adequately coined in the late 1980s by Professor Bell of the University of Birmingham. It states that *surface engineering involves the application of traditional and innovative surface technologies to engineering components and materials in order to produce a composite material with properties unattainable in either the base or surface material. Frequently, the various surface technologies are applied to existing designs of engineering components but, ideally, surface engineering involves the design of the component with a knowledge of the surface treatment to be employed.*

Historically, some of the earliest references on the subject of surface engineering are somewhat vague allusions to the metallurgical treatment of armour, notably quench hardening of weapons by the process of blinding the Cyclops in the *Odyssey*, by Homer. This Greek saga was written about 880 BC, but referred to an earlier Bronze Age period (1400–1200 BC). However, in the Iron Age the carburising of weapons by some form of case hardening, or more specifically carbonitriding, was widely practised. This heat treatment process was in reality an austenitic thermo-chemical procedure involving the diffusional addition of both carbon and nitrogen to austenite. This was a somewhat long-winded process and was followed by rapid and agitated vertical quenching to harden and improve the wear characteristics of the weapon, followed by tempering to enhance the toughness. Notably, in the eleventh century AD a German monk, Theophilus, in his work

Schedule diversarum artium, mentioned the word *temperamento*, meaning to quench and self-temper in goat urine – after certain complex pre-treatment of the goat! This prior management of the goat was necessary to increase the acidity level of the subsequent fluid media necessary for heat treatment and hence enhance its severity, enabling the liquid to be utilised for severe quenching applications. Although they did not know it at the time, such practical heat treatment enhanced the critical cooling velocities through the time–temperature transformation (TTT curve), obtaining the desired mechanical properties. In Japan the swords of the samurai were unsurpassed in their keen edge and cutting ability. Their success relied on very complex manipulation of the sword's grain structure (up-setting and folding by hammer blows of the weapon on the anvil) and intricate heat treatment processes, in order to obtain the desired metallurgical and mechanical properties. Contemporaneous with the Japanese were the early metallurgical developments by Chinese dynasties, whereby soya bean protein that had decomposed was employed to enrich the red-hot edges of steel swords with carbon and nitrogen, and in so doing enhance their mechanical properties.

In Britain a 1727 publication entitled *Husbandry and Trade Improvement* by John Houghton FRS made reference to surface hardening. In the West Midlands techniques for surface improvement "... as being done at Wolverhampton in a different manner; with burnt hoofs and horns, bag felt, old burnt leather and tartar all mixed together with urine" again seemingly found the latter ingredient essential! Pre-dating this time was the first scientific investigation of surface treatments undertaken by Réaumur in France in his *magnum opus* published in 1722. In fact, some rather dubious materials (burnt leather in particular) were being employed for case hardening, by pack methods, prior to the First World War. These natural forerunners of later surface engineering processing and treatments for enhancement of metallurgy, typified by techniques such as grain refining and surface quench hardening, could be utilised without reliance upon electrical heating/phenomena to induce modifications to surface properties. Only with the advent of an active research programme into electrical phenomena in the late nineteenth and early twentieth centuries was it possible to envisage surface technology applications, many of which are now becoming mature commercial technologies.

Contemporary surface engineering

If the subject of surface engineering was only partially developed, then the text would be voluminous and impracticable within the scope of the current work; however, only an overview can be given here. The level of technological development for surface engineering technologies across various manufacturing industries will be markedly different, although their natural sequence of events for component production may be summarised in the following manner:

Application ⇒ Properties (required) ⇒ Design ⇒
Materials (selection) ⇒ Engineer (substrate) ⇒
Engineer (surface) ⇒ Lubrication ⇒
Performance

(After Bell, 1990)

Recent and traditional surface technologies contribute to a multifarious assortment of treatments enabling design and manufacture of a diverse range of metallic-based composites. By modifying a component's structural characteristics/composition, it intrinsically changes its engineering properties. In simplistic terms, non-mechanical surface treatments can be classified into various groupings, as follows:

- thermal treatment;
- thermo-chemical treatment;
- plating and coating;
- implantation.

For an engineering designer, the technological selection of a means to manipulate and change a surface becomes an integral part of component design. The decisive step taken to enhance a component's material surface signifies the fact that one of the existing base material properties needs manipulation in some way, in order that it satisfactorily fulfils its functional performance, while at the same time being economically acceptable. An engineered surface can have its coating thickness varying from several millimetres, by weld overlays, to just a few micrometres – for either physical vapour deposition (PVD) or chemical vapour deposition (CVD) coatings – and surface depth modifications to approximately 0.1 μm – by ion implantation (Figure 195). In a similar manner the hardness of coatings spans a wide range; typically these might be:

- spray coatings – ranging from >250 to 350 H_V;
- nitrided steels – around 1000 H_V;
- detonation gun (D-gun) carbide-metal cermet coatings – ranging over 1300–1600 H_V;
- titanium nitride (TiN) coatings by the PVD process – up to 3500 H_V.

Initially when selecting a surface modification treatment, it is important to comprehend what the

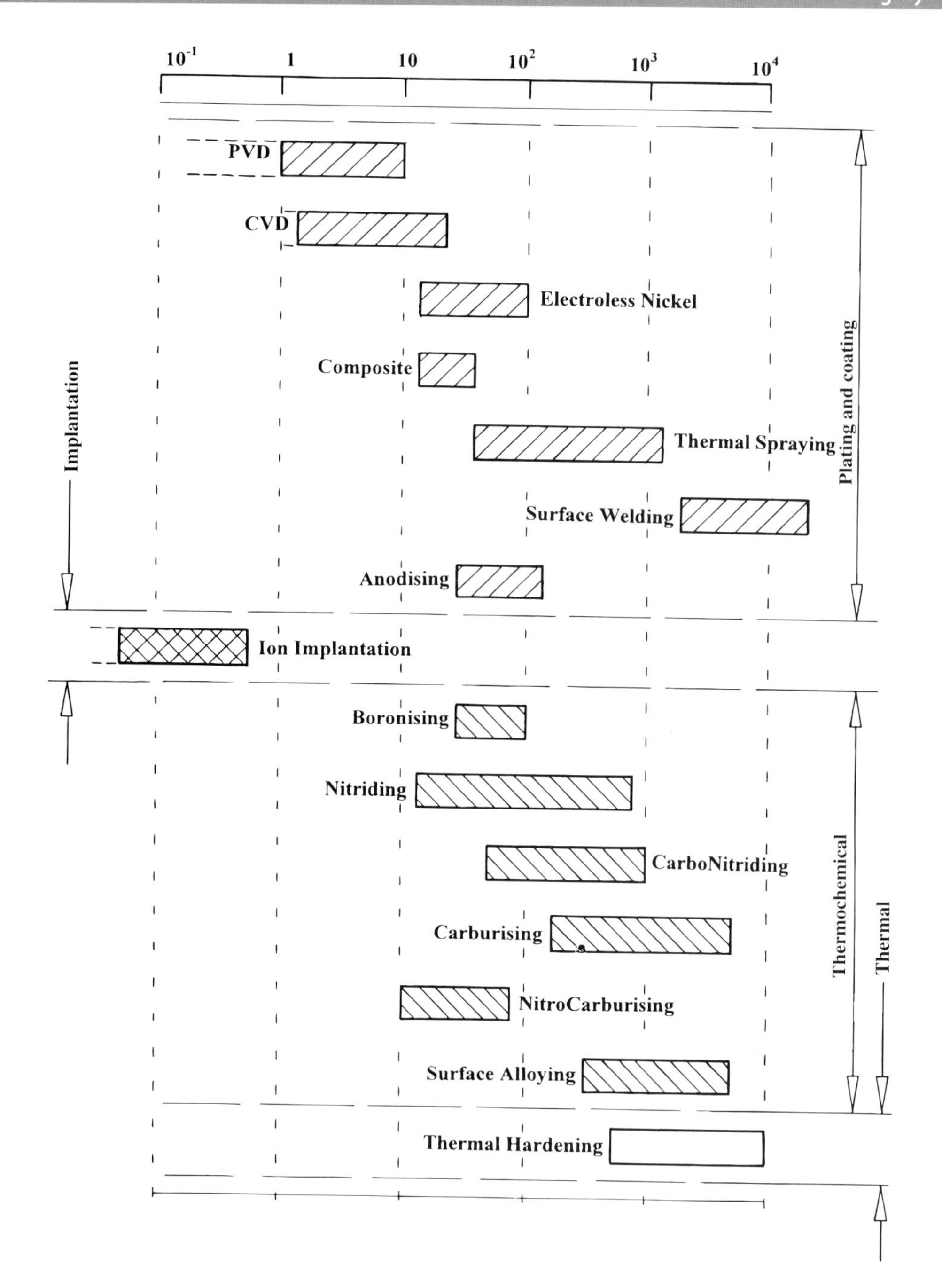

Figure 195. Typical thicknesses of surface layers for engineering components. [After Bell, 1990.]

anticipated surface and subsurface requirements might be for the component's in-service function. Often these decisions involve only one of the following set of properties, but may embrace several for complex surfaces:

- abrasive wear resistance – for either high or low compressive loading;
- scuffing resistance, or resistance to seizure;
- torsional, or bending strength;
- torsional fatigue/bending strength;
- resistance to surface contact fatigue (mechanical pitting);
- resistance to surface collapse (case crushing);
- corrosion, or erosion resistance.

In recent years, in order to assistance the designer in their choice of the correct decision related to the surface's functional performance, the development of databases utilising intelligent design systems has been of some success, although a comprehensive all-embracing design system is still some way off as a practical reality.

Future surface engineering applications

The widening market acceptance of the previous "first-generation" surface technological innovations and their assimilation into design and manufacture have allowed many technically enhanced engineering surfaces now to be considered as economically viable. It has been reported that at least eight widely diverse activities will dominate surface development in the future, excluding those currently in use for microelectronic surfaces and treatments; these areas are:

1. non-ferrous metal surface engineering;
2. polymer and composite surface engineering;
3. surface engineering of ceramic materials;
4. mathematical modelling of components for surface enhancement;
5. "second-generation" surface engineering technologies;
6. surface engineering in material manufacture;
7. statistical process control (SPC) in surface engineering;
8. non-destructive testing (NDT) of surface-engineered components.

Several of these surface-engineered activities are now becoming accepted across a wide field of specialisms, with others yet to become more than simply research topics as they are at present. It is undoubtedly true that the capability of producing complex duplex or multiple surface treatments, providing an intricate array of novel surface-engineered products, is almost unlimited in its scope. A typical example of this sophisticated surface treatment occurs when attempting to improve the machining process, this being provided by the multiple coating treatment of cutting tool inserts, where non-metallics and metallics are deposited as hard and soft-glide coatings that are utilised to significantly enhance a specific production process. Of late, surface engineers and designers must be encouraged to cooperate ever closer, in order to realise from a commercial viewpoint the combinations of surface technologies that currently exist and their relevant manner of exploitation.

This chapter has attempted to explain the importance of component surface integrity and how sometimes unwittingly, or intentionally a surface can be either enhanced or degraded by the actual production process. Moreover, some surface engineering treatments and processes have reached a significant level of sophistication, offering exciting challenges in the future for potential industrial applications. In the final chapter the reader's attention will be drawn to the reasons why, where and how it is important to ensure that instrumentation is appropriately calibrated at planned periodic intervals, thereby ensuring that any results obtained are both valid and traceable.

References

Journals and conference papers

Allen, A.F. and Brewer, R.C. The influence of machine tool variability and tool flank wear on surface texture. *Proceedings of the Sixth International Machine Tool Design and Research Conference*, 1965, 301–314.

Bell, T. Surface engineering: past, present and future, *Surface Engineering* 6(1), 1990, 31–40.

Bellows, G., Field, M. and Kahles, J.B. Influence of material and its metallurgical state on surface integrity. In *Influence of Metallurgy on Machinability*, ASM. 1975, 272–295.

Bonifácio, M.E.R. and Diniz, A.E. Correlating tool wear, tool life, surface roughness and tool vibration in finish turning with coated carbide tools. *Wear* 173, 1994, 137–144.

Bousfield, B. and Bousfield, T. Progress towards a metallography standard. *Metals and Materials* March 1990, 146–148.

Brinksmeier, E., Cammett, J.T., König, W., Leskovar, P., Peters, J. and Tönshoff, H.K. Residual stresses: measurement and causes in machining processes. *Annals of the CIRP* 31(2), 1982.

Chadda, Y. and Haque, M. Finish-turning surface roughness. *Cutting Tool Engineering* June 1993, 95–99.

Chou, K.Y and Evans, C.J. White layers and thermal modeling of hard turned surfaces. *International Journal of Machine Tools and Manufacture* 39, 1999, 1863–1881.

Davies, N., Spedding T.A. and Watson, W. Auto-regressive moving average processes with no normal residues. *Time Series Annals.* 2, 1980, 103–109.

El-Wahab, A.I. and Kishawy, H.A. A new method to improve surface quality during CNC machining. *International Journal of Production Research* 38(16), 2000, 3711–3723.

Fathailal, M., Danai, K. and Barber, G. Effect of flank wear on the topography of machined surfaces. *Tribology Transactions* 36(4), 1993, 693–699.

Field, M. and Kahles, J.F. Review of surface integrity of machined components. *Annals of the CIRP* 20(2), 1971, 153–163.

Ford, D.G. Machining to microns: error avoidance or compensation? *Proceedings of Lamdamap II. Computational Mechanics* 1995, 277–286.

Ford, D.G., Postlethwaite, S. and Morton, D. Error compensation applied to high precision machinery. *Proceedings of Lamdamap I.* Computational Mechanics, 1993, 105–121.

Galloway, D.F. Some experiments on the influence of various factors on drill performance. *Transactions of ASME* February 1957, 191–231.

Gillibrand, D. and Heginbotham, W.B. Experimental observations on surface roughness in metal machining. *Proceedings of the 18th International Machine Tool Design and Research Conference*, Las Vegas, 1977, 629–640.

Griffiths, B.J. Deficiencies in surface specifications. *Proceedings of Lamdamap III.* Computational Mechanics, 1997, 465–474.

Griffiths, B.J. Manufacturing surface design and monitoring for performance. *Surface Topography* 1, 1988, 61–69.

Griffiths, B.J. Mechanisms of white layer generation with reference to machining and deformation processes. *Transactions of ASME – Journal of Tribology* 109, July 1987, 525–530.

Griffiths, B.J. and Furze, D.C. Tribological advantages of white layers produced by machining. *Transactions of ASME – Journal of Tribology* 109, April 1987, 338–342.

Griffiths, B.J. Problems in measuring the topography of machined surfaces produced by plastic deformation mechanisms. *Wear* 109, 1986, 195–205.

Grum, J. and Oblak, A. Use of linear discriminating function for the description of the effects of microstructure on surface roughness after fine-turning. *Acta Stereol*, 18(3), 1999, 313–318.

Huynh, V.M. and Fan, Y. Surface-texture measurement and characterisation with applications to machine-tool monitoring. *International Journal of Advances in Manufacturing Technology* 7(2), 1992, 1–10.

Koster, W.P, Field, M, Fritz, L.J., Gatto, L.R. and Kahles, J.F. Surface integrity of machined structural components. *Airforce Materials Laboratory Technical Report AFML-TR-70-11*, MMP Project No. 721-8. Metcut Research Associates Inc., Cincinnati, OH, March 1970.

Lamb, A.D. Some aspects of the character of machined surfaces. *Metals and Materials* March 1987, 75–79.

Ng, E-G. and Aspinwall, D. Hard part machining AISI ((50 HRC) using Amborite AMB90: a finite element modelling approach. *Industrial Diamond Review* 60(587, part 4), 2000, 305–312.

Patz, M., Dittmar, A., Hess, A. and Wagner, W. Better hard turning with optimised tooling. *Industrial Diamond Review* 60(587, part 4), 2000, 277–282.

Rakhit, A.K., Sankar, T.S. and Osman, M.O.M. The influence of metal cutting forces on the formation of surface texture in turning. *International Journal of Machine Tool Design Research* 16, 1976, 281–292.

Rhoades, L. Understanding edmed surfaces. *Cutting Tool Engineering* April 1996, 22–31.

Shiraishi, M. and Sato, S. Dimensional and surface roughness controls in a turning operation. *Transactions of ASME* 112, 1990, 78–83.

Smith, G.T. Ternary manufacturing envelopes (TME's): a new approach to describing machined surfaces. *International Journal of Machine Tools and Manufacture* 40(2), 2000, 295–305.

Smith, G.T. Surface integrity. *Machinery Market* October 1998, 28–30.

Smith, G.T. Beauty is only skin deep?. *Metalworking Production* February 1998, 7–8.

Smith, G.T. The influence of insert wear on machined component surface texture whilst turning. *Proceedings of Lamdamap II*. Computational Mechanics, 1995, 355–366.

Smith, G.T. The surface topography resulting from boring powder metallurgy components. *Tribologia* 13(4), 1994, 44–62.

Smith, G.T. Hard-metal turning of crankshafts. *Proceedings of the Third International Conference on the Behaviour of Metals in Machining*, University of Warwick. IOM Communications, November 1994.

Smith, G.T. Some aspects in the surface integrity and tool wear associated with turning powder metallurgy compacts. *Wear* 151, 1991, 289–302.

Smith, G.T. Secondary machining operations and the resulting surface integrity: an overview. *Surface Topography* 3, 1990, 25–42.

Smith, G.T. The surface integrity of turned ferrous powder metallurgy components. *Powder Metallurgy* 33(2), 1990, 155–165.

Smith, G.T. and Maxted, P. Evaluating the high-speed machining performance of a vertical machining centre during milling operations. *Proceedings of Lamdamap II*. Computational Mechanics, 1995, 123–138.

Stout, K.J. and Spedding, T.A. The characterisation of internal combustion engine bores. *Wear* 83, 1982, 311–326.

Tönshoff, H.K. Eigenspannungen und plastische Verformungen im Werkstück durch spanende Bearbeitung. Dr.-Ing. dissertation, TH Hannover, 1965.

Turley, D.M., Doyle, E.D. and Samuels, L.E. A structure of the damaged layer on metals. *Proceedings of the International Conference on Production Engineering*, Tokyo. Japanese Society of Precision Engineering Part 2, 1974.

Vajpayee, S. Analytical study of surface roughness in turning. *Wear* 70, 1981, 165–175.

Watson D.W. and Murphy, M.C. The effect of machining on surface integrity. *Metallurgist and Materials Technology* April 1979, 199–204.

White, A.J., Postlethwaite, S.R. and Ford, D.G. An identification and study of mechanisms causing thermal errors in CNC machine tools. *Proceedings of Lamdamap IV*. WIT Press, 1999, 101–112.

Whitehouse, D.J. Surface measurement fidelity. *Proceedings of Lamdamap IV*. WIT Press, 1999, 267–276.

Whitehouse, D.J. Conditioning of the manufacturing process using surface finish. *Proceedings of Lamdamap III*. Computational Mechanics, 1997, 3–20.

Wu, D.W. and Matsumoto, Y. The effect of hardness on residual stresses in orthogonal machining of AISI 4340 steel. *Transactions of ASME – Journal of Engineering and Industry* 112, August 1990, 246–252.

Yeh, L-J and Lai, G-J. Study of the monitoring and suppression system for turning slender workpieces. *Proc. Instn. Mech Engrs.* 1995, 209, 227–236

Zhu, C. How to obtain a good quality surface finish in NC machining of free formed surfaces. *Computers in Industry* 23, 1993, 227–233.

Books, booklets and guides

American National Standard (ANS) B211.1 – Surface Integrity, 1986.

Behaviour of Materials in Machining. IOM Communications, November 1998.

Boothroyd, G. *Fundamentals of Metal Machining and Machine Tools*. McGraw-Hill, New York, 1975.

Cardarelli, F. *Materials Handbook*. Springer, Berlin, 2000.

Dieter, G.E. *Mechanical Metallurgy*. McGraw-Hill, New York, 1976.

Griffiths, B.J. *Manufacturing Surface Technology:* Penton Press, 2001.

Halling, J. *Introduction to Tribology*. Wykeham Engineering and Technology Series, 1976.

Höganäs Handbook for Sintered Components: Machining Guidelines. Höganäs AB, 1998.

Harris, W.T. *Chemical Milling*. Clarendon Press, Oxford, 1976.

Influence of Metallurgy on Hole Making Operations. ASM, 1978.

Influence of Metallurgy on Machinability. ASM, 1975.

McGeough, J.A. *Advanced Methods of Machining*. Chapman & Hall, 1988.

Momber, A. and Kovacevic, R. *Principles of Abrasive Water Jet Machining*. Springer, Berlin, 1998.

Metalworking Fluids. Verlag Moderne Industrie, 1991.

Modern Metal Cutting', Sandvik Coromant, 1994.

Rao, B.K.N. *Handbook of Condition Monitoring*. Elsevier Advanced Technology, 1996.

Finish Turning. Sandvik Coromant, 1995.

Shaw, M.C. *Principles of Abrasive Processing*. Oxford University Press, 1996.

Shaw, M.C. *Metal Cutting Principles*. Oxford University Press, 1996.

Smith, G.T. *CNC Machining Technology*. Springer, Berlin, 1993.

Smith, G.T. *Advanced Machining: The Handbook of Cutting Technology*. IFS/Springer, 1989.

Surface Technology. Verlag Moderne Industrie, 1990.

Trent, E.M. *Metal Cutting*. Butterworths, 1984.

Tlusty, G. *Manufacturing Processes and Equipment*. Prentice-Hall, Englewood Cliffs, NJ, 2000.

Williams, J.A. *Engineering Tribology*. Oxford Science, 1996.

6 Quality and calibration techniques

"Si parva licet componere magnis."

Translation
"If one may measure small things by great."
(Virgil, 70–19 BC)

Dimensional measurement: historical perspective

Archaic length measurement

Probably the oldest units of measurement employed in the ancient world were related to either body measurements or physical entities that were readily available in daily life. Notably, in Egypt around 3000 BC the basic form of measurement was the *cubit*, this being a measurement taken from the arm relating to the distance from the tip of the finger to the elbow. The cubit could then be subdivided into shorter units, such as the *foot* or the *hand* – this latter term today still being used for expressing the height of horses. Conversely, the cubit can be extended by adding several together to form longer units such as the *stride*. Unfortunately, these physical measurements would vary considerably owing to the difference in the size of people. In England, as early as 1000 the Saxon King Edgar kept a *yardstick* at Winchester as the official standard of measurement. However, it was not until the reign of King Richard I (the Lionheart) that any form of standardisation of units of measurement was first documented. In his Assizes of Measures in 1196 it was stated: "Throughout the realm there shall be the same yard of the same size and it should be of iron." This metallic artefact was the first permanent standard measure to be utilised in an attempt to control the vagaries of linear measurement. In 1215 in the reign of King John, the Magna Carta also considered units of standardised measurement – including those of wine and beer! The yard (or *ulna*, as it was sometimes known) together with its linear subdivisions and aggregated divisions, came into existence during the reign of King Edward I (1272–1307). Here, it was stated that "It is ordained that three grains of barley, dry and round make an inch, twelve inches make a foot, three feet make an ulna [yard], five and a half ulna make a perch (rod), and forty perches in length and four perches in breadth make an acre." The *perch*, or *rod* as it was known, was a traditional Saxon land measure that still survives into the twenty-first century, although it was originally defined as "the total length of the left feet of the first sixteen men to leave church on Sunday morning", this statement being somewhat open to variation in its determination of length!

In 1588 the *Elizabethan yard* or *ell* as it was otherwise known was once again based on a large man's stride. Alternatively, the measurement could be taken from the distance between a man's nose and his thumb when standing and stretching the arm – this technique often being used as a rudimentary form of measurement for the length of cloth. Artefacts such as the Elizabethan yard are termed "end standards", being engraved with an initial "E" and the Royal crown to denote universal acceptance in the country. However, until 1824 this and other measures remained legitimate until superseded by an Act of Parliament under King George IV, when the *imperial standard yard* was introduced, to minimise the inaccuracies generally associated with linear measurement at the time. This imperial yard had a relatively brief life as a comparative end standard because it was destroyed in 1834 in the fire that swept through the Houses of Parliament, where it resided. Restoration of this linear standard was undertaken from an alloy composition of copper (16 parts), tin (2.5 parts) and zinc (1 part). It was colloquially known as "Bronze No. 1" by its creator, the Rev. Sheepshanks, the principal design feature being that engraved gold plugs were recessed into the artefact at the plane of its neutral axis to minimise flexure and hence measurement inaccuracy. Duplicates of exacting dimensions were made and they were legalised by Parliament in 1855, one being presented to the Government of the United States of America in 1856.

International Bureau of Weights and Measures: the Metre

In 1875 in continental Europe parallel development was taking place creating the metric system, where at Paris the "Metre Convention" of participating nations established the International Bureau of Weights and Measures (BIPM). In 1889 the metallurgy and design of measurement artefacts had been improved from the rather crude designs and elementary metallurgical compositions to those of a more uniform cross-section and greater stability of artefact materials. In particular, of the number of artefacts produced, No. 6 manufactured from platinum–iridium, which replaced its more primitive flat cross-sectioned forerunner (*Métre des Archives*), was to become known as the *International Prototype Metre*. Of the remaining bars, they were distributed to the representative nations, Britain receiving bar No. 16. Britain did not actually sign the Metre Convention until 1884 and even then was reluctant to implement the clause referring to the introduction of metric measures into signatory countries. In fact, the use of the metric system of weights and measures by trade and industry only became law in Britain in 1897, but was not universally adopted – and then with certain resistance – until 1971.

It had been known for some time that a natural constant measurement value for the metre was required. The solution for its appropriate length was developed from a portion (quadrant) of the circumferential line of longitude through the Earth, theoretically at sea level. The French National Assembly meeting's 1791 definition of the metre was: "one ten millionth of the polar quadrant of the earth passing through Paris". In practical terms what this meant was that a line running from the North Pole to the Equator, passing through Paris, would represent a quarter of the Earth's circumference and, if this was broken down to a ten millionth part of this linear length, the amount would represent a metre. A team of surveyors was charged with the difficult task of establishing this geometrical/linear relationship and took six years to complete the work, during the war between France and Spain. In practice, only a relatively small portion of the geographical terrain could be surveyed, between Dunkirk (France) to Barcelona (Spain), minimising the affect of altitude whenever practicable. Many of these surveyors were charged as spies, nearly losing their heads in the process of this work, the outcome of the exacting and testing survey being the 1799 *Métre des Archives*. This artefact became the "master standard" for the first entity to be known as a means of global measurement classification, the so-called "metric system".

Optical and laser length measurement

Several major problems exist when using conventional artefacts as the basis of comparative measurement. Most notable among these problems are:

- the inability to manufacture linear-based artefacts to identical lengths;
- transferring these measurements precisely to employ them elsewhere;
- instability of the material resulting from movement/creep as internal stresses are released and gravitational and weight loss occurs with time;
- part accuracies and tolerances becoming tighter with chronological time, necessitating an even more accurate means of establishing length.

This latter point was well known but needed a new means to determine even greater accuracy and precision than was apparent using the International Prototype Metre and its contemporary artefacts around the world. In order to minimise the problems associated with metallic artefacts for standards, during the early to mid-twentieth century another natural standard was established for dimensional measurement, namely the wavelength of light. Once the metre had been defined in terms of the wavelength of light from an atomic discharge lamp, the necessary instrumentation could be reproduced in any well-equipped metrology laboratory. In order to get an accurate determination of the metre in terms of the wavelength of light, a comprehensive testing programme was initiated – nine times – between 1892 and 1940 at various laboratories, typically twice at the National Physical Laboratory (NPL) in 1932 and 1935. The mean of these nine results became the basis of a new definition for the metre: *the length equal to 1,650,763.73 wavelengths in a vacuum of the radiation ... of krypton-86*. In 1960 the metre was defined in this manner, thereby replacing the International Prototype Metre as the fundamental standard of measurement at the time.

At the same time (1960), the first laser was constructed and by the mid-1970s lasers were being used as length standards. Therefore, in 1983 the definition of the metre using krypton-86 source became: *the length of the path travelled by light in a vacuum during the time interval of 1/299,792,458 of a second*. This metre length is realised by employing iodine-stabilised helium–neon lasers having a reproducibility of $\geqslant \pm 2$ parts in 100,000,000,000, or alternatively to ±0.00000002 mm. Although as a wavelength source the iodine-stabilised helium–neon laser has excellent characteristics, in the near future it should be possible to construct lasers with vastly superior wavelength accuracy. The new laser instrumental design is necessary because the current iodine-stabilised helium–neon type is limited by the fact that the iodine molecules are at room temperature and collide into each other and fluctuate in and out of the laser beam. This relative motion causes them to absorb the laser light over a comparatively large range of wavelengths. The problem can be rectified by substituting the iodine molecules with a single atom of a metal such as ytterbium or strontium. This atom is given an electrical charge (changing its characteristics into an ion), which is held in an "ion trap". Once the ion has been captured in this manner it can be cryogenically treated (cooled down), so that its temperature creates an almost stationary atomic motion and either an ultraviolet light or blue laser illuminates the ion. This single atom source is a considerable improvement for the wavelength reference over that of current technique, as it absorbs light over a range of one millionth that of iodine, giving the potential for vastly improved performance in the future.

6.1 Size and scale

In recent times, some form of length measurement has been used in every sphere of life, to enable fair trading conditions and the development of new and improved products and processes to raise the standards of living of the populace. The range of precision products is vast and covers items such as nanometric electronic devices, with adjacent circuit dimensions of approximately 0.000006 mm. Conversely, at the other extreme similar measurements are required for geographical distances of many kilometres having a resolution of just a few millimetres – this latter measurement accuracy was necessary when railway tunnels were constructed and when the Channel Tunnel needed to meet in the middle between England and France. Accurate length measurement is a fundamental concept for the vast majority of technological products, ranging from video recorder heads to automotive fuel injector nozzles.

To gain a better understanding of the relative size and scale of measurement at the nano-and microtechnology end of ultra-high-precision manufacture, Figure 196 has been included and illustrates the generally accepted and logical divide between these two technologies, nominally quoted as around 10^{-8} m. In Chapter 3, describing notable technologies and features of surface microscopy, it was shown that instrumentation such as the atomic force microscope (AFM) could both resolve and, indeed, manipulate structures in the atomic resolution range (10^{-9} to 10^{-10} m). This atomic resolution sits at a level of ultra-high accuracy and resolution far below that of visible light, which is nominally quoted as ranging between the values of 10^{-6} and 10^{-7} m. In fact, until the middle of the twentieth century high resolution was considered by many as the thickness of a hair (around 10^{-5} m), people quoting a "hair's breadth" as being an exceptionally small measurement. If a group of hairs are pulled from several people's heads – the author's included – then it can be visibly appreciated (Figure 197a) that they vary considerably in diameter. The largest hairs in this group are around 100 μm and the smallest approximately 30 μm; therefore, this term for a hair's diameter is open to some degree of variability. What is apparent when viewing Figure 197 collectively is that although their individual diameters might vary in size, all of these hairs have one constant feature: diametral consistency. This natural sizing by our bodies is to be expected, as our evolutionary development over millions of years has led to hair follicles being of consistent size. The often quoted hair diameter is nominally agreed as being about 75 μm, which gives an indication of its apparent size, but is subject to a degree of *uncertainty* in its measurement, this being one of the main topics discussed later in this chapter.

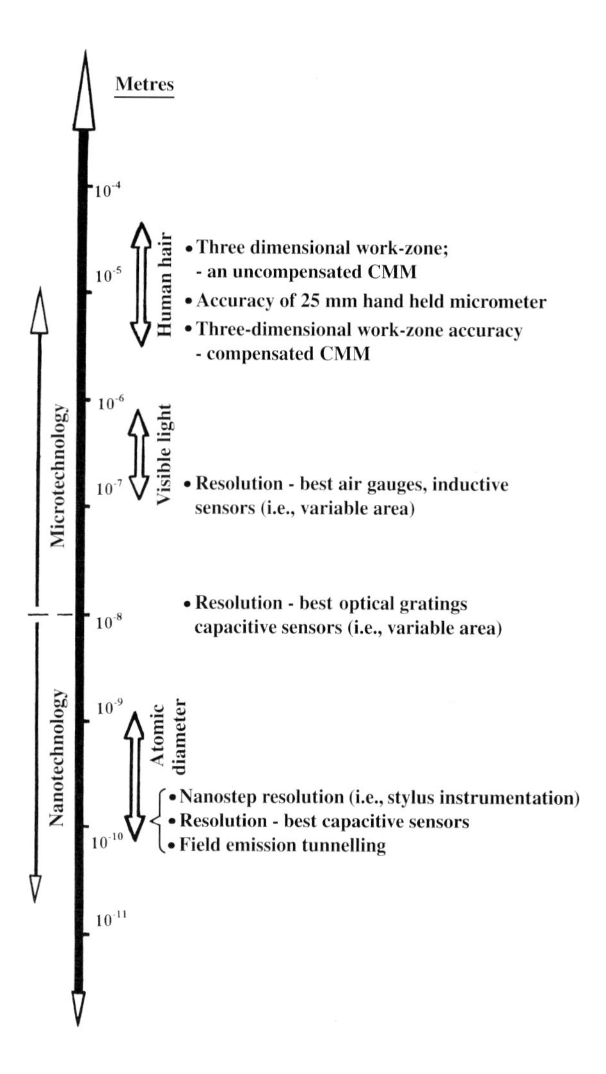

Figure 196. The relative size and scale of things. [After McKeown, 1986.]

6.2 Predictable accuracy: its evolution

At the end of the nineteenth century ultra-precision engineering meant machining to around 0.1 mm tolerances; such tolerancing had steadily improved to this level due to the earlier efforts of inventors in the Industrial Revolution. Pioneers such as John Wilkinson (1774) produced the first boring machines capable of producing a bored hole of 1270 mm diameter to an error of approximately 1 mm, while Henry Maudslay (1771–1831) invented many

(a) Hairs vary remarkably in size – often used as a standard of minute measurement (magnification ×270)

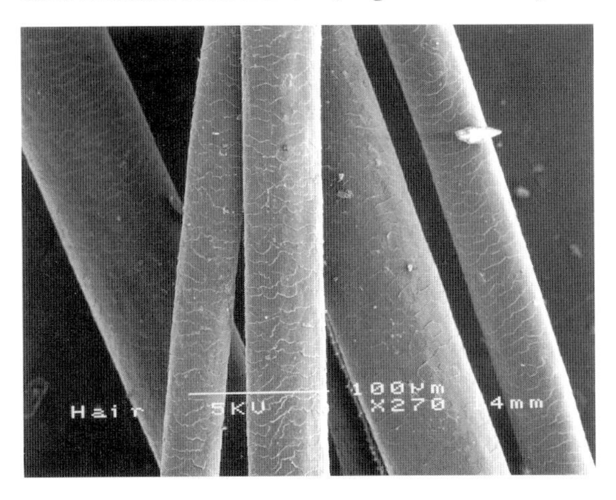

(b) Close-up of an individual strand of hair, illustrating good diametrical consistency (magnification ×1500)

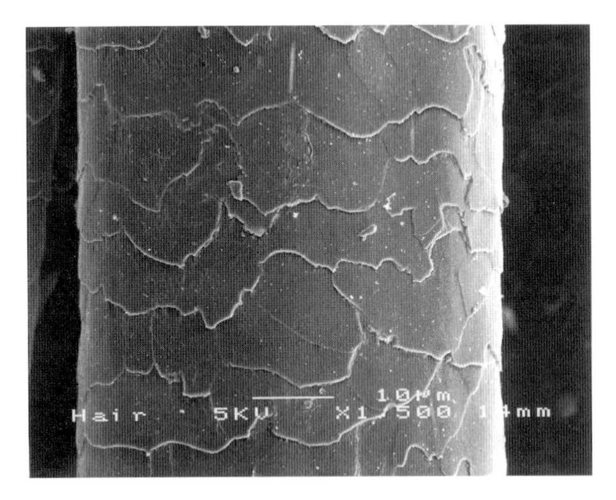

Figure 197. Hair size has often been selected to illustrate the magnitude of size. [Courtesy of Jeol (UK) Ltd.]

precision machine tools, but particularly the engine lathe. Notably others such as Sir Joseph Whitworth (1803–1887) developed the 55° included angle Vee-form screwthread, allowing precision feed motion to be accomplished through suitable gear trains on machine tools. These technological advances and fundamental work by others, for example Joseph R. Brown (1852), who designed a "dividing engine" for the production of precision engraving, further enabled a wider range of machine tools, such as that by Eli Whitney (milling machine) to be designed and developed, and then refined. With the considerable technological strides in precision engineering in the last hundred years, today the ability to reproduce consistent parts more economically enables significantly higher levels of precision and the attendant higher value-added tolerances, surface texture and other exacting features to accrue.

In Table 18 the currently achievable tolerances by contemporary precision CNC machine tools and specialised associated equipment are listed.

When it is considered that Taniguchi made his prediction of machined accuracy in 1974, he was remarkably accurate in his extrapolated data for normal, precision and ultra-precision machining accuracies (Figure 198). The advent of the term *nanotechnology*, coined by Taniguchi in 1974, heralded the concept of actually machining at nanometric accuracies (10^{-9} m), which is very close to the atomic lattice spacing. Today nanometric machining is a reality, with components of quite large dimensional size being held to diameters and surface textures of nanometric accuracies and finishes respectively, or a combination of both. Conversely, minute components can be manufactured to within nanoscale resolution and this technique is fast becoming increasingly popular, as part miniaturisation becomes essential in many industrial applications.

Nanotechnology instrumentation

The general rule of thumb that inspection instrumentation should be at least *ten times* more accurate than the equipment manufacturing the component is exceedingly difficult to achieve at present for parts made in the nanotechnology range. Nanotechnology instruments have been developed to both explore and measure surface properties in the vicinity of the atomic range, typically investigating surface topography, grain boundary coherence, atomic dislocations or similar – some of which were briefly touched on in Chapter 3. These nanometric requirements exceed the concept of performance of dimensional tolerances on a machine tool, or the attainable surface texture for a part. Notionally, the surface production and its accompanying measurement can be categorised as follows:

Functional performance for specific topographical and surface features

- surfaces of magnetic heads and compact disks for storage capacity;
- coatings on cutting inserts and tools, for wear properties;
- tribological investigation on bearings, or adhesion and adsorption problems;
- surface effects produced by non-conventional mechanisms (EDM, ECM, USM, laser and WJM, etc.);

Table 18. Acceptable tolerance errors in products

Tolerance (accuracy) (μm)	Mechanical components	Electronic components	Optical components
200	Normal machine and home-made parts	General-purpose electrical parts (switches, motors and connectors)	Camera and telescope bodies
50	General purpose mechanical parts (gears and threads), typewriter and engine parts	Packages (electronic parts) micro-motors, transistors, diodes and magnetic heads (video/type recorders)	Camera shutters, lens holders (for cameras and microscopes)
5	Mechanical watch parts, accurate gears, threads, machine tool bearings, ballscrews, rotary compressor parts, shaver blades	Electric relays, condensers, disk memory, silicon wafers, TV colour masks, video head and cylinder	Lens, prisms, optical fibres and connectors
0.5	Ball and roller bearings, precision drawn wires, flapper servo-valves, gyro bearings, air bearings, precision dies, roll thread dies, ink jet nozzle	Magnetic heads (video cassette recorders, CCDs, magnetic scales, quartz oscillator, magnetic bubble memory, ICs, magnetrons, thin film-pressure transducers, thermal printer heads, memory, electronic video disks	Precision lens and prisms, optical scales, elastic deflection mirrors, IC exposure masks (photo/X-ray), laser polygon mirrors, X-ray mirrors
0.05	Gauge blocks, diamond indentors, high-precision *X–Y* tables, high-precision stamping and dies, microtome cutters (diamond)	IC memory, electronic video disks, large-scale IC's, micro-vacuum tubes, TFT-LCD	Optical flats, precision Fresnel lens, optical diffraction gratings, optical video disk (CD)
0.005	Ultra-precision parts (plane, ball, roller, thread)	VLSI's, super-lattices (synthesis) thin films	Ultra-precision diffraction gratings
(Special features)	Shape (3-D) preciseness	Pattern (2-D) fineness	Mirror grating (1-D) accuracy

After Taniguchi (1994).

- damage monitoring of both pure and hybrid materials at the atomic scale;
- semiconductor and similar electrically based surface properties, with respect to either charge injection or storage;
- surface catalysts and reagents concerning chemical processes and reactions;
- surfaces of biological molecules in both liquids and membranes and their changes in real time.

Specific measurement requirements

- height, or topographical features;
- profile shapes and edge sharpness;
- position and relative position of features;
- volumetric analysis;
- flaw, or defect detection;
- structural characterisation of lattice parameters;
- atomic motion;
- time changes of atomic, or molecular structures.

Due to the comprehensive investigative and diagnostic assessments that are routinely required on such a diverse range of surface and subsurface topographical processes, a large number of instrumental techniques have been developed to encompass these investigative procedures. Each instrument in this large array of equipment needs to be calibrated at periodic intervals to ensure that data obtained is traceable to known international and national standards, so that some form of consistency in investigative procedure is assured. The following section will explore both the underlying philosophy and the route in achieving traceable measurements.

6.3 Traceability of measurement

In order to substantiate and validate any measurements taken on metrological instrumentation, it is vital to ensure that these measurements are correct and that they relate to international or national standards, or both. Prior to describing the mechanism for obtaining valid measurement from these instruments, it is worth pursuing the confusion that often

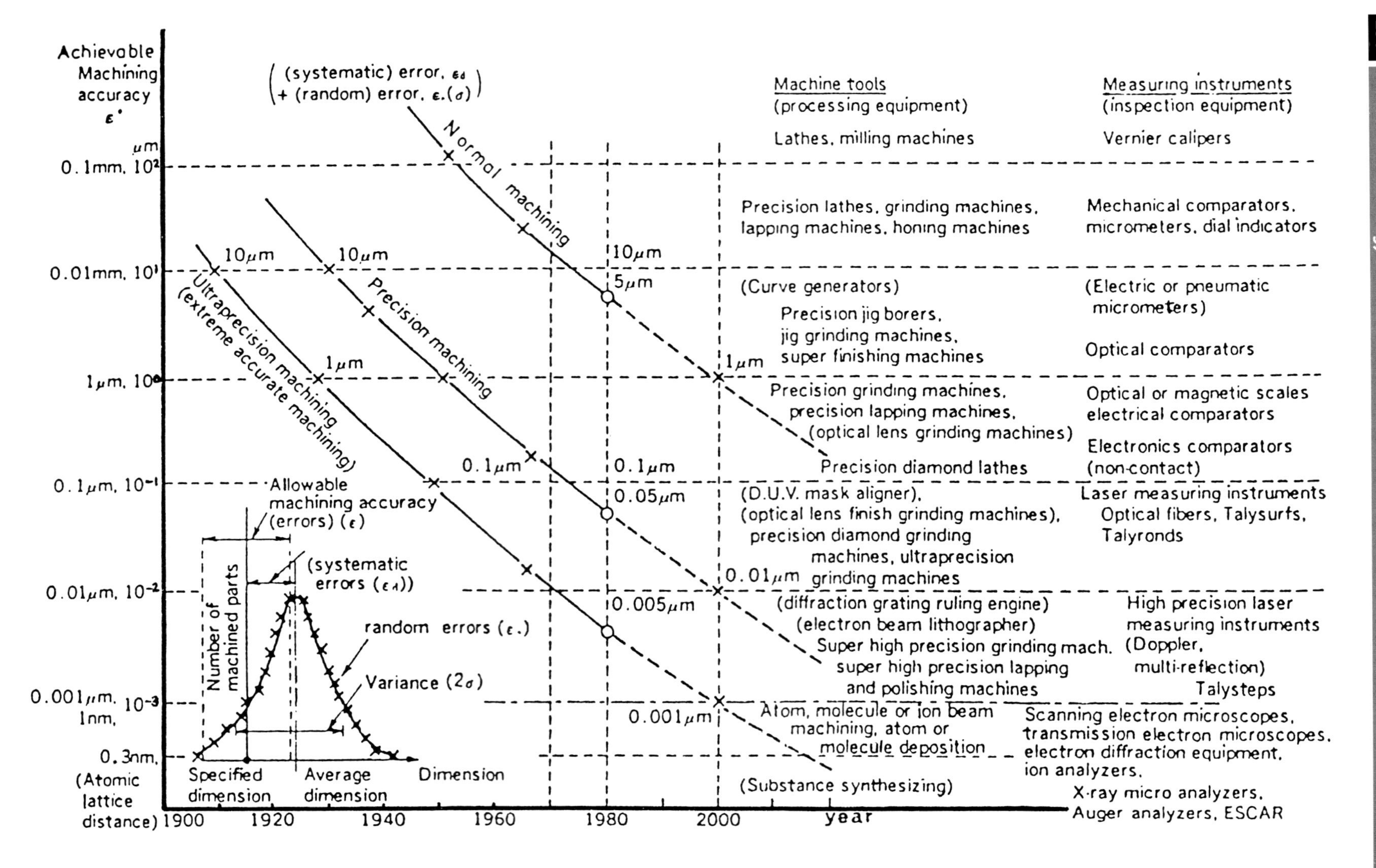

Figure 198. The prediction of achievable accuracy. [After Taniguchi 1974.]

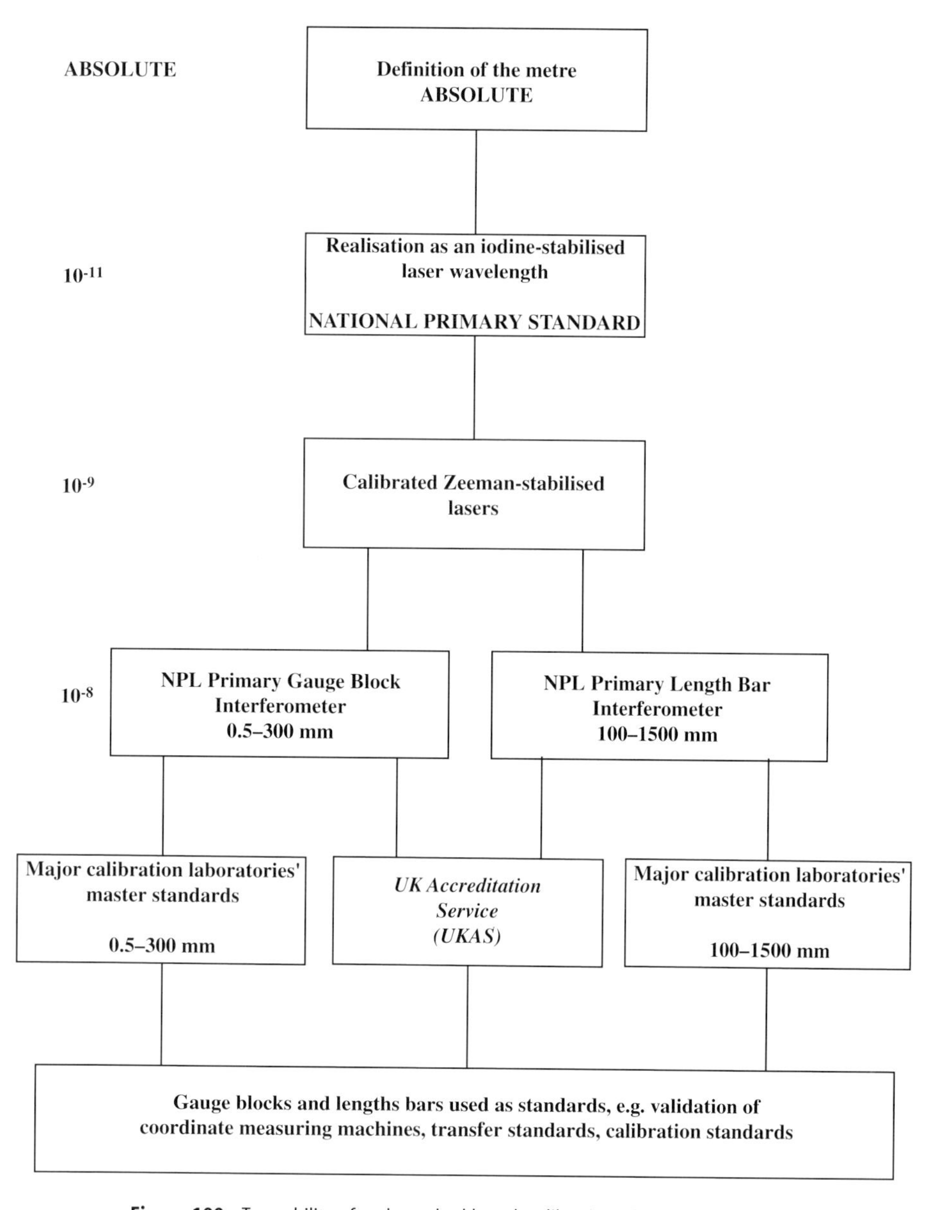

Figure 199. Traceability of end standard length calibrations. [Courtesy of NPL.]

arises over the subject of accuracy and repeatability, as they do not have the same origin; hence:

- *repeatability* is a function of the instrument's design;
- *accuracy* is a function of calibration.

For example, an instrument may repeat with great precision, but it is only by calibrating the equipment that we know whether to believe the results. This seems an obvious point to make, but it is often a widely misunderstood distinction that is very important whenever a new instrument is purchased. To take for granted the positional accuracies claimed by the manufacturer of the instrument is not wise, without ascertaining this fact for oneself. Returning to the current theme of traceable measurement, if the traceability chart illustrated in Figure 199 is considered, the route from metrological and inspection equipment in daily use, through the traceablity chain to the primary standard of dimensional measurement, can be shown. In each case, as calibration progresses toward the primary standard, more exacting measurements are made, thus ensuring that each piece of instrumentation in the "chain" is calibrated to a higher order of accuracy and precision. This process of calibration against known artefacts is both costly and, more important, a time-consuming process, taking valuable and often delicate instrumentation out of the metrological chain, but is essential to ensure traceability.

Today, many of the primary standards are now "derived" and do not exist as discrete entities; this is particularly true for many electrical and physical standards. As a result of such "pseudo-standards", a move in recent years in the United Kingdom has been an attempt to mitigate the losses in cost and time as essential calibration occurs, by employing *virtual gauging concepts*, which have yet to be introduced into conventional dimensional measurements. By way of illustration of the potential for this technique for dimensional measurement, that of an electrical calibration will be briefly reviewed. Here, users obtain calibration through on-line Internet access to the National Physical Laboratory (NPL), by way of its appropriate website – via encrypted security stages. It is important that in order to gain access to the calibration facilities the sensitive nature of the data to the site must be secure. Once access has been confirmed, users then enter information such as the type of equipment they wish to calibrate, together with the calibration kit and frequency ranges. This data is transferred to the NPL, where it is subsequently evaluated and the correction values are displayed on the PC's screen. Thereafter, a need to evaluate any imperfections that can arise via electronic noise and other outside approval by the user needs to be substantiated, with the values sent back to the instrument, which will automatically calibrate itself.

Although Figure 199 shows the traceability route to be relatively straightforward, in reality the development of the route to efficient calibration can involve some quite complex practical problems and philosophical interrelationships, as depicted in schematic form in Figure 200.

Surface description

Any given surface can be described in a variety of ways (Figure 200), the most obvious being where the height of each data point is plotted in relation to a reference plane. For example, in two-dimensional sectioning this is what the surface texture instrumentation would in the main be measuring. However, such information tends to be quite specific and relates to both the surface stylus probe interaction and the measuring conditions. These inspection parameters make it problematical to extract from the surface the required information to enable comparison of surfaces to each other. This problem has led to roughness parameters considerably multiplying in number, many with quite limited practical applications. In order to minimise this parameter proliferation an alternative option might be to describe the surface by way of its Fourier components. Anyone familiar with the complex waveforms existing in acoustics will be able to comprehend Fourier concepts. For instance, a musical instrument's range can be described by its relative properties at different frequencies in a harmonic series, in the same manner that a real surface is described in terms of a succession of sinusoidal corrugations, each having particular frequencies, amplitudes and phases. When a surface has a periodic profile, this can be described by sine waves of infinite series, whose actual periods are integral fractions of the fundamental; alternatively, if periodicity is not represented on the surface, then all spatial frequencies are probably present. If for simplicity, consideration is given to the cross-section of a random surface, two facts can be established; these are:

1 the surface can be represented by a graph of height-versus-position;
2 another graph can be made to represent amplitude versus spatial frequency.

The relationship between both graphs is in effect a Fourier transform. To obtain a complete description of this arbitrary surface, consideration must be given to the relative phases of the individual components. In effect this is the power spectrum, but as the power is proportional to the square of the amplitude it will not have any phase information. A major benefit in utilising the Fourier transform for surface description is that surface data can be simply extracted and as a consequence metrology equipment can be more easily assessed as the *instrumental response function*, or more notably on *Stedman diagrams*, yet to be described.

The traceable route from the concept of an ideal sinusoidal surface to its practical realisation, and from there to more complex surfaces to provide standards against which to compare real surfaces of functional performance, is depicted in Figure 200. Achieving this goal represents considerable scientific research in both fundamental and applied fields. Even when much practical metrological work is undertaken, it cannot be guaranteed that it will be possible to accomplish instrumental calibration to the level of accuracy one might expect, nor even that this accuracy will coincide with the sensitivities of typical instruments. However, it should be possible to establish uncertainty levels and, in the majority of instances, the causes of this uncertainty. Further, it cannot be certain that this exercise in establishing uncertainty will lead to improvements in measurement determination, but it enables the metrologist to obtain a clearer understanding of the present instrumental limitations. The following section

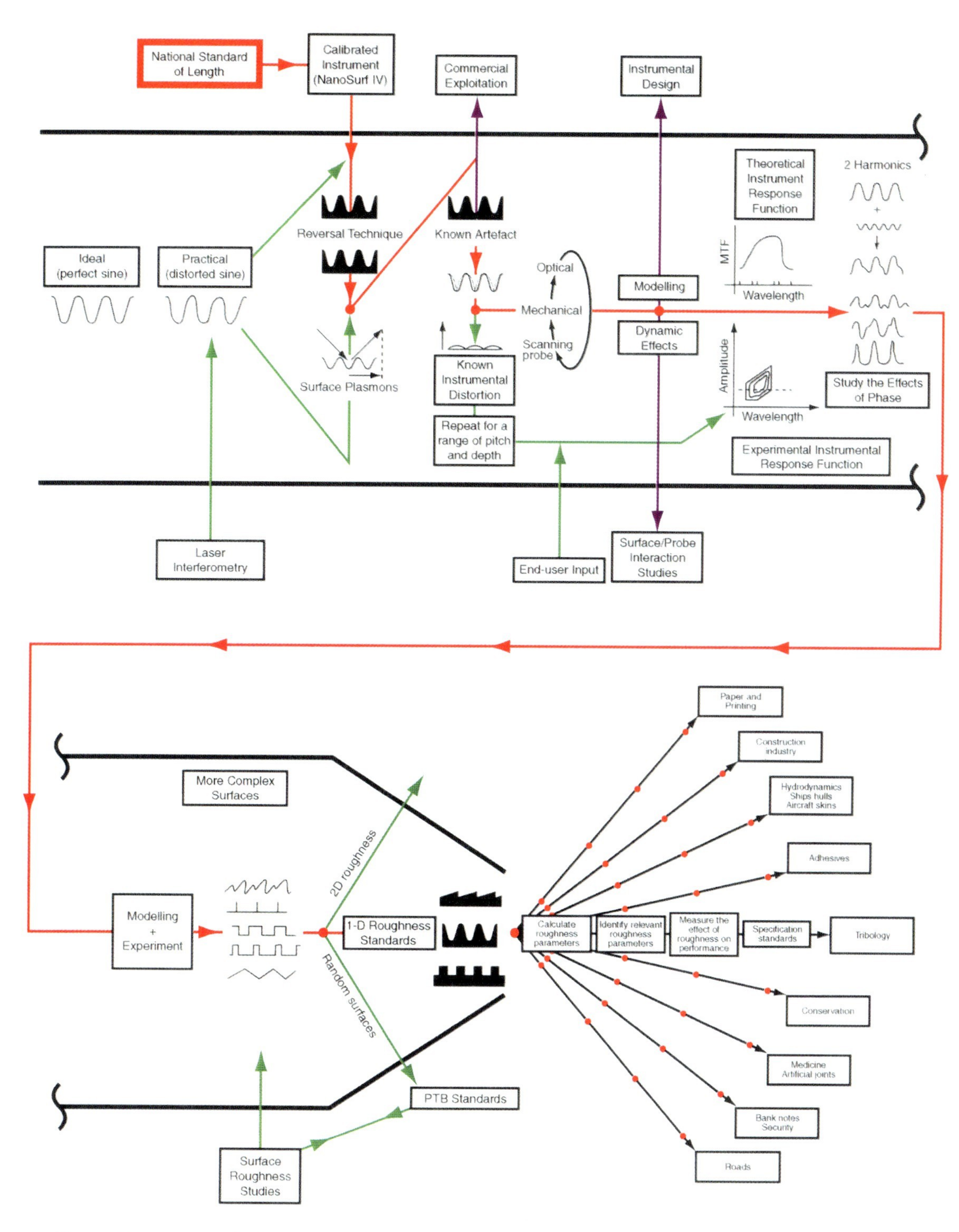

Figure 200. Route towards absolute standards for measurement of surface texture. [Courtesy of R.K. Leach & M.C. Hutley/NPL.]

briefly describes how, and in what manner, the influence of process capability on the production of components by industry is affected by calibration and recalibration of inspection equipment. Measurement uncertainty will then be described and how it can be determined, enabling instruments, artefacts and their users to establish the probable causes of metrological uncertainties.

Process capability factors

The output from a production process will differ according to the *stochastic* nature of the operation. A stochastic process can be defined as: *a process that has a measurable output and operating under a stable set of conditions causing the output to vary about a central value in a predictable manner.* Such an output is depicted in Figure 201, where the *process capability* (C_p) index is initially estimated at the beginning of a batch run to have a value of <1.0. Process capability is an extremely important measure of production performance and can be simply expressed in the following manner:

$$C_p = \frac{T_U - T_L}{6\sigma}$$

where: C_p is the process capability, T_U is the upper specification limit (tolerance), T_L is the lower speci-

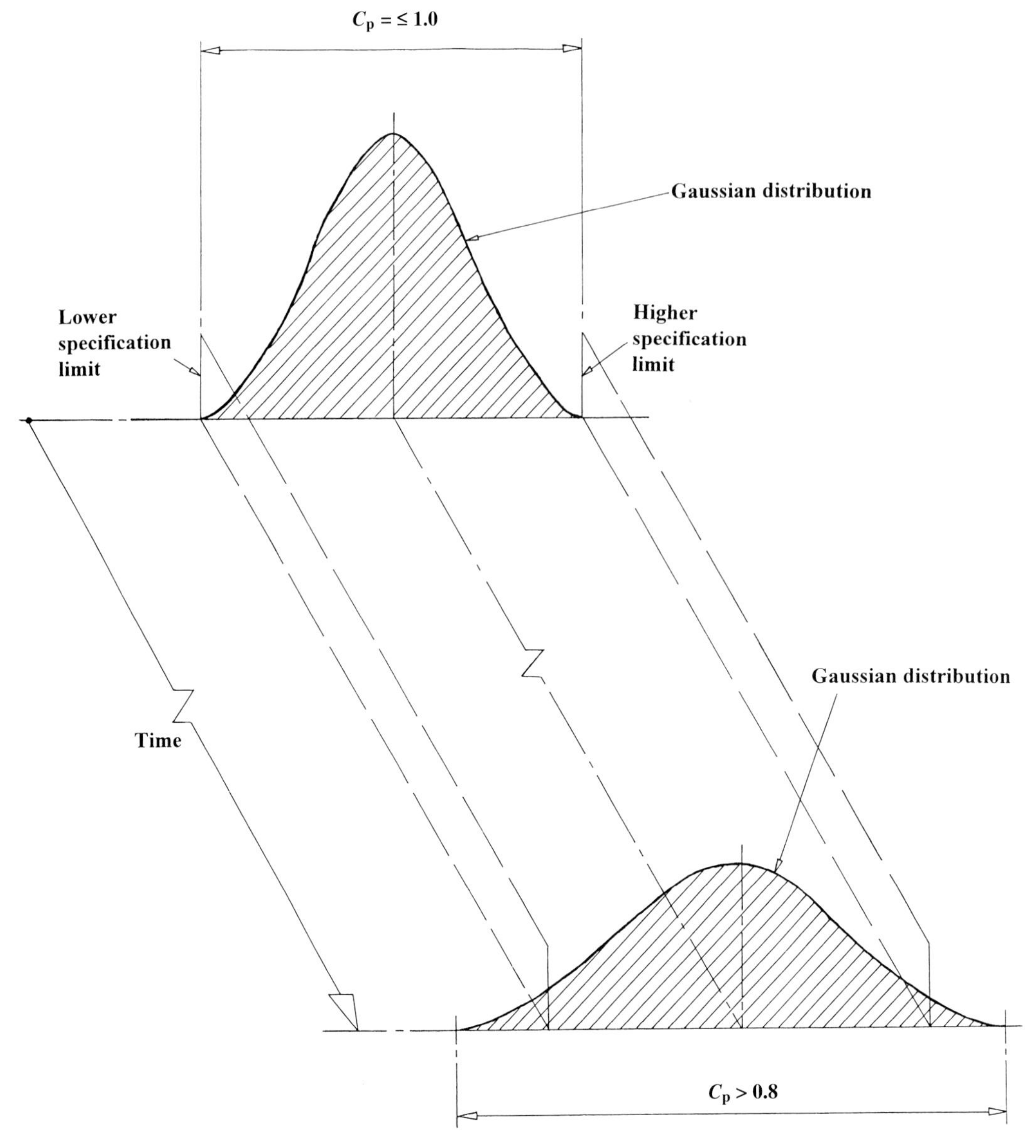

Figure 201. The process capability (C_p) index may change with time.

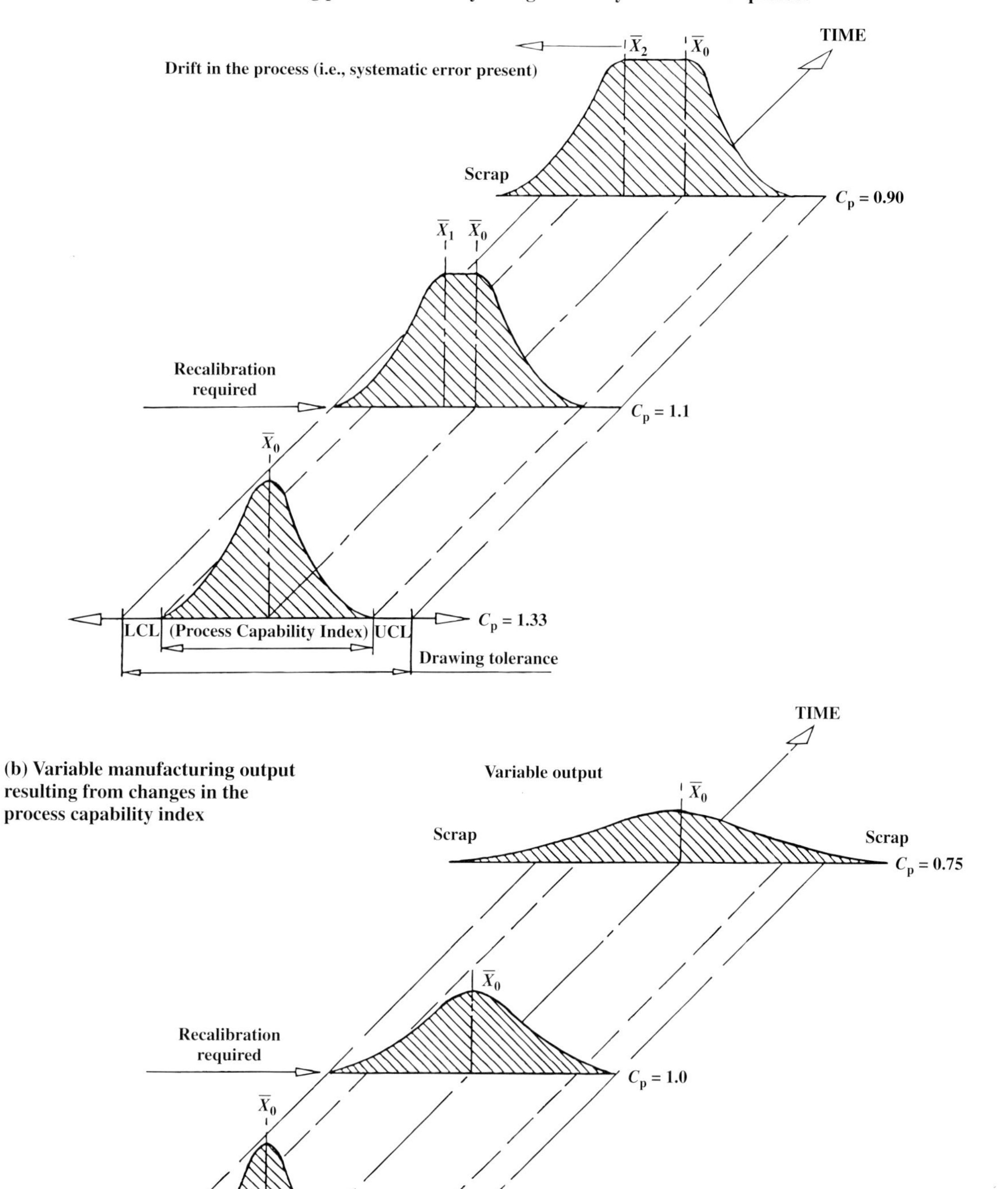

Figure 202. Illustration showing how (a) the manufacturing process mean changes with time; (b) process capability index changes with time; or how both can change with time.

fication limit (tolerance) and σ is the standard deviation for the process.

In this case the Gaussian (normal) distribution from the output at the beginning of the production run is centred on the mean value (x bar) and fills the complete statistical band (6σ). As the process continues to produce components the output is still centred on the mean, but the spread (variability) has increased considerably with respect to time. Here is an undesirable situation as the process capability has decreased to a value of C_p 0.8; this will inevitably introduce scrap into the process. Some reworking may be possible for certain parts outside one limit, depending upon whether this is a non-critical imposed limit in the process. However, the process capability has highlighted the fact that in this case there is an unacceptable variability occurring that needs drastic action if most of the production run is not to be lost. What was unusual about the production output in Figure 201 was the fact that the mean value had not changed, but the variability had increased. In the case of Figure 202(a) the mean value has changed with respect to time – it has introduced some "drift" as a systematic error in the production output has proportionally increased. This drift has caused the process capability to change from an initial value of C_p 1.33, being well in control, to an intermediate value of C_p 1.1, indicating that calibration was required *before* this point had been reached in the production run, as some scrap is now inevitably being produced. The situation is exacerbated with time until a C_p 0.90 production output occurs, indicating the whole process is out of control, because of this tendency to drift. In the case of Figure 202(b) once again the production run has the output situated at the mean, but with time the variability has increased from an initial level of C_p 1.33 to an intermediate value of C_p 1.0, where calibration is necessary. Finally, toward the end of the run, an undesirable situation occurs at a C_p value of 0.75, where scrap will be present.

If some form of in-process gauging and associated traceable calibration procedure had been introduced to limit the "process drift" or variability in output (or indeed both), then scrap parts are unlikely to be present in a batch production run. What has been said here for the control of production processes is equally true for the calibration of general metrology equipment, through acceptance of the fact that uncertainty will arise in any production or metrological process requiring appropriate action to minimise these effects.

6.4 Measurement uncertainty

Geometrical product specification (GPS)

The GPS procedure enables the designer to define the geometric shape, dimensions and surface characteristics of a component in a way that ensures optimum functioning of that workpiece. The procedure normally adopted includes a definition of the optimum value and its dispersion for its intended function, enabling it still to be satisfactory. The production process will in many circumstances produce parts that are not perfect, in that they may demonstrate some form of deviation from a defined

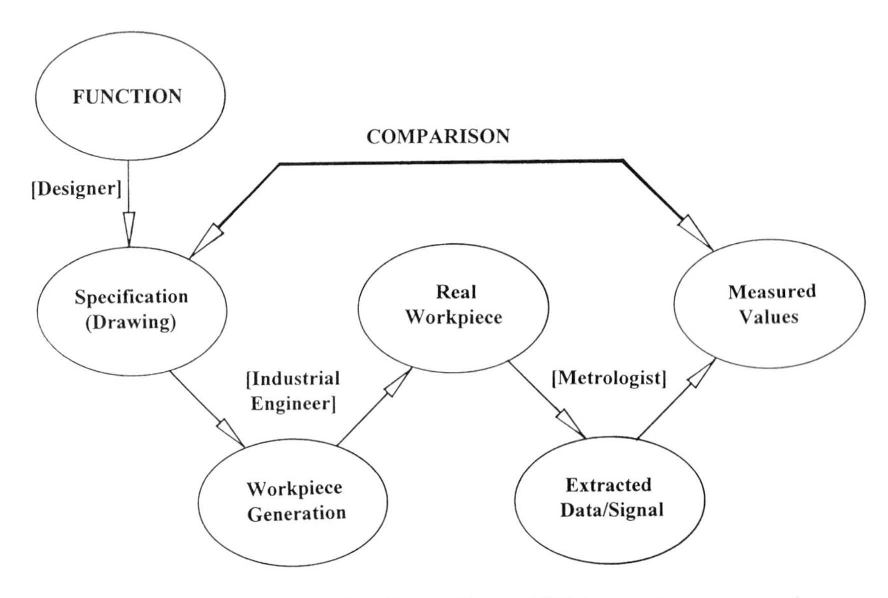

Figure 203. The basic geometrical product specification (GPS) framework: a new approach.

Table 19. The geometrical product specification (GPS) matrix model

Matrices: chains of standards:					
General GPS matrix		Complementary GPS matrix			
General GPS		A	Process specific, tolerance standards	B	Machine element, geometry standards:
1.	Size chain	A1	Machining chain	B1	Screwthread chain
2.	Distance chain	A2	Casting chain	B2	Gears chain
3.	Radius chain	A3	Welding chain	B3	Spline chain
4.	Angle chain	A4	Thermal cutting chain		
5.	Form of line (independent of a datum) chain	A5	Plastics moulding chain		
6.	Form of line (dependent on a datum) chain	A6	Metallic and inorganic coating chain		
7.	Form of a surface (independent of a datum) chain				
8.	Form of a surface (dependent on a datum chain				
9.	Orientation chain				
10.	Location chain				
11.	Circular runout chain				
12.	Total runout chain				
13.	Datums chain				
14.	Roughness profile chain				
15.	Waviness profile chain				
16.	Primary profile chain				
17.	Surface defects chain				
18.	Edges chain				

Source: BS 8888:2000 / ISO TR 14638.

optimum. If a comparison is undertaken between the component and its specification (Figure 203), it requires the following to be considered: that the workpiece (BS 8888: 2000):

- is conceived by the designer;
- is as manufactured;
- is as measured.

The standards introduced in the GPS field provide the fundamental rules for geometrical specification, such as the basic definition, symbolic representation and measurement principles (see ISO TR 14638 for an overview of these concepts). Several categories of the standard are concerned with concepts, dealing with fundamental rules of specification, others provide global principles and definitions, or various geometric characteristics, the latter including workpiece characteristics relating to differing production processes, together with specific machine elements, as shown in Table 19. GPS principles are necessary throughout a product's development: design, manufacturing, metrology and quality assurance.

In Figure 204 one of the "chains of standards" is depicted, illustrating a typical composition of a functional performance standard, such as the geometric tolerance for straightness. In boxes 1, 2 and 3 are shown the manner in which straightness and the associated geometric tolerances are defined. In box 1 the codification of the geometric tolerance symbol for straightness with its associated tolerance is depicted. Box 2 indicates the definition of this straightness tolerance and box 3 is the specification operator, indicating the manner in which this straightness deviation may appear within the tolerance. Box 4 illustrates the anticipated uncertainty associated with the upper and lower tolerance limits; more will be said on this subject later when discussing Figure 206. Boxes 5, 6 and 7 illustrate the metrological assessment of the real surface and the metrology instruments with their anticipated calibration requirements. In this visual manner these boxes depict the complete geometric product specification chain, this being presented in a logical and practical approach to the topic.

In the GPS philosophy the specification and verification operators are slightly more complex than first appearances would seem to indicate, as illustrated in the flow diagram in Figure 205. The linking of design intent and metrology is the objective here – the *duality principle*, with geometric features occurring across three disciplines. These are:

1 *specification* – this is where the designer conceives several representations of the workpiece;
2 *workpiece* – as it would appear in the physical world;

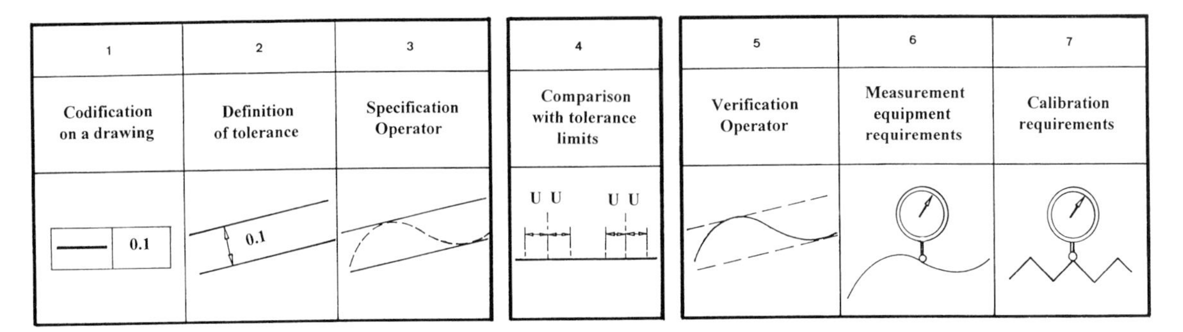

Figure 204. The chain of standards, e.g. straightness.

3 *inspection* – where a given workpiece representation is created by sampling of the part by measuring instruments.

To comprehend the nature of the relationship between these three disciplines is of great importance and Figure 205 depicts the parallel or "mirror" activities that connect those between the design intent and metrology. The design intent will result in specification characteristics – the conceptual perfect operator – with the metrology creating the evaluation of characteristics – the actual operator. These "operators" can then be compared with each other for comformity, as indicted in Figure 205.

Some of the key questions raised today by the manufacture of parts to tighter tolerances are:

- "What is the best manner to interpret metrology data, whenever there is a measurement uncertainty factor in a tolerance band?"
- "How to standardise on tolerances across the design, production and inspection functions, so that measurement uncertainty is accounted for throughout the creative procedure in the product's design and build life cycle?"

The term *uncertainty* has been mentioned here on numerous occasions and this will be the main theme from now on in this section.

Uncertainty issues

The question often asked in calibration-related tasks, is: "What is measurement uncertainty?" Uncertainty of measurement refers to the doubt that exists about any measurement; there occurs a margin of doubt for every measurement. This expression of measurement uncertainty raises other questions: "How large is the margin?" and "How bad is the doubt?" Hence, in order to quantify uncertainty two numbers are required, one being the width of the margin – its *interval* – the second is the *confidence level*; this latter value states how sure one is that the actual value occurs within this margin. Typically, a stylus length might be 100 mm plus, or minus 0.5 mm at the 95% confidence level. This uncertainty could be expressed as follows:

Stylus length = 100 mm ± 0.5 mm,

at a level of confidence of 95%.

In reality what this statement is implying to a metrologist is that they are 95% sure that the stylus length will lie between 99.5 mm and 100.5 mm in length. In this discussion it is essential not to confuse the term *uncertainty* with *error*. They can be distinctly classified thus:

- *error* – the difference between the measured value and the "true value" of the item being inspected;
- *uncertainty* – the quantification of the doubt existing with regard to the result of this measurement.

In most situations one tries to correct for any known errors; this is attempted by the application of *corrections* from certificates of calibration, although any error occurring that is not known will be a source of uncertainty. The estimation of measurement uncertainty is important because precise and accurate quality measurements enable the metrologist to comprehend the results without ambiguity. However, the expression of the measurement uncertainty is significant when taking any form of measurements, because they may be part of:

- *calibration* – where measurement uncertainty must be reported on appropriate certification;
- *test* – if measurement uncertainty is required to establish either a "pass" or "fail" condition.

Alternatively, information on measurement uncertainty is vital to hold a:

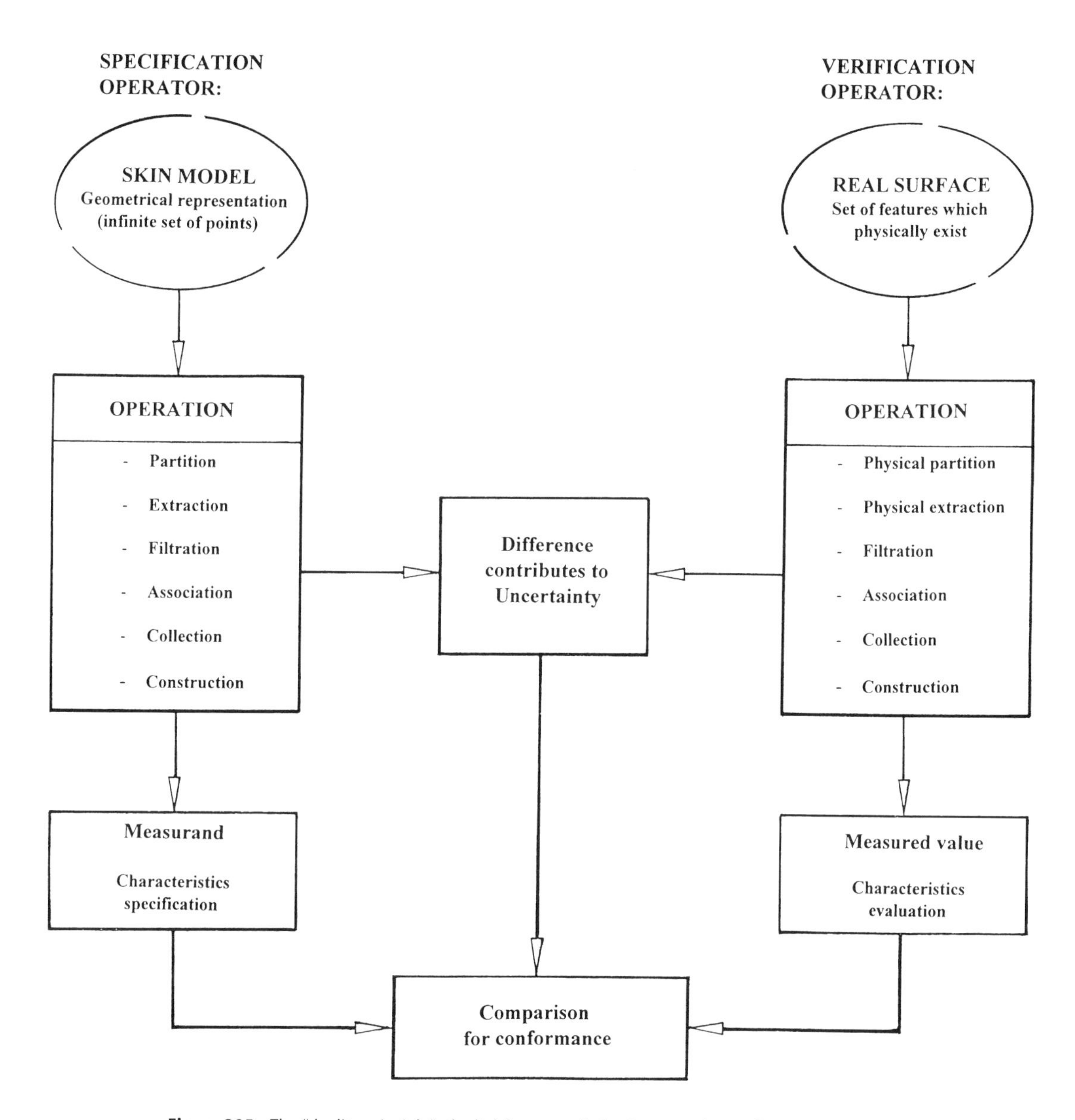

Figure 205. The "duality principle": the link between design intent and metrology [BS 8888: 2000.]

- *tolerance* – where the uncertainty had been previously established, prior to deciding whether the tolerance is met (see Figure 206).

Otherwise one might need to read and understand the calibration certificate or written specification for a particular test/measurement. It is worth noting, that a measurement is *not* traceable unless quoted with an uncertainty.

Statistical measures

In order to evaluate the test data from an inspection procedure, to determine whether adequate process control has been established, statistical measures are adopted. If the metrological process is not influenced by systematic or random errors, then the process is said to be behaving "normally" and any process output data is valid. Two statistically derived mathematical expressions are needed to define whether a process is behaving correctly: these are the *arithmetic mean* and its accompanying *standard deviation*. Normally, the arithmetic mean is conveniently shortened to *mean*, this being denoted by the symbol $\bar{x}$ (termed "x-bar"). This $\bar{x}$ value is the mean of all the values of x and can be derived from the following expression:

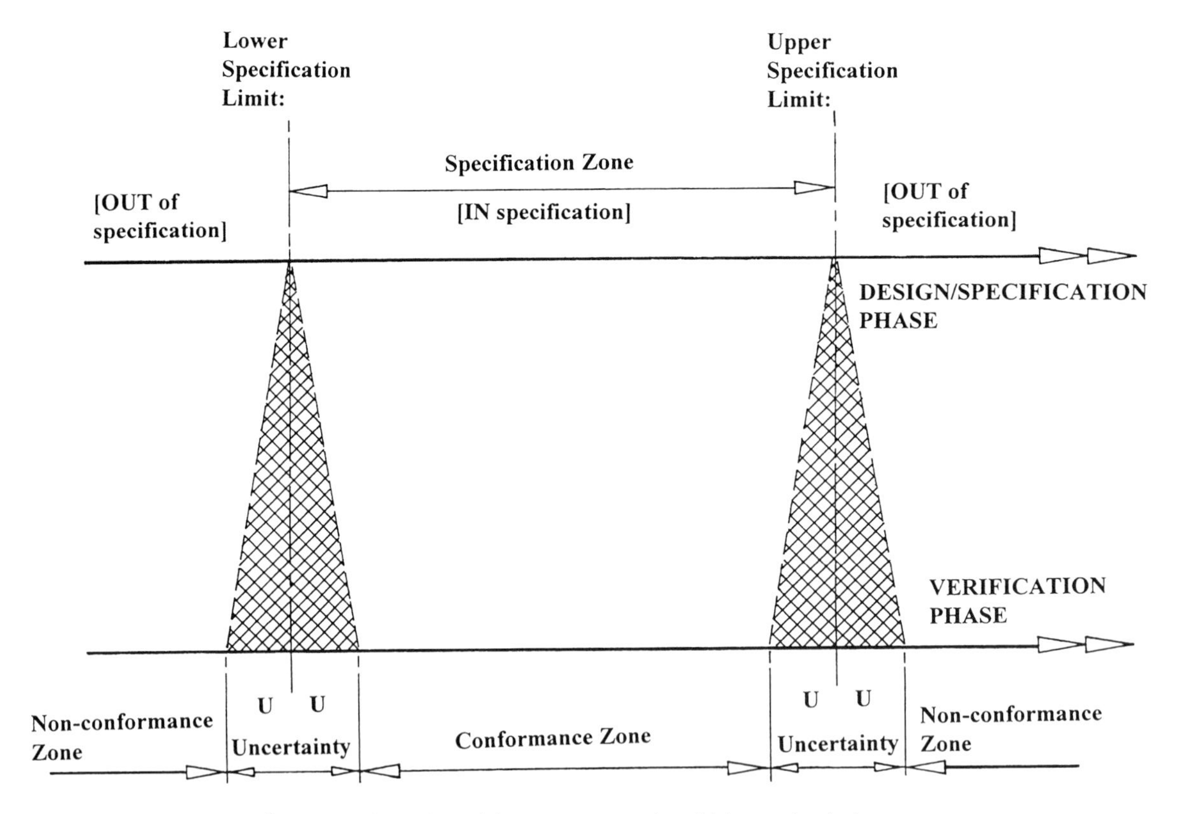

Figure 206. Comparison of the measurement value with its associated tolerance.

$$\bar{x} = \frac{1}{n}\sum_{i=1}^{n} x_i$$

where $\bar{x}$ = arithmetic mean, $\sum x$= sum of x and n = number of readings.

NB Sometimes several readings are taken, which modifies the calculation to

$$\bar{x} = \frac{1}{n}\sum_{i=1}^{n} f_i x_i$$

where $\sum fx$ = the sum of the frequencies of x.

For example in roundness measurement, if 10 LSC roundness polar plots were obtained in the course of an inspection procedure – 17, 18, 22, 18, 18, 19, 21, 19, 18 and 20 µm respectively, then the $\bar{x}$ value would be

$\sum x$ = 17 + 18, + 22 + 18 + 18 + 19 + 21 + 19 + 18 + 20 (µm)

= 190 µm

Therefore

$\bar{x}$ = 190/10 µm

= 19 µm

The *spread* of these readings will give an indication of the uncertainty of these measurements. Moreover, knowing how large this spread is enables one to form an opinion as to the quality of the inspection procedure or group of measured data. The customary way to quantify the spread is by utilising the *standard deviation* – often described by the Greek lower case letter σ (sigma). Usually only a moderate number of measured data is obtained and, in this case, an estimate of the standard deviation can be derived which is denoted by the letter s, meaning the *estimated standard deviation*. In principle the calculation of s for the previous series of LSC roundness polar plots would be

17 + 18 + 22 + 18 + 18 + 19 + 21 + 19 + 18 + 20 (µm) = 19 (mean)

The next step from the calculated mean is to find the difference between each successive reading; this is obtained in the following manner (by making the mean LSC value of 19 µm equate to zero):

−2 − 1 + 3 − 1 − 10 + 20 − 1 + 1

Then each of these values is squared as follows:

4 1 9 1 1 0 4 0 1 1

Next, sum the total, then divide by $n-1$; here, in the case of n this is 10, so $n-1$ is 9; hence:

$$(4+1+9+1+1+0+4+0+1+1)/9 = 22/9 = 2.44\ \mu m^2$$

The estimated standard deviation (s) can be obtained by taking the square root of the previous answer; therefore

$$s = \sqrt{2.44}\ \mu m^2$$
$$= 1.6\ \mu m \text{ (correct to one decimal place)}$$

The overall process of calculating the estimated standard deviation for a series of n measurements can be mathematically expressed in the following manner:

$$s = \sqrt{\frac{1}{n-1}\sum_{i=1}^{n}(x_i - \bar{x})^2}$$

where x_i is the result of the ith measurement and $\bar{x}$ is the arithmetic mean of n results considered.

To ensure that the production process was correctly operating, this estimated standard deviation s can show whether a process was in control or not. If plus or minus three s is utilised in conjunction with $\bar{x}$ (mean), then if the process is normally distributed all the LSC polar plot readings should fall within the upper and lower values, because this represents 99.7% of the overall population. Hence, the lower and upper limits for these LSC polar plots are

$$\bar{x} \pm 3s = 19 \pm 3 \cdot 1.6\ \mu m$$
$$= 19 \pm 4.8\ \mu m$$
$$= 14.2 \text{ (lower) and } 23.8 \text{ (upper) } \mu m$$

Therefore, with the lowest reading obtained being 17 μm and the highest reading 22 μm, these readings would fall within, say, some pre-selected tolerance set by the designer which might be an LSC of 14 μm and 24 μm, respectively.

One further way of expressing these values is to employ its associated *process capability index*, usually denoted by the C_p, which indicates the spread of the process. This can be derived in the following manner:

$$\text{Process capability } C_p = \frac{\text{overall tolerance}}{\pm 3s \text{ (that is } 6s \text{ or } 6\sigma)}$$
$$= 24 - 14/9.6$$

Therefore

$$C_p = 1.04$$

NB The tolerance represents the *specification limits* for the production process.

A term to denote the relationship between the tolerances and their associated process capability is the *relative precision index*' (RPI; see Figure 202). The relationship between the *tolerance* and 6s (that is, (3s), gives rise to three levels of precision for the process; these are:

1. *low relative precision* – where the tolerance band is less than or equal to 6s (representing a value of <1);
2. *medium relative precision* – where the tolerance band is greater than 6s (value of 1), but less than a value of 1.33 (a value of >1 but <1.33);
3. *high relative precision* – where the tolerance band is greater than 1.33 (representing a value of >1.33).

A distribution that equates to a high RPI that is "centred on the mean" will be unlikely to produce scrap parts, as any process drift introduced by systematic or random errors are unlikely to produce components out-of-tolerance. Today, with many companies artificially introducing RPIs that are close to or exceed values of 2, with these being "centred", there is quite a low probability that any components from such a production process will produce scrap – unless this process has become completely out of control through a fundamental change in its output.

Origins of uncertainties

There are numerous aspects that can contribute to poor measurements, many are easily established while others remain more difficult uncertainties to determine. Due to the fact that actual measurements are never taken under ideal conditions, any errors or uncertainties can be the product of a wide range of contributing factors (see Figure 207), such as:

- *metrological equipment* – the instrumental problems can accrue due to drift, wear, ageing and many more problems;
- *component inspection* – which may not be stable, an often quoted analogy of this problem being to attempt to measure the size of an ice cube in a warm room;
- *measurement process* – the practice of measuring a component may be difficult to achieve, this being particularly true if the subject of the measurement is in dynamic motion as readings are taken;

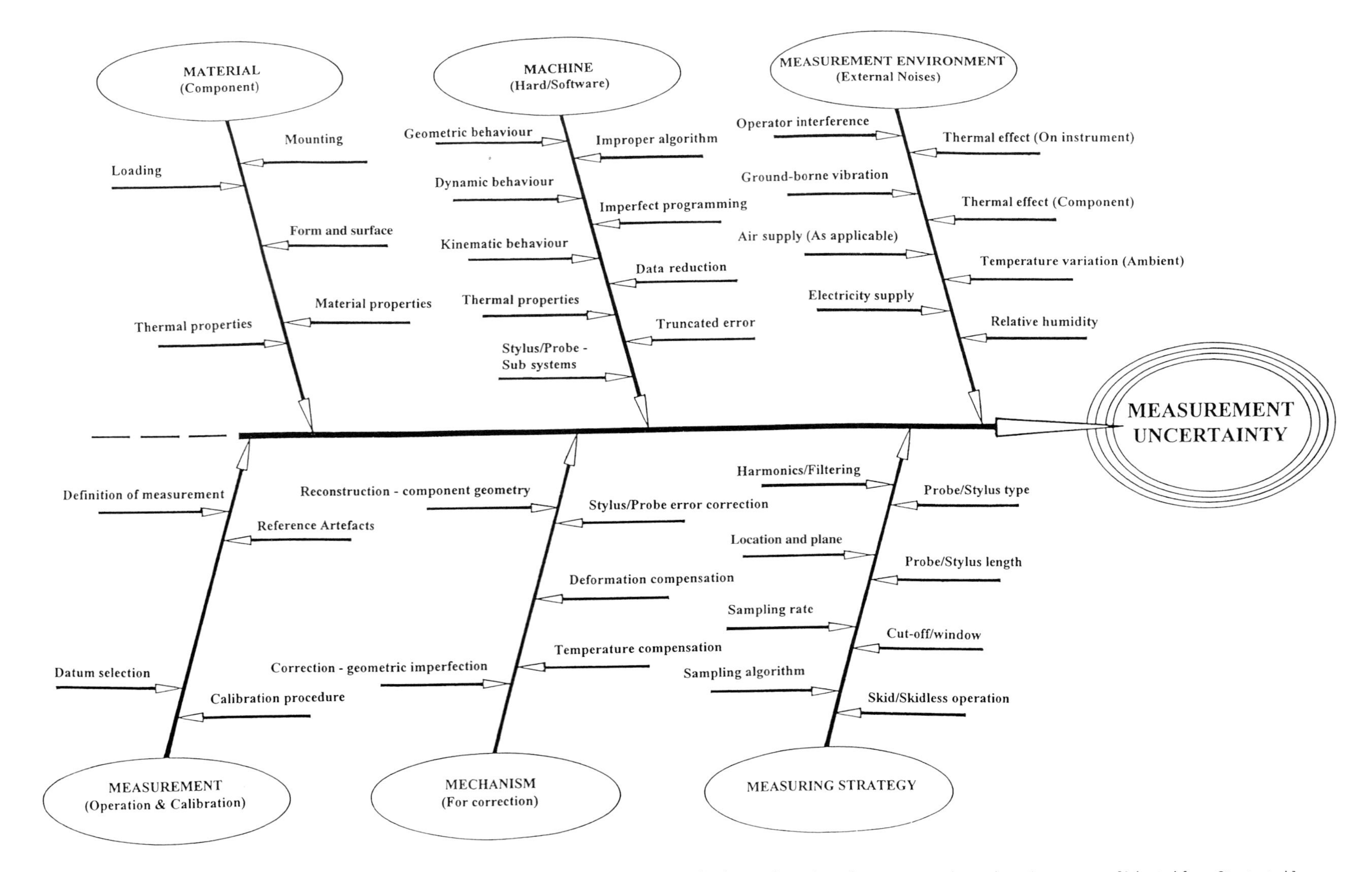

Figure 207. "Cause-and-effect" (6M) categorisation for measurement uncertainty – generalised causation – in surface texture and roundness instruments. [Adapted from Stout, et al.]

- *imported uncertainties* – any metrological instrument has its own uncertainty and this affects the subsequent measurements taken;

NB If an instrument has not been previously calibrated, then in this condition it will exacerbate any future measurements taken.

- *operator skill* – occasionally certain measurements rely on the operator's skill and judgement, with some operators being more efficient and reliable than others; for example, when timing a trial with a stop watch the level of uncertainty depends upon the operator reaction time;
- *sampling concerns* – any measurements obtained must be representative of the process being assessed.

NB As examples of these sampling issues, any temperature measurement should be assessed close to the item being monitored and not away from its local thermal environment. When selecting samples from the production line, ensure that they do not originate in the initial "start-up" batch run, nor that they are obtained from a mixed stillage from two separate production processes.

- *Metrological environment* – the measuring instrument can be significantly influenced by either the component being measured or variations in temperature, humidity, air pressure, possible vibrational effects and many more conditions.

NB The overall measurement uncertainty will be affected by the individual uncertainties, and where the error is known – that is, from a calibration certificate – a correction can be applied to the result of this measurement.

Calculation of measurement uncertainty

Prior to the calculation of measurement uncertainty, initially the sources of uncertainties need to be identified. Once this task has been performed then an estimate of the size of the uncertainty from each source must be established. Lastly, each of the individual and contributory uncertainties is then combined to produce an overall figure for the whole uncertainty. There are two methods used to estimate uncertainties: these are termed either *Type A* or *Type B* evaluations, and in many instances uncertainty evaluations of both types are required, their identification being according to the following:

- *Type A evaluations* – estimates of uncertainty by statistical techniques, normally from repeated readings;

Table 20. Determination of measurement uncertainty

Step	Procedure
1	Decide what you need to find out from your measurements. Decide what actual measurements and calculations are needed to produce the final result
2	Carry out the measurement needed
3	Estimate the uncertainty of each input quantity that feeds into the final result. Express all uncertainties in similar terms (see following text)
4	Decide whether the errors of the input quantities are independent of each other, but if not then additional calculations/information is required
5	Calculate the result of your measurement and include known corrections from previous calibration, etc.
6[a]	Find the combined *standard uncertainty* from each individual aspect
7[a]	Expressing uncertainty in terms of *coverage factor*, together with a size of uncertainty interval and, state confidence level
8[a]	Write down the measurement result and its uncertainty, stating how both were obtained

[a] To be discussed later in more detail in this section.

Source: Beginner's G.U.M. Bell/NPL, 2000.

- *Type B evaluations* – uncertainty estimates obtained from any other relevant information, such as from calibration certificates, manufacturers' specifications, published information, past experience or just common sense.

NB It is tempting to think that Type A uncertainties occur by random processes and Type B uncertainties result from systematic factors, but this logic is somewhat questionable.

Table 20 attempts to illustrate how to use information from both Type A and Type B evaluations for an overall estimate of uncertainty.

Any contributions to uncertainty must be expressed in the same units before they are combined, furthermore having matching levels of confidence. The contributing uncertainties and their respective confidence levels need to be uniform and this allows them to then be converted into standard uncertainties. Plus or minus one standard deviation is the margin for a *standard uncertainty*; moreover it also gives information on the uncertainty of an average and not just the spread of these values. It is normal to quote *standard uncertainty* using the symbol u, or as $u(y)$, this latter term being the *standard uncertainty in y*.

"Type A evaluation": standard uncertainty calculation

Once a series of measurement readings has been obtained for a "Type A evaluation", the $\bar{x}$ (mean) value and s (estimated standard deviation) can be calculated. From this set of readings the estimated standard uncertainty u of the mean can be calculated from the following:

$$u = s/\sqrt{n}$$

where n = number of measurements taken.

"Type B evaluation": standard uncertainty calculation

If information is somewhat scarce, typically for estimates of some Type B values, it might be possible to establish the upper and lower limits of uncertainty. If this action is taken, then it could be shown that these limits might equally fall anywhere between positions along the uniform distribution, or for that matter in a rectangular distribution. In this latter case the standard uncertainty for a rectangular distribution will be

$$u = a/\sqrt{(3)}$$

where a = the semi-range (half-width) between the upper and lower limits.

In most cases, for a calibration certificate of a measuring instrument it can usually be assumed that any uncertainty indicated here will have been derived from a normally distributed set of statistical data, although it is not uncommon to obtain data from either a uniform or rectangular distribution.

Combining standard uncertainties

Whether the individual standard uncertainties were calculated by either Type A or Type B evaluations is not an issue, as they can be validly combined by the mathematical techniques known as *summation by quadrature*, commonly termed either the *root sum of squares* or *root mean square*. The result is known as the *combined standard uncertainty* and is denoted by the notation u_c or $u_c(y)$.

Addition and subtraction: summation in quadrature

The most elementary case for this technique is where the sum of a series of measured values – by either addition and subtraction – is calculated. By way of illustration of the summation by quadrature method, if it is required to determine the total height of a group of "wrung-together" gauge blocks of differing individual sizes, then the standard uncertainty of each gauge block could be represented by the notation a, b, c, etc. The known (measured) values are then found by the *combined standard uncertainty* technique, by squaring the individual uncertainties, adding them together, then taking the square root of the total, as follows:

Combined standard uncertainty

$$\sqrt{(a^2 + b^2 + c^2 + \dots n^2)}$$

Multiplication: summation in quadrature

In cases of more complex uncertainty problems, an approach adopted to simplify the calculations is to work in terms of either the relative or fractional uncertainties. For example, in the simplest form of calculation, if a large rectangular surface area A has to be established, then it is simply determined by multiplying its length L by the width W, that is: $A = L \times W$. The fractional uncertainty in the area is established from the individual fractional uncertainties for the length and width. Therefore in the case of its length L with an uncertainty of $u(L)$, the relative uncertainty will be $u(L)/L$, likewise for the width W the relative uncertainty is $u(W)/W$. Hence, in the representative area the relative uncertainty $u(A)/A$ is given by

$$\frac{u(A)}{A} = \sqrt{\left(\frac{u(L)}{L}\right)^2 + \left(\frac{u(W)}{W}\right)^2}$$

Complicated functions: summation in quadrature

In many instances in the calculation of the final result, the value will be squared (Z^2); here the relative uncertainty due to this squared component takes the form $2u(Z)/Z$.

In some instances measurements taken are calculated using formulae that employ combinations of addition, subtraction, multiplication and division, etc. For example, in some "uncertainty budgets" (more to be said on this topic shortly) it might be necessary to measure the electrical resistance R and voltage V and then calculate the resultant power P, utilising the relationship:

$$P = \frac{V^2}{R}$$

In this case the relative uncertainty $u(P)/P$ in the value of power would be given by

$$\frac{u(P)}{P} = \sqrt{\left(\frac{2u(V)}{V}\right)^2 + \left(\frac{u(R)}{R}\right)^2}$$

In general, in any multi-step calculations the procedure for combining standard uncertainties in quadrature can be achieved in multiple steps, utilising the relevant form for addition, multiplication, etc., at each step. For complicated formulae the combination of standard uncertainties is more comprehensively discussed elsewhere (e.g., the UKAS Publication M 3003).

Correlation

The equations required to calculate the combined standard uncertainties above are only valid if the input values for standard uncertainties are not interrelated or *correlated*. *Correlation* refers to the question of whether all the uncertainty contributions are independent. Namely, could a large error in one input measurement cause a similar sizeable error in another? Further, might some outside environmental influence such as temperature create a similar effect simultaneously on several aspects of uncertainty, which are either easily established or require further investigation by to the metrologist? In many instances these individual errors are independent, but where they are not additional calculations are required.

Coverage factor k

In the previous discussion on uncertainty the components were invariably scaled to obtain the combined standard uncertainty. This result of the calculations for combined standard uncertainty may be thought of as equivalent to "one standard deviation", but it may be necessary to obtain the overall uncertainty expressed in terms of a different level of confidence, typically the 95% limit. Such rescaling can be achieved by using a *coverage factor* (k). Multiplying the *combined standard uncertainty* (u_c) by the coverage factor (k) results in what is termed the *expanded uncertainty* – normally denoted by U in the following manner:

$$U = k u_c$$

NB A specific value of the coverage factor (k) will introduce a particular confidence level for the expanded uncertainty.

In particular, if the overall uncertainty is scaled by utilising a coverage factor $k = 2$ to obtain an approximate level of confidence of 95%, the $k = 2$ value is acceptable when the combined standard uncertainty tends to be normally distributed. Some typical coverage factors for normal distributions are

$k = 1$ for a confidence level of around 68%;
$k = 2.58$ for a confidence level of around 99%;
$k = 3$ for a confidence level of around 99.7%.

Other less popular distribution shapes have a different range of coverage factors. Alternatively, whenever an expanded uncertainty is quoted, the standard uncertainty can be found by reversing the process, namely dividing this uncertainty by an appropriate coverage factor. On calibration certificates the quoted expanded uncertainty, if expressed properly, can be "interrogated" to obtain the required standard uncertainties.

When expressing the measurement uncertainty, it should be exacting in its definition, allowing no misinterpretation. To achieve accurate uncertainty definitions, several important factors need to be mentioned:

- *measurement result and uncertainty figure* – an artefact was 300 mm ± 1 mm;
- *coverage factor and confidence level* – recommended wording to minimise confusion is: "The reported uncertainty is based on a standard uncertainty multiplied by a coverage factor $k = 2$, providing a level of confidence of approximately 95%";
- *description of how uncertainty was estimated* – reference to the appropriate documentation.

Analysis of uncertainty: "uncertainty budgets"

To facilitate the process of calculating uncertainty, it is often helpful to summarise the uncertainty analysis – *uncertainty budget* – in a spreadsheet, as indicated in Table 21.

Another approach, which is similar in concept to that depicted in Table 21, is to establish the application of uncertainty budgets using spreadsheets, as depicted in Table 22. Assuming that the input quantities have been set, the next procedure is to calculate the variation limits (x_i estimates), uncertainty contributions, and then to set up an uncertainty budget. This can be achieved in several steps, as follows:

- *calculation of estimates* – based, in this case, on ambient conditions, material/product specifications and possible variation limits that are calculated for each input quantity;
- *application of distribution factor* – type of distribution is selected (depending on the probability that the result lies within the prescribed limits) for a rectangular distribution (100%) or normal distribution (99.73%).
  NB As an alternative to complex calculations, the following correction factors can be used:
  - *rectangular distribution*: multiply the limit by 0.7;
  - *normal distribution (3s)*: multiply the limit by 0.5.

Owing to the fact that in Table 22 there is no correlation between the uncertainty components, the combined standard uncertainty is expressed in the form of a standard deviation. In this case, the combined uncertainty (U_c), which was established at 2.29 μm, will need to be multiplied by a coverage factor of $k = 2$, to ensure a 95% probability that the measured result will lie within the uncertainty band. The metrology instrument in this case, which had a manufacturer's stated specified accuracy of 2 μm, here had a measurement uncertainty of 4.58 μm when used to inspect shafts under workshop conditions. In any uncertainty budget calculations the objective is to ascertain whether the measured result will be adequate for its intended purpose. For example, if the measurement uncertainty is small when compared to, say, the inspected diametral tolerance, then the metrology equipment can be used with confidence for the stated inspection procedure. In the case cited above for the uncertainty budget (Table 22) for the metrology instrument, the analysis illustrates that around 50% of this "budget" is accounted for by temperature factors alone. Because of the large impact that temperature influences have on the measurement uncertainty for any measurements on the shop floor, they must be "corrected back" to 20°C. However, correcting for temperature within the production shop may lead to unforeseen practical problems being encountered. The temperature on the shop floor could vary during the working day as dimensional measurements occur. The recorded thermal history may also not be correct during this time, because any thermometer readings may not be accurate, or even some uncertainty may exist concerning the actual value for the *coefficient of linear expansion* for the workpiece material, which leads to dubious inspection procedures. These uncontrolled factors of inadequate temperature correction mean that more generous limits are necessary to take into account the variability introduced because of such temperature influences. Here, the most obvious way to mitigate the lack of temperature control and its adverse effect on the measurement uncertainty is to undertake final inspection in a temperature-controlled environment that is as close as practicable to that of 20°C.

Thermal-induced measurement uncertainties during metrological inspection, notably the coefficient of linear expansion workpiece issues, become critical factors at the highest levels of accuracy. In calibration laboratories where thermal influences must be adequately controlled in order to minimise temperature-induced errors into measurement, the three greatest single contributors in uncertainty components are:

Table 21. "Uncertainty budget", in the form of a spreadsheet

Source of uncertainty	Value ± distribution	Probability	Divisor uncertainty	Standard
Calibration uncertainty	5.0 mm	Normal	2	2.5 mm
Resolution (size of division)	0.5 mm[a]	Rectangular	√3	0.3 mm
Artefact not lying perfectly straight	10.0 mm[a]	Rectangular	√3	5.8 mm
Standard uncertainty of mean 10 repeated readings	0.7 mm	Normal	1	0.7 mm
Combined standard uncertainty		Assumed normal		6.4 mm
Expanded uncertainty		Assumed normal ($k = 2$)		12.8 mm

[a]Here the (±) half-width divided by √3 is used.

Source: Beginner's G.U.M. Bell/NPL, 2000.

Table 22. A typical "uncertainty budget" for a metrology instrument

Quantity (x_i)	Type	Estimate (x_i)	Distribution factor	Uncertainty contribution (μm)
Repeatability	A			1.00
Variation of zero	A			1.00
Indication error	B	1.00	0.6	0.60
Flatness	B	0.60	0.5	0.30
Straightness	B	0.60	0.5	0.30
Parallelism	B	1.00	0.5	0.50
Ambient temp.	B	0.30	0.7	0.21
Temperature diff.	B	1.40	0.7	0.98
Comp. form error	B	2.00	0.6	1.20
Combined uncertainty	U_c			2.29
Expanded uncertainty	$U = 2 \times U_c$			4.58

Where

$U_c = \sqrt{(x_{1a}{}^2 + x_{2a}{}^2 + x_{1b}{}^2 + x_{1b}{}^2 + x_{2b}{}^2 + x_{3b}{}^2 + x_{4b}{}^2 + x_{5b}{}^2 + x_{6b}{}^2)}.$

Sources: Shankar (April 1999)/Beginner's G.U.M. Bell/NPL, 2000.

1. *calibration of material thermometers* – ensuring that these instruments (various types of temperature measurement devices) are thermally error-mapped;
2. *actual workpiece temperature is established* – as this will have a large influence on the coefficient of linear expansion (which must also be known);
3. *air temperature reading errors must be known* – due to air convection currents (lamellar air flow/air turbulence/hot-, or cold-spots within the room, etc).

Other environmental factors such as humidity and air pressure can also have a significant influence on the measurement uncertainty, particularly if the light path of laser-based equipment is employed in the calibration/measurement procedure.

In the previous tabulation and discussion of measurement uncertainty, it was important that the correct conclusions were drawn from these results, which is of great significance when deciding if the values fall within, on or outside the specification limit. If both the result and its accompanying measurement uncertainty fall inside the specification limits, then the process can be deemed to be *compliant*, in that the operation will be capable of sustaining efficient measurement and process control. For *non-compliance* to occur, neither the result nor any part of the uncertainty band will fall within the specified limits. In cases where neither the result nor part of the uncertainty band is either completely inside or outside the limits, then a dilemma occurs, as no firm conclusion about the compliance state can be made. Prior to stating the compliance with a specification, it is always advisable to check what the demands are for this specification. By way of illustration, sometimes a specification may incorporate factors such as "attributes" (non-measurable entities) – visible appearance, interchangeability, electrical connectivity, etc. – which have no actual bearing on the measurements previously taken, but can impinge on the whole uncertainty problem.

Reducing measurement uncertainty

It is essential to remember that it is just as important to attempt to minimise uncertainties as it is to quantify them. Good working practices will help to reduce these measurement uncertainties, which might include the following:

- *calibration of instruments* (or have them calibrated by a third party) – then apply those calibration corrections which appear on the certificate;
- *compensating for any corrections* – make adjustments for any other significantly known errors;
- *ensuring measurements are traceable to national standards* – employing calibrations that offer *traceability* via an unbroken chain of measurements to the relevant standard (see Figure 199);
- *selecting the optimum metrology equipment* – normally they would encompass a calibration facility with minimum uncertainty of measurement;
- *repeating and checking measurements* – occasionally allow someone other than the usual inspector to repeat these readings, or utilise a different method of assessment to ensure validity of the results;

- *checking calculations and transcription of results* – ensuring that results are substantiated and are correctly written;
- *utilising uncertainty budgets* – the "budget" will identify where the most significant uncertainties occur, so that they can be minimised;
- *awareness of problems in "calibration chains"* – at every step in the chain, uncertainty can increase.

NB Surface texture measuring instrument uncertainties covered in the *Measurement Good Practice Guide* no. 37 by the NPL (Chapter 8) – see References.

Utilise good working practices in measurements, such as by following equipment manufacturers' instructions operated by using experienced and relevantly trained personnel; checking and validating software (if employed) to ensure it operates satisfactorily; ensuring that any calculations in the "rounding" of measurement values is correct, ensuring that good records of both measurements and calculations are produced at the time of metrological inspection. Lastly, keep a written account of any other relevant additional information, as this may be useful if at a later stage these results are ever called into doubt.

Extra factors need to be considered in any dealings with uncertainty issues, particularly if they have any one or a combination of those listed below:

- when using statistics for very small data sets – normally considered to be less than about 10;
- if one uncertainty component is considerably larger than all the others involved in the "uncertainty budget";
- if just some inputs to the uncertainty calculation are correlated;
- when the distribution or spread has an unusual shape namely, that it is not "normally distributed";
- if the uncertainty has been obtained for other than a single result, that is, by fitting a curve or line to a number of points to obtain the required information.

This explanation is by no means an exhaustive account on the subject of measurement uncertainty, but attempts to discuss some of the relevant principles and practices behind its theoretical application to dimensional measurement.

6.5 Calibration: surface texture

The calibration of surface texture instruments principally relies on the use of secondary calibration artefacts. Due to the fact that most surface texture measuring instruments have a wide range of operating conditions and usage, the net result is that more than one type of artefact has been developed to cater for these demands. The current standard relating to surface texture calibration (ISO 5436-1: 2000) considers five general cases of artefact having several versions of each type - the first four initially being discussed below. A general consideration for any instrument calibration is that its profile should correspond to the surface that is to be measured. In more general terms, these artefacts fulfil the following assessment criteria, namely, Type A calibration standards will assess the instrument's vertical magnification factor, but do not give information with regard to instrument calibration in the scanning axis; this latter point of scanning is dealt with by the Type C artefact. Furthermore, it is important to check the instrument's overall calibration and that the Type D artefact has been produced to assess the ability to measure and calculate a surface texture parameter. Finally, the Type B artefact checks and ensures that the selected stylus performs to specification. In order to determine whether any surface texture instrument performs adequately, at least four previously calibrated and traceable types of artefact need to be available for this task. Hence, these Type A-to-E secondary calibration standards need to have the relevant current calibration certification to a known standard. Such secondary artefact calibration can be achieved by several means, either using interferometry or by an alternative stylus instrument, which itself has been calibrated to a traceable standard in an "unbroken and documented chain of calibration" to the primary standard (Figure 199), this traceability being a major requirement of ISO 9000: 2000. These surface texture calibration artefacts have a restricted range, relating to their own individual characteristics and to those of the instrument to be calibrated, as listed in Table 23.

Each type of surface texture artefact will now be briefly reviewed.

6.5.1 Surface texture artefacts (ISO 5436-1: 2000)

Depth measurement standard (Figure 208a–c)

Type A1: wide grooves with flat bottoms (Figure 208a–b)

The first artefact illustrated in this subgroup is Type A1 (Figure 208a), featuring a wide calibrated groove

Table 23. Surface texture artefact types, description and applications

Type	Artefact description	Application:
A	Depth measurement	Calibration of the vertical profile – having a known depth with wide grooves
B	Tip condition measurement	Calibration of the stylus tip condition – having various depths and widths and narrow grooves
C	Spacing measurement	Calibration of vertical profile – under certain conditions artefact can also be utilised calibrating horizontal profiles
D	Roughness measurement	Overall calibration of these instruments
E	Profile coordinate measurement	Calibration of instrument – for its profile coordinate system

Source: ISO 5436–1: 2000.

and a flat bottom, incorporating a ridge with a flat top. Alternatively, another approach is to have a number of separated features of equal or increasing depth/height, similar to that illustrated in Figure 213(I). On the artefact (Figure 208a), each calibrated feature has enough width to accommodate the stylus tip's shape or its condition, with at least five evenly distributed traces would be required over the window of measurement.

In a previous related standard (ISO 5436-1985, section 8) mention was made describing how to measure the depth of a Type A1 calibrated artefact groove with an appropriate stylus; the extract (slightly edited) has been reproduced below. The technique explains how and in what manner calibration can be achieved; this operational procedure for surface trace assessment is as follows (Figure 208b):

> A continuous straight mean line equal in length to three times the width of the groove is drawn over the groove to represent the upper level of the surface and another to represent the lower level, both lines extending symmetrically about the centre of the groove (see Figure 208b). To avoid the influence of any rounding of the corners, the upper surface on each side of the groove is to be ignored for a length equal to one-third of the width of the groove. The surface at the bottom of the groove is assessed only over the central third of its width. The portions to be used for assessment purposes are therefore those at A, B and C in Figure 208(b). The depth *d* of the groove shall be assessed perpendicularly from the upper mean line to the mid-point of the lower mean line. A significant number, not less than five, of evenly distributed traces shall be taken.

Type A2: wide grooves with rounded bottoms (Figure 208c)

Here (Figure 208c) the surface texture depth standard is comparable to Type A1, with the obvious difference in their visual appearance being the rounded groove, rather than flat bottomed. The radius must be sufficiently large enough to be insensitive to either the shape or condition of the stylus tip, or both. Once again, a significant number of surface traces is demanded – at least five – which are taken across the artefact and these are distributed evenly over the measuring window.

Tip condition measurement standards (Figures 208d, e and 209)

Type B1: narrow grooves (not illustrated)

The tip condition artefact (Type B1) can have either narrow grooves or a number of separated grooves designed to be progressively more sensitive to the stylus tip dimensions. The profile of these grooves is that they have rounded bottoms, the radii of each being sensitive to the stylus tip.

Type B2 (Figure 208)

These calibration standards normally have two or more groove patterns on a common base. In this calibration procedure a significant number of traces are demanded, the minimum being 18, which should be evenly distributed over the measurement window. Normally it is recommended that the stylus tip is sharp ((2 μm nominal radius) and has a filter with a λc cut-off according to the appropriate certification. In Figure 208(d) the Type B2 calibration artefact is illustrated, with the sensitive groove patterns being formed by the isosceles triangular groove geometry, having both sharp peaks and valleys. The recorded *Ra* from this calibration artefact is dependent on the stylus tip's size.

Alternatively, the Type B2 groove patterns that are insensitive to the stylus tip are shown in Figure 208(e), which occur by forming grooves on the artefact that are either sinusoidal (Figure 208e, left) or arcuate (Figure 208e, right). These types of calibrated groove profiles allow the value of *Ra* to be assessed independent of the stylus tip.

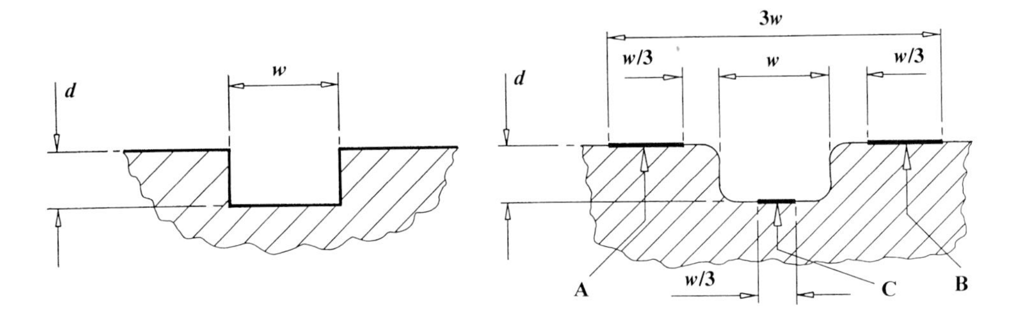

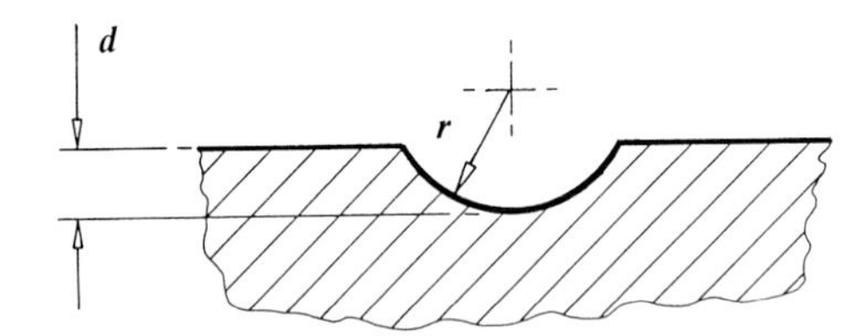

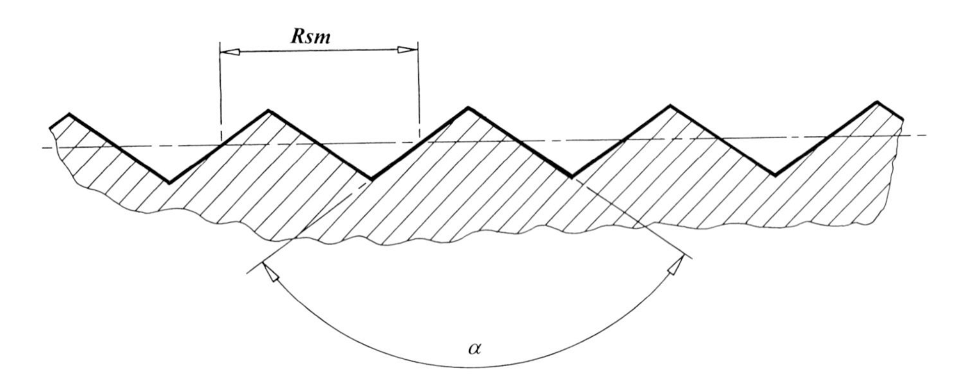

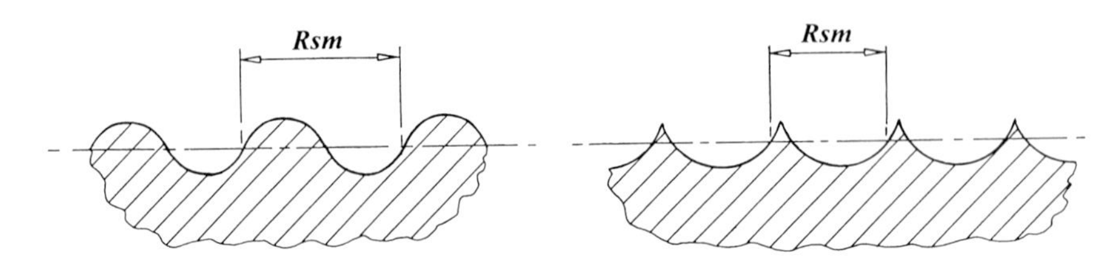

Figure 208. Depth measurement standards (Type A) and tip condition measurement standards (Type B). [Source: ISO: 5436–1: 2000.]

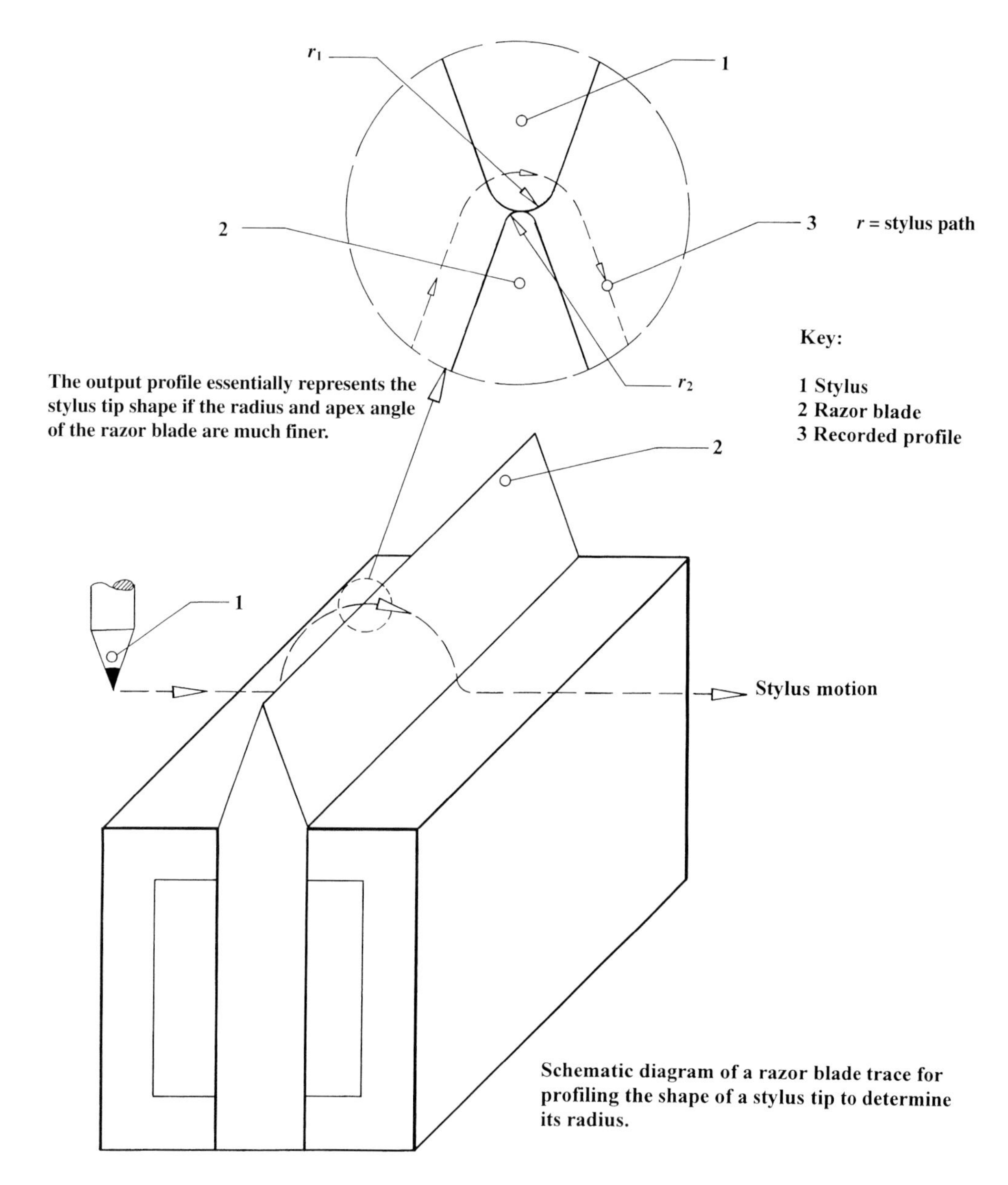

Figure 209. Calibration artefact Type B3. [Source: ISO 5436–1: 2000.]

Type B3 (Figure 209)

The artefact shown here represents a knife edge having a fine protruding edge, the radius and apex angle of which is shown, being both smaller and narrower than that of the stylus under test. The condition of the stylus can be established by traversing the stylus over the artefact and recording the surface profile, as schematically depicted in the main diagram. This type of stylus calibration technique can only be employed with low traversing speeds and for direct surface texture profiling instruments. The Type B3 artefact has a sharp razor blade, but even here a finite radius appears at the artefact's edge, as shown by the magnified view in Figure 209.

Spacing measurement artefacts (Figure 210a–d)

Artefacts of this type of design are mainly used to calibrate vertical profile components; however, an

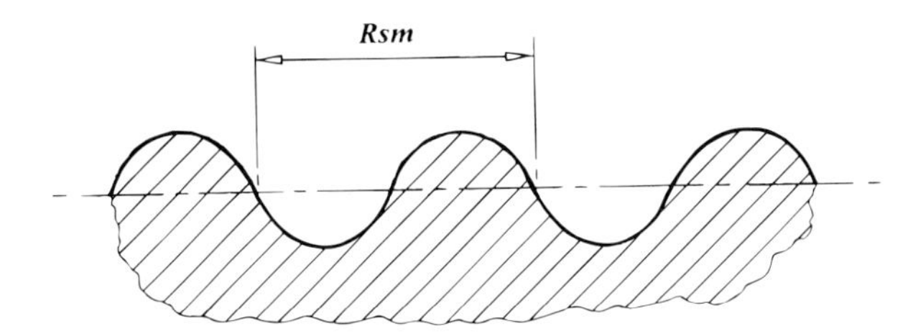

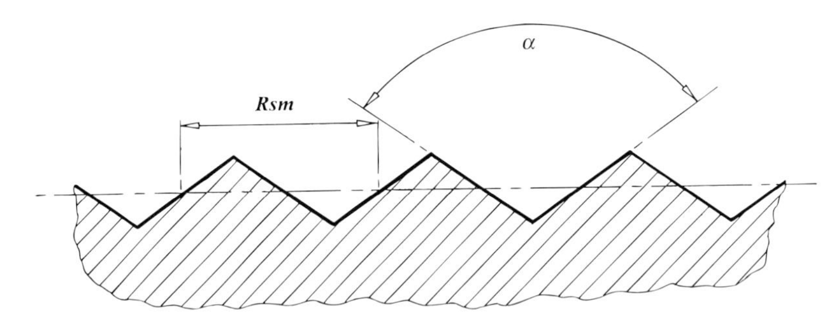

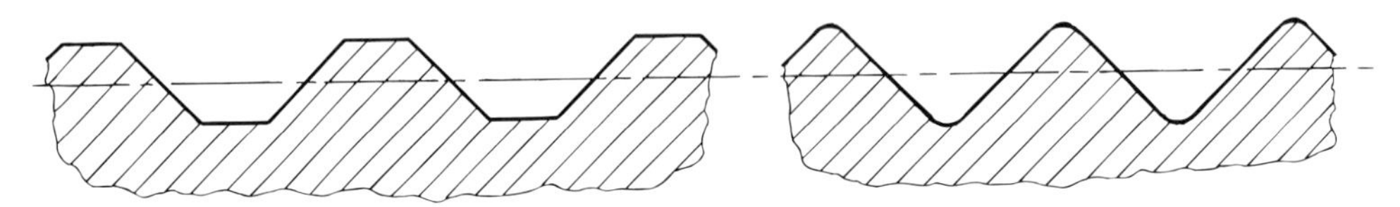

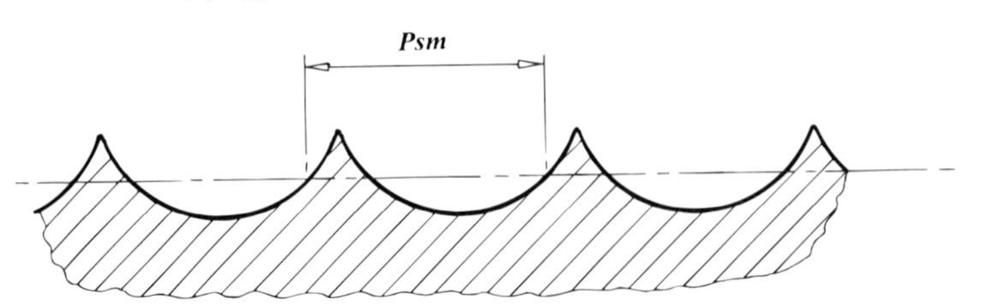

Figure 210. Calibration artefacts for spacing measurement. [Source: ISO 5436–1: 2000.]

alternative application for them is to calibrate the horizontal features based on the condition that the groove spacing will be maintained within acceptable limits of tolerance for this purpose. With spacing measurement artefacts, the grooves must be both consistent and repetitive and would normally come in a variety of profile shapes (Figure 210a–d).

Type C1 (Figure 210a)

The Type C1 artefact has sine wave profile grooves that are characterised by *Rsm* and *Ra*. The values for these parameters were selected to ensure that the attenuation by either the stylus or filter is negligible. As previously mentioned, a significant number of traces are advisable, the minimum number being 12, which are evenly distributed across the measurement window. This assessment strategy allows the parameters to be calculated according to the appropriate ISO standard.

Type C2 (Figure 210b)

This *Type C2* artefact has regularly spaced grooves based on the profile of an isosceles triangle, enabling both *Rsm* and *Ra* parameters to be assessed. The groove profile geometry values selected here are such that negligible attenuation occurs by either the stylus or filter. Once again, a significant number of traces are required, the minimum being 12, these being evenly dispersed over the measurement window. The surface texture parameters are calculated according to the relevant ISO standard.

Type C3 (Figure 210c)

With this *Type C3* artefact (Figure 210c), simulated sine wave grooves that feature either triangular profiles with rounded (right-hand side diagram), or truncated (left-hand side diagram) peaks and valleys may be present. At least 12 traces are required that are evenly distributed over the measuring window, having the parameters calculated according to the appropriate ISO standard.

Type C4 (Figure 210d)

In the case of this standard Type C4 artefact the grooves have an arcuate profile, being characterised by *Psm* and *Pa*. The values selected here should guarantee that negligible attenuation occurs by either the stylus or filter. As with the previously discussed assessment criteria, at least 12 traces are necessary that are evenly distributed across the measuring window. The parameters should be calculated according to the relevant ISO standard.

Roughness measurement artefact (Figures 211a, b)

Artefacts such as these are possibly the most commonly utilised for the overall calibration of surface texture instruments. To obtain the full benefit from such Type D standard artefacts it is normally necessary to average a statistically determined number of appropriately placed traces on the artefact.

Type D1 – undirectional irregular profile (Figure 211a)

This Type D1 artefact has an irregular profile in the direction of the stylus travel and is repeated for a given percentage of λc in the longitudinal direction, as shown schematically in the diagram. The profile shape is constant normal to the direction of measurement across the standard. The standard simulates workpiece surfaces that contain a wide range of spacing of crests, providing reassurance for an overall check on the instrument's calibration. This standard is characterised by the parameters *Ra* and *Rz*. As in most of the previous standards, the minimum number of traces required is 12, which are evenly distributed across the measuring window. The parameters should be calculated according to the appropriate ISO standard.

(a) Type D1 calibration artefact, undirectional irregular profile

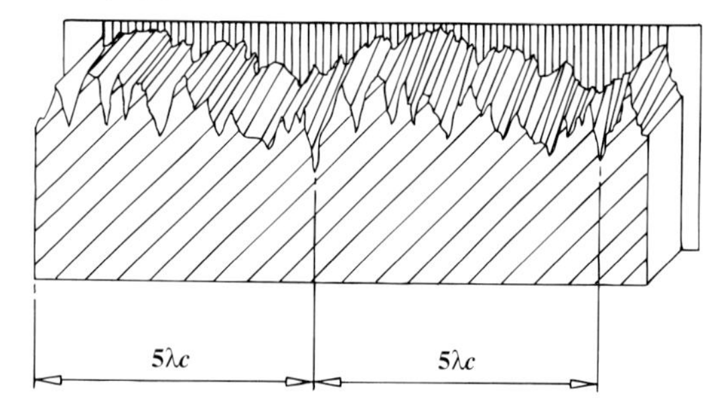

(b) Type D2 calibration artefact, circular irregular profile

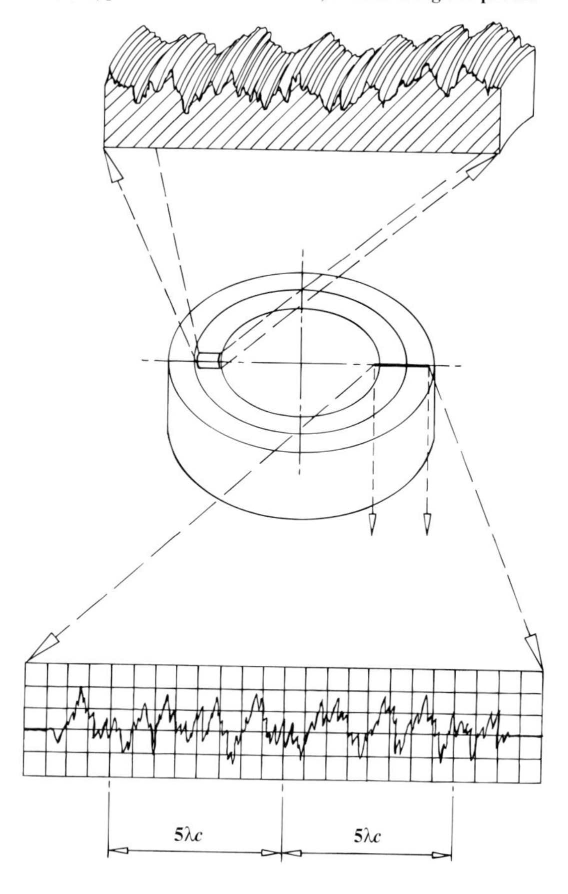

Figure 211. Roughness measurement artefacts. [Source: ISO 5436-1: 2000.]

Type D2 – circular irregular profile (Figure 211b)

The Type D2 artefact (Figure 211b) is characterised by *Ra* and *Rz* having irregular profiles repeated every $5\lambda c$ in the radial direction. The profile shape is constant normal to the measuring direction of the artefact, in the circumferential direction. At least 12 traces are required that are evenly distributed over the measuring window and, as before, the parameters should be calculated according to the appropriate ISO standard.

Profile coordinate measurement artefact (Figures 212 and 213)

Type E1 (Figure 212)

This Type E1 artefact measurement standard is characterised by radius and *Pt*. The radius of the sphere, or hemisphere as depicted in this instance, should be sufficient to allow the stylus tip to remain in contact with the convex surface profile and not foul on the stylus's stem during its measurement travel. Prior to calibration assessment the stylus tip should be positioned symmetrically either side of the highest point of the intended trace.

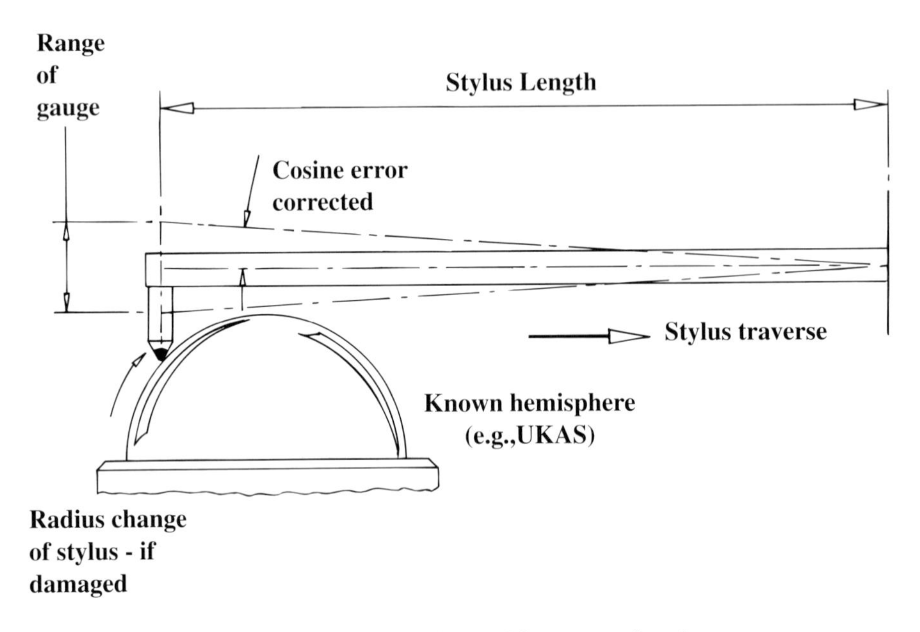

Figure 212. Type E1 profile coordinate measurement artefact for gauge and stylus calibration using a "qualified" hemisphere. (Courtesy of Taylor Hobson/ISO 5436-1: 2000)

Type E2 (Figure 213)

The Type E2 artefact is a precision prism characterised by angles between the surfaces and *Pt* on each surface. The size and shape of the standard should be such that, as in the case of the previous artefact (Type E1), the stem does not foul during stylus travel across it and the stylus should be symmetrically set for ramping up, over and down the adjacent prism faces. The length of the top plane should be long enough to enable the standard to be levelled in a stable manner.

For any calibration procedure, the surface texture instrument should be calibrated at and in its operating environment and across the anticipated ambient conditions that could influence the in-service use of the equipment. Generally, adoption of the following calibration procedure will minimise unexpected problems:

- the artefact should be aligned to within 10% of the measuring range of the instrument;
- selection of the appropriate measuring conditions, such as
  - sampling length,
  - evaluation length,
  - cut-off wavelength;
- undertake the specified measurements on each artefact distributed across the measurement window as shown in the measurement location pattern.

Type F standards (not illustrated)

ISO 5436: 2000 Part 2 describes the use of reference software and its associated reference data sets – termed "softgauges" – in facilitating calibration tasks for surface texture measuring instruments. At present these software gauges are not yet commercially available and, as such, are outside the current discussion. However, they will no doubt be featured in future calibration applications by industry at large.

6.5.2 Stylus damage

Because of the potential and possibly fragile nature of most surface texture styli, calibration is an important diagnostic/condition monitoring aid for the current state of the tip. In particular, the conical diamond styli are particularly susceptible to unintentional damage as the forces on the point are high, because of the very small contact area on the workpiece under test. Normally, wear is not the main contributor to stylus rejection via calibration, but the more likely scenario is as a result of tip fracture. The tip condition of both a perfect and unacceptable pair

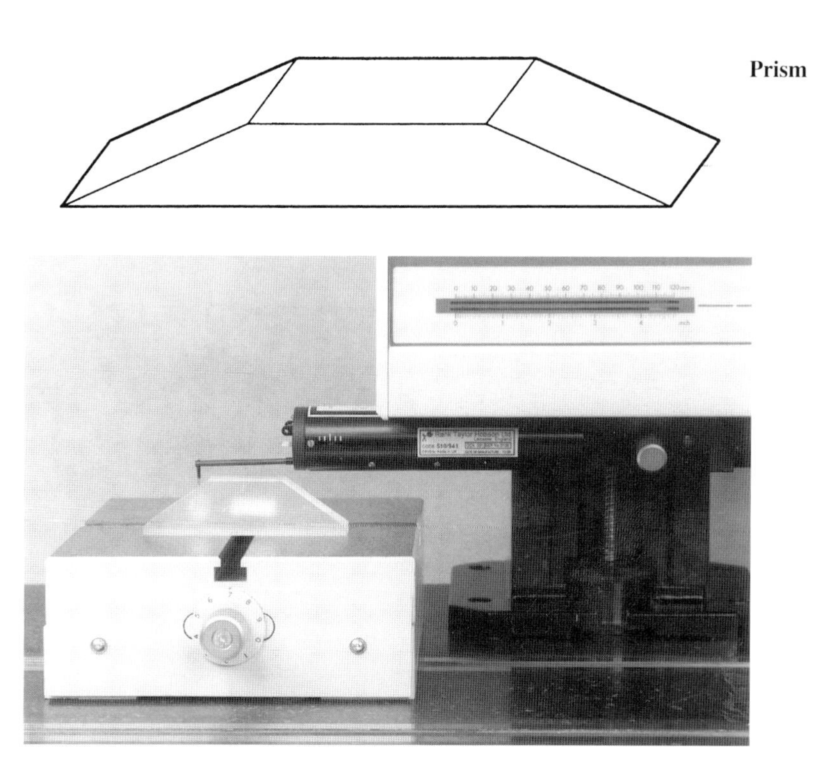

Figure 213. The profile coordinate measurement artefact.

Figure 214. Photomicrograph taken on an electron microscope of a conical diamond stylus with a radius tip of 5 μm. [Courtesy of Hommelwerke GmbH.]

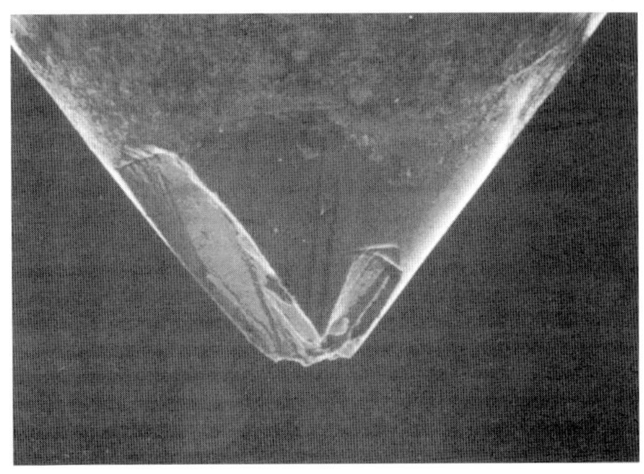

Figure 215. SEM photomicrograph illustrating catastrophic damage to the tip of a stylus. [Courtesy of Hommelwerke, GmbH.]

of styli is depicted in Figures 214 and 215, respectively. In Figure 214 the tip is in excellent condition and the 5 μm tip radius is intact. Conversely, in Figure 215 the tip has suffered catastrophic damage, through misuse or perhaps because of inappropriate protection – left unprotected without the skid and its accompanying cover – resulting in a potential condition for accidental damage. The uninformed user may not have noticed this damage, which may not be visually apparent until high optical magnification is employed. Therefore with the stylus tip in this poor state, the operator is currently unaware that false readings are being introduced into the measurement cycle. This stylus tip condition (Figure 215) is clearly unacceptable, which is why it is important to periodically check – via calibration artefacts – that the tip is in perfect condition. This calibration procedure should, as a minimum, be undertaken at least once per day. Ideally, calibration needs to be part of the measurement strategy and its frequency adjusted according to:

- *conditions of inspection frequency* – many readings and components to be assessed, necessitates more frequent calibration routines to be undertaken;
- *critical nature of these components* – high accuracy, more calibration and vice versa;
- *other unusual metrological conditions* – which might degrade the stylus performance during the measurement routine.

The ramifications of a damaged stylus on the recorded surface texture result can be gleaned from the schematic diagrams shown in Figure 216. Here, both a perfect and damaged stylus is illustrated traversing across a calibrated surface. In reality, this standard should have a wider included angle between the peaks (see Type B2, Figure 208d) to avoid the potential "worst case" situation that is shown in Figure 216, where contact could be made along the stylus flank, rather than at the tip. The tip damage as shown in the lower diagram in Figure 216 will not fully enter the valley and will foreshorten the peak height as it traverses across the surface, leading to a lower recorded *Ra* reading under this condition. This unintentional error in the required surface texture parameter could have major repercussions in service, or at best will allow substandard/scrap parts to enter the supply chain. Clearly, there is a requirement to prevent these unanticipated and undesirable measurement/quality problems from occurring; with systematic and regular calibration their likelihood is significantly diminished.

6.6 Calibration: roundness

Roundness-testing instruments are generally of two basic configurations, as depicted in Figure 112. Instrument calibration consists of some general testing procedures and the use of several artefacts to assess the equipment's capabilitity. The types of testing regime to be undertaken might normally include the following:

- *spindle* – checking with a previously calibrated glass sphere or hemisphere;
- *stylus deflection* – using an appropriate set of calibrated gauge blocks, or utilising the so-called "flick-standard";
- *stylus-to-component alignment* – using the "cresting standard";
- *column straightness* – employing a calibrated cylindrical square;

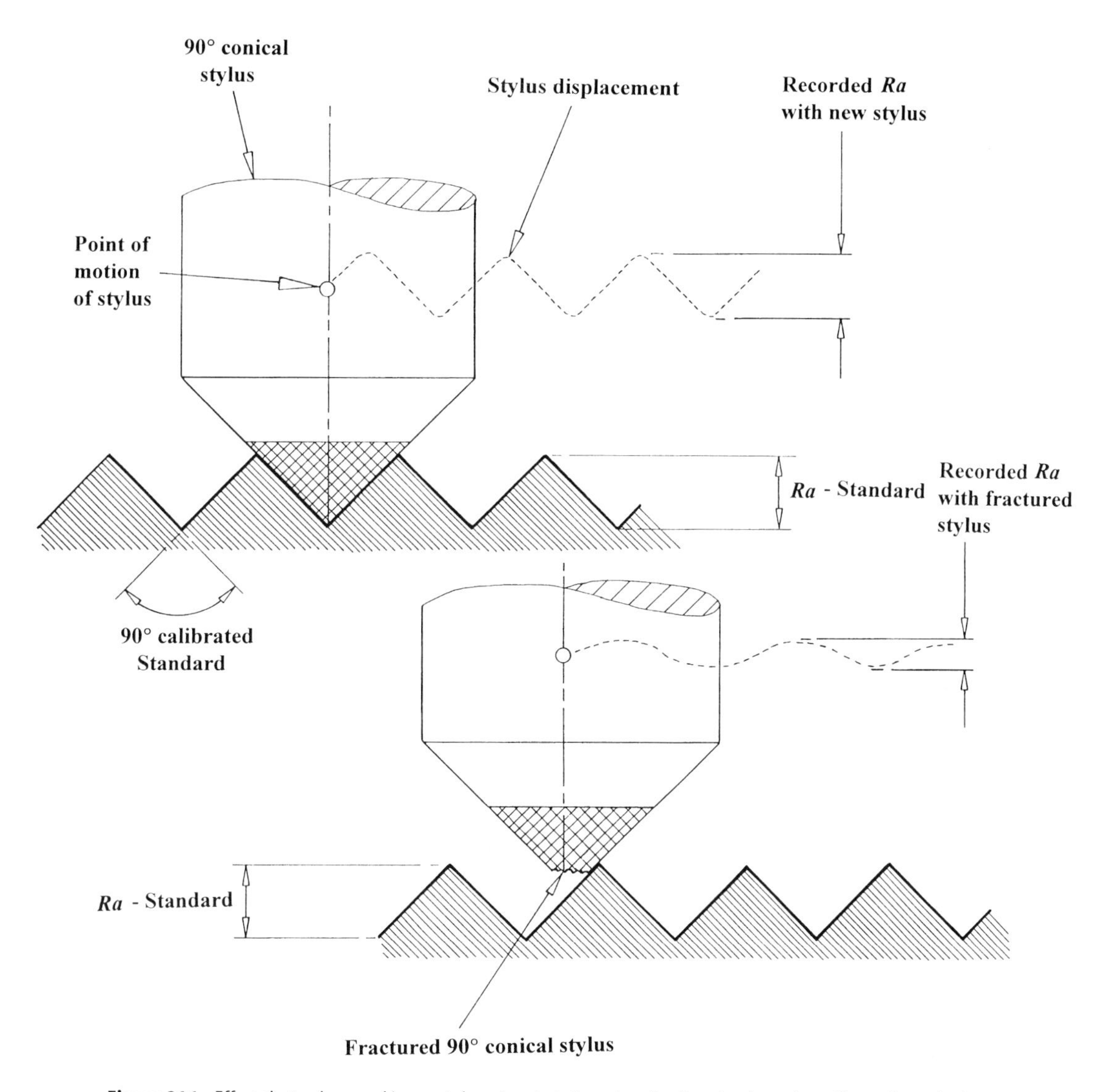

Figure 216. Effect that a damaged/worn stylus plays in influencing the *Ra* value from the calibrated standard.

- *column squareness* – utilising the same calibrated cylindrical square.

NB In the case of the latter two column checks, these are only undertaken if the instrument has an appropriate alignment column.

Some of the known sources of uncertainty in roundness measurement are the result of the following factors:

- horizontal and vertical fluctuations of the instrument's spindle path from one trace to the next;
- variability due to stylus type and its length;
- general misalignments in the stylus column to that of the spindle axis;
- non-uniformity of recorded displayed profile;
- variability in the user's interpolation of trace position on polar graph (for manual "fitting" techniques only).

NB This latter uncertainty is a human-induced error and is not strictly speaking an instrument calibration error.

Spheres and hemispheres: spindle assessment

A hemisphere can be employed to check the roundness measuring instrument in one of three ways, these are:

(i) *Verification* – a previously calibrated hemisphere is measured on the instrument being tested. If the differences fall within the limits, then the instrument is accepatable for further use;

NB Prior to such calibration work, the instrument must have previously been calibrated for gain by using a "flick-standard" – see following pages.

(ii) *Secondary calibration* – a previously calibrated hemisphere is measured on the instrument being tested. The result from the instrument is compared to the known result. These differences are stored in the instrument and are used to correct subsequent measurements (software correction);

NB The instrument must have previously been calibrated for gain by the "flick-standard".

(iii) *Primary calibration* – of spindle errors.

Regardless of the equipment configuration prior to any instrument calibration, the artefact should be centred; on some instruments they utilise automatic procedures to align the work axis to the table spindle, while others require manual centring and levelling. In order to ensure consistency and repeatability in the calibrated data, it is necessary to orient the artefact to that of a known angular position on the instrument. This artefact orientation can be undertaken by its "witness mark" which denotes the zero angular position; this is made to coincide with a similar mark on the machine, equating to both their datum (zero) positions. On sophisticated instruments, this known error at specific angular orientations means that a software correction can be applied to minimise such errors. Spheres and hemispheres used to assess the performance of roundness measuring instruments, are usually round to better than 250 nm and have a calibration unccertainty of $\pm$10 nm. To achieve sphere calibration, a special precision indexing fixture has to be manufactured (Figure 217; see the appendix for a schematic view). A displayed profile is made with the spherical artefact in this position averaged over four complete revolutions), then via the indexing fixture it is indexed clockwise by $360/n$ (where n in this case is 10). Once again, four artefact revolutions are undertaken and recorded, with a total of 11 (36° interval) indexings being carried out to return to the original angular datum fixture position (Reeve method). It is normal practice to obtain three orthogonal planes on the calibration sphere at 90°. Rather than a glass sphere artefact being supplied with the instrument (Figures 217 and 218), the more usual item is a glass hemisphere, which has its roundness calibrated at a 3 mm offset from the equator. In any calibration assessment it is important to record the angle (δ) between the stylus arm and the tangent to the artefact at the point of contact. This angle is normally set at 0° for a sphere measured at its equator but for

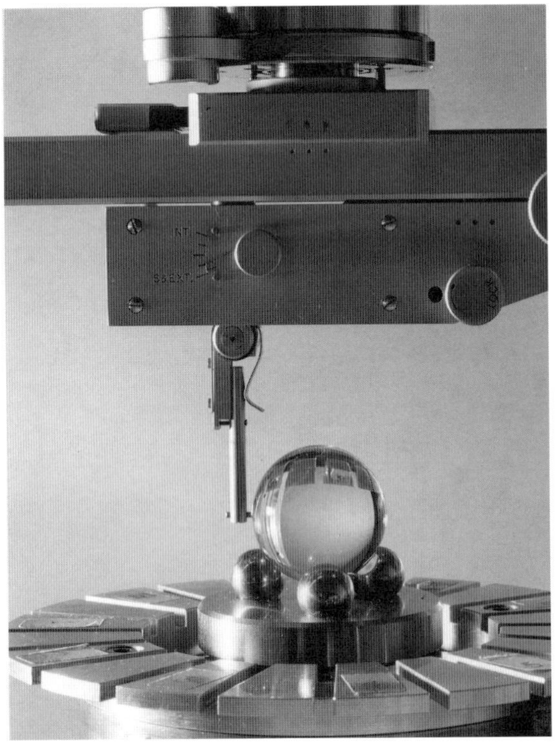

Figure 217. Calibrating a "test sphere" with a special-purpose indexing fixture on a precision roundness-testing instrument. (Courtesy of Taylor Hobson.)

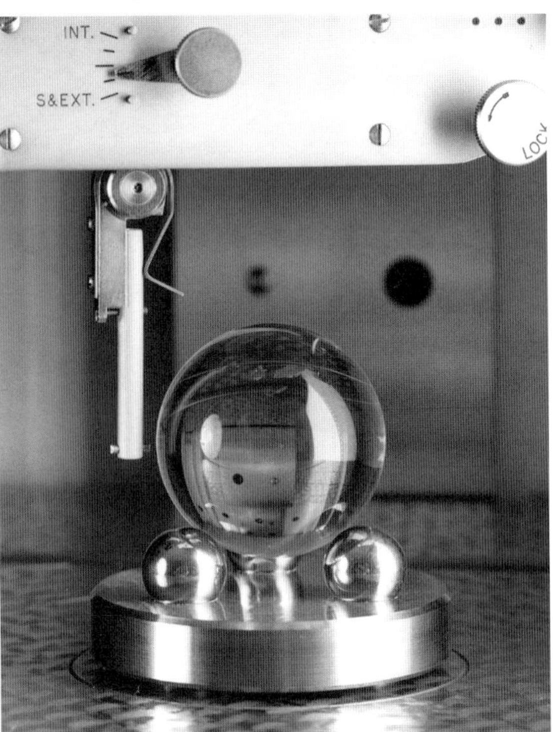

Figure 218. Calibrating a roundness-testing machine using a glass sphere, to 28 nm departures from roundness. (Courtesy of Taylor Hobson.)

hemispheres it can be both angled and offset (3 mm) with respect to the equator (see Appendix C, Figure 242). (Figure 218).

This offset correction for δ is a useful tip to note when assessing a precision workpiece on a roundness instrument that has not been recently calibrated. If a significant "error spike" appears on the polar trace, its origin can be established by simply rotating it through 90°; if it does not disappear then the error is in the instrument's spindle.

Gauge blocks/"flick-standard": stylus deflection

Stylus deflection calibration, for accuracy and sensitivity of the probe and its associated lever arm length, can be established by utilising either a set of calibrated gauge blocks and their accompanying optical flat (Figure 219, inset photograph), or by using a "Flick-standard" (Figure 220). In the former

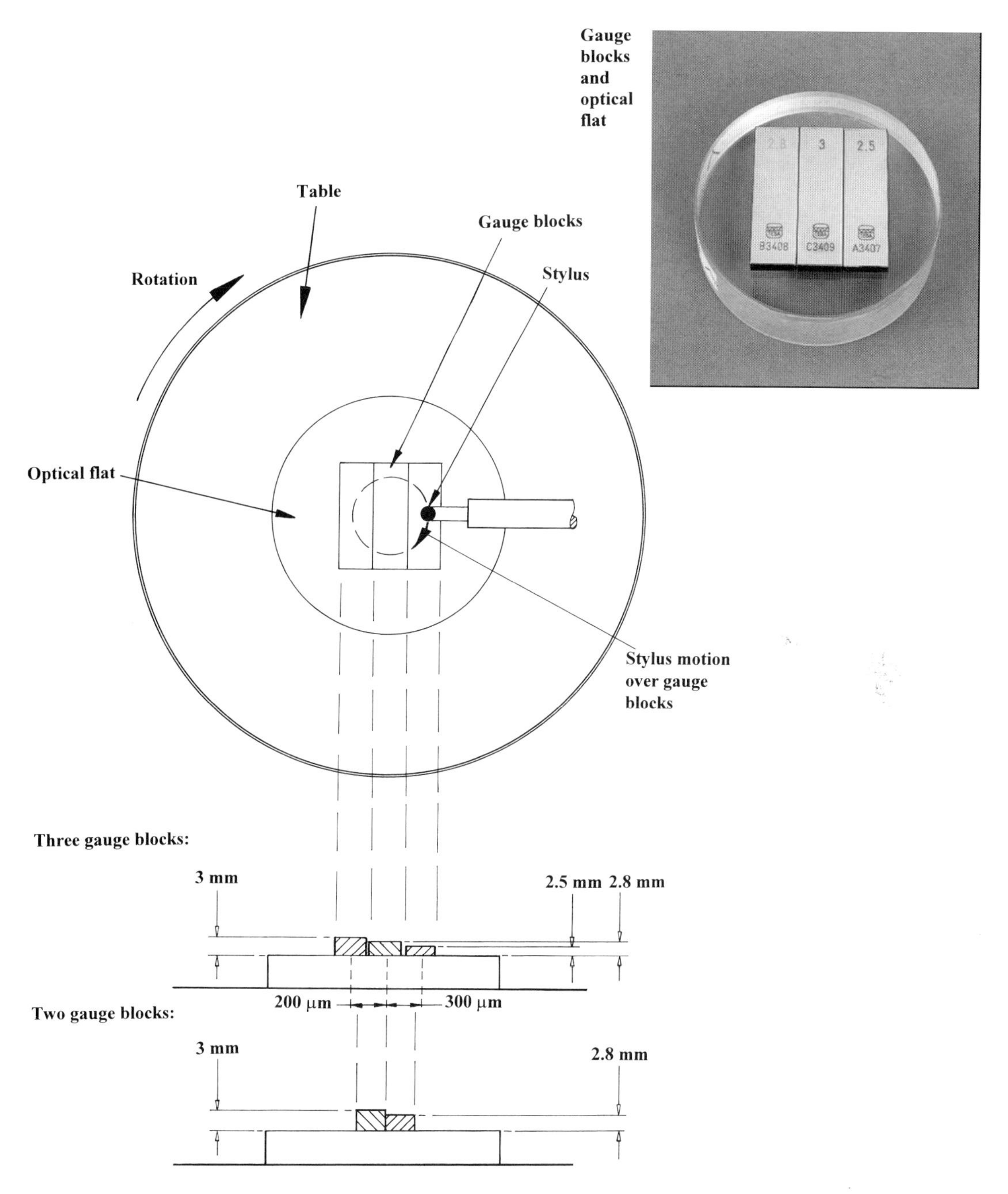

Figure 219. Calibration of roundness instrument employing optical flat with gauge blocks. [Photograph courtesy of Taylor Hobson.]

case the optical flat is positioned on the instrument's table, then initially centred and accurately levelled, enabling the calibrated precision gauge blocks to be "wrung" onto the flat's surface, eliminating the "air gap" which would otherwise occur. The stylus tip is brought down to contact these gauge blocks at a radius offset and the table is rotated, as shown in Figure 219. If three gauge blocks are used, then a two-step height can be accurately established, giving a recorded series of step heights which can then be compared to the gauge block calibration certification for establishing the stylus deflection/sensitivity. The correction figures can be entered into the instrument's software and such correction factors play an essential role in minimising uncertainty in the measurement (traceability) chain. For example, the calibration factor for a two-gauge block step height could be found in the following manner:

Measured step $= X - Y = Z$

Gauge block step $= C$

Calibration factor $= C/Z = F$

The general artefact configuration of gauge blocks with their associated optical flat arrangement for the calibration of stylus deflection can be enhanced by employing the *reduction lever principle* are illustrated in Appendix C, Figure 241. For ultra-high precision instruments the conventional gauge blocks and optical flat arrangement have to be mechanically magnified by a reduction lever (Appendix C), typically obtaining magnifications of 20:1. This arrangement can produce a total uncertainty budget on gauge block steps to <0.016%, this being equivalent to <3.2 nm on a step height of 2 μm.

For calibration of stylus deflection on either longer stylus arms or with earlier roundness testing equipment featuring analogue-based instrumentation, the flick-standard can be successfully utilised (Figure 220). This artefact is designed around a precision cylinder with a small flat on the peripheral face (Figure 220b). The standard artefact has a calibrated flat on its periphery, which when rotated in contact with the stylus tip (Figure 220b), gives a notable "flick" on the displayed profile (Figure 220a), which equates to the calibration value – if instrument calibrated is within its range. It has been reported that such flick-standards have a major disadvantage such as a small dynamical content for larger wave numbers. For example, the amplitude of the signal of a typical "flick" having a ϕ20 mm cylinder with a flatness depth of 12 μm has no significant contributions from larger wave numbers other than at 75 upr (undulations per revolution) (see Figure 220a). For this calibrated artefact all the wave numbers less than 75 upr contribute to the signal, while the individual wave number amplitudes are quite low if compared to the 12 μm form deviation. Any calibration work has to be undertaken in a suitable measurement range for the full form deviation, although the actual calibration is dependent upon the individual wave amplitudes, hence the signal-to-noise ratio (SNR) tends to be quite low for such flicks.

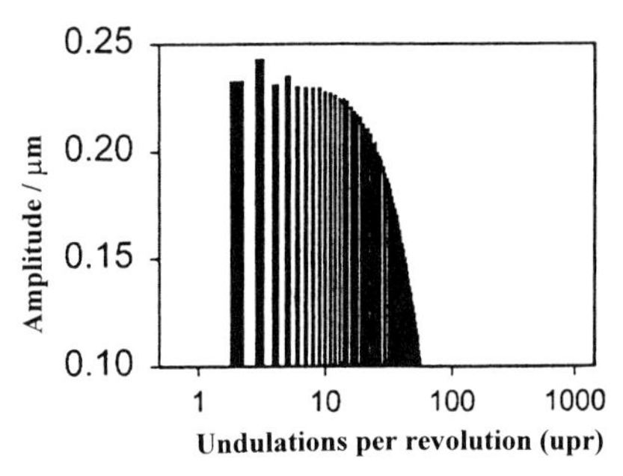

(a) Flick standard amplitude spectrum [Courtesy of PTB, Braunschweig, Germany.]

(b) Flick standard and centring iris on instrument

Figure 220. A "flick standard" and its amplitude spectrum, being used to assess a roundness instrument.

It has been suggested that a new form of calibration standard will improve data capture and response, based on a spatial embodiment of the superposition of several sinusoidal waves, termed a multi-wave standard (MWS). This waveform artefact can be reproduced around the cylindrical circumference, whereas the generatrices should be straight lines without form deviations as these act to lower the influence of potential variance of differing axial sections on the prospective result (Figure 221). Such artefact geometry has been termed "R-type MWS" – the "R" denoting roundness, with the waves being reproducible both for internal and external cylindrical artefacts. Hence, these internal "R-MWS" artefacts are the first internal calibration standards.

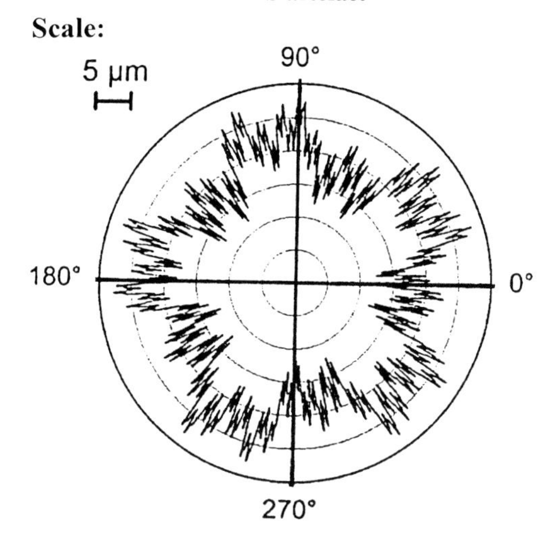

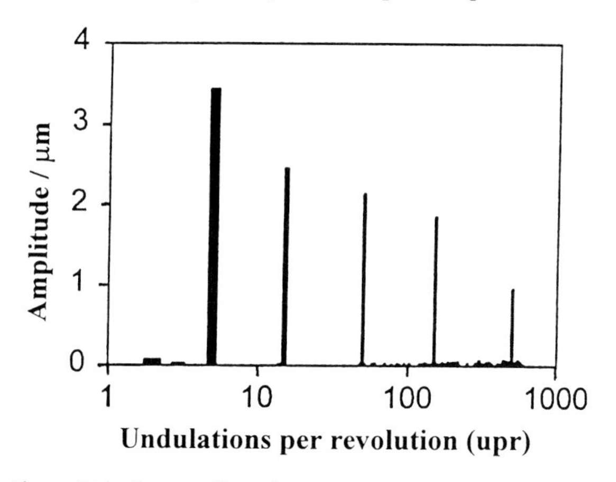

Figure 221. Form profile and spectrum of a multi-wave standard (MWS). [Source: Jusko and Lüdicke, 2000/PTB.]

Another variant of these waves machined along the generatrices of a cylinder and their nearly perfect circles as roundness profiles are termed "S-type MWS" – the "S" in this case denoting straightness. Additionally, such a waveform could be machined into a nominally flat surface, which then can perform its task as a calibration artefact for surface roughness instruments. The 80 mm approximate diameter of the R-MWS artefact has its wave profile machined into a chemically applied nickel layer on an aluminium substrate, by diamond turning coupled to a fast tool servo-mechanism. A reference mark denotes the 0° (datum) position or witness mark. The embodied wavelengths were selected to match the limit wavenumbers of filters defined in ISO/FDIS: 12181, namely, 5, 15, 50, 150 and 500 upr. This evolved multi-wave standard artefact and its integrating frequency-space analysis, when compared to the conventional flick-standard, exhibits a much higher stability under test, inferring improved transfer stability of calibration to the end-user.

Cresting standard

Yet another artefact that is utilised for assessment of angular offsets (arcuate uncertainty) produced by spherical stylus tips on roundness instruments is the so-called "cresting standard". "Cresting" is alignment of the stylus to the axis of rotation of the instrument (Figure 222a–c). An incorrect stylus/component contact causes a certain amount of *noise*, which is included with the measurement data. However, this stylus and component geometric relationship will not normally significantly influence the filtered results.

The results of the measurement taken can be influenced in the following three ways:

1 *Errors in peak-to-valley measurements* (Figure 222d) – this is probably the least significant error introduced by poor "cresting" and its value is of the order of: $1 - \cos\alpha$ (where α is the angular offset) . Typically, with diameters of components of approximately 10 mm and the instrument "crested" within 0.5 mm, α is 5.7° and this will relate to $(1 - \cos\alpha) = 0.005$, that is, introducing a 0.5% error in peak-to-valley values. If the diameters are larger than 10 mm, the cresting problem becomes even less significant, although for very small diameters the use of the bar-type stylus is recommended. Normal cresting accuracy of a roundness instrument is typically about 0.1 mm, being set at installation. If customer-manufactured styli are produced, then any cresting may be adjusted to suit.
2 *Errors in absolute radii* (Figure 222e) – when calibrating at a larger diameter than the measured diameter, the condition can arise where $E_{rad} = E_2 - E_1$. For example, if $E_{crest} = 0.5$ mm, $R_1 = 50$ mm and $R_2 = 25$ mm, then $E_{rad} = (5\ \mu m - 2.5\ \mu m) = 2.5\ \mu m$. This introduces a 2.5 μm error in absolute R.
3 *Errors in centring and levelling* – as the absolute radius information is used in order to undertake automatic centring and levelling operations (on some sophisticated instruments), then the previous error criteria apply. Such factors will not prevent accurate centring and levelling from occurring, but will have the effect of increasing the number of operations required, particularly as the component approaches its theoretical centre.

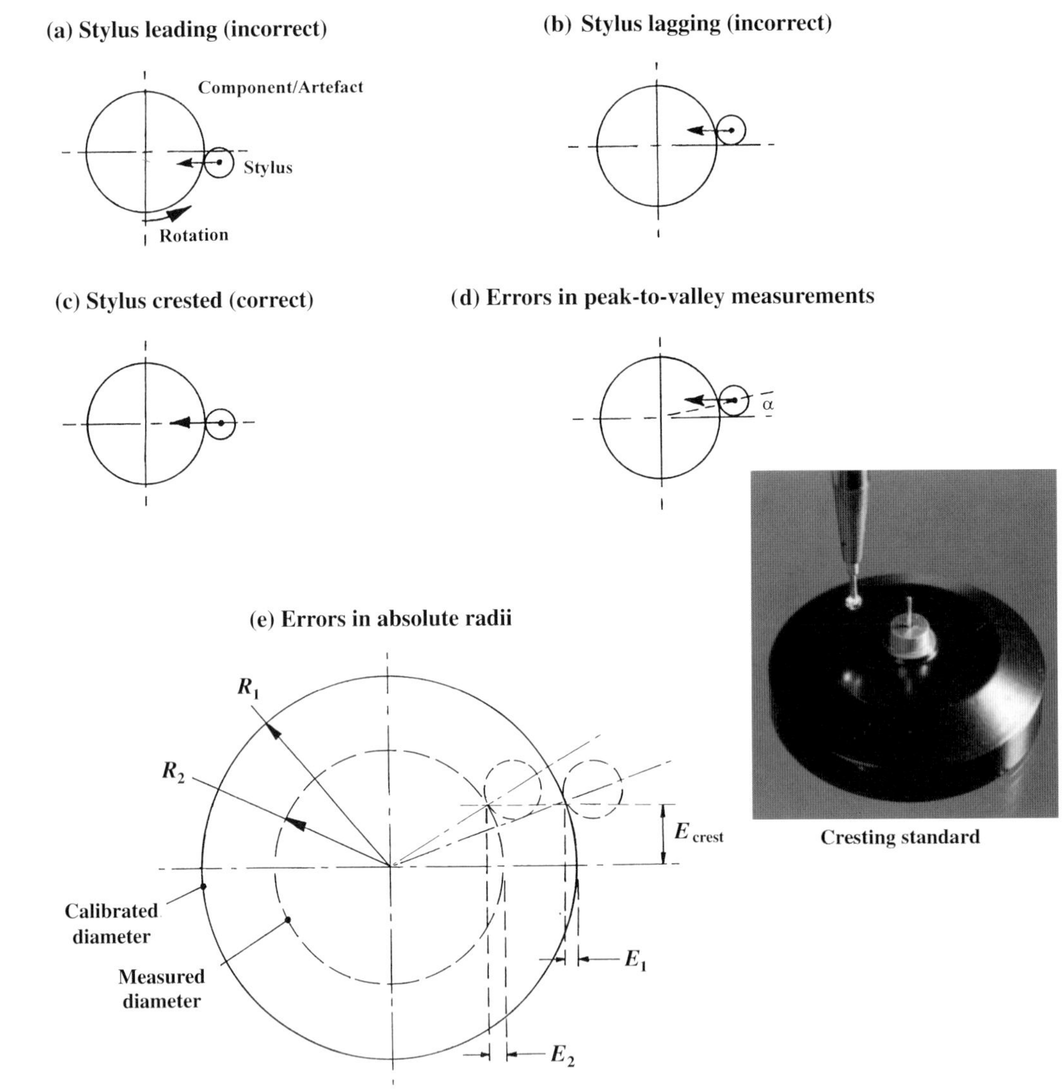

(a) Stylus leading (incorrect)

Component/Artefact

Stylus

Rotation

(b) Stylus lagging (incorrect)

(c) Stylus crested (correct)

(d) Errors in peak-to-valley measurements

α

(e) Errors in absolute radii

R_1

R_2

E_{crest}

Calibrated diameter

Measured diameter

E_1

E_2

Cresting standard

Figure 222. Stylus fine crown (cresting). (Courtesy of Taylor Hobson.)

Column straightness and squareness

Many of the more sophisticated roundness instruments are equipped with a straightness column (see Figures 113, 114 and 122 for typical machine configurations). In order to calibrate the alignment (straightness and squareness) of this column with respect to the spindle axis, most notably for rotating table instruments, a cylindrical square is employed. These "squares" can be purchased in a number of lengths, typically 500 mm, 1000 mm and 2000 mm. The smallest of these artefacts has an LSC roundness of better than 0.25 μm and the largest less than 0.5 μm. In Figure 223 is depicted one of the shortest cylindrical squares.

One major problem when assessing any straightness/squareness alignment is termed "coning error". This coning error is principally the result of the initial set-up.

6.7 Probing uncertainty: roundness and form

One method to obtain more discrimination in roundness assessment and at the same time reduce measuring uncertainty is to be able to calculate an estimate of how much a measured result may vary

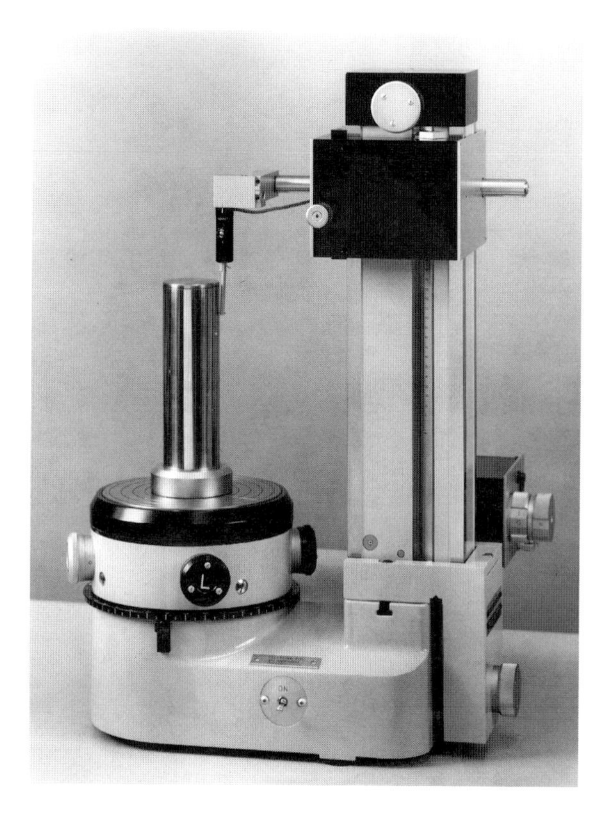

Figure 223. A precision artefact – "cylindrical square" – being employed to calibrate column straightness/squareness on rotating-table roundness instrument. (Courtesy of Taylor Hobson.)

from the true value. This uncertainty may be the consequence of a number of factors, not least of which are operator influence and environmental conditions. For example, if collecting a greater data density – perhaps acquiring 800 instead of eight data points on a roundness profile – an industrial company can reduce its measuring uncertainty, as illustrated in Figure 224a.

At one extreme, today's scanning instruments can read hundreds or thousands of points in the time it takes a coordinate measuring machine (CMM) with its associated touch-trigger probe to check around six data points. By way of illustration, a high-speed scanning probe can measure a ϕ150 mm bore in around 10 seconds, taking over 800 measuring points in the process. To press the point still further, if a 30-second scan of a ϕ50 mm boss is taken this results in approximately 3000 measuring points on the component feature. However, if a CMM is equipped with a scanning head, it can achieve equivalent levels of operational performance, to that of high-speed scanning probes. If the measurement involves the assessment of a near-perfect component, the number of probing points and their locations become somewhat irrelevant to the calculation of the result (see Figure 224b). Both conventional roundness instruments and scanning instrumentation comes into their own, as component accuracy and precision increase. Higher density of data points means improved process control. Gauge repeatability and reproducibility (GR&R) tests become significantly enhanced when sampling over several hundred data points, by eliminating dependency on where the actual data was taken, from one trial to the next. In any GR&R work component alignment is critical. If an operator removes and then replaces the same part, its initial set-up datum will be lost, resulting in different measurement results. Scanning instruments can successfully mitigate this set-up problem if the part has to be disturbed for whatever reason, as repeatability and consistency in data capture can be more readily controlled, via the software algorithms and the manner in which geometric data is extracted from the component.

6.8 Nanotechnology instrumentation: now and in the future

The term *nanotechnology* was first introduced in 1974 by Professor Taniguchi to describe production tolerances and finishes within the nanometric range. His graphical representation of achievable machining accuracies (Figure 198) was extrapolated into the future and has been shown to be a remarkably accurate prediction of what came to pass (Table 18). Taniguchi quite correctly concluded that by the mid-1990s some metrological instrumentation would be consistently operating in the region of between 100 nm and 1 nm. This has meant that highly sophisticated instruments would need to be developed and that further exploitation of surface and roundness technologies would be even more exacting in the future. Nanotechnology encompasses more than simply highly accurate and precise metrology equipment, as many micro-miniaturised components and assemblies today are fabricated as integral nanometric devices. Physics, chemistry, biology, engineering, materials science and other disciplines meet at the atomic scale.

Instrument performance

Instrument characterisation falls into a number of categories, although it is possible to list them by three variables:

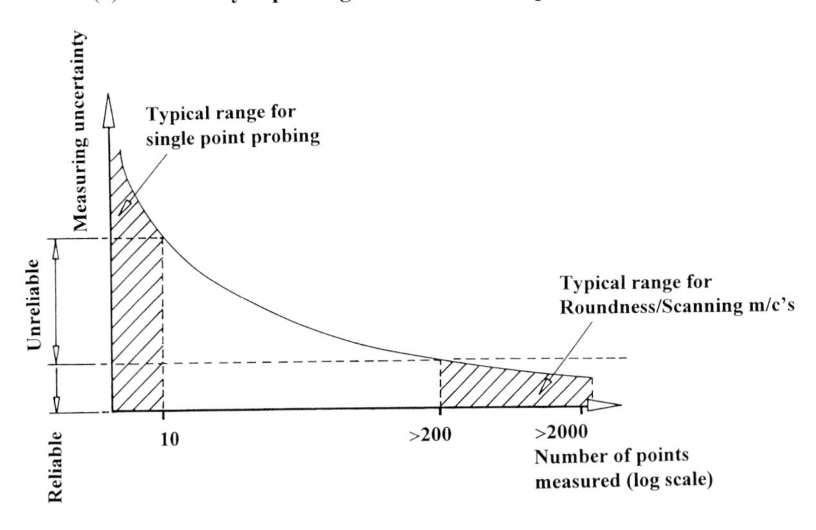

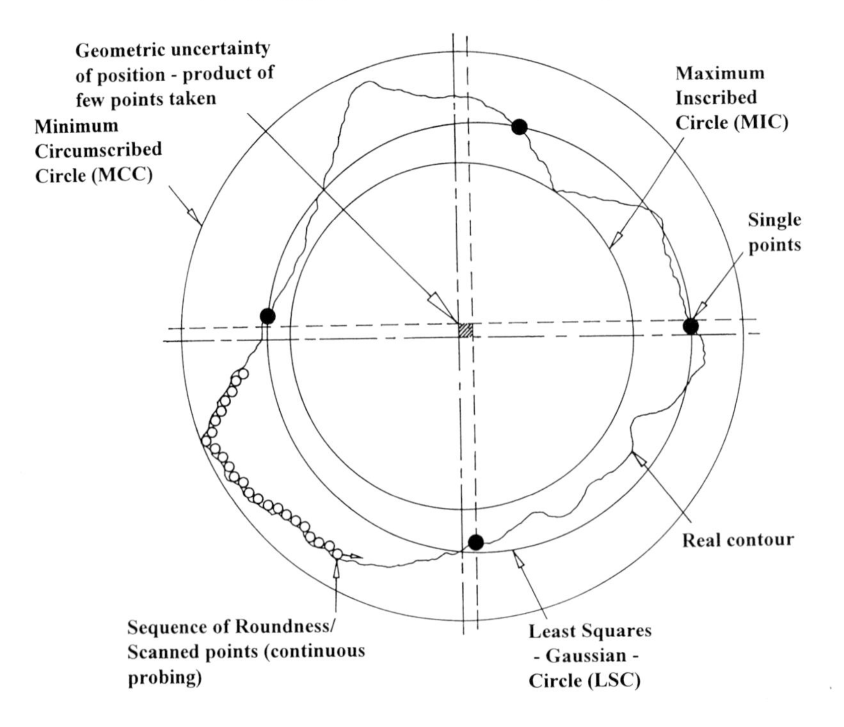

Figure 224. Probing uncertainty arises if too few point are taken on roundness parts. (After Knebel/American Machinist, 1999.)

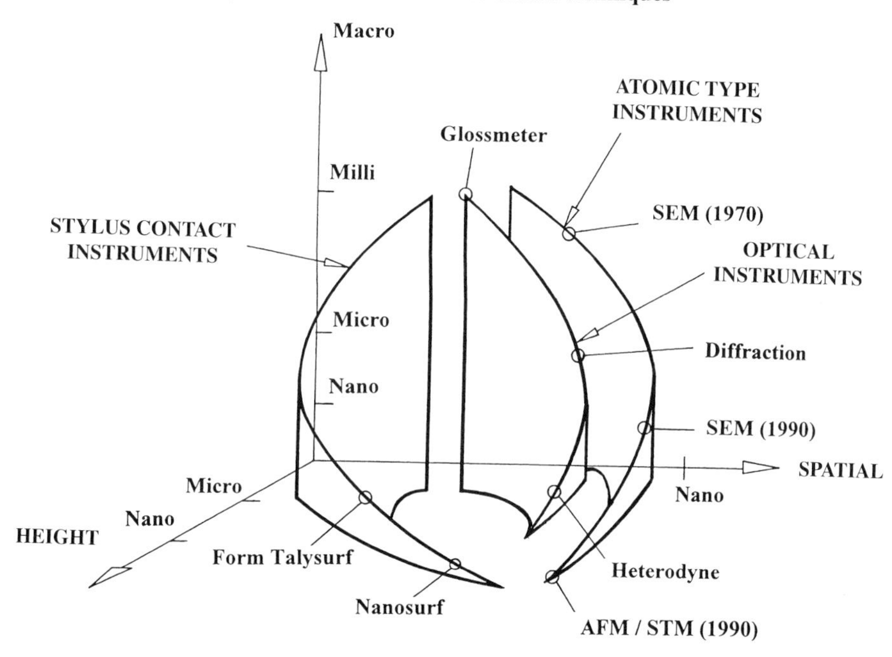

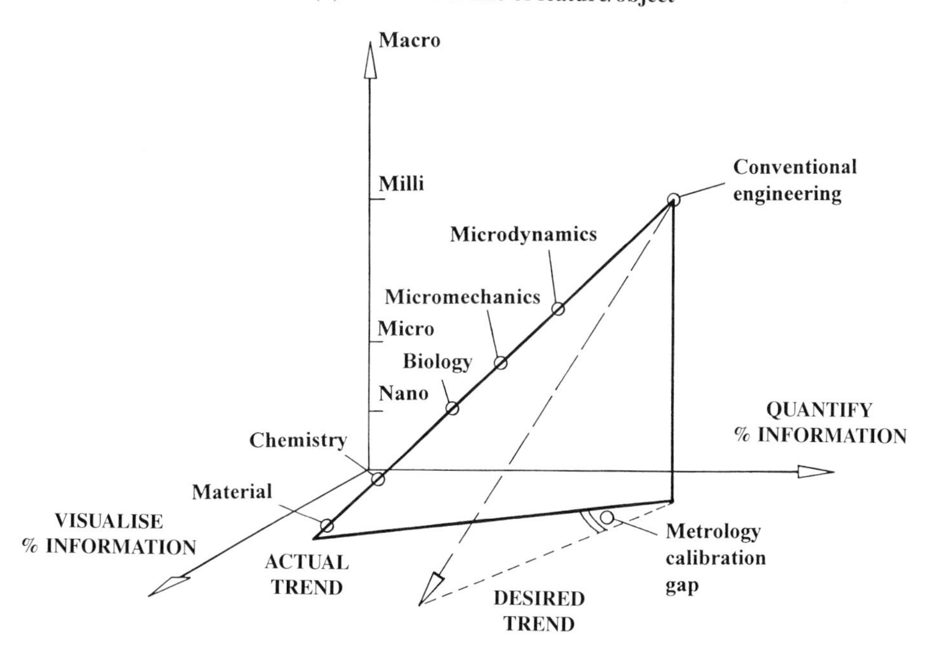

Figure 225. Future instrument trends. [After Whitehouse 1991]

- *spatial bandwidth* – by lateral range/resolution;
- *vertical bandwidth* – by height range/resolution;
- *frequency response* – or response time.

Any of the above instrumental characteristics is important, because they directly affect spatial and vertical bandwidths. They can also indirectly affect the instrument in the case of frequency response, being related to measurement fidelity. The relative weighting of each of these instrument characterisation categories will change from that of conventional instrumentation to those at the nanometric scale. Significant progress has been made recently in vertical scale calibration in the atomic region; this was previously difficult to achieve because there are many natural spacings that emerge on the surfaces of a boundary. By utilising X-rays in the form of an interferometer, it is possible to achieve height calibration to better than 10^{-2} nm. However, as instrumentation applications are spread wider from basic technological work to that of cross-disciplines where physics, chemistry and biology merge, it is spatial information which becomes paramount. This emphasis toward spatial recognition characterisation explains the great emphasis being made on nanomechanistic movements, the general trends in such development being addressed in Figure 225; the graphical relationship shown in Figure 225(b) indicates how requirements change with discipline as a function of the scale of size.

Instrumentation: the future

The previous comments concerning nanotechnology, and to some extent their associated enabling technologies, will in the future be expanded in their operational capabilities to embrace flexible and multi-purpose applications. Such instruments will be fast in operational and data capture and in this way will minimise the effects of environmental changes, most notably negating the influences resulting from thermal drift and vibrational effects. More will be said shortly concerning minimisation of environmental effects when discussing the newly developed instrument depicted in Figure 226. One approach to increasing instrumentation versatility is to enable the equipment to perform many different functions with the same set-up and at this time, but across a wider range of resolution. For example, it should be possible to assess various critical aspects of the surface topography, including its associated frictional forces and coefficients, together with surface micro-/nano-hardnesses. Such assessment could be undertaken in a single vacuum chamber, which would have the ability to utilise the same software for a range of operational functions, such as precise and accurate specimen positioning and surface manipulation and image enhancements. In the early 1990s it was shown at the nano-scale that both friction and surface topography could be successfully measured simultaneously. Moreover, *scaling theory* becomes a critical factor due to the manner in which physical phenomena change with their accompanying scale. By way of illustration, the influence of system inertia dramatically reduces in importance in relation to damping factors and stiffness, as equipment is employed when approaching nano-resolution levels. Furthermore, sensor technology dictates that in future metrological applications single atomic particles might be utilised to explore surface detail at the atomic level, perhaps using deflection techniques. In Chapter 3, concerning the uses and applications of atomic force/tunnelling microscopes and their derivatives, it was mentioned that it is now standard operational procedure to regularly manipulate the atomic structure by moving atoms to specific and preferential sites on a specimen's surface. The manipulation and quantification of surfaces within the atomic range is set to continue and diversify, with a new range of instrumentation being developed for the twenty-first century, offering exciting times ahead for the metrologist.

Nanometric instrument: design and operation

The instrument illustrated in Figure 226 is one of the latest research developments from the National Physical Laboratory in Britain. This "NanoSurf IV" surface texture measuring instrument offers traceable measurement based on sound metrological and kinematic principles, by measuring displacements in two orthogonal axes, having a measurement uncertainty of approximately 1 nm. This "NanoSurf IV" is a stylus-based instrument that has been primarily manufactured from Zerodur™, comprising three basic elements:

- *base plate* – to support the entire structure of the instrument;
- *slideway* – to translate the specimen for measurement;
- *bridge* – which carries a probing system and a vertical stage to position the probe over the specimen.

The base plate is lapped and polished flat. The bridge section of the instrument forms part of a metrology frame, being kinematically located on the base plate via ball contacts and three associated vee-grooves. The bridge carries the probing system,

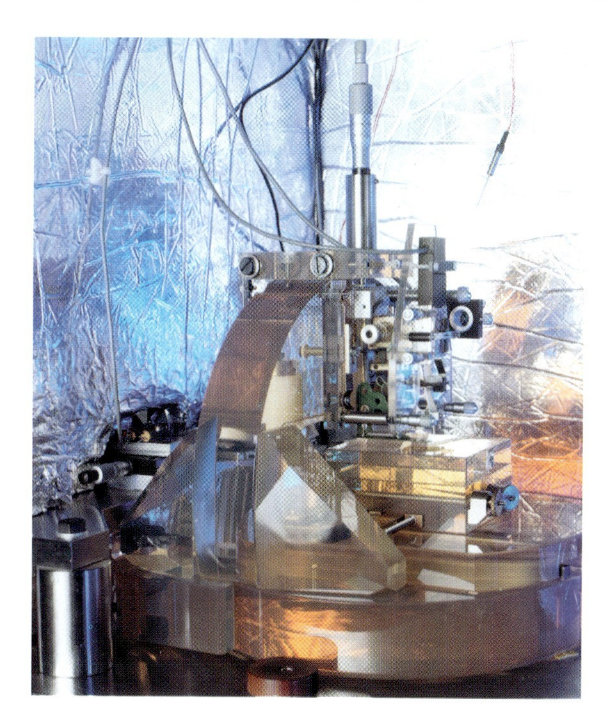

Figure 226. The ultra-precision "NanoSurf IV" instrument, having an uncertainty of measurement of 1 nm. (Courtesy of NPL.)

a Z-axis slideway and a mount for the Z-axis interferometry.

The "NanoSurf IV" slideways are of the dry-bearing prismatic design, similar to those currently available on the "Nanostep" instrument. The bearing surfaces for the X-slideway are the top surface of the base plate and a straight edge that is cemented to this base plate. Two magnets are fixed to the underside of the carriage, being attracted to a steel rod to pull two bearing pads into contact with the straight edge. Three further pads contact the base plate, ensuring the full complement of five constraints, allowing linear motion in one axis only.

The probe on the "NanoSurf IV" has its origin in that developed by Taylor Hobson for their "Nanostep" instrument (Figures 58–60). So that the laser interferometer could be incorporated as the sensing element, the Taylor Hobson "Nanostep" probe was modified to allow the polished and aluminised end of a Zerodur™ rod to act as the plane mirror of a cat's-eye retro-reflector for the Z-interferometer. The diamond-tipped stylus is kinematically mounted onto the other end of this Zerodur™ rod.

The two interferometers on the "NanoSurf IV" are based on an NPL-designed retro-reflecting Twyman-Green type, which does not have polarisation optics and therefore avoids the errors caused by polarisation leakage. Such interferometers have been shown to produce sub-nanometre accuracy. The light source for the interferometer is a 633 nm frequency-stabilised He–Ne laser, with an output power of approximately 1 mW. The beam from the laser is split into two components to be utilised by the X- and Z-interferometers. A modified version of the method developed by Heydemann is used to correct the quadrature signals from the interferometers for DC offsets and differential gains. Computer simulations by Birch were used to obtain a correction method to predict a fringe-fractioning accuracy of 0.1 nm.

The uncertainty analysis of the "NanoSurf IV" has been calculated according to the ISO G.U.M. and gives a combined standard uncertainty of measurements in the X- and Z-axes at 95% confidence of ±1.3 nm. The traceability to the metre is via the calibrated frequency of its laser source. An instrument such as the "NanoSurf IV" can be used to measure ultra-precision parts, such as sinusoidal gratings, depth standards and random roughness samples; conversely, the instrument can be employed in characterising other surface texture measuring equipment.

Progress will continue to refine and develop new instruments for present and future applications of a metrological nature, with yet wider scope and range being presented for surface and roundness exploitation in combination with work of a fundamental and applied scientific nature.

References

Journals and conference papers

Birch, K.P. Optical fringe subdivision with nanometric accuracy. *Precision Engineering* 12, 1990, 195–198.

Bosse, H., Krüger, R., Lhr, W. and Lüdicke, F. Oberflächensimulator für die Dynamische Kalibrierung von Längenmeßtastern. *Technisches Messen* 2, 1994, 82.

Donaldson, R. How good are your measurements? *Quality Today* July 1993, 16–20.

Downs, M.J., Birch, K.P., Cox, M.G. and Nunn, J.W. Verification of a polarization-insensitive optical interferometer system with subnanometric capability. *Precision Engineering* 17, 1995, 1–6.

Downs, M.J. and Nunn, J.W. Verification of the subnanometric capability of an NPL differential plane mirror interferometer with a capacitance probe. *Measurement Science and Technology* 9, 1998, 1437–1440.

Eigler, D.M. and Schweiger E.K. Positioning single atoms with a scanning tunnelling microscope. *Nature* 344, April 1990, 524.

Evans, C.J., Hocken, R.J. and Estler, W.T. Self-calibration: reversal, redundancy, error separation and absolute testing. *Annals of the CIRP* 45, 1996, 617–623.

Franks, A. Nanometric surface metrology at the National Physical Laboratory. *Nanotechnology* 2, 1991, 11–18.

Frennberg M and Sacconi A. International comparison of high-accuracy roundness measurements. *Metrologia* 33, 1996, 539–544.

Garratt J.D. and Bottomley, S.C. Technology transfer in the development of a nanotopographic instrument. *Nanotechnology* 1, 1990, 28–43.

Haitjema, H. Uncertainty analysis of roughness standard calibration using stylus instruments. *Precision Engineering* 22, 1998, 110–119.

Haitjema, H. and Kotte, G.J. Dynamic probe calibration up to 10 kHz using laser interferometry. *Measurement* 21(3), 1997, 107–111.

Haitjema, H., Bosse, H., Frennberg, M., Sacconi, A. and Thalmann, R. International comparison of roundness profiles with nanometric accuracy. *Metrologia* 33, 1996, 67–73.

Hall, C. and Griffiths, B. Comparing 2D and 3D stylus surface measurement. *Quality Today* July 1999, 30–34.

Hawthorn, J. The practical use of SPC charts. *Quality Today* September 1994, 38–40.

Heydemann, P.L.M. Determination and correction of quadrature fringe measurement errors in interferometers. *Applied Optics* 20, 1981, 3382–3384.

Hillman, I.W. Calibration of contact stylus instruments within the Deutscher Kalibrierdienst (DKD = German Calibration Service). *EUROMET: Workshop Traceable Measurement of Surface Texture* 289, 1993, 23–30.

Hundry, B. A brief history of quality. *Manufacture and Engineering* September 1991, 49–52.

Jørgensen, J.F., Garnaes, J., Thulstrup, D. and Rasmussen, S.E., Calibrated surface characterization, problems and solutions. *Seventh International Conference on Metrology and Properties of Engineering Surfaces* 1997, 358–365.

Jusko, O. and Lüdicke, F. Novel multi-wave standards for the calibration of form measuring instruments. *Physikalisch-Technische Bundesanstalt*, Braunschweig, Germany.

Kellock, B. Getting the measure of calibration. *Machinery and Production Engineering* December 1997, 47–48.

Knebel, R. Scanning for better results. *American Machinist* November 1999, 76–78.

Leach, R., Haycocks, J., Jackson, K., Lewis, A., Oldfield, S. and Yacoot, A. Advances in traceable nanometrology at the National Physical Laboratory. *Nanotechnology* 12, 2001, R1–R6.

Leach, R.K. Traceable measurement of surface texture in the optics industry. *Large Lenses and Prisms Conference*, University College London, March 2001.

Leach, R.K. Traceable measurement of surface texture at the National Physical Laboratory using NanoSurf IV. *Measurement Science Technology* 11, 2000, 1162–1172.

Leach, R.K. Telling the rough from the smooth. *Materials World* 8(2), 2000, 18–19.

Lee, G.L., Mou, J. and Shen, Y. An analytical assessment of measurement uncertainty in precision and machine calibration. *International Journal of Machine Tools and Manufacture* 37(3), 1997, 263–276.

Lee, G.L. and Mou, J. A method to enhance the quality and reliability of inspection data for precision manufacturing applications. *Proceedings of ASPE (10th Annual Meeting)* 1995, 96–99.

Litsikas, M. The do's & don'ts of roundness measurement. *Quality Magazine* November 1996, 1–6.

Locke, J.W. Quality standards for laboratories. *Quality Progress* July 1993, 91–94.

Nakajima, N. et al. Study on microengines. *Sensors and Actuators* 20, 1989, 75–82.

Petzing, J. and Joy, M. Metrology, measurement and calibration, assessment and training. *Engineering Technology* December 1996, 21–23.

Pope, C. Made to measure. *Professional Engineering* March 2001, 29.

Reeve, C.P. *The Calibration of a Roundness Standard*. US Department of Commerce, National Bureau of Standards – NBSIR 79-1758, June 1979.

Saundry, L. Calibration and the user. *Quality Today* March 1999, 27–30.

Schneider, U. and Hübner, G. Dynamic calibration and testing of roundness measuring devices by means of a waviness standard. *Proceedings of the Fourth International Symposium on Dimensional Metrology in Production and Quality Control*, Tampere, Finland, 1992, 394–414.

Schobinger, J.P. Investigation into the production of roughness standards. PhD thesis, Federal Technical College, Zürich, 1959.

Shankar, N. Reducing uncertainty in measurement. *Quality Today* April 1999, 36–40.

Sharman, H.B. Calibration of surface texture measuring instruments. *Proceedings of the Institute of Mechanical Engineers* 182, 1967, 319–326.

Sheer, B.W. and Stover, J.C. Development of a smooth-surface microroughness standard. *Proceedings of SPIE* No. 3141, 1998, 78–87.

Shen, Y.L. and Duffie, N.A. An uncertainty analysis method for coordinate referencing in manufacturing engineering. *Transactions of the ASME: Journal of Engineering for Industry* 117(1), 1995, 42–48.

Smith, G.T. Why the need for calibration? *Metalworking Production* September 1998, Q5–Q8.

Stout, K.J. and Johnson, A. Nano-tronics: the role of the engineer in nanotechnology. *Korean Journal of Precision Engineering* October 1998.

Taniguchi, N. The state of the art of nanotechnology for processing of ultraprecision and ultrafine products. *Precision Engineering* 16(1), 1994, 5–24.

Taniguchi, N. Current status in and future trends of ultra precision machining and ultrafine materials processing. *Annals of the CIRP* 32(2), 1983, 573.

Taniguchi, N. On the basic concept of nanotechnology. *Proceedings of ICPE*, Tokyo, 1974.

Teague, E.C. Uncertainties in calibrating a stylus type surface texture measuring instrument with an interferometrically measured step. *Metrologia* 14, 1978, 39–44.

Tsukada, T. and Kanada, T. Evaluation of two- and three-dimensional surface roughness profiles and their confidence. *Wear* 109, 1986, 69–78.

Tyler Estler, W., Evans, C.J. and Shao, L.Z. Uncertainty estimation for multiposition form error metrology. *Precision Engineering* 21, 1997, 72–82.

Vasseur, H., Kurfess, T.R. and Cagan, J. Use of the quality loss function to select statistical tolerances. *Transactions of the ASME* 119, August 1997, 410–416.

Whitehouse, D.J. Nanotechnology instrumentation. *Measurement and Control* 24, March 1991, 37–46.

Whitehouse, D.J. Nano-calibration of stylus-based surface measurements. *Journal of Physics E: Scientific Instruments* 21, 1988, 46–51.

Books, booklets and guides

Guide to Uncertainty Measuurement. BIPM, IEC, IFCC, ISO, IUPAC, IUPAP annd OIML, 1995.

Bell, S. *A Beginner's Guide to Uncertainty of Measurement*. National Physical Laboratory (NPL), 2000.

Birch, K., Estimating Uncertainties in Testing. Measurement Good Practice Guide no 36, MTA, March 2001.

Bergman, B. and Klefsjö, B. *Quality*. McGraw-Hill, New York, 1994.

Dietrich, C.F. *Uncertainty, Calibration and Probability* (2nd edn). Adam Hilger, 1991.

EAL-R2-S1. *Supplement 1 to EAL-R2 Expression of the Uncertainty of Measurement in Calibration*. European Co-operation for Accreditation of Laboratories, Rotterdam, November 1997.

EAL-R2. *Expression of the Uncertainty of Measurement in Calibration*. European Co-operation for Accreditation of Laboratories, Rotterdam, April, 1997.

Farago, F.T. *Handbook of Dimensional Measurement*. Industrial Press, New York, 1968.

ISO 5436-1985. *Calibration of Stylus Instruments.*

Jusko, O. and Lüdicke, F. *Novel Multiwave Standards for the Calibration of Form Measuring Instruments*, Physikalisch-Technische Bundesanstalt (PTB), Braunschweig, Germany.

Leach, R.K. *Calibration, Traceability and Uncertainty Issues in Surface Texture Metrology.* NPL Report CLM-7, June 1999.

Leach, R.K. *The Measurement of Surface Texture using Stylus Instruments.* Measurement Good Practice Guide no 37, NPL, July 2001.

Moore, W.R. *Foundations of Mechanical Accuracy.* Moore Special Tool Co., 1970.

Munro-Faure, L. and Munroe-Faure, M. *Implementing Total Quality Management.* Pitman, 1994.

Manual of British Standards in Metrology. Hutchinson, 1984.

Nakazawa, H. *Principles of Precision Engineering.* Oxford University Press, 1994.

Engineering Dimensional Metrology. National Physical Laboratory, 1996.

Oakland, J.S. *Statistical Process Control.* William Heinemann, 1986.

Ozeki, K. and Asaka, T. *Handbook of Quality Tools.* Productivity Press, 1990.

Price, F. *Right First Time.* Gower, 1984.

Reeve, C.P. *The Calibration of a Roundness Standard.* National Bureau of Standards (USA), Report No 79-1758, June 1979.

Smith, S.T. and Chetwynd, D.G. *Foundations of Ultraprecision Mechanism Design.* Gordon & Breach, London, 1993.

Stout, K.J. *From Cubit to Nanometre.* University of Huddersfield, December 1998.

Stout, K.J. *Quality Control in Automation.* Kogan Page, 1985.

Swanson, R.C. *The Quality Improvement Handbook.* Kogan Page, 1995.

The National Measurement System Programme for Length. National Physical Laboratory (NPL), October 1999.

Taniguchi, N. *Fundamentals and Applications of Nanotechnology.* Kogyo Chosakai (in Japanese), 1988.

UKAS pub. no. M 3003. *The Expression of Uncertainty and Confidence in Measurement.* December 1997.

Appendices

"Begin at the beginning," the King said gravely, "and go on till you come to the end: then stop."

(*Alice's Adventures in Wonderland*, Lewis Carroll, 1832–1898)

Appendix A

Previous and some current surface texture parameters

In Chapter 1 the current parameters were discussed relating to ISO 4287: 1997; here their immediate predecessors and some current ones are mentioned at some length, as many companies still require information relating to these previously valid parameters for surface texture assessment.

Prior to a discussion on the previous methods of obtaining roughness data from a surface texture trace, a brief review of these surface texture parameters will allow a greater understanding of how, where and why they are applied to gain an insight into methods of classifying surface conditions. In essence, surfaces that are manufactured can be produced in a myriad of ways, with surface roughness parameters being historically developed to fulfil specific commercial and industrial needs. This has meant that a considerable number of surface descriptors, as they are sometimes known, have occurred, resulting in a frequently quoted term, "parameter rash", coined some years ago by Professor David Whitehouse. Many of these parameters are utilised only by particular specialist industries. Generally, they can be broken down into distinct groups:

- *amplitude parameters* – which are measures of the *vertical* characteristics of the surface deviations;
- *spacing parameters* – measures of the *horizontal* characteristics of the surface deviations;
- *hybrid parameters* – which are some *combination* of amplitude and spacing parameters;
- *statistical parameters* – amplitude heights treated as statistical data sets.

Amplitude parameters

Ra: arithmetical mean roughness

The classification of the relative roughness of surfaces was developed in England and was termed *centre line average* (CLA), while in the USA its equivalent term was the *arithmetic average* (AA). The derivation of *Ra* is graphically illustrated in Figure 18(b) (see Chapter 1). Mathematically, *Ra* is the arithmetic average value of the departure of the profile from the centre line throughout the sampling length. Determination of *Ra* is normally computed by the software but can be derived using the following formula:

$$Ra = \frac{1}{1r}\int_0^{1r} z(x) \mid \mathrm{d}x \quad \text{(units of m)}$$

The *Ra* parameter would be primarily used in applications to monitor a production process, where a gradual change in the surface finish was anticipated, for example, due to tool wear in the cutting process.

There are a number of conditions that must be met for *Ra* to be utilised satisfactorily:

- the *Ra* value over one sampling length represents the *average* roughness. The effect of a single spurious non-typical peak or valley within the profile's trace will be averaged out and, as such, has only a small influence on the *Ra* value;
- it is usual to take assessments over several consecutive sampling lengths and then to accept the average as the *Ra* value. The evaluation length contains several sampling lengths (Figure 18a); this ensures that the *Ra* value is typical of the surface under test;
- an *Ra* value alone is meaningless, unless quoted with the metre cut-off (λc), or if this can be assumed. Repeatability of the *Ra* value will only occur at an identical cut-off length;
- where a dominant surface texture pattern occurs (lay), then *Ra* measurement is undertaken at 90° to this direction;
- that *Ra* does not provide information as to the shape of either the profile or its surface irregularities. Different manufacturing processes will produce divergent surface finishes; therefore it is normal procedure to quote the actual production process along with its expected *Ra* numerical value(s);
- *Ra* offers no distinction between peaks and valleys on the surface trace;
- assuming that the *Ra* numerical value has not been computed, requiring determination from a graph previously made of the surface, this can be simply undertaken using suitable software, or previously in a more basic form by a planimeter.

Figure 227 graphically depicts this *Ra* problem and illustrates the fact that the workpiece's *Ra* numerical value alone (as has already been mentioned) is not only meaningless but can, under certain conditions of incorrect interpretation, cause catastrophic consequences. This identical *Ra* value shows three potential in-service conditions (Figure 227: (a) bearing surface; (b) intermediate surface; and (c) locking surface; all having an *Ra* of 4.2 μm in these examples. In case (a) this surface might prove ideal for a "light" bearing application, perhaps where some local oil retention was necessary to reduce its frictional characteristics; whereas case (c)

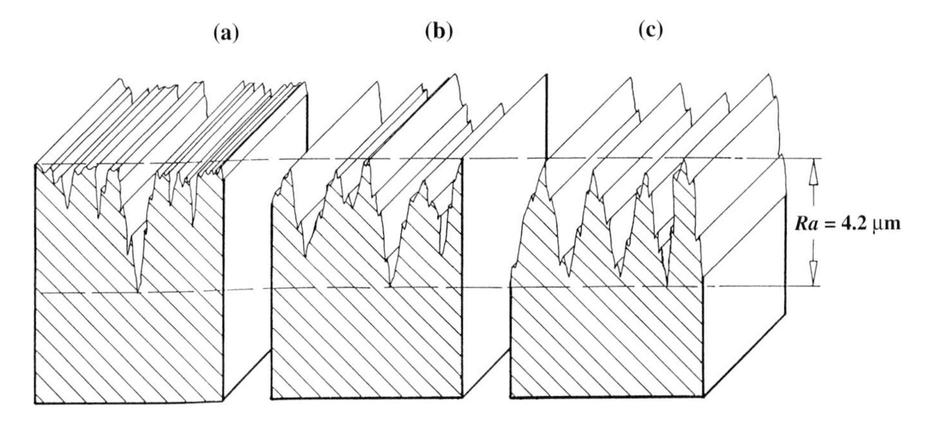

Figure 227. Differing profiles producing identical *Ra* values.

would prove somewhat useful if surfaces were needed to be locked together for either a permanent or semi-permanent state. Case (b) might be more appropriate if the in-service application required neither a locking nor a bearing condition.

What can be conclusively stated is that if a designer required a critical and highly stressed part to have bearing condition (a), but the metrologist had incorrectly assessed the surface just using *Ra* alone – thus giving a false interpretation of the surface condition and leading to the selection of a surface with locking state (c) – this would drastically shorten the component's life under highly stressed states. Therefore, using such a parameter and only quoting its numerical value in isolation, without further specifying the other mutually related descriptors to quantify the surface condition, can later lead to disastrous consequences. Incorrect use or interpretation by a designer or metrologist of any surface descriptor could result in, at best, the part being totally inappropriate for its current application – causing poor reliability and, at worst, leading to catastrophic failure.

Rq: root mean square

An alternative method of calculating the average roughness of a surface is to use statistical techniques, such as the root mean square (rms). The rms can be obtained by squaring each value, then taking the square root of the mean. For example, the arithmetic average of four values, a, b, c and d, would be:

$$\text{rms} = \frac{\sqrt{(a^2 + b^2 + c^2 + d^2)}}{4} \quad \text{(m)}$$

When compared to the arithmetic average, rms has the effect of giving extra weight to higher values, as shown by the following three groups of values:

$3, 4, 5 \quad 2, 4, 6 \quad 1, 4, 7$

In each case, the arithmetic average is 4, the successive increase of 1 in the higher value being balanced by a decrease of 1 in the lower value. The rms values are $\sqrt{16.6}$, $\sqrt{18.6}$ and $\sqrt{22}$, respectively, illustrating that the increase in the highest number in the group outweighs the decrease in the lowest, or, for a more mathematical treatment of the rms value:

$$Rq = \sqrt{\frac{1}{1r} \int_0^{1r} z^2(x)\,\mathrm{d}x} \quad \text{(m)}$$

The reason why both *Ra* and *Rq* roughness average parameters were adopted was, in the main, historical. The *Ra* surface descriptor is simpler to estimate from the profile recording and as a result was initially adopted before roughness-measuring instruments became more popular. Instrumentally, *Rq* has the advantage of simplicity, as phase effects in the electrical filters become of comparatively less importance and can be disregarded, although they influence the arithmetic average somewhat and cannot be ignored.

What can be seen in many of the amplitude parameters is that there is no fixed relationship allowing conversion from *Ra* to *Rq*, as it would depend solely on the shape of the profile. For many production processes the *Rq* value of the surface would be at least 10% greater than its equivalent *Ra*. In general, a reasonable approximation of the ratio *Ra*:*Rq* = 1:1.1, but care must be taken with this conversion factor as some surface topographies can introduce an error. As an example of the misleading ratio effect, the case of a lapped surface has an asymmetrical profile (i.e., with deep valleys and rounded crests) – the *Ra*:*Rq* ratio can increase to 1:1.5.

Rt, Rtm, Rv, Rz, Rmax, Rp and *Rpm*: peak and valley heights

Occasionally, it is useful to specify the *maximum* roughness (peak-to-valley) height, rather than *Ra*, which gives the mean height. Typical of one of the maximum roughness descriptors is *Rt*; this parameter correlates to a tactile assessment of the surface when handled. *Rt* equates to the German term for roughness depth – *Rauhtiefe* – being the highest and lowest points of the profile within the evaluation length, measured perpendicular to the ordinate lines on the profile chart (see Figure 228). Some points worth mentioning relating to the *Rt* parameter are that

- with the definition of *Rt* the centre line does not occur;
- the value of *Rt* is determined over the evaluation length;
- *Rt* is directly affected by the presence of dirt and scratches, because there is no "averaging effect";
- if the slope of the profile is ignored, then there is no ambiguity as to which are the highest and lowest slopes.

The symbol *Rmax* (Figure 228) is the vertical distance between the highest and lowest points of a profile within the *sampling length*, but because of its sensitivity to scratches, etc., it is more usual to take the mean – *Rtm* – of the five individual (consecutive) sampling lengths (Figure 228). In Germany, *Rtm* equates to $Rz_{(DIN)}$, whereas elsewhere around the world $Rz_{(ISO)}$ is a different parameter, as described later in this section.

Rp is illustrated in Figure 228, which is another height term, being the maximum height of the profile above the centre line within the sampling length. As the *Rp* value tends to be extremely sensitive to large profile peaks within the sampling length, the information is not too reliable and as a

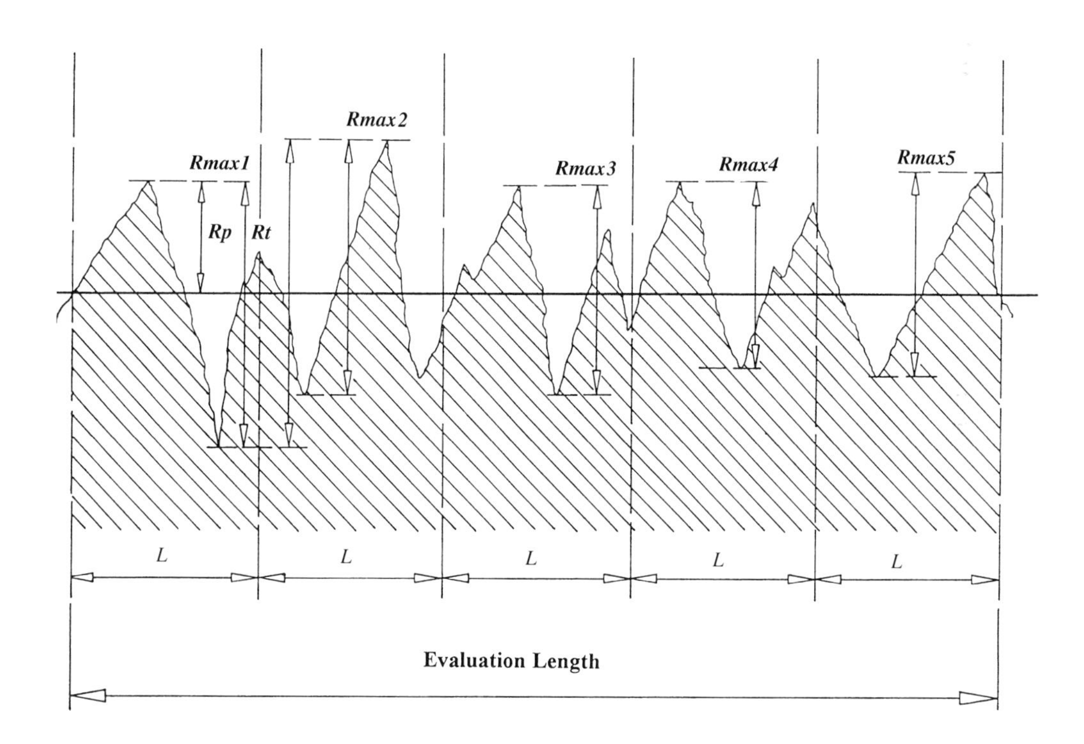

Rmax = maximum peak-to-valley height within the sampling length *L*.
Rt = vertical height between the highest and lowest points of the profile within the evaluation length
Rtm = Mean value of the **Rmax** of five consecutive sampling lengths.
Rp = Height of the highest point of the profile above the centre line within sampling length *L*.
Rpm = Mean value of the **Rp** of five consecutive sampling lenghs.

Figure 228. Derivation of some peak parameters.

result should be used with care. Due to this uncertainty introduced by "stray" dominant peaks using Rp, the "mean levelling depth" parameter Rpm offers a more consistent alternative. This latter surface roughness descriptor is the mean value of the levelling depths of five consecutive sampling lengths, quantified by the following equation:

$$Rpm = \frac{R_{p/1} + R_{p/2} + R_{p/3} + R_{p/4} + R_{p/5}}{5}$$

or

$$Rpm = \frac{1}{5}\sum_{i=1}^{5} p_i$$

This Rpm term gives reliable information on the profile's shape and can indicate whether the part in question might be suitable for wear-resistant applications. Due to Rpm being used to determine the functionality of a surface, it is often used in tribological applications, for example, bearings, for establishing interference/shrink fit component behaviour, sliding surfaces, or surface treatment analysis – prior to coatings. Small values of Rpm are characterised by surface topography having wide peaks and narrow valleys, while large Rpm values indicate a sharp-pointed profile (i.e., "spiky surfaces"), this latter surface obviously having considerably worse wear resistance than the former topography.

$Rz_{(JIS)}$: 10-point height

Although the $Rz_{(JIS)}$ or $Rz_{(ISO)}$ parameter has now been deleted from the latest ISO standard (ISO 4287: 1997), it has been included in this discussion as there are still many engineering drawings in existence that utilise this surface texture descriptor. The $Rz_{(JIS)}$ parameter averaged the peak-to-valley values, which had the effect of reducing the influence of either a spurious irregularity or a scratch. As its name suggested, it was derived from 10 points within the sampling length (i.e., being a German-based standard the letter "z" was derived from the German; Zehn Punkt Höhe which equates to "10-point height") and is the mean distance between the five highest and lowest peaks and valleys, respectively, measured perpendicularly to a base line (see Figure 229). $Rz_{(JIS)}$ was determined using the following equation:

$$Rz_{(JIS)} = (Zp_1 + Zp_2 + Zp_3 + Zp_4 + Zp_5) - (Zv_1 + Zv_2 + Zv_3 + Zv_4 + Zv_5) \div 5$$

$$Rz_{(JIS)} = \frac{1}{5}\left[\sum_{i=1}^{5} p_i + \sum_{i=1}^{5} v_i\right]$$

The five peaks and valleys can be located anywhere across the sampling length, unlike $R_{(DIN)}$, allowing it to be utilised for short surface lengths, whereas the alternative $R_{(DIN)}$ would be inappropriate because significant peaks or valleys per cut-off length are not present. $Rz_{(JIS)}$ suffers from the

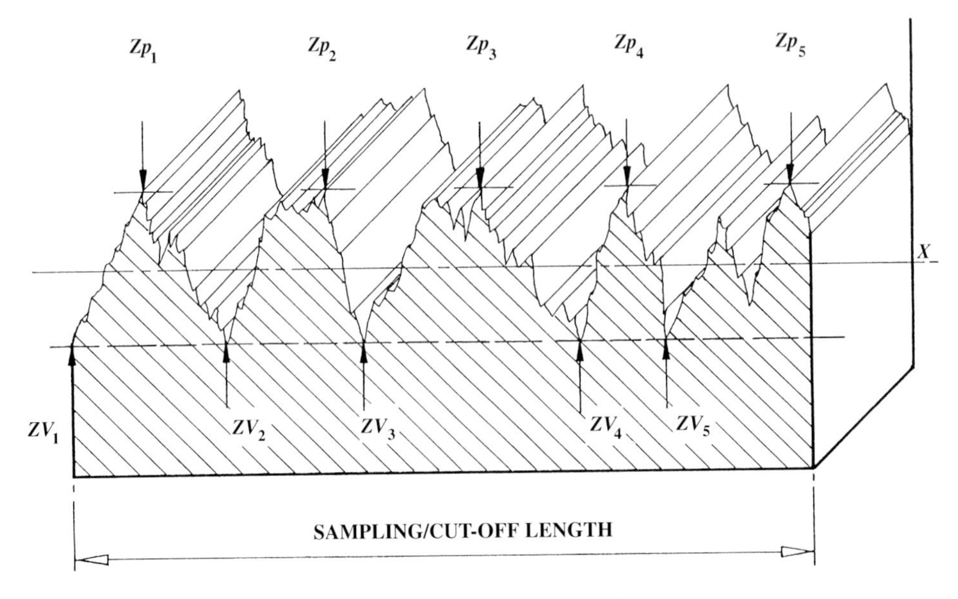

Figure 229. The 10-point *Rz* parameter derivation.

limitation that if a number of peaks and valleys occurred condensed within the cut-off length they would be unrepresentative of the whole surface and will have influenced the average condition over the profile. The centre line (i.e., Figure 229) is only used as a reference direction from which peak and valley measurements are made, although the same result can be achieved if measured from any line inside or outside the profile, but parallel to the centre line. $Rz_{(JIS)}$ lent itself to assessment from the graph, needing only 10 direct linear measurements. Further, there was no ambiguity in defining the maximum peak, or valley (Zp_1 or Zv_1); however, prior to determining which were the peaks, Zp_2 to Zp_5, and valleys, Zv_2 to Zv_5, it was necessary to define what constituted a peak or valley. A slight digression here might be in order to establish how and in what manner peaks and valleys can be established.

Peaks and peak counting

From a profile graph, if an attempt is made to measure a parameter such as the example shown in Figure 229, or to count the number of peaks and valleys within the sampling length, the question arises: "What constitutes a peak or valley?" Figure 230(a) illustrates the dilemma associated with peaks; in this example two areas of material project above the mean line, area A between points C and D, plus area B between points E and F – these can be termed profile peaks. In this example within area A are superimposed a number of minor peaks a–f, termed "local peaks". Are these to be regarded as separate peaks and included within the assessment? It might be said that such local peaks, being of low amplitude, can be discounted because they are non-repeatable and are not really part of the surface topography, resulting from factors such as instrument instability, vibration or debris on the surface. Whatever the cause for these minor peaks, once the surface is in service and a tribological action occurs, then various smaller local peaks will be worn away before any significant effect occurs to the surface's performance. As a result of the last statement, the question raised is: "Should local peaks be neglected?" Taking the case of local peak f in Figure 230(a), it is almost of comparable height to the highest peak and, as such, should not be ignored. Guidelines based on this criterion for inclusion of local peaks have been proposed: when a peak's height measured from the smallest adjacent local valley is not less than 1% of the maximum height of the profile peak to the valley, measured from the mean line (i.e., height h in Figure 230a must not be less than 1% of $Rp + Rv$).

Hence by this criterion local peaks e and f (Figure 230a) would be accepted; conversely, the smaller peaks a, b, c and d would be ignored. This decision to ignore smaller local peaks is concurrent with the practical significance of these minor profile irregularities, although it does ignore the fact that the crest of a profile peak should be retained in the surface assessment. What has been said for local peaks is also true for profile valleys and their associated local valleys.The roughness amplitude is important for surface assessment for many applications. The spacing of these roughness peaks has importance too, particularly on rolled sheet steels, etc., where the control of surface texture is essential to obtain uniform lubrication as the sheet is either pressed or drawn to minimise scoring and its influence on the final paint or plating finishes. The value of the *Ra* parameter alone is not sufficient to specify the differing types of texture obtainable from the rolling process. Peak spacing control gives improved bonding of finish treatments with better uniformity while reducing the tendency to cracking during drawing or forming operations. With our eyes being very sensitive to differences in appearance, some plastic parts are plated to enhance their cosmetic appearance. When identical parts have been produced at two factories, if their average peak spacings (peak counts) are not similar they may not be a visual match. Moreover, peak spacing is often a significant factor in the frictional performance of surfaces for automobile brake drums.

Spacing parameters

To the eye the visual presentation of a surface is a combination of its profile depth and peak spacing. Even when a surface has the same *Ra* value, the appearance can be markedly different. The number of peaks within the sampling length becomes a controlling factor where visual appearance is critical. Establishing the type of peak from the surface is important. Peaks can be established in a number of ways:

1. profile peaks are normally established if their profile crosses the centre line twice; thus the actual peaks are half these numerical values. An example of counting peaks is shown in Figure 230(b), where a, b, c and d represent two peaks;
2. establishment of "local peaks" provided they exceed a certain height; in this example there were seven, namely A, B, C, D, E, F and G;
3. the number of peaks similar to profile peaks projecting above a reference level at height *Y* can be determined by halving the number of times the

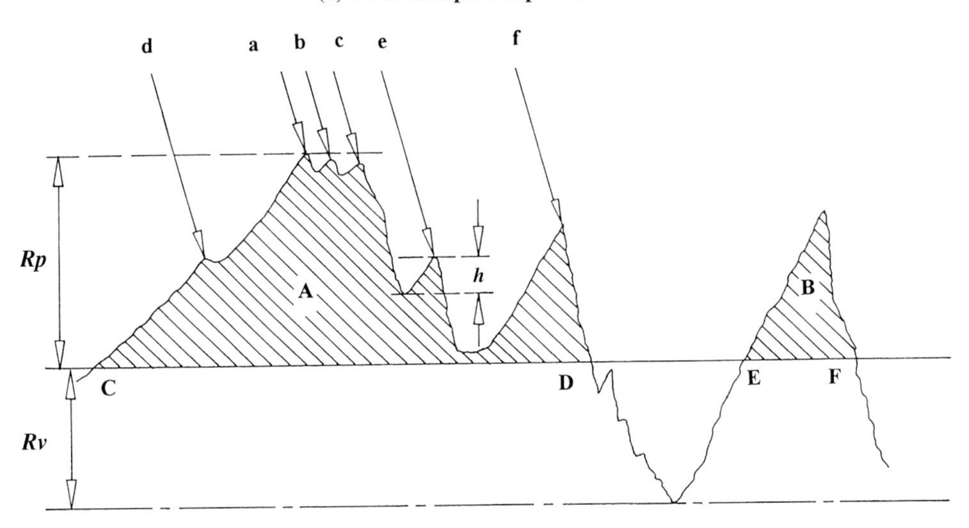

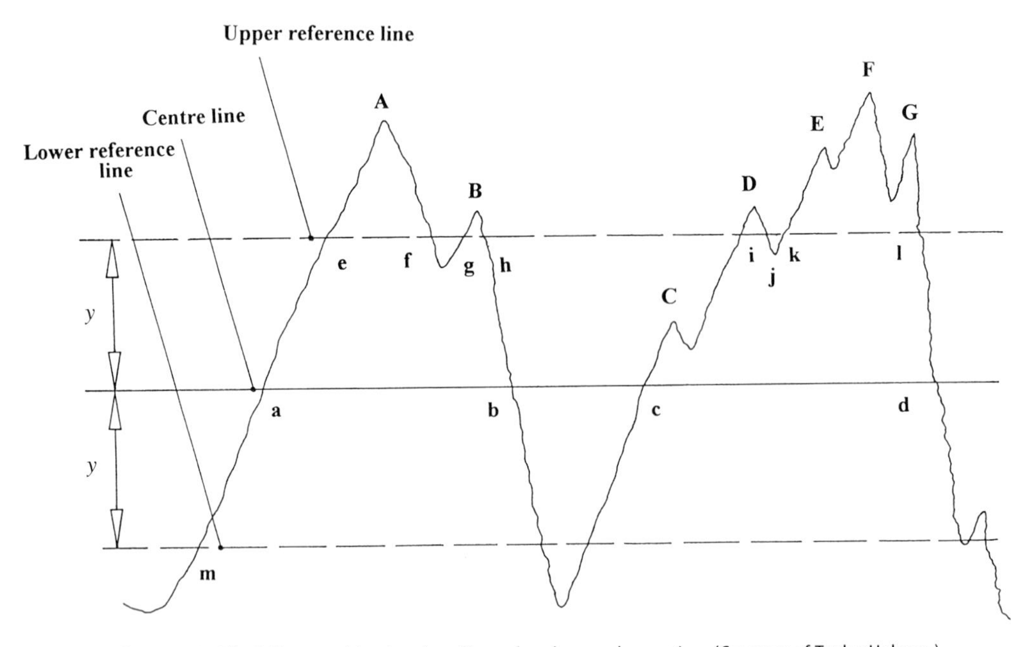

Figure 230. The influence of local and profile peaks when peak counting. (Courtesy of Taylor Hobson.)

profile crosses this reference line; in this case crossings at e, f, g, h, i, j, k and l give a count of four.

4 The number of peaks which project through a selectable band centred around the mean line can be established; typically, when the profile projects below the reference level at m, then above the upper level at e, crossing this level twice at f and h, this represents just one peak.

In the first case, the counting of these profile peaks is termed *high spot count* (HSC; see Figure 231a), whereas the peaks counted by the second, third and fourth cases are known as *peak counts* (*Pc*), but to minimise any ambiguity it should be stated what represents a peak's definition in our case. *Pc* is normally determined over the greatest assessment length possible; that is:

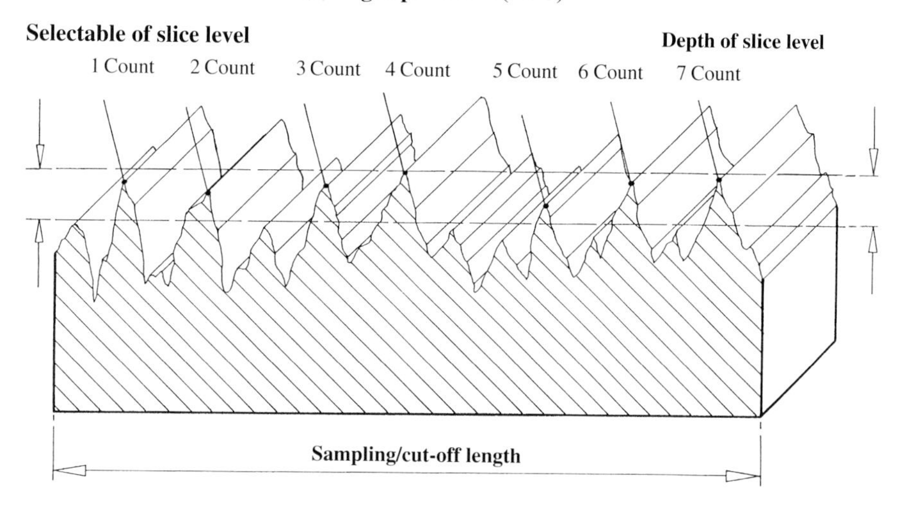

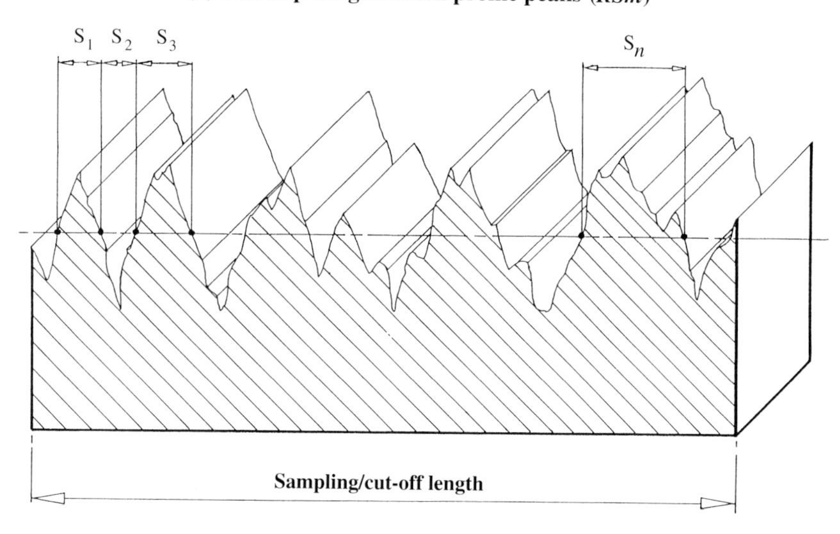

Figure 231. Spacing parameter derivation: high spot count and mean spacing.

$$Pc = \frac{\text{No. of counts}}{\text{Assessment length (m)}}$$

$$= \text{Peaks/m}$$

Another factor that should be noted is that at the beginning and end of the evaluation length, if peaks occur in these regions do we count them? This needs to be defined as the count could otherwise vary by ±2. When there are an odd number of crossings of the reference line within the evaluation length, then do we halve this count or round it up? This paradox can be resolved, since there is every chance that at the beginning of the evaluation length a peak probably existed; therefore it might be prudent to count up rather than down. As a predetermined evaluation length is utilised the count is determined over this distance, but in order to ensure uniformity it can be converted to count per unit length, this being independent of the evaluation length.

RSm: mean peak spacing

The *RSm* parameter is concerned with the mean peak spacing of the profile and is the average value of the length of the centre line section containing a profile peak and an adjacent valley, or in other words between adjacent crossings – in the same direction – of the centre line (see Figure 231b; i.e., S_1, S_2, S_3 . . . S_n). It can be calculated as follows:

$$RSm = \frac{1}{n}\sum_{i=1}^{n}\frac{(S_i = S_1 + S_2 + S_3 \ldots S_n)}{n}$$

RSm is particularly useful as it can be used to represent (normally) the feedrate of a machining pass for an anisotropic surface – that is, one having predominant machined cusps; conversely, *RSm* can be used to determine the feed associated with surfaces having directional lay.

The more sophisticated instruments can utilise additional geometrical parameter terms, such as *WSm* and *PSm*; these are the corresponding parameters established from the waviness and primary profiles (form errors), respectively. These parameters are defined in the relevant standard – Clause 4 (ISO 4287: 1997) – and can be calculated from the profile.

In fact, these *P*-, *R*- and *W*-parameters are derived as follows:

- *P*-parameter is calculated from the primary profile;
- *R*-parameter from the roughness profile;
- *W*-parameter from the waviness profile.

NB The first capital letter in any of these parameter symbols designates the particular type of profile being evaluated. For example, *Pt* is calculated from the primary profile and likewise *Ra* from the roughness profile.

Hybrid parameters

A further refinement from the amplitude and spacing parameters are those that can be classified as neither, hence the term "hybrid" parameters, derived from differentials in the surface texture profile. This hybrid group of parameters tended to be somewhat unreliable initially, because of the relative inability to measure the differentials correctly with analogue-based instrumentation, as they are naturally "noisy". Digital techniques have enabled them to be utilised more readily, but care must be taken in their selection and usage. Hybrid parameters have been particularly favoured by users in the tribological and optical industries, and a brief mention of how and why they are favoured might well be illuminating to the reader.

Pmr, Rmr and *Wmr*: material ratio/Abbott–Firestone curve (ISO 4287: 1997)

The interaction of engineering surfaces in contact with each other determines their ability to perform satisfactorily in service. If enhanced surface contact is desirable, such as in the case of bearing applications, adhesive situations, sealing properties, etc., then the surface topography must be established to suit such conditions. Considering the case of the bearing, the interface between adjacent surfaces would normally be separated via some form of lubricating film. Hence any surface asperity – peak – on

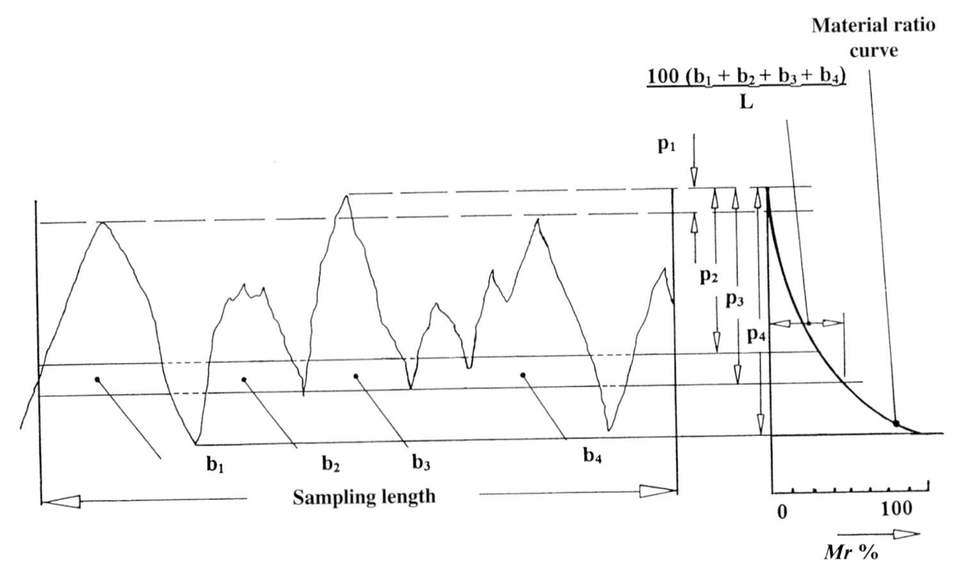

Figure 232. Representation of the material ratio/Abbott–Firestone curve. (Courtesy of Taylor Hobson.)

either of the contact surfaces that is higher than the film of lubrication will penetrate through this medium, resulting in physical contact with its associated mating surface. Such contact will introduce friction, thermal increases and premature bearing wear. Thus any surface having too many inappropriate peaks will not be as effective as a bearing, when compared to one having low peaks and deep valleys.

The material ratio curve, namely *Pmr*, *Rmr* and *Wmr*, for primary, roughness and waviness profiles, respectively, or the Abbott–Firestone curve (as it was previously known and denoted as T_p, but often written as t_p), can be expressed as a percentage of the evaluation length at a predefined depth below the highest peak. It can be determined by the following expression (see Figure 232):

$$Pmr\,(\%) = \frac{(b_1 + b_2 + b_3 + b_4 \ldots b_n)}{L} \times 100$$

$$= \left(\frac{100}{L}\right)\sum_{i=1}^{n} b_i$$

Bearing applications enable the kinematic movement of mating component surfaces, either as they slide, or produce rotational motions, or a combination of both relative to each other. Through tribological studies (from the Greek *tribos* meaning "to rub") and the associated wear, this being a natural consequence of bearing surface contact. Is significantly reduced when a lubricating film is present, thus the bearing ratio, which simulates wear effects, has progressively been shown to be of increasing use by industry. In the early days bearing ratio was manually established from a profile graph; today many surface texture measuring instruments incorporate this parameter. Practical workshop examples of artificially induced wear are relevant here, such as when scraping slideways, or the bedding-in of large bore of white metal bearings. In the latter example, the use of "engineer's blue" in combination with a "master diameter reference surface" – this being coated in the substance and manipulated in a controlled fashion by the fitter over the surface to be scraped – allows telltale "witness marks" to be left by the blued "master surface" and these high spots are then scraped off by the fitter. Then the "master" is reapplied and further scraping occurs until the desired amount of "high spots" are uniformly distributed over the newly created surface, giving the estimated fractional bearing area – being the criterion for the quality of the bearing surface.

The material/bearing ratio T_p designation (now *Pmr*, *Rmr* and *Wmr*) was originally derived from the German word *Tranganteil*, which equates to bearing fraction and *p*, the profile depth. This T_p parameter, superseded by *Pmr* (although it simulates wear effects) is only an approximate substitute for true running-in tests, this is because:

- the bearing ratio is only a fraction of the actual length and not representative of a surface area;
- estimation of the bearing area is obviously established from a predetermined and short cut-off length – ignoring gaps that could result from waviness and form errors;
- *Pmr*, being a theoretical value, by its very nature relates to an unloaded surface, whereas in reality the actual surface may undergo elastic deformation;
- practical in-service conditions mean that two contacting surfaces are involved and each specific surface's features contribute to overall wear;
- induced wear is often accompanied by bulk or plastic flow of bearing material, with the result that the geometrical concept of crests being neatly truncated by an arbitrary line drawn through them is impractical and unrealistic.

Despite these reservations, *Pmr*, *Rmr* and *Wmr* are very useful surface descriptors, finding numerous applications which have been shown to be successfully correlated with their in-service performance. The converse of these parameters can also be applied, namely the depth at which a certain bearing ratio is obtained (see Figure 233). For example, if experience has shown that a bearing ratio of 65% is necessary for an application, the depth *p* at which the crests must be truncated to reach this value can be easily established.

The material/bearing ratio can be determined from the graph by establishing a line at the selected depth, parallel to the centre line, then computing the lengths of the intercepted profile. Examples of preselected depths on the profile's trace are indicated in the material ratio curve given in Figure 232.

For differing slices or depths through the profile a specific ratio of air-to-material occurs (see Figure 233a), its calculation being dependent upon the preselected depth being expressed as a unique and continuous curve for that profile. Figure 233 depicts three such profiles and their associated curves. This Abbott–Firestone curve (named after the two Americans credited with its development), alternatively known as the material ratio curve – the latter term being the preferred name – gives meaningful information on the profile's shape.

The top of the highest peak within the profile, have been evaluated, establishes the reference or zero percentage line for the material ratio curve. Calculation of this curve is influenced by the highest peak's height in relation to the others, although in

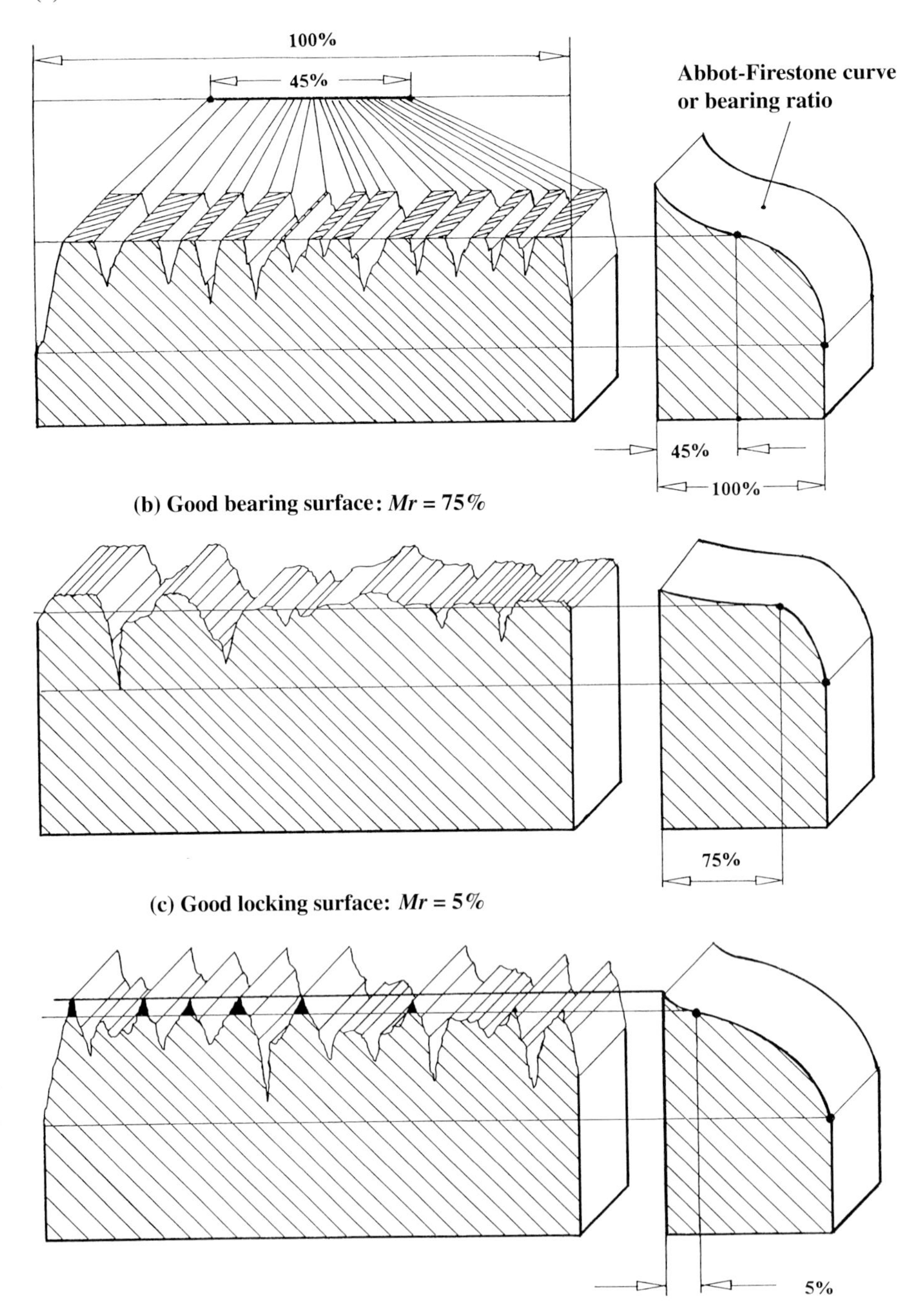

Figure 233. Hybrid surface texture parameters: spacing and depth.

reality the effect of a single peak on a surface's in-service function has little significance. In order to minimise the effect of a single peak on the material ratio curve, an artificially induced reference line is chosen to shift the line below this highest peak as depicted in Figure 233(a and b), this being expressed as a material ratio percentage. Specifying, for example, a 5% reference line (Figure 233c) testifies that the top 5% of the profile is not included as part of the calculation for the material ratio. Selection of the zero line beneath the highest measurable peak will be dependent on the topography of the associated peaks in the profile, but industrial practice suggests the reference level is set between 2% and 5%.

The influence of filtering on the preselected depths through the profile will be an important factor in modifying the material ratio curves, as will the measuring length selected. Terms such as *micro-material* and *macro-material ratios* relate to the filtered or unfiltered (total) roughness profiles, respectively. Normally the filtered parameter is preferred, because it eliminates the effects of waviness in the measured profile. Filter distortion of the measured profile can introduce erroneous results; this is particularly a problem when plateau-type surfaces occur. Typical of these are sintered, honed, plasma-coated, ceramic and chrome finishes. Some apparent confusion has arisen when measuring these plateau-type surfaces, as to whether a mechanical form of filtering is preferred, via the use of a skid to minimise the effect that waviness can play in influencing the shape of the material ratio curve. Some of the literature categorically states that skids should be utilised for these plateau-type surfaces, while others are equally adamant that a skidless mode of operation is preferable, particularly when quantifying secondary machined powder metallurgy surfaces. Powder metallurgy and many scored surfaces are affected by filtering, particularly via "overmodulation", necessitating the use of a "phase-corrected" filter (ISO 11562: 1996). Regardless of which skid technique is employed, topographical filtering of plateau-type surfaces to remove waviness effects is important; further, as has been mentioned earlier, it is also strongly influenced by the selected measuring length for the surface under test. These conditions of surface determination should appear with the surface texture symbol (at the appropriate position – see Figure 3).

Previously, no standardised measurement conditions for the assessment of T_p existed; this parameter has of late become of less importance with the advent of newer parameters such as; *Rk*, *Mr*1 and *Mr*2 – this will be discussed in more detail shortly.

From the idealised profile traces and their accompanying material ratio curves illustrated in Figure 233, it is possible to visualise whether a surface is either a "bearing type" or "locking type" of surface. Furthermore, pre-selecting a zero reference line at some predefined depth below the highest peak will show the anticipated wear pattern or likely asperity (peak) collapse under predefined in-service conditions. This level of peak attrition by tribological action or through plastic deformation may be an important factor relating to its functional operation, anticipated useful life and subsequent reliability.

Parameters derived from Abbott–Firestone/material ratio curve (ISO 13565: 1996 – Parts 1 to 3)

The standard denotes several surface parameters relating to the functional behaviour of highly stressed surfaces, namely lubricating, sealing and rolling faces. Being derived from the material ratio curve (Figure 234), these parameters not only characterise the profile's roughness but also the shape of the profile. These derived parameters include:

- *Rk* – core roughness depth;
- *Rpk* – reduced peak height;
- *Rvk* – reduced valley depth;
- *Mr*1 – peak material ratio;
- *Mr*2 – valley material ratio;
- *A*1 – material filled profile peak area;
- *A*2 – lubricant filled profile valley area;
- *Vo* – oil retention volume.

The material ratio curve is broken down into a series of straight lines comprising three parts to calculate the "*Rk* family" of parameters – the *k* represents the word "kernel", meaning core:

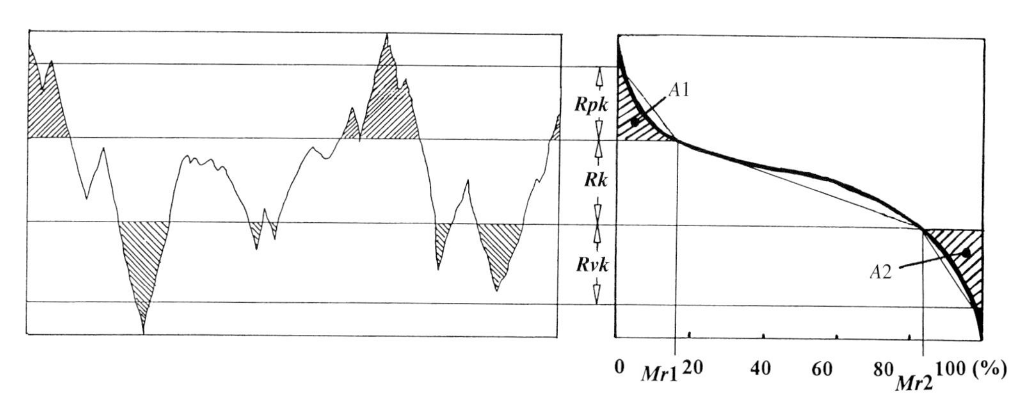

Figure 234. Derivation of the kernal roughness depth, i.e., depth of roughness core profile, *Rk*. (Courtesy of Taylor Hobson.)

- profile peak area – *Rpk*;
- core roughness area of the profile – *Rk*;
- profile depth area – *Rvk*.

Such parameters are exclusively utilised for obtaining the roughness profile of plateau-like (asymmetrical) surfaces with the digital valley suppression filter (ISO 11562: 1996). The derivation of the *Rk* parameter is depicted in Figure 234. It is necessary to find the tangency points for the minimum slope (the turning point) from the "S-shaped" material ratio curve, delineated here by the vertical lines at *Mr*1 and *Mr*2. These tangency points relate to the 0% and 100% material ratio lines, being equatable to heights *Rpk* (for peaks) and *Rvk* (for valleys), respectively. The *Rk* zone represents the depth of core roughness, as it is in the vicinity where the profile has the densest bearing region. The locations of both *Mr*1 and *Mr*2 relate to the material ratios at the top and bottom of the roughness core, respectively. These values can be used to determine still other parameters, such as the "oil retention volume", derived from the *Rvk* and *Mr*2 relationship (*Vo*, which relates to the shaded triangular area A_2 in Figure 234), principally used by the automotive industry. The material ratio curve *Vo* can be described in terms of the area below the core roughness (*Mr*2) and the demarcation of the 100% material line. The numerical value of *Vo* can be expressed by

$$Vo = \frac{Rvk(100 - Mr2)}{200}$$

Rpk – reduced peak height – is illustrated in Figure 234, being the height of the top portion of the surface profile positioned above the core profile. If a small value of *Rpk* occurs, this characteristic would represent good wear resistance or its "running-in" behaviour. The magnitude of *Rpk* is indicated by triangular area $\mathbf{A}_1$, whose height is equal to *Rpk* and with a base of *Mr*1. By utilising such areas the effect of a spurious peak in the profile trace becomes minimised. At the lower portion of the material ratio curve the height *Rvk* represents the depth of the lowest part of the profile, with valleys extending below the core profile. This parameter has an oil-retaining capacity for lubricated faces, such as cross-honing of cylinder liners, hence the need under this condition for a high numerical *Rvk* value.

Porous components produced via powder metallurgy processing cannot be accurately measured, due to the fact that the pores tend to be interconnected internally (in most cases), so their exact depths are not readily apparent, due to the stylus tip's size and geometry not being able to measure them with sufficient accuracy. Therefore, when considering valleys, they are normally discussed separately from peaks. Plateau-like surfaces, when measured, can be represented by peak height, whereas the valleys' depths are only a means of establishing the character of its respective roughness profile. The functional behaviour of a component is not only decided by its relative "global" roughness, as indicated by an equivalent *Ra* value, but also by the profile's surface structure.

The average ($R\Delta a$) and RMS profile ($R\Delta q$) slopes (ISO 4287: 1997)

Figure 235 illustrates how the parameters $R\Delta a$ and $R\Delta q$ are derived. Expressly, the parameter $R\Delta a$ is derived from the average slope of a filtered profile relative to the mean line of the trace. Its derivation can be established by dividing the profile peak in question into smaller portions, then numerically averaging the slope for each portion, as indicated in the enlarged detail of the peak shown in Figure 235. Calculation of $R\Delta a$ can be found from the following formula:

$$R\Delta a = \frac{1}{n}\sum_{i=1}^{n}\left|\frac{\Delta y_i}{\Delta x_i}\right|$$

where n represents the number of ordinates.

Rather than taking individual numerical average values for profile slopes within the sampling length (as in the case of $R\Delta a$), the root mean square "RMS profile slope" ($R\Delta q$) can be calculated. A bonus of utilising the RMS – $R\Delta q$ – value is that it offers increased sensitivity to extreme values, unlike that of the numerical method of determination using $R\Delta a$, which tends to minimise their influence.

$$R\Delta q = \sqrt{\left(\frac{1}{1r}\right)\int_0^{1r}[\theta(x) - \bar{\theta}]^2\,\mathrm{d}x}$$

where

$$\theta = \left(\frac{1}{1r}\right)\int_0^{1r}\theta(x)\,\mathrm{d}x$$

and θ is the slope of the profile at any given point.

An advantage gained from using these hybrid parameters, namely $R\Delta a$ and $R\Delta q$, is that they can detect modifications in the profile's geometry, both prior to and after processing. Typically, if peaks on the surface topography become somewhat rounded through tribological action – wear – this will have the effect of diminishing the $R\Delta a$ and $R\Delta q$ values, which could be used as a means of quality control for a particular set of processing conditions.

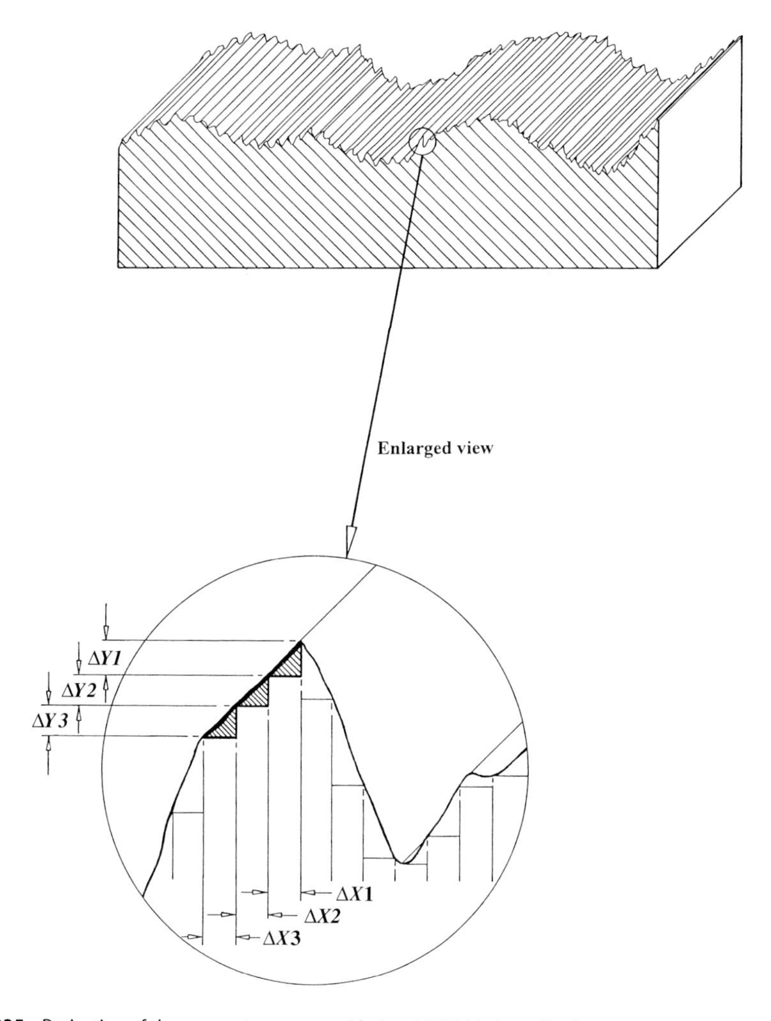

Figure 235. Derivation of the parameters average (Δa) and RMS (Δq) profile slopes. (Courtesy of Taylor Hobson.)

The average (λa) and RMS profile (λq) wavelengths

Figure 236(a) indicates how λa is derived, which is the average wavelength of the surface profile. The parameter λa should not be confused with the "spacing parameter" *Sm* – this being the mean peak-to-peak spacing within the profile's length when peaks exceed a certain threshold, whereas λa considers all the "components" of the surface profile, namely, the wavelength in conjunction with the amplitude (its peak-to-valley height) –although if a profile has a significant sinusoidal profile of predictable periodicity, under these conditions λa will approach that of the peak-to-valley spacing represented by *Sm*.

The average wavelength (λa) surface descriptor can be determined as follows:

$$\lambda a = \frac{2\pi Ra}{R\Delta a}$$

Whenever a complex profile occurs – this being an amalgamation of several different wavelengths – then the average wavelength of these "weighted components", according to their amplitude, can be represented by λa. In Figure 236(a) the roughness profile is illustrated for three surfaces. If a large average roughness (Ra) is present, but having a small slope ($R\Delta a$) – as in the case of the top trace – this indicates a wavy type of surface having a large λa value. Conversely, if the magnitude of roughness decreases but the slope increases, the surface becomes rougher but less wavy, as indicated by the middle trace, or as indicated in the bottom trace where the roughness has been superimposed onto waviness. The visual appearance of a surface may

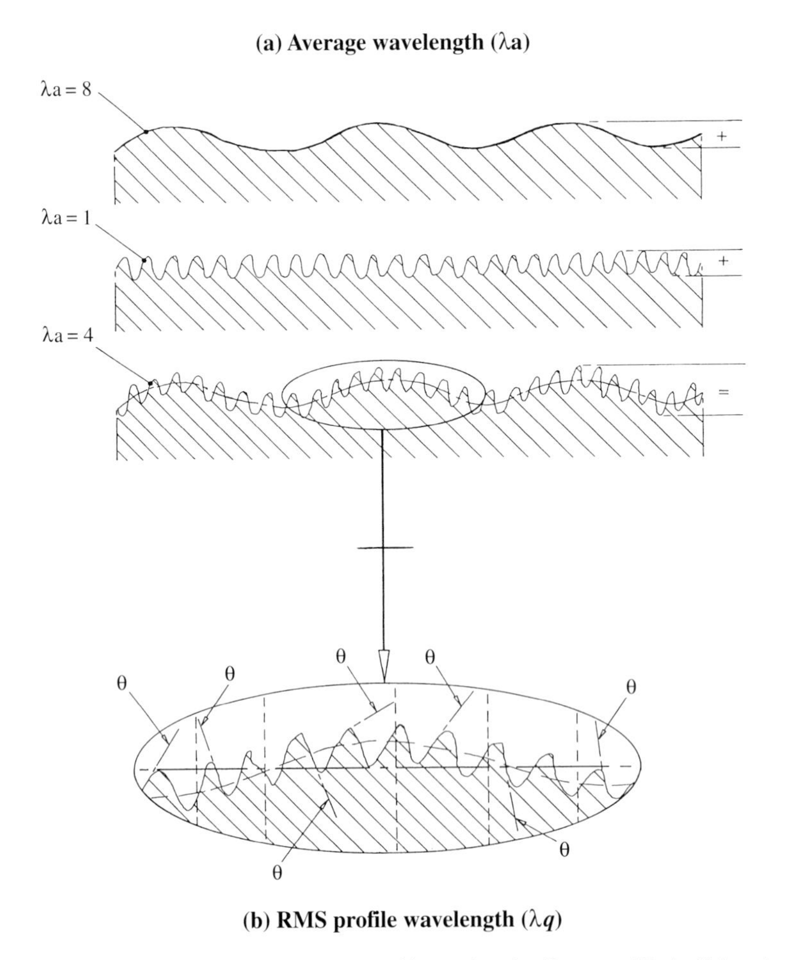

Figure 236. Average (λa) and RMS (λq) profile wavelengths. (Courtesy of Taylor Hobson.)

vary from that of a highly polished and reflective one to matt surfaces, but they can be compared using the λa parameter.

The RMS profile wavelength (λq) can be visualised in Figure 236(b) and is a measure of the spacings between local peaks and valleys, taking into account their relative amplitudes and individual spatial frequencies. Its derivation can be established from

$$\lambda q = \frac{2\pi Rq}{R\Delta q}$$

As in the case of the slope parameter $R\Delta q$, this equivalent wavelength parameter (λq) is more sensitive to extreme values than either the $R\Delta a$ or λa, these being for average slope and wavelengths, respectively.

The λa parameter is particularly useful in sheet steel applications where the average wavelength is a measure of the openness or closeness of the texture and correlates well with the cosmetic assessment of surfaces – an important factor for automobile bodies. λq can be utilised in tribological situations where very closely spaced roughness irregularities of the surface tend to be of relatively small amplitude and rapidly wear away in service – such as for a bearing application. Since these irregularity amplitudes are quite small, any changes in the overall Ra value during running-in are also small, with the shorter wavelengths disappearing and resulting in more pronounced changes in the average wavelength. Of more practical importance is the use of the average wavelength as a means of directly monitoring the manufacturing process. For example, in a continuous outside-diameter turning operation of high-quality finishing, the average wavelength directly relates to the feedmarks promoted by the residual effects of the partial tool geometry as it passes along the workpiece. If the machine tool settings are incorrect, the values of both λa and λq will dramatically change, even though the Ra or similar height parameter may not alter significantly. Similarly, for most grinding operations the para-

Table 24. Comparison of actual λa and Ra values from measurements made on a series of specimens

Production process	λa (μm)	Ra (μm)
Planning	400	17.8
Fine cut planing	440	5.1
Multiple-tooth milling	1300	5.1
Fly-cut milling	300	2.2
OD turning	100	1.7
Diamond turning	37	0.3
Electro-erosion	340	10.1
Electro-sinking	57	1.8
Surface grinding	77	0.95
Circumferential grinding	33	0.55
Plunge grinding	43	0.17
Honing	19	0.07

meter λa or λq has a direct relationship to the average grit size, hence by monitoring the average wavelength this will provide relevant information relating to when the wheel must be redressed to maintain its surface quality.

In Table 24 are shown the λa values taken using samples from a range of production processes. This table illustrates that surfaces generated by a variety of production methods having similar Ra values (fine cut planing and multiple tooth milling, or alternatively turning and electro-sinking) can exhibit considerably differing values of λq. This enables the metrologist or manufacturing engineer to assess a surface under test with considerably more discrimination than might otherwise be the case if only the Ra value alone had been used.

Statistical parameters

The amplitude distribution curve (Figure 237) represents the comparative total heights over which the trace achieves any selected range of heights above or below the centre line. One can deal with this amplitude or height data statistically, in the same manner as one might physically measure anthropometric data such as a person's average stature, within a specific population range. As with the statistical dispersion of population stature, engineering surfaces can exhibit a broad range of profile heights.

For example, a boring operation with a relatively long length-to-diameter ratio may cause deflection (elastic deformation of the boring bar) and occasion the cutting insert to deflect, producing large peak-to-valley undulations along the bore (waviness). Superimposed onto these longer wavelengths are small-amplitude cyclical peaks (periodic oscillations) indicating vibrations resulting from the cutting process. The resultant surface profile for the bored hole would depict the interactions from the boring bar deformations and any harmonic oscillations. This boring bar motion reflected in the profile trace would exhibit low average profile height, but with a large range of height values.

Surface texture data can be statistically manipulated, beginning with the profile trace's height or amplitude distribution curve – this being a graphical representation of the distribution of height ordinates over the total depth of the profile.

The characteristics of amplitude distribution curves can be defined mathematically by several terms called "moments"; these are:

- *arithmetic average of the profile* – the first moment ($m1$);
- *profile height variance* – the second moment ($m2$);
- *skewness* – the third moment ($m3$);
- *kurtosis* – the fourth moment ($m4$).

Arithmetic average of the profile

The first moment ($m1$) can be found from the expression

$$m1 = \bar{y} = \left(\frac{1}{n}\right)\sum_{i=1}^{n} y_i$$

where y_i is the height ordinate and n is the number of ordinates.

NB By definition, the mean height of a filtered profile equates to zero.

Profile height variance

The formula for the second moment ($m2$) is

$$m2 = Rq^2 = \left(\frac{1}{n}\right)\sum_{i=1}^{n} (y_i)^2$$

NB The variance gives an indication of the profile range heights, its square root being the standard deviation of the amplitude distribution curve or its roughness parameter Rq.

Skewness (Rsk) – ISO 4287: 1997

Skewness is a measure of the asymmetry of the amplitude distribution curve; alternatively, this can be expressed another way, as the symmetry about the mean line (see Figure 237a). If symmetry occurs then *Rsk* equates to zero, but an unsymmetrical profile gives rise to a skewed curve (see Figure 237b and c). The bias of *Rsk* depends upon whether the bulk of the material is above or below the mean line. In the case of Figure 237(b), the bulk of material occurs below the mean line – promoting a "locking" type of surface. Conversely, in the case of Figure 237(c), the majority of material is present above the mean line – representing a "bearing" or plateau surface. Such skewness bias enables a metrologist to distinguish between two profiles having identical *Ra* values but divergent shapes and hence considerably different in-service applications. A numerical value can be given to *Rsk* and in the case of Figure 237(b) the "positive skewness" may eventually obtain an adequate bearing surface, although it is unlikely to have oil-retaining abilities. This type of surface can typically be exploited for adhesive bonding applications. The surface characterised by Figure 237(c) might occur in the cases of porous, sintered or cast iron surface topography, having comparatively large numerical values of negative skewness. The surface is sensitive to extreme ordinate values within the profile under test; this is due to *Rsk* being a function of the cube of the ordinate height. As a result of this peak sensitivity, it is a hindrance when attempting to inspect plateau-type surfaces.

Skewness is the third moment ($m3$) of the amplitude distribution curve and can be expressed in the following manner:

$$m3 = Rsk = \left(\frac{1}{(Rq)^3}\right)\left(\frac{1}{n}\right)\sum_{i=1}^{n}(y_i)^3$$

The shape or "spikeness" of the amplitude distribution curve can also relay useful information about the dispersion or "randomness" of the surface profile which can be quantified by means of a parameter known as kurtosis (*Rku*).

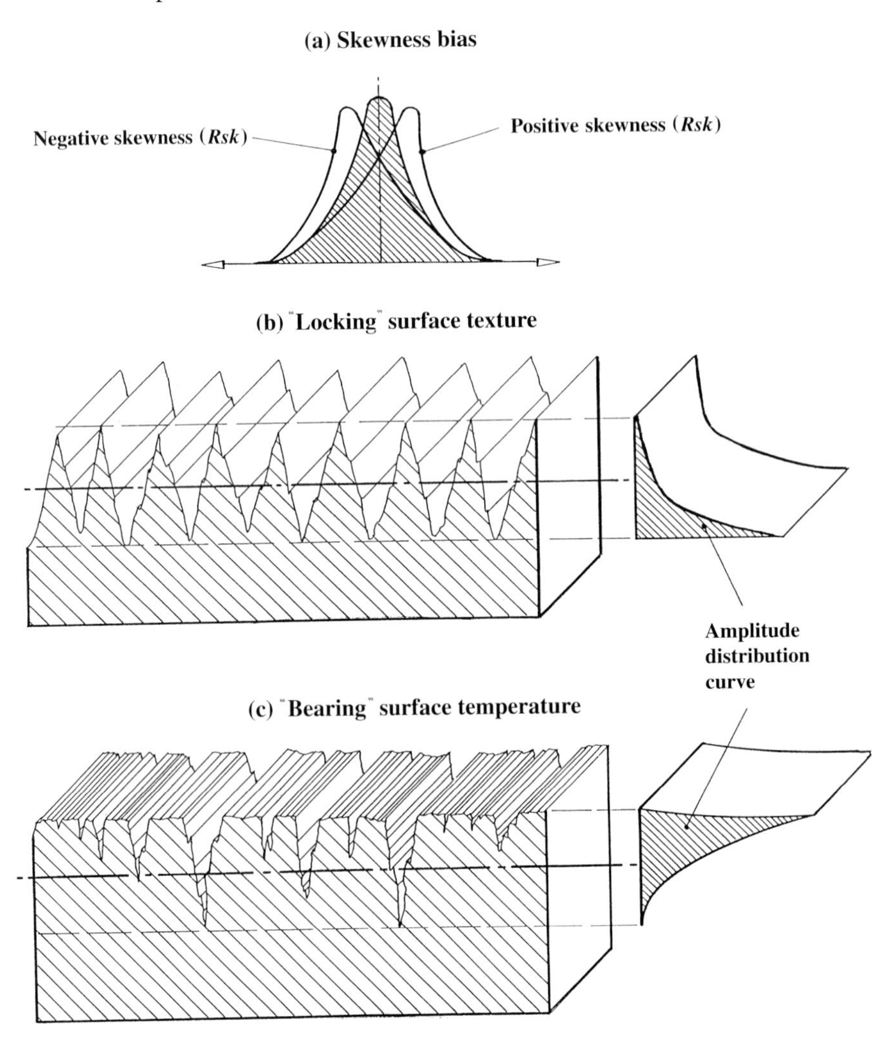

Figure 237. How surface texture topography influences the amplitude distribution curve.

(a) Kurtosis is influenced by the distribution shape

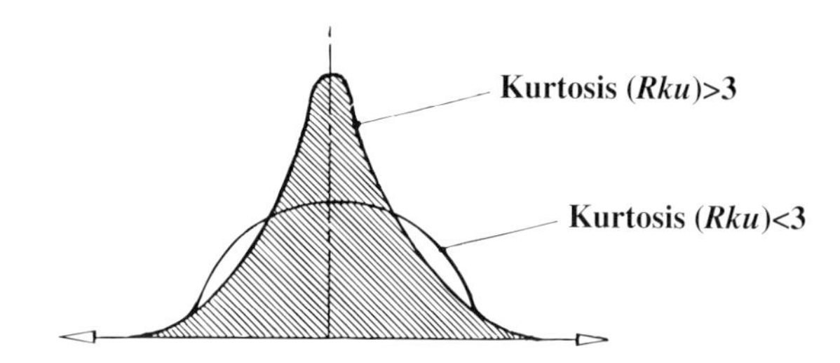

(b) Material distributed evenly across the whole of the surface topography

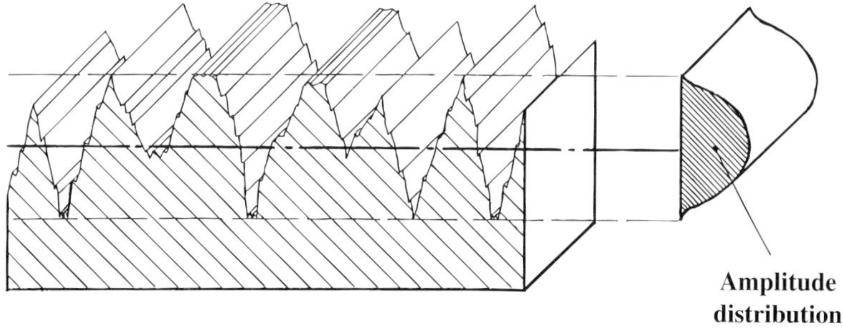

(c) Material distributed about the centre line

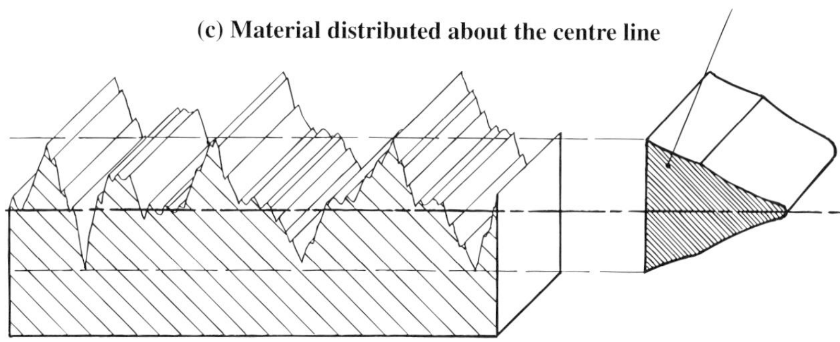

Figure 238. Variation in surface topography influences the shape and height of the amplitude distribution curve.

Kurtosis (Rku) – ISO 4287: 1997

The surface parameter skewness (*Rsk*) is not sensitive if the profile's "spikes" are distributed evenly both above and below the mean line. Another term, kurtosis (*Rku*), has this ability (see Figure 238a). Kurtosis provides a means of measuring the *sharpness* of the profile, with a "spiky" surface exhibiting a high numerical value of *Rku* (Figure 238c); alternatively, a "bumpy" surface topography will have a low *Rku* value (Figure 238b). As a consequence of this ability to distinguish variations in the actual surface topography, *Rku* is a useful parameter in the prediction of in-service component performance with respect to lubricant retention and subsequent wear behaviour.

By definition, *Rku* is the fourth moment of the amplitude distribution curve (*m*4), being the relative "sharpness" of the curve. The amplitude distribution curve shown in Figure 238(b) indicates smaller and less pronounced condensed peaks and valleys than that of the larger, more widely spaced "spiky" peaks and valleys illustrated in Figure 238(c). As a result of these topographical profile trace differences in peak and valley height and their density distribution, the *Rku* shape changes, as will their respective numerical values, as indicated in Figure 238(a).

Due to the fact that *Rku* is a function of the fourth power of its profile ordinates, it has extreme sensitivity to variations in peaks and valleys within the profile trace under test. Due to *Rku* sensitivity, in practice it can be difficult to apply to the analysis of surface texture. As has already been mentioned, kurtosis – being the fourth moment of the amplitude distribution curve ($m4$) – can be calculated in the following manner:

$$m4 = Rku = \left(\frac{1}{(Rq)^4}\right)\left(\frac{1}{n}\right)\sum_{i=1}^{n}(y_i)^4$$

NB Higher levels of moments for the amplitude distribution curve can be calculated, but they are of little practical relevance for analysis of surface texture.

This completes the review of just some of the previous and current surface texture parameters, but it is by no means an exhaustive list of those previously utilised, as many industries have developed their own parameters for specific industrial or research-based requirements.

Appendix B

Amplitude–wavelength (AW) analysis: "Stedman diagrams"

Introduction

A myriad of different types and models of profilometers and associated instrumentation exists for use by the surface metrologist, ranging from the basic (portable) stylus contact instruments, to sophisticated non-contact equipment. Each type of instrument has its own distinct characteristics and capabilities; these instruments can be compared by predicting their performance from information supplied by the manufacturer in terms of their stylus radius, ranges and resolutions, slide quality, etc. Techniques have been developed that are based on how such instruments or probes will respond to measuring sinusoidal waves of varying amplitudes. The method employed is termed *amplitude–wavelength* (AW) *analysis*, which determines charts of instrument performance in AW space (see Figures 239 and 240).

AW space: the concept

In essence, the concept provides a topographical characterisation of surface texture that can be modelled in two-dimensional space, by combinations of surface wavelengths and amplitudes of sinusoidal profiles enabling mapping in *AW* space. Individual points in this *AW* space correspond to a unique amalgamation of amplitude (A) and wavelength (W) for a specific topographical character. For example, the sine wave $y = A \sin(2\pi x/W)$ has a maximum slope (S_{max}) given by $2\pi A/W$ and a minimum radius of curvature (R_{min}) given by $W^2/4\pi^2 A$. The particular values of S_{max} and R_{min} can be plotted as straight lines of slopes 1 and 2 respectively in a logarithmic AW space (see Figure 239a). This technique allows for four topographical

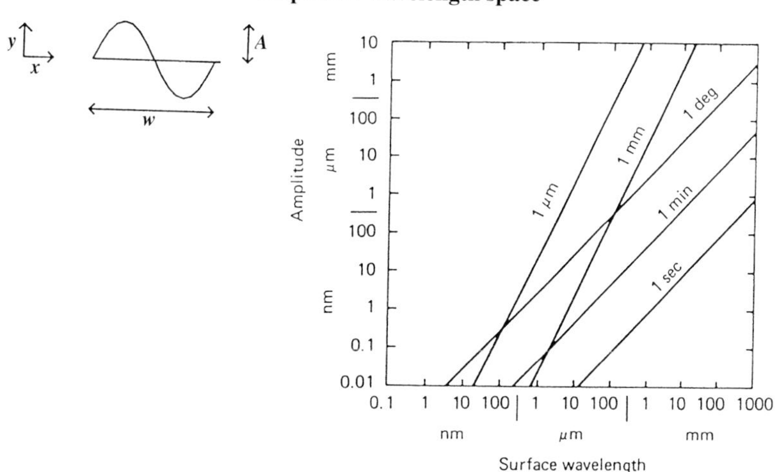

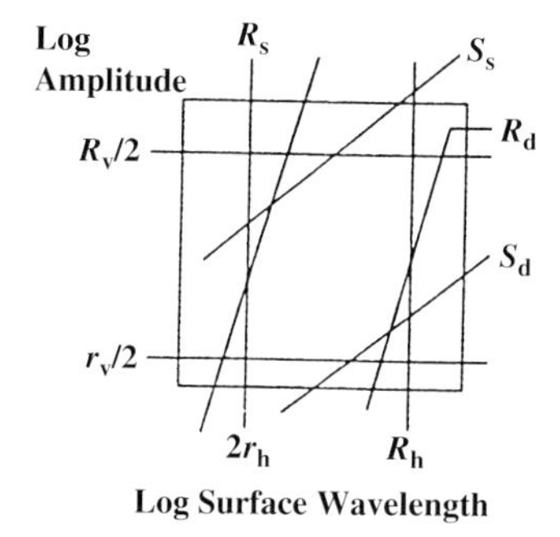

Figure 239. Basis for the Stedman diagrams. [Sources: Stedman, 1987, 1990; Franks, 1991.]

(a) Stedman diagram: range and resolution

Cmax :sharpest curvature; **Zmin** :smallest height difference;
θmax :steepest slope; **λmin** :shortest wavelength;
***R*p** :largest height difference; **λmax** :longest wavelength

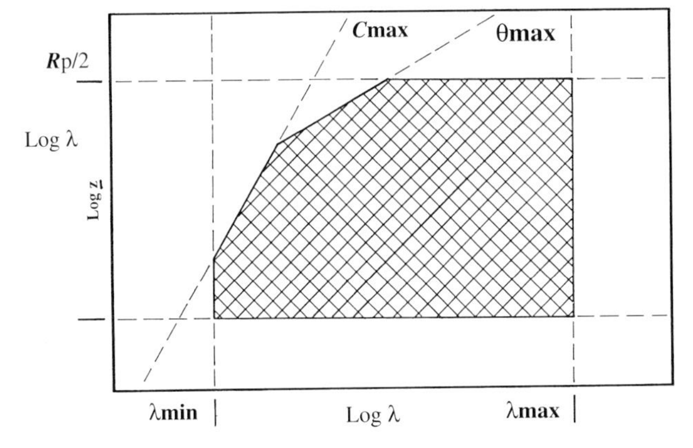

(b) Stedman diagram: stylus instrument [After Thomas, 1999]

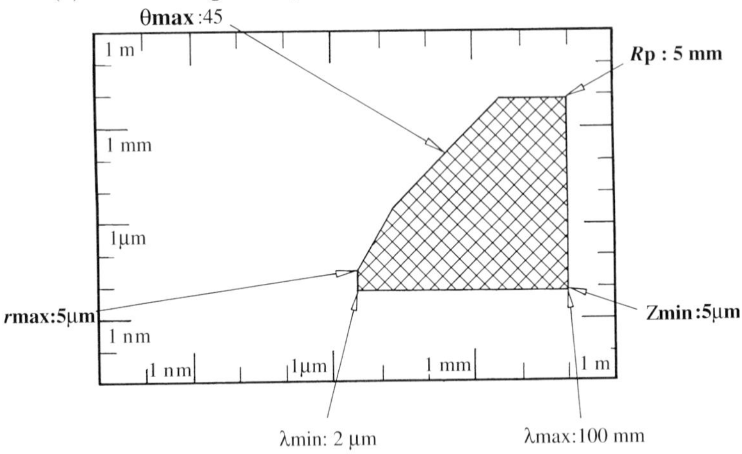

Figure 240. Stedman diagrams for instrument assessment of comparative performance. [Source: Stedman (b), 1987.]

features of the particular instrument in question to be found, namely height, surface wavelength, slope and curvature, which can be represented at any potential scales within an individual uncomplicated chart of AW space.

If one considers the instrument or probe interaction with the sinusoidal profile, it follows that most instrument parameters enforce limits on the sine wave ranges that can be reliably measured. By way of illustration, the tip radius (R_s) of a stylus must not exceed R_{min} of a sine wave that is to be drawn. R_s can be plotted as a limit line of slope 2 in low AW space, with only sine waves to the right of the limit line being measurable. The slope errors (S_d) of a datum or reference slideway should ideally not be greater than S_{max} of any sine wave to be measured. The S_d plots as a limit line of unit slope, with values above this allowing acceptable measurements to be made. In a similar manner the vertical range (R_v) and resolution (r_v) restrict amplitudes, with the horizontal range (R_h) and resolution (r_h) restricting the wavelengths that can be measured. For other classes of profilometers typified by those based on optical or other non-contact probes, analogous limit lines can be derived. For example, the numerical aperture of an objective through which a beam is focused to form an optical probe will limit the maximum slope that can be sensed.

In a collective sense, these limit lines will define a polygonal zone, within which all the conditions for acceptable measurement are satisfied (see Figure 239b). This bounded profile can be taken as the working zone for the instrument in question, being referred to as its *AW map*.

Comparison of instruments in AW space

It is possible to compare a diverse range of instruments in AW mapping space. In a single scan it is possible to show the preset magnification and scan length for a selected probe or stylus that is constrained within the area of AW space, being only sensed to a particular location of the overall AW map. Moreover, a more in-depth AW mapping application might show the effects of setting different magnifications and filters, together with other operating conditions for a single instrument. Such detailed maps can be an important insight to the user and can be incorporated in the instrument's manual. Further, it can display the appropriate conditions currently set up for the instrument in the computer's software.

AW space: surface function and quality assurance

Yet another facet of this important technique of AW mapping is that surface texture requirements can be expressed by appropriate analysis of the function of a surface. For example, a mirror's optical specification could be converted to AW limits by consideration of diffraction theory and geometrical optics, plus the dimensions of the mirror would define AW space relevant to the function of the mirror. The application of surface metrology within the AW window can be a route to quality assurance. The capabilities of the production processes – grinding, lapping and polishing – can also be mapped, with AW analysis being utilised as the unifying concept that links both specification and function to its manufacture and metrology.

AW analysis: higher-order assessment

The AW analysis for instrument performance previously described can suggest limit lines that delineate sharp boundary attributes between a "go" and "no go" situations. Often in practice, such boundaries tend to be "fuzzy" and this anomaly can be resolved by higher-order analysis that can quantify this boundary fuzziness as error contours. This strategy will enable AW maps to be realised, having closed contours for different levels of uncertainty. Thus, AW maps can provide a unifying link between the workpiece specification, its surface function, production processes and subsequent metrological assessment, hence playing a vital role in quality assurance. Moreover, these AW analysis tools are a vehicle in presenting complex information in a succinct manner.

References

Franks, A. Nanometric surface metrology at the National Physical Laboratory. *Nanotechnology* 2, 1991, 11–18.

Stedman, M. How to compare different surface texture measuring instruments. *Quality Today*, April 1990, 35–36.

Stedman, M. Basis for comparing the performance of surface measuring machines. *Precision Engineering* 9(3), 1987a, 149–152.

Stedman, M. Mapping the performance of surface-measuring instruments. *Proceedings of SPIE*, 803, 1987b, 138–142.

Thomas, T.R. Rough Surfaces (2nd Edn.). *Imperial College Press*, 1999.

Appendix C

Surface texture and roundness: calibration diagrams and photographs

*** Major lever length can be accurately set and measured to ±0.05 mm. This could produce a 0.04% uncertainty in step at the stylus**

****Minor lever length can be set using gauge blocks to an accuracy of ± the uncertainty in the blocks (typically <±0.0001 mm), producing a possible error of 0.001%.**

NB: Total uncertainty budget on actual step generated is ±0.16% (±0.16% (±3.2 nm on a 2mm step height on gauge blocks).

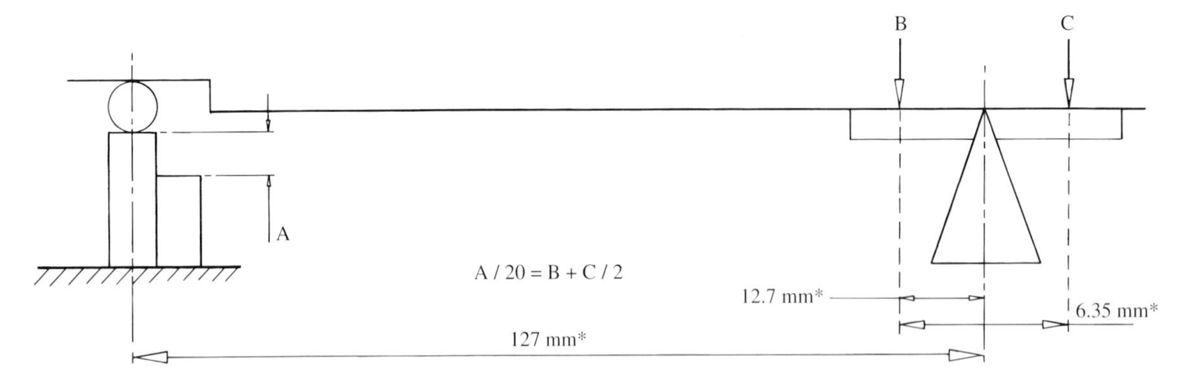

Figure 241. The traceable magnification and calibration on surface texture/roundness instruments using the *"reduction lever principle"*. (Courtesy of Taylor Hobson.)

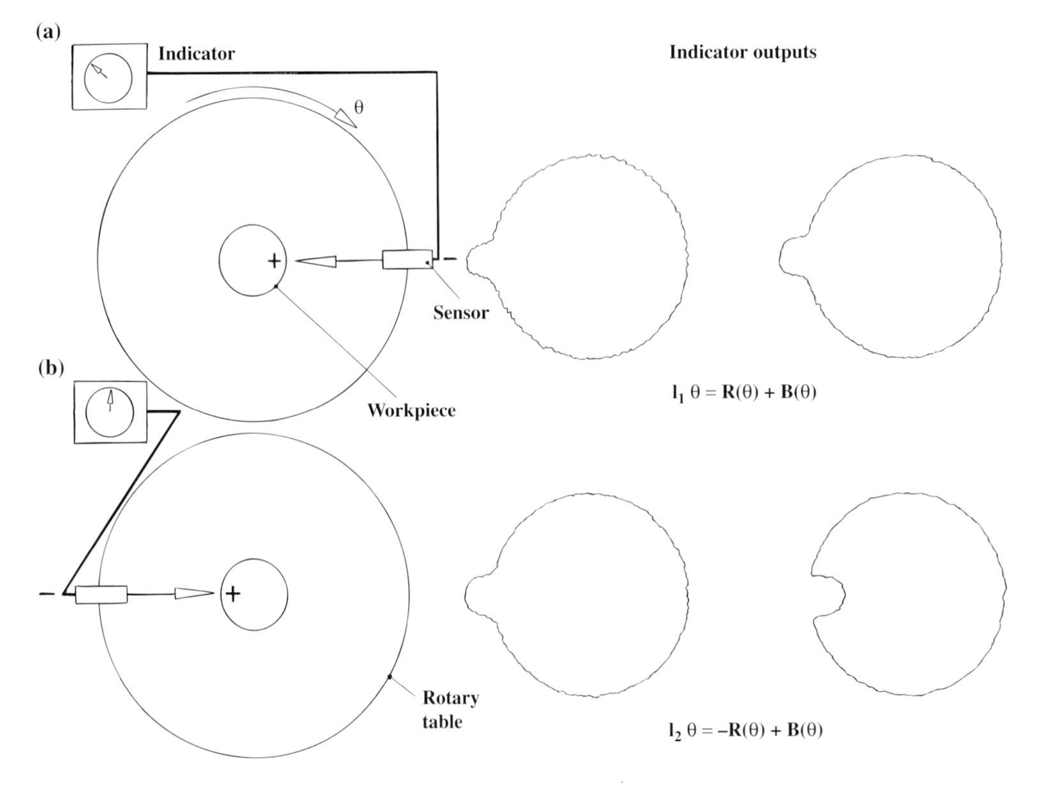

Figure 242. The Donaldson Ball Reversal Technique.

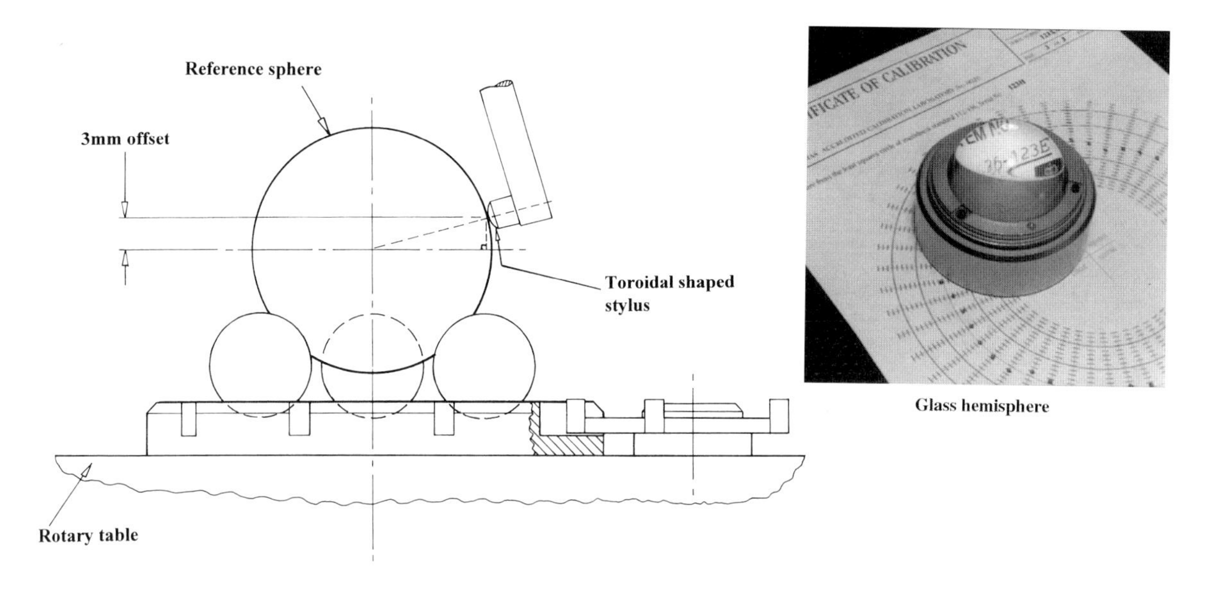

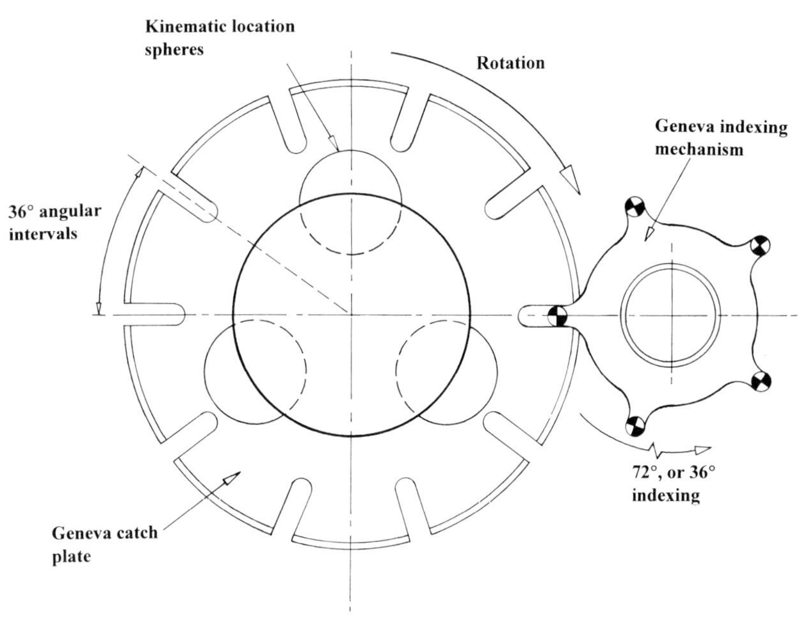

Figure 243. Calibration of a glass reference sphere. (Courtesy of Taylor Hobson.)

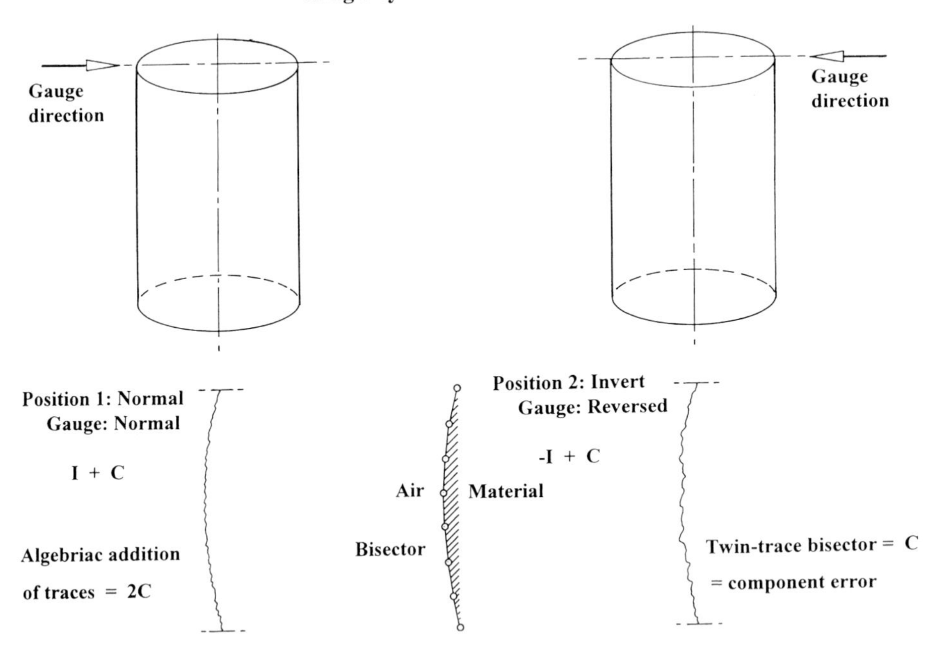

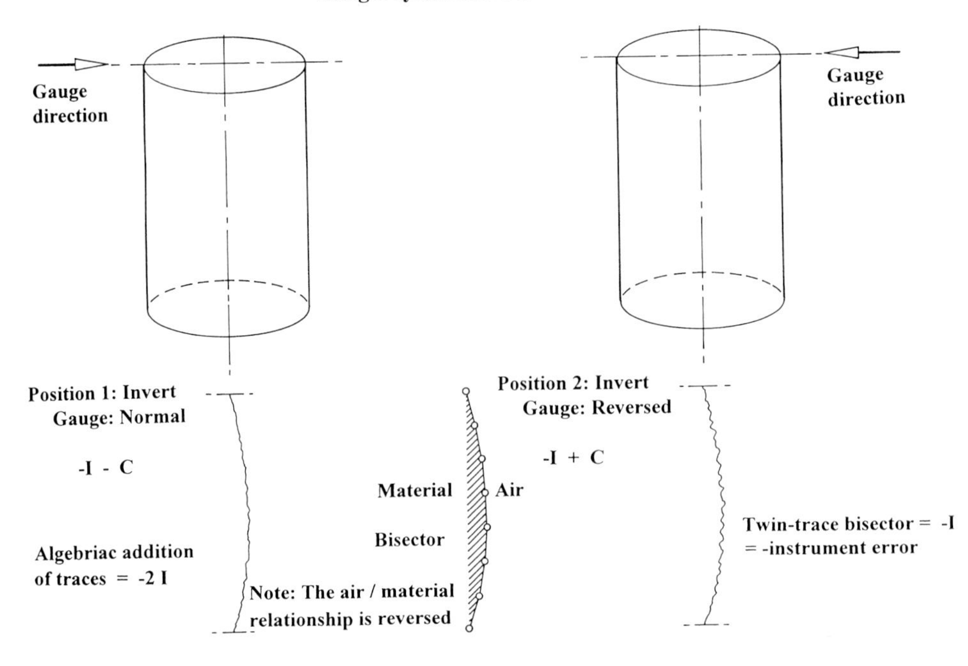

Figure 244 . Calibration a roundness instrument for component or instrument uncertainty. (Courtesy of Taylor Hobson.)

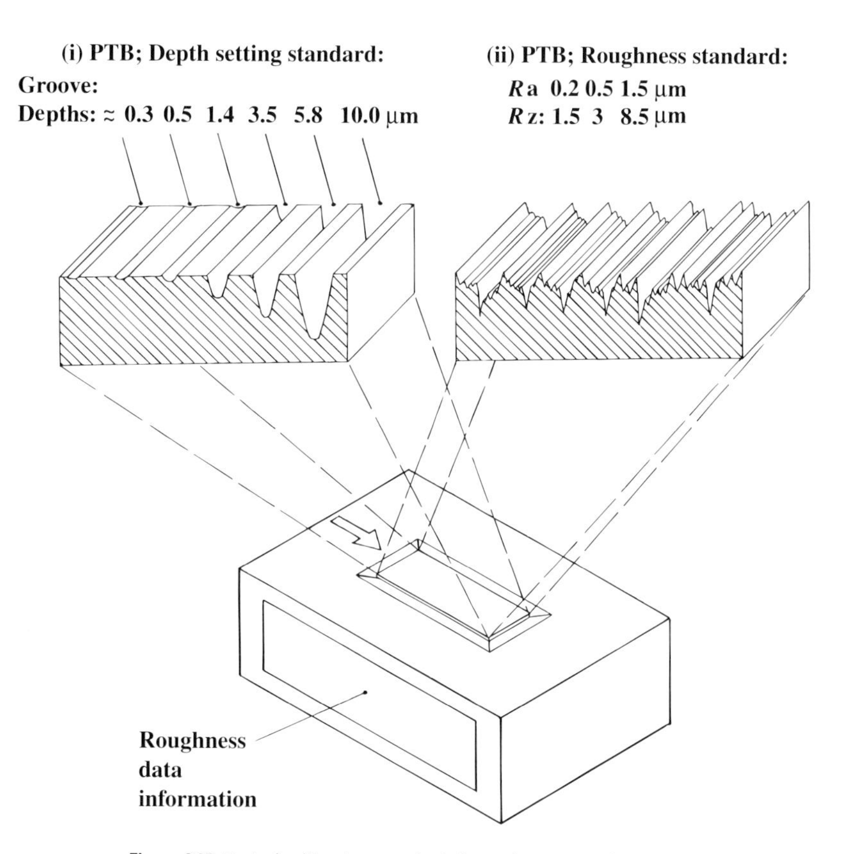

Figure 245. Typical calibration standards for surface texture instruments.

Figure 246. A range of three calibration standards for checking electronic stylus surface texture instruments [Courtesy of Rubert & Co.].

Figure 247. Calibrating the 3-D surface scanning instrument to an accredited artefact. (Courtesy of Taylor Hobson.)

Figure 248. A non-contact laser diffractometer being calibrated with a suitable series of roughness comparison blocks. (Courtesy of Taylor Hobson.)

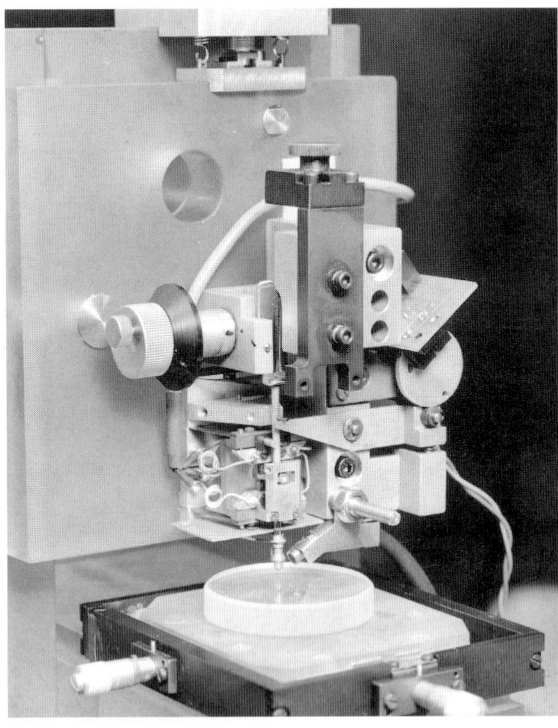

Figure 249. Calibrating the "Nanostep" instrument in conjunction with an optical flat. (Courtesy of Taylor Hobson.)

Appendix D

Hardness conversion chart (Courtesy of Wilson/Rockwell)

Relationships between values determined on "ROCKWELL", "ROCKWELL" Superficial and TUKON Hardness Testers and values determined on other testers.

Although conversion tables dealing with hardness can only be approximate and never mathematically exact, it is of considerable value to be able to compare different hardness scales. This table is based on the assumption that the metal tested is homogeneous to a depth several times as great as the depth of the indentation.

The indentation hardness values measured on the various scales depend on the work hardening behaviour of the material during the test and this in turn depends on the degree of previous cold working of the material. The B-scale relationships in the table are based largely on annealed metals for the low values and cold worked metals for the higher values. Therefore, annealed metals of high B-scale hardness such as austenitic stainless steels, nickels and high nickel alloys do not conform closely to these general tables. Neither do cold-worked metals of low B-scale hardness such as aluminium and softer alloys. Special correlations are needed for more exact relationships in these cases. All values are consistent with ASTM E-140 Tables 1 and 2, and ASTM A-370 Tables 3A and 3B, where applicable. Microficial number values were developed in the *Wilson* Standards Laboratory.

Note 1: A 10mm steel ball is used for 450 BHN and below. A 10mm carbide ball is used above 450 BHN.

Note 2: The tensile strength relation to hardness is inexact, even for steel, unless it is determined for a specific material.

Hardened steel and hard alloys.

C 150 Kg Brale ROCKWELL HARDNESS TESTER	A 60 Kg Brale ROCKWELL HARDNESS TESTER	D 100 Kg Brale ROCKWELL HARDNESS TESTER	15-N 15 Kg N Brale ROCKWELL SUPERFICIAL HARDNESS TESTER	30-N 30 Kg N Brale ROCKWELL SUPERFICIAL HARDNESS TESTER	45-N 45 Kg N Brale ROCKWELL SUPERFICIAL HARDNESS TESTER	Vickers Hardness VICKERS (DPH) DIAMOND PYRAMID - 136° Apex Angle	Knoop Hardness 500 Gr. & Over KNOOP ELONGATED DIAMOND	Brinell Hardness 3000 Kg 10mm Ball BRINELL HULTGREN 10 mm. Ball	G 150 Kg 1/16" Ball ROCKWELL HARDNESS TESTER	Tensile Strength Approx Only Thousand lbs. per square inch
80	92.0	86.5	96.5	92.0	87.0	1865	-	↑	↑	↑
79	91.5	85.5	96.3	91.5	86.5	1787	-	↑	↑	↑
78	91.0	84.5	96.0	91.0	85.5	1710	-	↑	↑	↑
77	90.5	84.0	95.8	90.5	84.5	1633	-	↑	↑	↑
76	90.0	83.0	95.5	90.0	83.5	1556	-	↑	↑	↑
75	89.5	82.5	95.3	89.0	82.5	1478	-	↑	↑	↑
74	89.0	81.5	95.0	88.5	81.5	1400	-	↑	↑	↑
73	88.5	81.0	94.8	88.0	80.5	1323	-	NOTE 1	↑	NOTE 2
72	88.0	80.0	94.5	87.0	79.5	1245	-	↑	↑	↑
71	87.0	79.5	94.3	86.5	78.5	1160	-	↑	↑	↑
70	86.5	78.5	94.0	86.0	77.5	1076	972	↑	↑	↑
69	86.0	78.0	93.5	85.0	76.5	1004	946	↑	↑	↑
68	85.6	76.9	93.2	84.4	75.4	940	920	↑	↑	↑
67	85.0	76.1	92.9	83.6	74.2	900	895	↑	↑	↑
66	84.5	75.4	92.5	82.8	73.3	865	870	NA	↑	↑
65	83.9	74.5	92.2	81.9	72.0	832	846	739	↑	↑
64	83.4	73.8	91.8	81.1	71.0	800	822	722	↑	↑
63	82.8	73.0	91.4	80.1	69.9	772	799	706	↑	↑
62	82.3	72.2	91.1	79.3	68.8	746	776	688	↑	↑
61	81.8	71.5	90.7	78.4	67.7	720	754	670	↑	↑
60	81.2	70.7	90.2	77.5	66.6	697	732	654	↑	NA
59	80.7	69.9	89.8	76.6	65.5	674	710	634	↑	351
58	80.1	69.2	89.3	75.7	64.3	653	690	615	↑	338
57	79.6	68.5	88.9	74.8	63.2	633	670	595	↑	325
56	79.0	67.7	88.3	73.9	62.0	613	650	577	↑	313
55	78.5	66.9	87.9	73.0	60.9	595	630	560	↑	301
54	78.0	66.1	87.4	72.0	59.8	577	612	543	↑	292

Soft steel, grey and malleable cast iron and most non-ferrous metal.

B 100 Kg 1/16" Ball ROCKWELL HARDNESS TESTER	F 60 Kg 1/16" Ball ROCKWELL HARDNESS TESTER	G 150 Kg 1/16" Ball ROCKWELL HARDNESS TESTER	15-T 15 Kg 1/16" Ball ROCKWELL SUPERFICIAL HARDNESS TESTER	30-T 30 Kg 1/16" Ball ROCKWELL SUPERFICIAL HARDNESS TESTER	45-T 45 Kg 1/16" Ball ROCKWELL SUPERFICIAL HARDNESS TESTER	E 100 kg 1/8" Ball ROCKWELL HARDNESS TESTER	K 150Kg 1/8" Ball ROCKWELL HARDNESS TESTER	A 60Kg Brale ROCKWELL HARDNESS TESTER	Knoop Hardness 500 Gr & Over KNOOP ELONGATED DIAMOND	BRINELL Hardness 500 Kg 10 mm Ball STANDARD BRINELL	BRINELL Hardness 3000 Kg Vickers 10 Kg SQUARE BASE DIAMOND PYRAMID - 136° Apex Angle	Tensile Strength Approx. Only Thousand lbs. per square inch
100	↑	82.5	93.1	83.1	72.9	↑	↑	61.5	251	201	240	116
99	↑	81.0	92.8	82.5	71.9	↑	↑	60.9	246	195	234	114
98	↑	79.0	92.5	81.8	70.9	↑	↑	60.2	241	189	228	109
97	↑	77.5	92.1	81.1	69.9	↑	↑	59.5	236	184	222	104
96	↑	76.0	91.8	80.4	68.9	↑	↑	58.9	231	179	216	102
95	↑	74.0	91.5	79.8	67.9	↑	↑	58.3	226	175	210	100
94	↑	72.5	91.2	79.1	66.9	↑	↑	57.6	221	171	205	98
93	↑	71.0	90.8	78.4	65.9	↑	NA	57.0	216	167	200	94
92	↑	69.0	90.5	77.8	64.8	↑	100	56.4	211	163	195	92
91	↑	67.5	90.2	77.1	63.8	↑	99.5	55.8	206	160	190	90
90	↑	66.0	89.9	76.4	62.8	↑	98.5	55.2	201	157	185	89
89	↑	64.0	89.5	75.8	61.8	↑	98.0	54.6	196	154	180	88
88	↑	62.5	89.2	75.1	60.8	↑	97.0	54.0	192	151	176	86
87	↑	61.0	88.9	74.4	59.8	↑	96.5	53.4	188	148	172	84
86	↑	59.0	88.6	73.8	58.8	↑	95.5	52.8	184	145	169	83
85	↑	57.5	88.2	73.1	57.8	↑	94.5	52.3	180	142	165	82
84	↑	56.0	87.9	72.4	56.8	↑	94.0	51.7	176	140	162	81
83	↑	54.0	87.6	71.8	55.8	↑	93.0	51.1	173	137	159	80
82	↑	52.5	87.3	71.1	54.8	↑	92.0	50.6	170	135	156	77
81	↑	51.0	86.9	70.4	53.8	↑	91.0	50.0	167	133	153	73
80	↑	49.0	86.6	69.7	52.8	↑	90.5	49.5	164	130	150	72
79	↑	47.5	86.3	69.1	51.8	↑	89.5	48.9	161	128	147	70
78	↑	46.0	86.0	68.4	50.8	↑	88.5	48.4	158	126	144	69
77	↑	44.0	85.6	67.7	49.8	↑	88.0	47.9	155	124	141	68
76	NA	42.5	85.3	67.1	48.8	↑	87.0	47.3	152	122	139	67
75	99.6	41.0	85.0	66.4	47.8	↑	86.0	46.8	150	120	137	66
74	99.1	39.0	84.7	65.7	46.8	↑	85.0	46.3	147	118	135	65

B 100 kg 1/16" Ball ROCKWELL HARDNESS TESTER	F 60 Kg 1/16" Ball ROCKWELL HARDNESS TESTER	15-T 15 Kg 1/16" Ball ROCKWELL SUPERFICIAL HARDNESS TESTER	30-T 30 Kg 1/16" Ball ROCKWELL SUPERFICIAL HARDNESS TESTER	45-T 45 Kg 1/16" Ball ROCKWELL SUPERFICIAL HARDNESS TESTER	E 100 Kg 1/8" Ball ROCKWELL HARDNESS TESTER	H 60 Kg 1/8" Ball ROCKWELL HARDNESS TESTER	K 150kg 1/8" Ball ROCKWELL HARDNESS TESTER	A 60 Kg Brale ROCKWELL HARDNESS TESTER	Knoop Hardness 500 Gr. & Over KNOOP ELONGATED DIAMOND	Brinell Hardness 500 Kg 10mm Ball STANDARD BRINELL
50	85.4	76.9	49.7	22.7	87.0	↑	64.5	35.0	107	83
49	84.8	76.6	49.0	21.7	86.5	↑	63.5	34.6	106	82
48	84.3	76.2	48.3	20.7	85.5	↑	62.5	34.1	105	81
47	83.7	75.9	47.7	19.7	85.0	↑	61.5	33.7	104	80
46	83.1	75.6	47.0	18.7	84.5	↑	61.0	33.3	103	80
45	82.6	75.3	46.3	17.7	84.0	↑	60.0	32.9	102	79
44	82.0	74.9	45.7	16.7	83.5	↑	59.0	32.4	101	78
43	81.4	74.6	45.0	15.7	82.5	↑	58.0	32.0	100	77
42	80.8	74.3	44.3	14.7	82.0	↑	57.5	31.6	99	76
41	80.3	74.0	43.7	13.6	81.5	↑	56.5	31.2	98	75
40	79.7	73.6	43.0	12.6	81.0	↑	55.5	30.7	97	75
39	79.1	73.3	42.3	11.6	80.0	↑	54.5	30.3	96	74
38	78.6	73.0	41.6	10.6	79.5	↑	54.0	29.9	95	73
37	78.0	72.7	41.0	9-6	79.0	NA	53.0	29.5	94	72
36	77.4	72.3	40.3	8.6	78.5	100	52.0	29.1	93	72
35	76.9	72.0	39.6	7.6	78.0	99.5	51.5	28.7	92	71
34	76.3	71.7	39.0	6.6	77.0	99.0	50.5	28.2	91	70
33	75.7	71.4	38.3	5.6	76.5	98.8	49.5	27.8	90	69
32	75.2	71.0	37.6	4.6	76.0	98.5	48.5	27.4	89	69
31	74.6	70.7	37.0	3.6	75.5	98.0	48.0	27.0	88	68
30	74.0	70.4	36.3	2.6	75.0	97.8	47.0	26.6	87	67
29	73.5	70.0	35.6	1.0	74.0	97.5	46.0	26.0	87	66
28	73.0	69.3	34.5	NA	73.5	97.0	45.0	25.5	86	66
27	72.5	69.5	34.0		73.0	96.5	44.5	25.0	85	65
26	72.0	69.0	33.0		72.5	96.3	43.5	24.5	84	65
25	71.0	68.8	32.5		72.0	96.0	42.5	24.3	83	64
24	70.5	68.5	32.0		71.0	95.5	41.5	24.0	82	64

53	77.4	65.4	86.9	71.2	58.6	560	594	525		283
52	76.8	64.6	86.4	70.2	57.4	544	576	512		273
51	76.3	63.8	85.9	69.4	56.1	528	558	496		264
50	75.9	63.1	85.5	68.5	55.0	513	542	481		255
49	75.2	62.1	85.0	67.6	53.8	498	526	469		246
48	74.7	61.4	84.5	66.7	52.5	484	510	455		238
47	74.1	60.8	83.9	65.8	51.4	471	495	443		229
46	73.6	60.0	83.5	64.8	50.3	458	480	432		221
45	73.1	59.2	83.0	64.0	49.0	446	466	421		215
44	72.5	58.5	82.5	63.1	47.8	434	452	409		208
43	72.0	57.7	82.0	62.2	46.7	423	438	400		201
42	71.5	56.9	81.5	61.3	45.5	412	426	390		194
41	70.9	56.2	80.9	60.4	44.3	402	414	381		188
40	70.4	55.4	80.4	59.5	43.1	392	402	371		182
39	69.9	54.6	79.9	58.6	41.9	382	391	362		177
38	69.4	53.8	79.4	57.7	40.8	372	380	353		171
37	68.9	53.1	78.8	56.8	39.6	363	370	344		166
36	68.4	52.3	78.3	55.9	38.4	354	360	336		161
35	67.9	51.5	77.7	55.0	37.2	345	351	327		156
34	67.4	50.8	77.2	54.2	36.1	336	342	319		152
33	66.8	50.0	76.6	53.3	34.9	327	334	311		149
32	66.3	49.2	76.1	52.1	33.7	318	326	301		146
31	65.8	48.4	75.6	51.3	32.5	310	318	294	NA	141
30	65.3	47.7	75.0	50.4	31.3	302	311	286	92.0	138
29	64.6	47.0	74.5	49.5	30.1	294	304	279	91.0	135
28	64.3	46.1	73.9	48.6	28.9	286	297	271	90.0	131
27	63.8	45.2	73.3	47.7	27.8	279	290	264	89.0	128
26	63.3	44.6	72.8	46.8	26.7	272	284	258	88.0	125
25	62.8	43.8	72.2	45.9	25.5	266	278	253	87.0	123
24	62.4	43.1	71.6	45.0	24.3	260	272	247	86.0	119
23	62.0	42.1	71.0	44.0	23.1	254	266	243	84.5	117
22	61.5	41.6	70.5	43.2	22.0	248	261	237	83.5	115
21	61.0	40.9	69.9	42.3	20.7	243	256	231	82.5	112
20	60.5	40.1	69.4	41.5	19.6	238	251	226	81.0	110

73	98.5	37.5	84.3	65.1	45.8	■	84.5	45.8	145	116	132	64
72	98.0	36.0	84.0	64.4	44.8	NA	83.5	45.3	143	114	130	63
71	97.4	34.5	83.7	63.7	43.8	100	82.5	44.8	141	112	127	62
70	96.8	32.5	83.4	63.1	42.8	99.5	81.5	44.3	139	110	125	61
69	96.2	31.0	83.0	62.4	41.8	99.0	81.0	43.8	137	109	123	60
68	95.6	29.5	82.7	61.7	40.8	98.0	80.0	43.3	135	107	121	59
67	95.1	28.0	82.4	61.0	39.8	97.5	79.0	42.8	133	106	119	58
66	94.5	26.5	82.1	60.4	38.7	97.0	78.0	42.3	131	104	117	57
65	93.9	25.0	81.8	59.7	37.7	96.0	77.5	41.8	129	102	116	56
64	93.4	23.5	81.4	59.0	36.7	95.5	76.5	41.4	127	101	114	NA
63	92.8	22.0	81.1	58.4	35.7	95.0	75.5	40.9	125	99	112	
62	92.2	20.5	80.8	57.7	34.7	94.5	74.5	40.4	124	98	110	
61	91.7	19.0	80.5	57.0	33.7	93.5	74.0	40.0	122	96	108	
60	91.1	17.5	80.1	56.4	32.7	93.0	73.0	39.5	120	95	107	
59	90.5	16.0	79.8	56.7	31.7	92.5	72.0	39.0	118	94	106	
58	90.0	14.5	79.5	55.0	30.7	92.0	71.0	38.6	117	92	104	NOTE 2
57	89.4	13.0	79.2	54.4	29.7	91.0	70.5	38.1	115	91	103	
56	88.8	11.5	78.8	53.7	28.7	90.5	69.5	37.7	114	90	101	
55	88.2	10.0	78.5	53.0	27.7	90.0	68.5	37.2	112	89	100	
54	87.7	8.5	78.2	52.4	26.7	89.5	68.0	36.8	111	87	NA	
53	87.1	7.0	77.9	51.7	25.7	89.0	67.0	36.3	110	86		
52	86.5	5.5	77.5	51.0	24.7	88.0	66.0	35.9	109	85		
51	86.0	4.0	77.2	50.3	23.7	87.5	65.0	35.5	108	84		
50	85.4	2.5	76.9	49.7	22.7	87.0	64.5	35.0	107	83		

23	70.0	68.0	31.0		70.5	95.3	41.0	23.5	82	63
22	69.5	67.8	30.5		70.0	95.0	40.0	23.0	81	62
21	69.0	67.5	29.5		69.5	94.5	39.0	22.5	81	62
20	68.5	67.3	29.0		68.5	94.3	38.0	22.0	80	61
19	68.0	67.0	28.5		68.0	94.0	37.5	21.5	79	61
18	67.0	66.5	27.5		67.5	93.5	36.5	21.3	78	60
17	66.5	66.3	27.0		67.0	93.0	35.5	21.0	78	60
16	66.0	66.0	26.0		66.5	92.8	35.0	20.5	77	59
15	65.5	65.5	25.5		65.5	92.5	34.0	20.0	76	59
14	65.0	65.3	25.0		65.0	92.0	33.0	NA	75	59
13	64.5	65.0	24.0		64.5	91.8	32.0		75	58
12	64.0	64.5	23.5		64.0	91.5	31.5		74	58
11	63.5	64.3	23.0		63.5	91.0	30.5		73	57
10	63.0	64.0	22.0		62.5	90.5	29.5		72	57
9	62.0	63.8	21.5		62.0	90.3	29.0		71	57
8	61.5	63.5	20.5		61.5	90.0	28.0		71	56
7	61.0	63.0	20.0		61.0	89.5	27.0		70	56
6	60.5	62.8	19.5		60.5	89.3	26.0		69	55
5	60.0	62.5	18.5		60.0	89.0	25.5		69	55
4	59.5	62.0	18.0		59.0	88.5	24.5		69	55
3	59.0	61.8	17.0		58.5	88.0	23.5		68	54
2	58.0	61.5	16.5		58.0	87.8	23.0		68	54
1	57.5	61.0	16.0		57.5	87.5	22.0		68	54
0	57.0	60.5	15.0		57.0	87.0	21.0		67	53

Index